Behavior of Radionuclides in the Environment II

Alexei Konoplev • Kenji Kato •
Stepan N. Kalmykov
Editors

Behavior of Radionuclides in the Environment II

Chernobyl

Springer

Editors
Alexei Konoplev
Institute of Environmental Radioactivity
Fukushima University
Fukushima, Japan

Kenji Kato
Faculty of Science
Shizuoka University
Shizuoka, Japan

Stepan N. Kalmykov
Chemistry Department
Lomonosov Moscow State University
Moscow, Russia

ISBN 978-981-15-3567-3 ISBN 978-981-15-3568-0 (eBook)
https://doi.org/10.1007/978-981-15-3568-0

This Springer imprint is published by the registered company Springer Nature Singapore Pte Ltd.
The registered company address is: 152 Beach Road, #21-01/04 Gateway East, Singapore 189721, Singapore

Preface

The Chernobyl accident occurred on 26 April 1986 at 1:24 a.m. local time during the scheduled outage of the fourth unit reactor when the experiment on testing a turbine generator was being carried out (IAEA 1986). The experiment was to investigate reactor safety in the event of failure of the mains electricity supply to the plant. The test program, however, was not properly designed and concurred, with no additional safety measures stipulated and the emergency core cooling system switched off. The experiment led to a steam explosion which blew the lid of the reactor and resulted in the largest accidental release of radioactivity into the environment in the history of nuclear power production. The exposed reactor core continued to burn for approximately 10 days with continued releases of radioactivity into the atmosphere over this period (IAEA 1986; Alexakhin et al. 2001; Beresford et al. 2016).

This was the worst nuclear disaster in history in terms of the amounts of released radioactive materials and extent of contaminated territories. Because of high risk to human health from high-level radiation exposure, about 44,000 inhabitants of Pripyat town located 3 km from the Chernobyl Nuclear Power Plant (NPP) were evacuated on 27 April 1986, and on 6 May 1986 the entire population of 30-km zone around the NPP was evacuated. Totally 350,000 people were resettled from the affected areas of Ukraine, Belarus, and Russia. Many of these evacuated areas remain abandoned today (IAEA 2006; Beresford et al. 2016).

The Chernobyl accident led to renewed research on the environmental behavior of radionuclides, considering that methods were required to model and predict the dynamics of environmental contamination, systems of emergency response and decision support systems were to be established, and strategy and recommendations for remediation and rehabilitation of contaminated areas were to be developed and implemented. Prior to the Chernobyl accident, data on the environmental behavior of radionuclides had been primarily obtained from studies of global fallout following nuclear weapons tests and radionuclides emerging as a result of accidents (Chalk River, Canada, 1952; Windscale, UK, 1957; Kyshtym accident at PA Mayak, USSR, 1957; Three Mile Island, USA, 1979, etc.), as well as radionuclides occurring in radioactive waste discharged to the environment (Hanford, USA; PA Mayak,

USSR/Russia; Sellafield, UK, etc.) (Prister et al. 2013). With more than 30 years passed after the Chernobyl accident, we have an opportunity to review major achievements in understanding the environmental behavior of radionuclides and, more specifically, their dynamics immediately after the accident and in the long term.

The volume is designed to review major achievements in research of the behavior of Chernobyl-derived radionuclides, ranging from air transport and resuspension, mobility and bioavailability in soil-water environment, vertical and lateral migration in soils and sediments, soil-to-plant and soil-to-animal transfer, and water-to-aqueous biota transfer. Special attention is given to long-term dynamics of radionuclides in rivers and lakes, including the heavily contaminated cooling pond of the Chernobyl NPP, which is currently being decommissioned. The focus on long-term aspects of environmental behavior of radionuclides enables a better understanding of environmental contamination mechanisms and developing methods to model and predict long-term consequences of radioactive contamination. This knowledge can also be used for predicting evolution of environmental contamination due to the Fukushima Dai-ichi NPP accident or other sources of man-made radionuclides.

Over the 30 years since the accident, a number of large-scale international programs and projects have been implemented to investigate in detail and gather data about Chernobyl-derived radionuclides in environmental compartments. One of the first projects was International Chernobyl Project on Radiological Consequences in the USSR of the Chernobyl Accident: Assessment of Health and Environmental Effects and Evaluation of Protective Measures. It was launched in October 1989 by International Atomic Energy Agency (IAEA) in response to the request from the USSR Government to organize and coordinate assessment of Soviet authorities' guidance to persons living in radiologically contaminated areas and evaluation of measures to safeguard the health of population (IAEA 1991). In June 1992 the European Commission and relevant ministries of Belarus, Russia, and Ukraine signed Agreement for International Collaboration on the Consequences of the Chernobyl Accident. This Program supported 16 research projects, involving 80 research groups in Western Europe and around 120 research groups in the three Republics (Karaoglou et al. 1996). From 1997 until the end of 2003 the French-German Initiative for Chernobyl (FGI) was implemented including, among other things, Project on the Radioecological Consequences of the Accident, the main goal being collection and harmonization of data on the behavior of radionuclides in the environment (FGI 2006).

It should be stressed that a lot of information and data obtained after the Chernobyl accident were published in Russian, Ukrainian, and Belorussian languages only and up to now is unavailable in English language literature. What is more, in the first years after the accident actual data on environmental contamination levels was not always publicly available. In this respect, the chapters containing original results, many of which were not previously published, are written by experts from Ukraine, Russia, and Belarus having 30 years' experience of investigating Chernobyl-derived radionuclides in the environment. It may be worth noting that at the time of the Chernobyl accident considerable expertise in emergency response and study of the environmental behavior of accident-related radionuclides was available from the so-called Kyshtym accident which occurred due to explosion of a tank with high-level radioactive waste at Mayak plant in 1957 and led to the formation of East-Urals Radioactive Trail (EURT) (Alexakhin et al. 2001). Results of those studies were practically unavailable in open literature and unknown for the world scientific community at large, but then immediately after the Chernobyl accident scientists with the EURT research track record became engaged, among them academicians Alexakhin R.M. and Prister B.S., professors Tikhomirov F.A. and Pavlotskaya F.I., and many other holders of unique knowledge, who took the lead in respective scientific research areas. One of them, academician Prister B. S., is the author of the chapter about Chernobyl-derived radionuclides in agricultural ecosystems in the present volume. Studies of the behavior of Chernobyl-derived radionuclides in the environment, contrary to the Kyshtym accident, were broadly covered in literature, including English language scientific literature, starting from the 1990s. In this context, this volume addresses the knowledge acquired through the legacy of the two unfortunate accidents.

The Chernobyl studies have clearly demonstrated that the behavior of accidentally released radionuclides in the environment is governed by radionuclide speciation in fallout and site-specific environmental characteristics. With this in view, Chaps. 1 and 2 deal with atmospheric transport of released radionuclides, their deposition, resuspension, and secondary atmospheric transport. Main chemical forms of radionuclides, their transformation, and accordingly mobility and bioavailability in soil and sediments are the subject of Chap. 3. Chapter 4 is dedicated to the wash off of radionuclides from contaminated catchments and their redistribution due to erosion and redeposition of contaminated soil particles. Chapter 5 concerns the behavior of Chernobyl-derived radionuclides in agricultural ecosystems. In Chap. 6, the behavior of radionuclides and radiation effects in forest ecosystems of the Chernobyl NPP near zone is discussed. Chapter 7 addresses the long-term dynamics of radionuclides in contaminated rivers and lakes, underlying mechanisms, and process level modeling. Chapter 8 focuses on the behavior of radionuclides in a unique water body of the Chernobyl NPP cooling pond which is being currently decommissioned, with the water level significantly reduced and part of the contaminated bottom exposed to the air. Chapter 9 provides an overview of data and modeling of radionuclide accumulation by aqueous biota in different stages after the accident, as exemplified by the Chernobyl NPP cooling pond, Dnieper River (Ukraine), and Lake Kozhanovskoe (Russia).

The book may be of interest to researchers of environmental contamination, radioecology, and radiation safety, as well as to students and all those majoring in environmental radioactivity and radioecology. We would be gratified if the main messages of this volume could be used for the benefit of future generations of scientists.

Fukushima, Japan | Alexei Konoplev
Shizuoka, Japan | Kenji Kato
Moscow, Russia | Stepan N. Kalmykov

References

Alexakhin R, Buldakov L, Gubanov V, Drozhko E, Il'in L, Kryshev I, Linge I, Romanov G, Savkin M, Saurov M, Tikhomirov F, Holina Yu (2001) Severe radiation accidents: consequences and protective measures. Moscow, Izdat, p 751 (in Russian)

Beresford N, Fesenko S, Konoplev A, Skuterud L, Smith JT, Voigt G (2016) Thirty years after the Chernobyl accident: what lessons have we learnt? J Environ Radioact 157:77–89

FGI (2006) The French-German initiative for Chernobyl. Program 2: study of the radioecological consequences. GRS/IRSN, p 109

IAEA (1986) Summary report on the post-accident review meeting on the Chernobyl accident. Safety series no. 75-INSAG-1. IAEA, Vienna

IAEA (1991) The International Chernobyl Project. Technical report. Assessment of radiological consequences and evaluation of protective measures. IAEA, Vienna, p 640

IAEA (2006) Environmental consequences of the Chernobyl accident and their remediation: twenty years of experience. IAEA, Vienna, p 166

Karaouglou A, Desmet G, Kelly GN, Menzel HG (Eds.) (1996) Proceedings of the first international conference "The radiological consequences of the Chernobyl accident", Minsk, Belarus, 18–22 Mar 1996. ECSC-EC-EAEC, Brussels-Luxemburg, p 1192

Prister B, Kluchnikov A, Shestopalov V, Kuhar V (2013) The safety problems of nuclear power. The lessons of Chernobyl: monograph. Chernobyl, p 200 (in Russian)

Contents

Part I
Atmospheric Transport of Radionuclides and Resuspension

Chapter 1
Atmospheric Transport of Radionuclides Initially Released as a Result of the Chernobyl Accident

Mykola Talerko, Evgeny Garger, Tatiana Lev, and Anatolii Nosovskyi

Abstract A brief overview of the characteristics of the Chernobyl release is provided here, including the radionuclide release activity, radionuclide and physicochemical composition of the release, and the available estimates of the temporal dynamics of the released materials and the release initial height. An overview of the meteorological conditions of emission transport is given. The main results of the measurements of radioactive contamination due to the Chernobyl accident on different spatial scales are presented, including within the Chernobyl exclusion zone and over Europe. The results of modeling the radionuclide atmospheric transport and their deposition on the underlying surface, including the results of the reconstruction of the emission source parameters, are given. The review of the measurement of radioactive contamination of air and the underlying surface with iodine isotopes during an initial stage of the accident is given, the same as the results of a reconstruction of the iodine contamination fields in the territory of Ukraine, Belarus, and Russia.

Keywords Chernobyl accident · Source term · Atmospheric dispersion · Hot particles · Modeling · Iodine

1.1 Introduction

On April 26, 1986, an accident occurred at the fourth unit of the Chernobyl nuclear power plant (ChNPP), which, according to its scale and consequences, became the largest accident in the history of nuclear power. The accident resulted in the release of a large amount of radionuclides that determined the radionuclide contamination of the territory of Ukraine, Belarus, Russia, and other states, as well as the radiation dose of the population in these territories. As the subsequent events showed, neither

M. Talerko (✉) · E. Garger · T. Lev · A. Nosovskyi
Institute for Safety Problems of Nuclear Power Plants, National Academy of Sciences of Ukraine, Kyiv, Ukraine

A. Konoplev et al. (eds.), *Behavior of Radionuclides in the Environment II*,
https://doi.org/10.1007/978-981-15-3568-0_1

specialists in the field of nuclear energy and radiation protection nor the human society as a whole was ready for an adequate response to such a technogenic catastrophe of a planetary scale. It took time to realize its true scope and threats to human health for the people responsible for operational decisions on emergency response against the accident. During the period after the Chernobyl accident, specialists from all over the world performed a detailed analysis of its technical causes. The results of this analysis are published elsewhere (GKIAE 1986; Velikhov et al. 1991; IAEA 1992; Nosovsky et al. 2006). A short but capacious conclusion was given in the report of the Nuclear Energy Agency (NEA) within the Organisation for Economic Co-operation and Development (OECD): "In summary, the Chernobyl accident was the product of a lack of "safety culture". The reactor design was poor from the point of view of safety and unforgiving for the operators, both of which provoked a dangerous operating state. The operators were not informed of this and were not aware that the test performed could have brought the reactor into explosive conditions. In addition, they did not comply with established operational procedures. The combination of these factors provoked a nuclear accident of maximum severity in which the reactor was totally destroyed within a few seconds" (NEA 2002).

On many issues related to the study of the long-term consequences of the Chernobyl accident for the environment and the health of the population inhabited (or living now) in radioactively contaminated areas, research continues to this day. More than 30 years after the Chernobyl accident, the problem of atmospheric transport of emissions from the emergency unit 4 of the Chernobyl nuclear power plant and the formation of fields of radioactive contamination of air and the land surface in its acute phase of the accident can be considered one of the most thoroughly studied topics. This chapter presents a brief review of the results of such studies and refers to the scientific and practical suggestions made for increasing the emergency preparedness of nuclear power enterprises and public authorities responsible for protecting public health in the event of possible happened radiation accidents in the future.

1.2 General Characteristics of the Chernobyl Release

1.2.1 Methods to Estimate the Radionuclide Release Activity

The long-term process of forming the radioactive fallout fields due to the prolonged release of radionuclide from the emergency reactor resulted in radioactive contamination of large territories in the former USSR and in most European countries. The traces of the Chernobyl releases were detected throughout the Northern Hemisphere.

Timely and adequate emergency response is possible if complete and reliable information is available on the radioactive contamination of natural environments and associated radiation doses. Such information about the current state can ideally be obtained from the data of operational radiation monitoring. Emergency monitoring should be carried out in conjunction with the use of atmospheric transport modeling, which enables us to estimate the release transport and predict the level of radioactive contamination of the environment. Thus, in the context of this chapter, a correct analysis of the processes that took place in the fourth unit of the Chernobyl nuclear power plant in the first days after the accident is important for determining the parameters of the release source of radioactive materials into the atmosphere.

The first calculations of the Chernobyl release transport in the atmosphere had great difficulty due to a high degree of uncertainty or almost complete lack of reliable data on most of the parameters of the release source: the release activity amount, its phase and isotope composition, the parameters of the size distribution of aerosol particles, the initial release height, and so on, required as input information for the atmospheric transport modeling.

Several independent methods were employed to estimate the radionuclide release in the Chernobyl accident:

1. *Using measurement data of the nuclide activity concentration in the air directly above the reactor.*
   This method used the data obtained from aircraft and helicopter sampling, which was systematically conducted over the reactor and the industrial site of the Chernobyl nuclear power plant. Due to the unsteadiness of the release itself and the meteorological conditions, the accuracy of the method proved to be low. Gavrilin (2001) stated that the uncertainty of estimates based on this method could reach three orders of magnitude. Nevertheless, the data obtained were used by Soviet specialists in 1986 for the first estimates of the emission value (IAEA 1986). The most important result of the measurements carried out in the radioactive plume was the qualitative conclusion that during the active stage of the accident, the release of radionuclides, with the exception of volatile elements, occurred in the form of finely dispersed nuclear fuel particles (FPs; Borovoi and Gagarinskii 2001).
2. *Reconstruction of the source term according to deposition measurements in different territories after finishing the active phase of the accident.*
   The emission reconstruction has been made by integrating the field of the fallout density over the entire contaminated area if it is assumed that the total activity deposited within the area where the measurements were made, they are equal to the released activity. The accuracy of the estimates for such an approach is determined both by the detailing of the deposition field measurements and the total area of the territory, which is taken into account. A few years after the accident, when detailed maps of radioactive contamination of the territory were compiled, this method of estimating the release became the principal one. With its help, the most accurate estimates of the emission of all long-lived nuclides were made, including the release of strontium and transuranium elements (Kashparov et al. 2003). The main drawback of this method is the impossibility of its direct use for estimating short-lived elements, primarily iodine, due to insufficient

volume of measurement data obtained in the initial period of the Chernobyl accident (especially in the territory of the former USSR).

3. *Comparison of calculations of nuclide accumulation in the active zone before the accident with estimates of their amount in the destroyed unit.*
   Long-term studies of the state of fuel-containing materials in the "Shelter" object enable, in principle, the possibility of estimating the amount of fuel in it after the accident and thereby assess its release. However, the accuracy of such estimates is very low. Nevertheless, Borovoi and Gagarinskii (2001) noted that fuel research in the "Shelter" made it possible to make an independent estimate of the emission of volatile nuclides ^{137}Cs and ^{131}I from the depletion degree of the fuel with cesium and long-lived ^{129}I in the destroyed unit. The obtained estimates have poor accuracy, but on the whole, they do not contradict the results obtained by other methods.
4. *Evaluating the inventory of radionuclides in the reactor core separately at the time of the accident and the fraction of the inventory of each radionuclide that was released into the atmosphere; the products of those two quantities are the amounts released* (UNSCEAR 2008).
5. *Solving the inverse problem of atmospheric transport of radionuclides using appropriate atmospheric transport models.*

Unlike the second method, for the reconstruction of the radionuclide release value, it can use measurement data of the air activity concentration or depositions, obtained only in a part of the contaminated area or at its individual points (for example, at the points of the radiation monitoring network) (see Sect. 1.5.5).

Besides the values of radionuclide activity, characterization of the Chernobyl source term should include, in addition, an estimate of the effective release height (or the initial vertical distribution of radioactivity in the atmosphere over the destroyed reactor), a description of the phase composition of the release for individual nuclides (gases and aerosol particles), and the size distribution of aerosol particles in release (or at least the mean particle size). Complicated nuclear and thermal processes that took place in the emergency reactor during the first days of the accident resulted in time variability of the mentioned release parameters. This introduces additional uncertainty in estimates of emission parameters, especially for some individual days of intensive releases.

1.2.2 Radionuclide Inventory in Unit 4 Reactor Core at the Time of the Accident

The radionuclide inventory value at the time of the accident is a key point for using third and fourth methods mentioned above. The first assessment of the radionuclide inventory in Unit 4 reactor core at the time of the accident was carried out at the Moscow Engineering Physics Institute at the request of academician V.A. Legasov. Data on the average fuel burnup in the reactor were used for the analysis, which led to errors in calculating the amount of radionuclides, the accumulation of which is nonlinearly related to the burnup value. The results of these calculations were used in the preparation of the report made by the Soviet delegation to the IAEA (GKIAE 1986). In 1987–1989, similar calculations were made in a number of countries, differing from each other by the selected fuel burnup value and the code used for computations. A review of used initial assumptions and results of these studies was made by Güntay et al. (1996).

Subsequently, the group of A.A. Borovoi from the National Research Center "Kurchatov Institute" (Russia) carried out a refined core inventory calculation that incorporates the results of the fuel burnup for each of the 1659 assemblies in the reactor before the accident (Begichev et al. 1990). Among the more recent studies, one can distinguish the work of Sich (1994), which performed calculations of the accumulation for 65 nuclides, considering the significant nonlinear accumulation of transuranics (actinides) as a function of the burnup in more detail.

The various estimates of core inventories obtained after the first assessment in 1986 are consistent with each other in general. The values of inventories obtained by Begichev et al. (1990) are considered to be the most reliable since they were calculated with the most detailed fuel history. They were used as a main estimate of the radionuclide inventory in the Unit 4 reactor core for the analysis of the effects of the Chernobyl accident in the UNSCEAR (2000) report. According to them, the activity of ^{137}Cs was 260 PBq, ^{90}Sr was 230 PBq, and ^{238,239,240}Pu was 3.4 PBq. The iodine and tellurium nuclides inventory were estimated as 3200 PBq for ^{131}I, 4300 PBq for ^{132}I, 4700 PBq for ^{133}I, and 4200 PBq for ^{132}Te (Gavrilin et al. 2004).

1.2.3 Estimations of Radionuclide Activity Released from the Reactor

The first assessment of releases was made by the Soviet scientists and presented at the IAEA Post-Accident Assessment Meeting in Vienna in August 1986 (IAEA 1986). The initial release estimation was conducted based on air sampling over the destroyed reactor with integration of the ground deposition data on the territory of the former Soviet Union (Table 1.1). Materials that went outside the borders of the USSR were not included. It was estimated that 100% of the core inventory of the noble gases (xenon and krypton) was released. The amount of volatile elements released to the environment was estimated from 10–13% of ^{134}Cs and ^{137}Cs to 20% of ^{131}I. The early estimate for refractory elements in the fuel particles released was $3 \pm 1.5\%$. The release amount of intermediate volatility nuclides was estimated to be the same to or slightly more than the fuel material—2.9% of ^{103}Ru and ^{106}Ru, 4% of ^{89}Sr and ^{90}Sr, and 5.6% of ^{140}Ba. The total release of fission products (excluding radioactive inert gases) was estimated to be approximately 1900 PBq (decay-corrected on 6 May 1986) or about 3.5% of the total inventory of radioisotopes in the reactor at the time of the accident. The estimates in Table 1.1 are decay-corrected

Table 1.1 Estimates of activities of the principal radionuclides released in the Chernobyl accident

Radionuclide	Activity of release (PBq)			
	IAEA (1986)	Buzulukov and Dobrynin (1993)	Devell et al. (1996)	UNSCEAR (2008)
Inert gases				
^{85}Kr	33	33		33
^{133}Xe	6300	6500	6500	6500
Volatile elements				
^{129m}Te		240		240
^{132}Te	410	1000	~1150	~1150
^{131}I	630	1200–1700	~1760	~1760
^{133}I		2500		910
^{134}Cs	19	44–48	~54	~47
^{136}Cs		36		36
^{137}Cs	37	74–85	~85	~85
Intermediate				
^{89}Sr	93	81	~115	~115
^{90}Sr	8	8	~10	~10
^{103}Ru	140	170	>168	>168
^{106}Ru	60	30	>73	>73
^{140}Ba	270	170	240	240
Refractory (including fuel particles)				
^{95}Zr	155	170	196	84
^{99}Mo		210	>168	>72
^{141}Ce	130	200	196	84
^{144}Ce	90	140	~116	~50
^{239}Np	850	1700	945	400
^{238}Pu	0.03	0.03	0.035	0.015
^{239}Pu	0.024	0.03	0.03	0.013
^{240}Pu	0.03	0.044	0.042	0.018
^{241}Pu	5.2	5.9	~6	~2.6
^{242}Pu	0.00007	0.00009		4×10^{-5}
^{242}Cm	0.78	0.93	~0.9	~0.4
Total (excluding noble gases)		8000	5300	~5300

The activities are decay-corrected to April 26, 1986

back to 26 April 1986. Decay correction to the beginning of the accident allows more short-lived radionuclides to be included, giving a higher estimate of total release, which, however, is a probable overestimate since many of these radionuclides would have decayed inside the damaged core before any release to the atmosphere could occur (IAEA 1986).

During the first 10 years after the accident, several refined estimates of the release were made on the basis of new data on radionuclide inventory in the reactor core and results of global fallout measurements. The main differences in the results of these studies stemmed from the assessment of volatile radionuclide activity in the release. In various studies, the ^{137}Cs release was increased to 25–60%, and estimations for ^{131}I have ranged from 20% to 80%. Buzulukov and Dobrynin (1993) reviewed new estimations made by this time and proposed the most reliable estimates based on analyses of the core inventory of Borovoi (1992). The release of cesium isotopes was estimated to be 33±10% release fraction, and for iodine, it was 50–60%. The very similar results for most of the radionuclides were obtained by Devell et al. (1996) (Table 1.1) and used in the UNSCEAR (2000) report.

In studies of this period, the estimates of releases of refractory and intermediary volatile nuclides (and the nuclear fuel itself) remained almost unchanging—about 3.5% of the core inventories. Later, the last important and widely recognized change in the assessment of the activity of this nuclide group was made by Kashparov et al. (2003). The released fraction of strontium, europium, plutonium, and americium deposited outside the industrial site of the Chernobyl nuclear power plant was estimated to be only 1.5±0.5%, which is about 2 times lower than the previous estimates. The revised estimation data have been used in the UNSCEAR (2008) report (Table 1.1).

In general, it can be concluded that in cases of a large radiation accident leading to radioactive contamination of a large area, the assessment of the source parameters is an extremely difficult task. None of the methods mentioned in Sect. 1.2.1 alone can solve this problem with acceptable accuracy. To this end, it is necessary to carry out joint assessments using different methods, using the available data collected throughout the radioactively contaminated area.

1.2.4 Physical and Chemical Properties of Released Radionuclides

The physicochemical forms of radionuclides in the Chernobyl release largely determine their environmental behavior, solubility, bioavailability, and ecological and health effects (Salbu et al. 1998). After the accident, volatilization of some radionuclides took place during oxidation and dispersal of the nuclear fuel. After entering into the atmosphere, the more volatile of the released radionuclides remained in the gas phase, while the less volatile condensed on particles of construction materials and atmospheric aerosol particles. Thus, the chemical and physical forms of radionuclides in the Chernobyl release were determined by the volatility of their compounds and the conditions inside the reactor (EGE 2005).

Constituents of radionuclides released from the accidental reactor to the environment are summarized as follows (IAEA 2011): (1) gases (isotopes of inert gases and iodine in different chemical forms); (2) particles formed as a result of the condensation of volatile fission products (radioisotopes of I, Te, and Cs, which were released during the high-temperature annealing of the nuclear fuel) on the surfaces of different carriers; (3) fuel particles (i.e., particles of finely dispersed nuclear fuel) formed by mechanical disintegration; and (4) monoelemental (or bielemental) particles.

Isotopes of iodine were released in the form of condensation particles, and in gaseous elemental and organic forms, as well as they were found in the fuel particles. In more detail, this issue will be considered in Sect. 1.6.

Isotopes of refractory elements (e.g., cerium, zirconium, niobium, and plutonium) were released into the atmosphere primarily in the form of fuel particles. Other radionuclides (isotopes of cesium, tellurium, etc.) were found in both fuel and condensed particles.

The specific activity of radionuclides in condensation particles was determined by the duration of the condensation process and the process temperature, as well as particle characteristics. Volatile fission products (cesium, iodine, and tellurium) were found in very small particles (0.5–1 μm) with nuclide composition different than fuel. In measurements of airborne nuclides in Switzerland from 30 April to 13 May 1986, Jost et al. (1986) found that an activity distribution on particle size for ^{132}Te, ^{103}Ru, and ^{137}Cs has the maximum at 0.71 μm whereas most of ^{131}I was associated with smaller particles (the distribution maximum at 0.35 μm). Georgi et al. (1988) and Kauppinen et al. (1986) obtained similar results for measurements in Germany, Austria, and Finland. More than 80% of the activity of cesium, ruthenium, and tellurium radionuclides is determined by particles with less than 1 μm. In contrast, a ^{131}I distribution is shifted to smaller aerosol diameters: more than 50% of the activity was found on particles smaller than 0.5 μm. The condensation particle type was the major components of long-range transport and deposition.

The fuel particles are fragments of uranium oxide fuel containing the range of fission products found in the irradiated fuel, including ^{239}Np, ^{141}Ce, ^{144}Ce, ^{103}Ru, ^{106}Ru, ^{95}Zr, ^{95}Nb, ^{140}La, and ^{99}Mo and the more volatile ^{131}I, ^{132}Te, ^{134}Cs, and ^{137}Cs (Sandalls et al. 1993). More than 90% of ^{89,90}Sr activities were also released in fuel particles (EGE 2005). The chemical and radionuclide composition of fuel particles was close to those estimated for the reactor fuel burnup of Unit 4, but fuel particles are often depleted in gases and volatile elements such as ^{131}I, ^{132}Te, ^{137}Cs, and ^{134}Cs, sometimes also in Ru and Ba isotopes. The depletion in some volatile elements may be ascribable to evaporation since the fuel temperature was likely to exceed 2000 °C (IAEA 2011). The different ^{144}Ce/^{141}Ce and ^{106}Ru/^{103}Ru ratios indicated that the fuel fragments originated from the fuel of various degrees of burnup (Devell 1988). The activity distribution of the particle sizes for the fuel component of the release in May 1986 was found to be a log-normal function with an activity median aerodynamic diameter (AMAD) of more than 10 μm (Kutkov et al. 1995).

The oxidation of nuclear fuel was the basic mechanism of fuel particle formation. Less oxidized fuel particles were formed as a result of the mechanical destruction of the fuel during the explosion in the first stage of the Chernobyl accident. These are pieces of nuclear fuel with a well-formed "grain" structure, deposited as separate grains (crystallites) of UO_2 (Salbu et al. 2001). After the initial core disruption event on 26 April 1986, a fragmentation of the fuel continued during the next days of the release. A possible mechanism for fuel fragmentation late in the accident is air oxidation of relatively cool (<1000 °C) fuel. Air oxidation of uranium dioxide to form U_3O_8 is known to cause decrepitation. Fine fuel fragments, depleted in volatile radionuclides but still containing nearly the initial concentrations of low volatility radionuclides such as zirconium and cerium, could have been dispersed in the core (Güntay et al. 1996). As a result, more oxidized and soluble fuel particles were prevailing in the remaining release and in the following depositions.

Monoelemental particles are often called ruthenium (Ru) particles since, for measurements made after the decay of short-lived isotopes, their total activity was due to ^{103}Ru and ^{106}Ru isotopes in most cases. Some parts of these particles revealed measurable amounts of other elements (mostly Cs). Some other radionuclides could also be detected with lower activities, for example, ^{125}Sb, ^{60}Co, and some volatile fission nuclides (Pöllänen 1997). Additionally, these types of particles showed the presence of a number of nonradioactive elements such as Fe, Ni, Mo, Tc, Rh, and Pd. These particles contained little or no uranium.

For the measurements made during the first days after the accident, Devell et al. (1986) and Pietruszewski et al. (1990) reported about particles composed almost entirely of ^{140}Ba and ^{140}La found in Sweden and Poland, respectively. These particles were referred to as bielemental particles. Elemental analysis proved that these particles contain U.

Sandalls et al. (1993) discussed various suggested possible mechanisms for the formation of Ru particles, including their formation by condensation processes in the air above the reactor, and their dispersion as solid particles already presented in the fuel as a result of the explosion. Salbu et al. (2001) and Kashparov et al. (1995) stated that spherical Ru particles detected in Western Europe could have been present in the nuclear fuel long before the accident (so-called "white inclusions"), whereas large and irregularly shaped Ru particles deposited near the plant were formed at the rate of Ru oxidation during the accident and its subsequent condensation on the nonradioactive particles.

Particles with features "intermediate" between the fuel and ruthenium particle types were also found. They contained the same isotopes as fuel particles but revealed a considerable higher relative activity of Ru. Osuch et al. (1989) assumed that they could be formed as a result of coagulation of U and Ru particles or condensation of the Ru vapors on U particles.

1.2.5 Release Dynamics

In order to assess comprehensively the accident consequences for the environment, it is necessary to know not only the total activity but also the time dynamics of emissions from the accidental reactor. The radionuclide release took place with varying release rates over an extended period of time (mainly over 10 days). In most of the studies, estimates of the time dynamics of the radionuclide release were made taking into account the following four stages of the release from the reactor (GKIAE 1986):

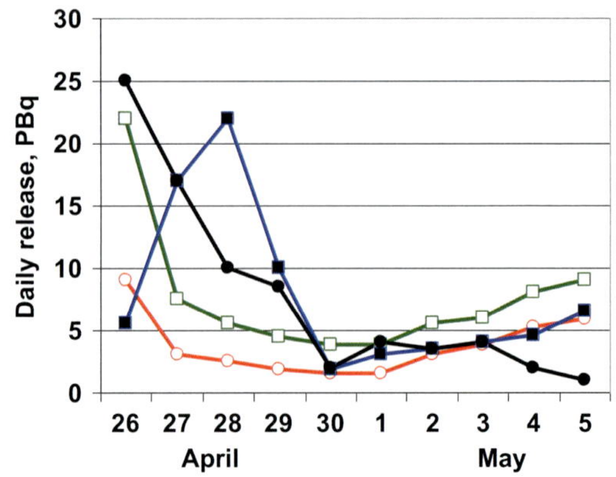

Fig. 1.1 The ^{137}Cs emission intensity (PBq per day) depending on time for the first 10 days of the Chernobyl accident according to GKIAE (1986) (white circles); Izrael et al. (1990)—first version (white squares); Izrael et al. (1990)—second version (black squares); Talerko (2005a) (black circles)

First stage, 26 April 1986. An initial release of dispersed fuel on the first day was caused by mechanical discharge as a result of the heat explosions in the reactor. The radioisotopic composition corresponded roughly to that of the irradiated fuel but was enriched by volatile isotopes of iodine, tellurium, cesium, and inert gases.

Second stage, from 26 April to 2 May 1986. The rate of release from the unit decreased as a result of the measures taken to stop the graphite burning and to filter the releases. During this period, the composition of the radioisotopes being released was again similar to that in the fuel. During this stage, finely dispersed fuel was being carried out of the reactor by a flow of hot air and the graphite combustion products. During this stage, a variability of the values of daily releases could reach more than one order of magnitude (Fig. 1.1).

Third stage, from 2 May to 5 May 1986. An increase in the rate of release of fission products from the unit took place. In the initial phase of this stage, the release was composed mainly of volatile components, especially iodine, but then the radioisotopic composition once more became similar to that of the irradiated fuel (on 6 May 1986). The reason for this was the heating of the fuel in the core to a temperature exceeding 1700 °C as a result of the reactor after-heat. The temperature caused the migration of fission products and the chemical transformation of uranium oxide, which, in turn, led to an escape of fission products from the fuel matrix and their release in the aerosol form on the graphite combustion products.

Fourth stage, after 5 May 1986. A rapid drop in releases took place due to the formation of more infusible fission product compounds as a result of their interaction with the materials introduced, and the stabilization and subsequent lowering of the fuel temperature.

In the framework of such ideas, the well-known reconstructions of the dynamics of the radioactive product release were proposed—in the report submitted to IAEA (GKIAE 1986) and later versions of Izrael et al. (1990) (Fig. 1.1). According to the USSR report (GKIAE 1986), the daily ^{137}Cs release in the atmosphere was maximal on April 26 and then decreased over the next 4 days. After that, for the same relative dynamics of daily emissions, the estimate of total ^{137}Cs emissions was increased approximately twofold (version 1 from Izrael et al. (1990)) (Fig. 1.1). In the later version of the daily release scenario (Fig. 1.1, the second version of Izrael et al. (1990)), a qualitatively different dependence of the release dynamics was proposed: unlike previous versions, the maximum of daily radioactive emissions fell on the third day of the accident (April 28), and the difference in these estimates of daily emissions reaches four times. In all the mentioned scenarios, the qualitative trend of the release intensity in the following days is the same: decrease of the release intensity by April 30–May 1, then the next increase until May 5, and a sharp decrease in the emission intensity after May 5. For comparison, in Fig. 1.1, estimates of the ^{137}Cs emission dynamics during the first 10 days after the accident made by Talerko (2005a) are also given (see below in Sect. 1.5).

The primary estimates of ^{137}Cs emissions from the ChNPP were obtained using measurement data of the gamma radiation dose rate near an emergency reactor. However, a more detailed analysis of available data on radioactive contamination of air and soil, especially for the near zone and on mesoscale distances, has demonstrated a number of problems that are difficult to explain quantitatively (and in some cases qualitatively too) within the framework of these scenarios. This led to the appearance of a series of works in which, based on the solution of the inverse problem of atmospheric transport, alternative reconstruction of the Chernobyl source was carried out (Izrael et al. 1990; Borzilov and Klepikova 1993; Vakulovskii et al. 1993).

Constructing databases of measured data on the radionuclide activity in air and soil after the Chernobyl accident (for example, Evangeliou et al. 2016) enables us to reconstruct the accidental source term (including the release temporal dynamics) on the basis of inverse atmospheric transport modeling techniques (see Sect. 1.5.5). Largely, these works were made possible due to the fact that weather conditions during the initial period of the accident significantly changed (Izrael et al. 1990; Borzilov and Klepikova 1993). In this case, by using modeling, one can try to connect deposition values in certain areas with an emission rate during some time periods.

Significant discrepancies in the results of these studies are the result of the exceptional complexity (and ambiguity of the solution) of the task of restoring a prolonged emission source, which, in turn, is due to a large number of source parameters that need to be known to determine the further distribution of radioactivity in the atmosphere.

1.2.6 *Initial Height of the Release*

From the moment the accident began, a convective plume of hot air transported radionuclides from the destroyed reactor into the atmosphere. GKIAE (1986) stated that according to aircraft monitoring data, the height of the plume on 27 April exceeded 1200 m. During the following days, the height of the plume did not exceed 200–400 m. Subsequently, many studies used this scenario for modeling, sometimes with minor modifications (Sofiev et al. 2007).

However, the results of measurements of radionuclide activity in the air and in fallouts, conducted during the period of the release dispersion in different countries, showed that the height of the radionuclide rise during the Chernobyl release could be much larger.

As is generally known, the transport of most of the release in the first hours after the accident was determined by the wind in the lower part of the atmosphere and was directed to the northwest to Belarus and further to Sweden and Finland. Makhon'ko et al. (1996) presented measurement data of daily iodine-131 deposition conducted on the network of meteorological stations of the former USSR during the period of emissions from the Chernobyl NPP (see Sect. 1.6). In particular, the value of ^{131}I deposition in a meteorological station in Odesa (550 km to the south of the ChNPP) for the period from 08 h on 25 April to 08 h on April 26, 1986 was equal to 2849 Bq/m^2. For comparison, during this period, ^{131}I deposition in Kyiv (110 km to the south of the ChNPP) was only 152 Bq/m^2. A significant increase in daily iodine deposition in Kyiv to a similar level began only during the measurements on April 29–30 (2458 Bq/m^2), when the wind in the atmosphere boundary layer changed to the northern one and the transport of release from the Chernobyl reactor directly to Kyiv began. The explanation for these data is that part of the radioactive release on 26 April was raised to a height of 5 km or more and was transported along the northwest and west periphery of the high-altitude cyclone to the south with a velocity of about 25 m/s (Borzilov and Klepikova 1993). Then, an activity was deposited to the ground due to the rain in Odesa (5.4 mm of precipitations during the first half of 26 April).

Conclusions about the possibility of the rise of radionuclides released from the Chernobyl emergency unit were obtained in a number of works on the modeling of radionuclide atmospheric transport. Gudiksen et al. (1989) suggested that some of the radioactivity must have been transported to heights well within the middle troposphere. They have reconstructed the source term for the Chernobyl accident by comparing the results of atmospheric dispersion modeling with the available radiological measurement data over the Northern Hemisphere. By optimizing the agreement between the measurements and the model calculations, they have suggested the initial vertical distribution of radioactivity in the atmosphere, which includes the "upper-level" part (to be centered at 4500 m with a vertical extent ranging from 1500 to 7500 m) and 'low-level" part (centered at 1300 m and extended from the surface to 1500 m). They obtained the initial radioactive release during the first day, 26 April, became segmented (see Sect. 1.5.2).

Talerko (1990) developed the model of a convective hot plume formed in a temperature-stratified atmosphere in a shear layer with an arbitrary wind velocity profile. Using this model, he calculated trajectories of the active part of the plume formed over the accidental unit of the Chernobyl NPP for the initial 10 days of the release. He obtained that the maximal height of the plume rise significantly varied in time, increasing to 2500 m. Thermal stratification of the atmospheric boundary layer in this period (including its diurnal variability) was revealed as the main factor that determined the plume rise.

Buzulukov and Dobrynin (1993) tried to solve the contradiction between the estimate of Izrael et al. (1990) and the above results stating that this estimate of the release height is accurate only for the nonvolatile radionuclides. They meant, apparently, that volatile nuclides in the form of gases or finely dispersed aerosol particles could reach higher altitudes in the convective plume compared with large fuel particles. But at an initial (active) phase of a plume, within air updraft with the vertical velocity of about 1 m/s or more that was formed above the reactor, the trajectories of a volatile particle with aerodynamic diameter of 1 μm and a relatively large fuel particle of 20 μm (the sedimentation velocity is about 3 cm/s) are not different considerably. Most likely, the separation between nuclides with significantly different sedimentation velocity could take place later, at a diffusion (passive) phase of a plume. In addition, Vakulovskii et al. (1989) indicated that, according to the aircraft measurements of the nuclide aerosol composition that were used by Israel to estimate the initial height of the plume, the ratio of $^{144}Ce/^{137}Cs$ activities in the plume over the reactor was approximately 2 and $^{95}Zr/^{137}Cs$ was approximately 4. Thus, these measurements were carried out in the plume that consisted mainly of condensation rather than fuel particles. So, this explanation cannot be regarded as sufficiently convincing.

Moreover, there is an additional problem in modeling of atmospheric transport of the Chernobyl release. A somehow estimated initial release height is used often as the effective release height value in different tasks of pollutant dispersion modeling. In the Chernobyl case, the release source should be treated not as a "point" elevated source but as a volume one distributed over height in the range from the height of the Unit 4 building (about 60 m) up to the estimated value of the initial release height. The release from this source takes place due to a radionuclide extrainment, that is, their entrainment out of a plume into the surrounding atmosphere due to the ambient turbulence (Netterville 1990). Besides the mentioned work of Gudiksen et al. (1989), attempts to estimate the initial vertical distribution of the Chernobyl release were made by Talerko (2005a) and Davoine and Bocquet (2007) (see Sect. 1.5.5).

1.3 Meteorological Conditions of Radionuclide Transport in the Vicinity of the Chernobyl Nuclear Power Plant

The weather during the release atmospheric transport changed in a complex manner, which resulted in the spread of radionuclides practically throughout Europe and their transport to America and Asia. Rains in several European countries (especially in Sweden, Finland, Austria, and Great Britain) in this period led to the formation of spots of radioactive fallout. Meteorological conditions over Europe during the release were analyzed thoroughly in many studies (Smith and Clark 1988; Persson et al. 1987; ApSimon et al. 1989; Puhakka et al. 1990).

During the acute phase of the accident (from 26 April to 05 May 1986), atmospheric transport of radioactive substances occurred under complicated

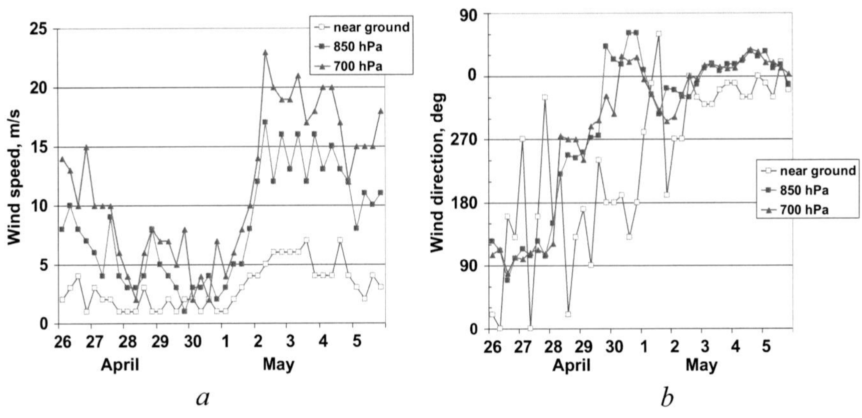

Fig. 1.2 The wind speed (**a**) and direction (**b**) in the layer up to 700 hPa for the period April 26 to May 05, 1986 in Kyiv (according to near-surface and radiosounding measurements)

meteorological conditions that represented various natural synoptic periods. In the initial period of the accident (April 25–26), the weather was determined by a large anticyclone with a center located in the south of the Ural Mountains and a low-pressure region between Iceland and Northwest Europe. The area of the Chernobyl nuclear power plant was located in the region of a low-gradient pressure field with a weak wind of variable directions. According to meteorological measurements in Kyiv, in the first days after the accident, the wind direction in the lower atmosphere was mainly east and southeast (Fig. 1.2), which caused the transport of the release from the Chernobyl NPP in the layer from the earth surface to 700 hPa toward Belarus and the Baltic states. According to the radiosounding data of the atmosphere in the nearest to the ChNPP stations in Kyiv and Gomel at 0 h 26.04.1986, there was stable temperature stratification in the air layer up to 300–400 m in this region, which collapsed in the morning.

The southern, eastern, and central regions of Ukraine were influenced by a small cyclone with a center over the Crimea (Izrael et al. 1990). The nearest atmospheric cold front passed over Western Europe; it was inactive, with small clouds and precipitations. In Ukraine and Belarus, there were no frontal precipitations on the day of the accident, with the exception of small rainfall in the southern regions of Ukraine.

In the second natural synoptic period from 27 to 30 April, the meteorological situation was determined by the pressure field located between the two high- and low-pressure regions, that is, baric col. The field of high pressure slowly moved southeastward, and the low-pressure field, including some of the poorly defined pressure subsystems, was spreading over most of the European territories of the former USSR (Borzilov and Klepikova 1993). One of the pressure subsystems was a minor near-surface cyclone located south of Gomel on April 27. The direction of transport in the vicinity of the ChNPP gradually changed in a layer up to 700 hPa clockwise from the northwest on April 27 to the northeast on April 28 and further to the southeast and south on April 29–30 (Fig. 1.2). During this period, precipitation was observed on the territory of Ukraine and eastern Belarus, associated with the movement of warm air masses from the south to the northwest and an intensive advection of cold air from the north. Maximum precipitation (more than 10 mm per day), coinciding with the period of passage of radioactively contaminated air masses over this territory, was observed in the northeastern Ukraine (Chernihiv and Sumy regions) and in the east of Belarus (Gomel and Mogilev regions).

During the third synoptic period (May 1–6, 1986), an anticyclone had formed over Central Europe and began moving to Scandinavia (Albergel et al. 1988). This changed the wind direction in much of Europe and determined the transport of cold arctic air southward into Ukraine. The wind speed at an altitude of 700 hPa increased to 15–20 m/s, and the radioactivity transport direction changed to the south and remained until 3 May 1986. In the final period of intense emissions on 4–5 May, the wind direction continued to rotate clockwise to southwest direction. Because of a cyclone formation in the southeast of Ukraine, small precipitation was observed in the entire eastern part of Ukraine, which was recorded in some meteorological stations to 5–10 mm per day or more.

Within the entire period of the Chernobyl release atmospheric transport (26 April to 7 May), it had been raining almost throughout Ukraine except in some western regions. At more than 50% of stations of the rainfall gauge network, the amount of precipitation for this period did not exceed 10 mm. Heavy rains and maximum rainfall amounts were noted in the northeast (Kharkiv region) and east (Donets'k region) of Ukraine. It should be noted that according to available data, in most cases, heavy rains (more 15 mm per 12 h) took place in those regions of Ukraine where there was no intensive atmospheric radionuclide contamination caused by the Chernobyl release.

1.4 Measurement Data on the Formation of Air and Ground Contamination in the Initial Period

The values of activity concentration in the air and depositions during the initial stage of the Chernobyl accident, the same as radionuclide composition in them, were dependent greatly on the distance from the release source. The near contaminated zone (up to 100 km) could be distinguished, which includes the Chernobyl exclusion zone (ChEZ) of area 4300 km^2. This territory was contaminated mainly with a fuel component of radioactive fallout, that is, with finely dispersed particles of the nuclear fuel. For the rest of the contaminated territory, ^{131}I and ^{137}Cs were the most important radionuclides in the Chernobyl release from the radiological point of view, because they are responsible for most of the population radiation exposure doses. After ^{131}I decay, ^{137}Cs became a major marker of the radioactive contamination caused by the Chernobyl accident, so main efforts in many countries were aimed at determining the levels of the cesium-137 contamination of the territory.

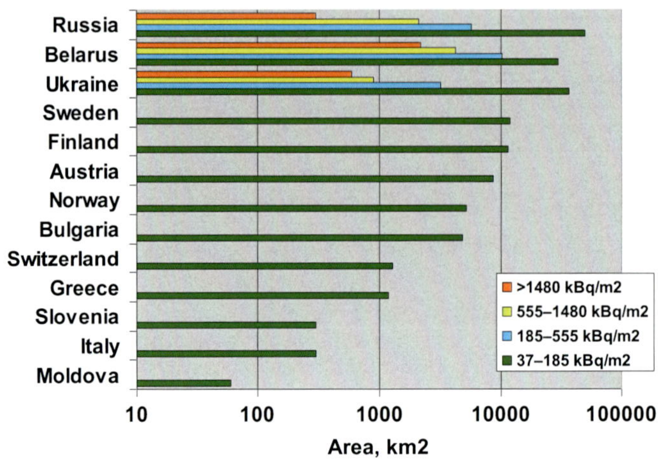

Fig. 1.3 Areas in European countries contaminated with ^{137}Cs (according to De Cort et al. 1998)

Radioactive contamination of the ground was found to some extent in practically every country of the Northern Hemisphere. The largest radioactive contaminated areas were found mainly in Belarus, Russia, and Ukraine (Fig. 1.3).

1.4.1 Measurements in the Near Zone of the Chernobyl NPP

Due to time-varying emission characteristics (total activity, radionuclide composition, and physicochemical forms of radionuclides) and meteorological conditions, a complicated picture of radioactive contamination of the near zone of the Chernobyl nuclear power plant was formed. Near-field contamination of the ChNPP was mainly caused by radionuclides associated with the irradiated nuclear fuel particle matrix (Kashparov et al. 1999). Certain radionuclides such as ^{95}Zr, ^{95}Nb, ^{99}Mo, $^{141,144}Ce$, $^{154,155}Eu$, $^{237,239}Np$, $^{238–242}Pu$, $^{241,243}Am$, and $^{242,244}Cm$ were released from the accident unit in the fuel particle (FP) matrix only. More than 90% of $^{89,90}Sr$ and $^{103,106}Ru$ activity was also released in the FP form (Kashparov et al. 2003). The cesium contamination inside the 30-km zone was due to the combination of fuel and condensation particles. Contamination of the territory with the condensed component has a spotted nature, and on the whole, the relative input of condensed radiocesium increases with the distance from the ChNPP.

In the initial period of the accident, a very limited number of measurements of radioactive contamination in the air of the near zone of the Chernobyl nuclear power plant were obtained. The concentration of radionuclide activity in the air in the vicinity of the Chernobyl NPP and in the plume of releases was regularly measured by the airplanes and helicopters of the State Committee on Hydrometeorology and Environmental Protection of the USSR equipped with sampling and gamma

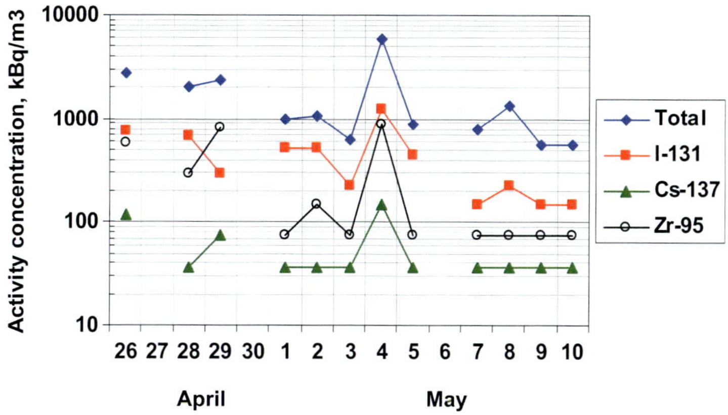

Fig. 1.4 The activity concentration in the air over the fourth block of the ChNPP in April to May 1986 sampled with a helicopter according to data from Whitehead et al. (1988) for 26 April 1986, and Dobrynin and Khramtsov (1993) for following days

spectrometry equipment. Helicopter measurements above the reactor were begun from 27 April 1986 for nine gamma-emitting nuclides, including ^{131}I, ^{103}Ru, ^{106}Ru, ^{144}Ce, ^{134}Cs, ^{137}Cs, ^{95}Zr, and ^{95}Nb (Dobrynin and Khramtsov 1993). In Fig. 1.4, the results obtained in the period of intensive releases for the total activity and three nuclides are shown. These data are completed with results of activity measurements on 26 April 1986 presented in the USSR first report (GKIAE 1986) and recalculated by Whitehead et al. (1988). The maximum total concentration of gamma-emitting nuclides according to the data of helicopter sampling was measured on May 4, 1986 and amounted to 5900 kBq m^{-3}. The relative contribution of ^{131}I to the total activity in this period ranged from 12% on April 29 to more than 50% on 1 May. The contribution of ^{137}Cs to the measured activity ranged from 2% to 7%. As pointed out by Ogorodnikov et al. (2008), the obtained changes in the concentration of radionuclides in the air can be associated not only with the variability in the intensity of radioactive release from the reactor but also simply due to the poor representativeness of the samples.

The first maps of the distribution of the depositions density of main dose-forming nuclides in the near zone of the Chernobyl NPP were created in the first months after the accident, mainly using data on gamma radiation dose rate measurements and calculation results of the relative contribution of individual nuclides to the doses. However, a detailed picture of the radioactive contamination of the near zone of the Chernobyl nuclear power plant by the main radiological long-living nuclides was finally clarified only a few years after the accident. In 1992–1994, a detailed map of ^{137}Cs contamination of this territory (scale 1: 25,000) was created on the basis of aerogamma-spectrometric survey data, conducted by the Institute of Radioecology (Kyiv, Ukraine) under the guidance of E.K. Garger (Shestopalov 1996). In 1997–2000, based on measurements of radionuclide content in 1300 soil samples, the Institute of Agricultural Radiology (Kyiv, Ukraine) under the guidance of

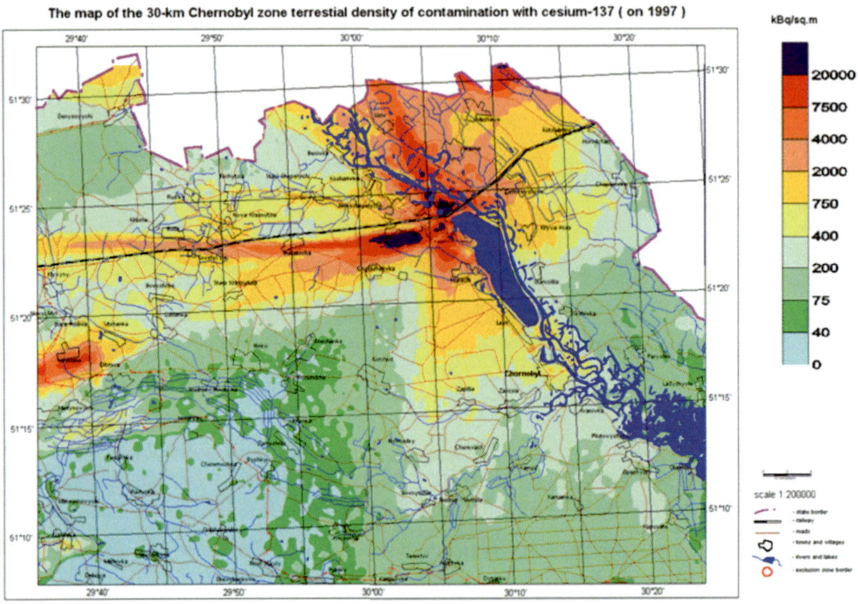

Fig. 1.5 The map of the ^{137}Cs soil contamination density in the Chernobyl exclusion zone (taken from Kashparov et al. 2001)

V.A. Kashparov created detailed maps of contamination of the 30-km zone of the ChNPP with ^{137}Cs (Fig. 1.5), ^{90}Sr (Kashparov 2001), and $^{239+240}$Pu (Kashparov et al. 2003).

The contamination field in the near zone of the Chernobyl NPP consists of three well-marked traces: western, northern, and southern ones. They contain about 10–15%, 50–60%, and 20–25%, respectively, of the fuel particles that have been deposited in the Chernobyl exclusion zone outside the ChNPP industrial site during the accident (Kashparov 2001).

As a result of the initial explosion at the fourth unit on 26 April 1986, a large number of fuel particles in the form of crystallites (or grains) of the uranium dioxide fuel with a median diameter of 5–6 μm (Kashparov et al. 2003) were released. The radioactive cloud of fuel particles was transported in the lower atmosphere to the western direction. These presumably nonoxidized fuel particles deposited quickly on the ground, formed a narrow (a width of only 1–1.5 km) western trace of fallout extending 100 km of the ChNPP (Fig. 1.5). Additionally, the U–Zr–O particles were formed as the result of the interaction between nuclear fuel and zircaloy during a meltdown at the time of the accident, and then they were deposited within the western trace (Zhurba et al. 2009; Savonenkov et al. 2009). Particles with maximal sizes up to 1.0 mm were deposited in the immediate vicinity of the ChNPP, at distances of up to 0.5 km from the ChNPP. However, the radiological importance of particles with size over 50–100 μm is low outside the 2–5 km zone, and their contribution to the radioactive contamination of the territory as a whole is much lower than the contribution of FP of micron range (Kuriny et al. 1993; Kashparov et al. 1999). In the western trace at distances from 1 to 13.5 km from the fourth unit, the average sizes of fuel particles decrease from 750 to 20 μm. Radionuclide ratios in particles on the western trace were close to the ratios in the initial fuel—the fractionation coefficient of ^{106}Ru relative to ^{144}Ce in soil samples taken in 1986–87 at a distance from 1.0 to 13.5 km varied within 0.8–1.0, and of ^{137}Cs 0.82–1.1 (Savonenkov et al. 2009).

Further to the middle of the day on 26 April, the direction of transport in the lower atmosphere changed rapidly to the southwest. The fuel heating in the reactor due to the remaining heat generation caused the increase of the volatile fission product release, in particular, iodine and cesium radioisotopes. As a result of the release of the condensation component and its further atmospheric transport and deposition, a so-called "cesium spot" was formed in this period in the near zone close to the settlements of Vesniane and Polesskoye at a distance of 32–55 km southwest from the nuclear power plant (Fig. 1.5). Vakulovskii et al. (2012) published the results of the first measurements of radioactive soil contamination conducted on this territory near the village of Termahovka on April 30, 1986, that is, during the period of intensive emissions from the reactor. By this day, the ^{137}Cs deposition density was about 40% of the ultimate value, and the fractionation coefficient of ^{137}Cs relative to the refractory ^{95}Zr was 57.4, that is, radioactive fallout in this spot virtually did not contain fuel particles.

In the second half of April 26, the direction of the transport began to change to the west again, and by the end of the first day of the accident, the transport was already in the northwest direction. In this time and later, the main mechanism of fuel particle formation was the oxidation of uranium dioxide to U_3O_8 in air at high temperatures, and fuel particles with a median radius between 2 and 3 μm were formed (Kashparov et al. 1996). The fuel particles of this type were depleted with volatile nuclides. The relative release of fission products of uranium oxide at high temperatures decreases with the increase of their binding energy with oxygen in the following sequence: Kr > Xe > I > Ag > Cs > Te > Sr > Ru > Ba > Zr > Ce (Andriesse and Tanke 1984). In the period from 27 to 29 April in the near zone, the northern trace of deposition was formed, which contained a high proportion of condensed particles, and the fraction of fuel particles in depositions decreased with distance from the reactor. Unlike the western trace, the contamination of the northern trace consists of particles of oxidized fuel with a high degree of dispersion. In the same period, a spot of fallout was formed in the territory of Belarus near the villages Radin and Kryukovo, in which the condensation component enriched with ^{137}Cs predominates. The mechanism of formation of this spot with a deposition density of more than 37,000 kBq m^{-2} remained unclear. The most likely explanation for the causes of intense deposition of fine particles is wet removal with atmospheric precipitation. However, there is no data on rain measurements in this area during the period of the release atmospheric transport (the nearest meteorological station in Bragin, about 60 km away, did not detect any rains in this period).

In the period of the south trace formation (from 30 April to 6 May 1986), the temperature of the reactor decreased and the release of volatile fission products decreased. However, the lower temperatures intensified the oxidation of the exposed uranium dioxide fuel (a process which is most efficient at between 600 and 1200 K) resulted in its dispersion (Kashparov 2016). Thus, the southern trace was formed mainly of finely dispersed oxidized fuel particles. Southward and southeastward from the reactor, more than 50% of the radiocesium was in the form of fuel particles, the fraction of fuel particles in the fallout decreases with increasing distance from the NPP (Kuriny et al. 1993; Ivanov et al. 1996).

At the present time in the Chernobyl exclusion zone (ChEZ), ^{90}Sr is the most radioecologically significant radionuclide after the decay of short- and medium-lived radionuclides. Initially, ^{90}Sr incorporated within the fuel particle matrix had low migration mobility compared with the condensation component of the Chernobyl fallout. In 1986–1987, in sod-podzolic sandy soils of the 30-kilometer exclusion zone, only 10–30% ^{90}Sr was in the exchangeable form (Konoplev et al. 1988). Weathering of the fuel particles deposited in the Chernobyl exclusion zone causes increasing the radionuclide mobility in soil with time after deposition (Konoplev 2020). Now the major radioactive flow-out from the territory of the exclusion zone is caused by ^{90}Sr leaching from fuel particles and its migration to surface and ground waters. The portion of exchangeable ^{90}Sr in soil increases, and it results in an increase of plant contamination with ^{90}Sr. It was found that along with the acidity of the soil solution, the degree of oxidation of the deposited fuel particles determines the rate of their dissolution considerably (Kashparov et al. 1999). The U–Zr–O particles in the Chernobyl release were found as chemically extra-stable. Between 2% (in the northern and the southern traces) and 21% (for the western trace) of ^{90}Sr activity is associated with these weathering-resistant fuel particles (Kashparov et al. 2004). For the rest of fuel particles, the mentioned differences in the part of nonoxidized and oxidized fuel particles within three deposition traces determine the different values of their dissolution rate and the exchangeable form fraction of ^{90}Sr in soil.

Kashparov et al. (2001) created a map of the contamination density of the ChNPP near zone with $^{239+240}$Pu and estimated that approximately 70–80% of the total inventory of the fuel component radionuclides in the upper 30-cm soil layer are located in only 10% of the 30-km zone territory in its central part. So, contamination of the ChEZ with transuranium elements rapidly decreases with the distance from the ChNPP. At the periphery of the exclusion zone, the density of contamination with $^{239+240}$Pu is lower than 1 kBq m^{-2}, further decreasing to the background levels caused by global fallout.

1.4.2 Measurements Outside of the Near Zone of the Chernobyl NPP

The Chernobyl accident became the test for the capabilities of national monitoring networks to early detect and to predict the large-scale nuclear accident consequences for the environment. During the initial stage of the accident, the monitoring networks in many European countries the same as many different laboratories throughout Europe measured the major radionuclide concentration activity in the air and daily depositions (Steinhauser et al. 2014). There were some degrees of inconsistency in these data because of different sampling and measurement techniques and various sampling periods. But, nevertheless, the obtained data became very valuable information for both the reconstruction of the large-scale radioactive contamination dynamics and the estimation of the Chernobyl accident consequences. Additionally, these data were used for the validation of long-range transport models (Klug et al. 1992; Evangeliou et al. 2016).

During the Chernobyl accident in April to May 1986, the main source of radiological information in the territory of the former USSR was measurement data of the USSR Goskomgidromet (State Meteorological Service). The particulate ^{131}I and ^{137}Cs daily depositions on collectors were sampled in 71 meteorological stations, including 24 in Ukraine and 8 in Belarus. In some of them, the daily-averaged ^{137}Cs and ^{131}I activity concentration in the air was also measured. The gaseous iodine was not measured in any of these posts. In Fig. 1.6, locations of the Goskomgidromet measurement stations in the European part of the former USSR are shown. The daily-averaged ^{137}Cs and ^{131}I activity concentrations are given in Fig. 1.7 for some measurement points. As can be seen, the maximal daily-averaged air activity concentration decreases with increasing distance from the source.

The regular measurement data carried out during the initial period of the Chernobyl accident in 17 Central and Western European countries were collected lately in the REM (Radioactivity Environmental Monitoring) Data Bank in the Joint Research Centre, Ispra (Raes et al. 1990; JRC 2015). The aim of this activity was to establish a historical record of the accident and its consequences, as well as provide the scientific community with data sets useful for checking models on radionuclide behavior in the environment. The databank includes data of ^{131}I, ^{134}Cs, and/or ^{137}Cs activity concentrations in 94 measurement points, depositions in 24 points, and total ^{131}I (gaseous + aerosol fraction) in 13 points (Fig. 1.6). On the whole, monitoring stations all over the world began intensive sampling programs only after the first Swedish communication about detecting the increased radioactive contamination. So, in most of the European monitoring points, the measurement data in the first 3–5 days are absent. The largest monitoring networks were activated in this period in Germany (22 points), France (20), Italy (8), Austria (7), and former Czechoslovakia (10). In three Scandinavian countries, measurements were made in 11 points, but they start regular samplings ahead of others (on 28 April, as a rule), and in most points, the measuring data were obtained with a high resolution in time—up to 1 h. The monitoring network in different countries was managed to detect periods of Chernobyl release arrival in Central Europe (29 April to 1 May) (Fig. 1.8). The maximal values of the ^{131}I activity concentration in air were obtained to be 70 Bq m^{-3} in Prague on 29 April and 58 Bq m^{-3} in Vienna on 30 April, and the values of ^{137}Cs were 23 Bq m^{-3} in Prague and 11.1 Bq m^{-3} in Brotjacklriegel, Germany, on 30 April 1986.

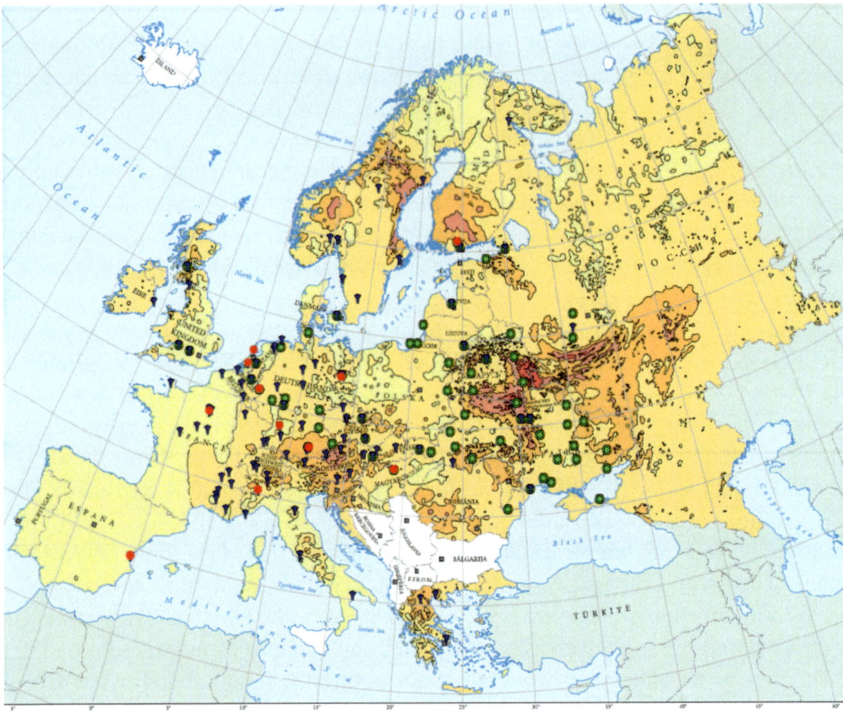

Fig. 1.6 Map of the ^{137}Cs surface ground deposition in Europe after the Chernobyl accident (from De Cort et al. 1998). The monitoring sites are added where the aerosol ^{131}I and ^{137}Cs air concentration (▼), the total (gaseous + aerosol) ^{131}I air concentration (●), and the ^{131}I and ^{137}Cs deposition (◉) were measured during the initial stage of the Chernobyl accident

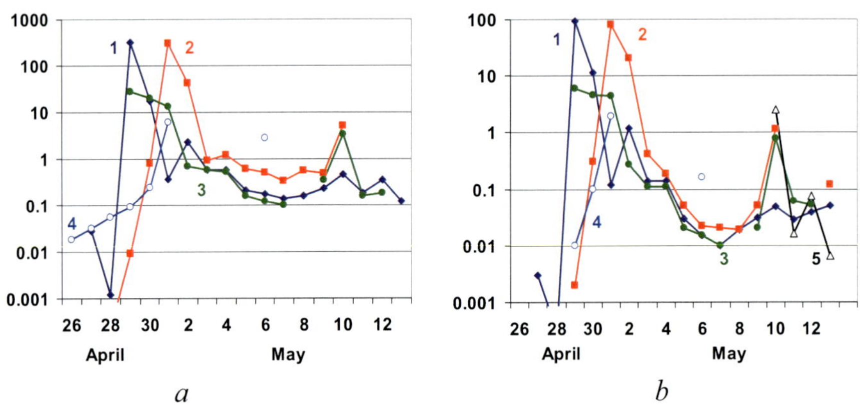

Fig. 1.7 The daily-averaged activity concentration of ^{131}I (**a**) and ^{137}Cs (**b**) in the air in Minsk (1), Baryshivka (Ukraine, 150 km to the southeast from the ChNPP) (2), Vilnius (3), Odesa (4), and Kyiv (5)

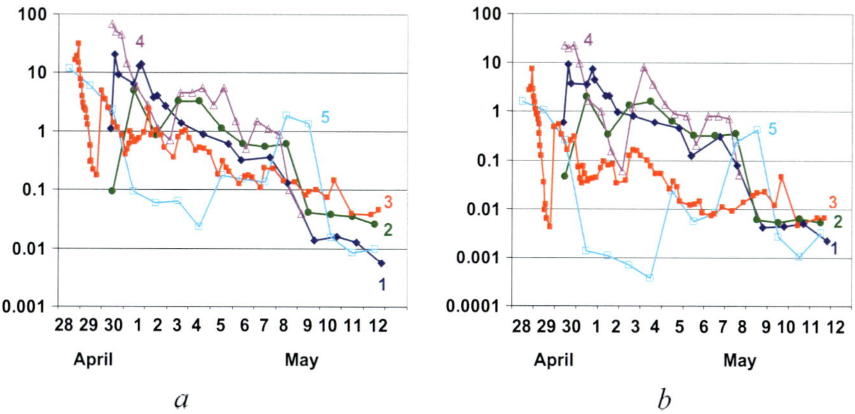

Fig. 1.8 The ^{131}I (**a**) and ^{137}Cs (**b**) activity concentration in the air in Neuherberg (1), Salzburg (2), Helsinki (3), Prague (4), and Stockholm (5)

Recently, the Norwegian Institute for Air Research (NILU) formed the updated database of air concentration and deposition measurements over Europe after the Chernobyl accident (Evangeliou et al. 2016). It includes both the data of the REM database and additional data collected over the years following the accident and make them publicly available, mainly ^{137}Cs deposition data obtained in Ukraine and Belarus.

In several countries, it was possible to measure the activity of radionuclides of the Chernobyl release not only at ground stations but also at various heights using aircrafts equipped with air samplers.

In Finland, the aircraft measurements of radioactivity were started on April 28 (Sinkko et al. 1987). The Chernobyl radionuclide debris was found to be concentrated at an altitude of 1300–1500 m above southwest Finland. The largest activity concentration was detected on 29 April at a height of 1500 m east of Helsinki: 420 Bq m^{-3} of ^{131}I, 167 Bq m^{-3} of ^{137}Cs, and 65 Bq m^{-3} of ^{140}Ba. The concentrations of radionuclides in the cloud were about 20 times and more than at the ground level on April 28 and 29 (Fig. 1.8). The next day, the measured high-altitude concentrations were obtained to be about two orders of magnitude lower.

Jaworowski and Kownacka (1988) reported about measuring ^{131}I and $^{134,136,137}Cs$ activity concentration in the atmosphere over Poland at the ground level at two stations and at heights from 1 to 15 km during 29 April to 20 May 1986. The maximal near-surface ^{131}I activity concentration was measured in Warsaw on 29 April, 11.49 Bq m^{-3} with three orders of magnitude lower values in the troposphere between 3 and 12 km. In the following days, the maximum ^{131}I concentration was observed at upper heights 1–9 km (up to 1 Bq m^{-3}). The upward transport of radioactivity through the tropopause (about 11 km in this study) resulted in rather high contamination of the lower stratosphere (on 5 May, the iodine concentration at 15 km was only ten times lower than at the ground level). The reasons for the stratosphere contamination with the Chernobyl radionuclides could be both the effect of the initial explosive of the accidental fourth unit of the ChNPP and upward movement of activity up to high altitudes during the following atmospheric transport. The importance of the latter mechanism was confirmed by measurements on 5–11 May. During this period, a local maximum of the concentration was obtained at levels from 3 to 9 km. On 5 May, the iodine concentration at 9 km was 3.5 higher than in the ground level layer, and on 11 May at 3 km altitude, it was 4.5 higher than near the ground. On the other hand, the high-altitude levels were a source of additional radioactive contamination of the surface. A sharp increase in the deposition of the gross beta activity was observed between 8 and 10 May at some of the measurement stations in Poland due to the activity washout of tropospheric masses with rain.

Shortly after the accident, surveys were conducted in most European countries using aircraft spectrometers to measure the deposition of ^{137}Cs and other radionuclides on the earth's surface. In addition, many soil samples were collected and analyzed in radiological laboratories. Thus, large data sets were collected, and on this basis, the detailed maps of contamination with cesium-137 (De Cort et al. 1998) and other nuclides, including ^{90}Sr and plutonium isotopes (Kashparov et al. 2001; BNR 2011; UNR 2011), were created. Based on the obtained measurement data, meteorological analysis, and subsequent atmospheric transport modeling, the picture of the formation of radioactive contamination over the territory of the former USSR, Europe, and the Northern Hemisphere was described in detail elsewhere (UNSCEAR 2000; Borzilov and Klepikova 1993; De Cort et al. 1998; NEA 2002).

On 26–27 April, the released nuclides were transported predominantly in the west and northwest directions, resulting in contamination of Northwestern Ukraine and Southwestern Belarus (Pinsk region), and further to Kaliningrad and Lithuania. Over Lithuania, the plume separated into two main paths. At lower altitudes (750–1000 m). the plume continued toward Sweden and Norway. At higher altitudes (1500–2500 m), the plume turned toward the north. The plume arrived in Southwestern Finland on 27 April at 12 UTC for a release height of 2000 m. Then the plume went across the country northeastward and back to Soviet Union (Paatero et al. 2011).

During 27–28 April, the wind direction near the ChNPP rotated clockwise, and on 29 April, the release was transported to the east. Outside of the near zone of the Chernobyl NPP, the releases in this period resulted in the formation of two large cesium contamination spots in Belarus and Russia due to rainfalls in these areas on 28–29 April at the time the plume passed over (Fig. 1.9). In Belarus and Russia, the Gomel–Mogilev–Bryansk spot was formed at a distance of about 200–250 km to the north-northeast of the ChNPP. The maximum level of ^{137}Cs soil contamination in this spot was found in the village of Chudyany, Mogilev region: 59,000 kBq m^{-2} (BNR 2001). The Kaluga–Tula–Orel spot in Russia, 500 km northeast of the ChNPP, was formed from the same radioactive cloud. However, the levels of deposition of ^{137}Cs were lower, usually less than 600 kBq m^{-2} (Balonov et al. 1996). Linnik et al. (2016) made detailed analysis of the spatial inhomogeneity of the radioactive fallout fields in Russia and showed the role of various factors in the formation of deposition spots at various space scales, including impact of

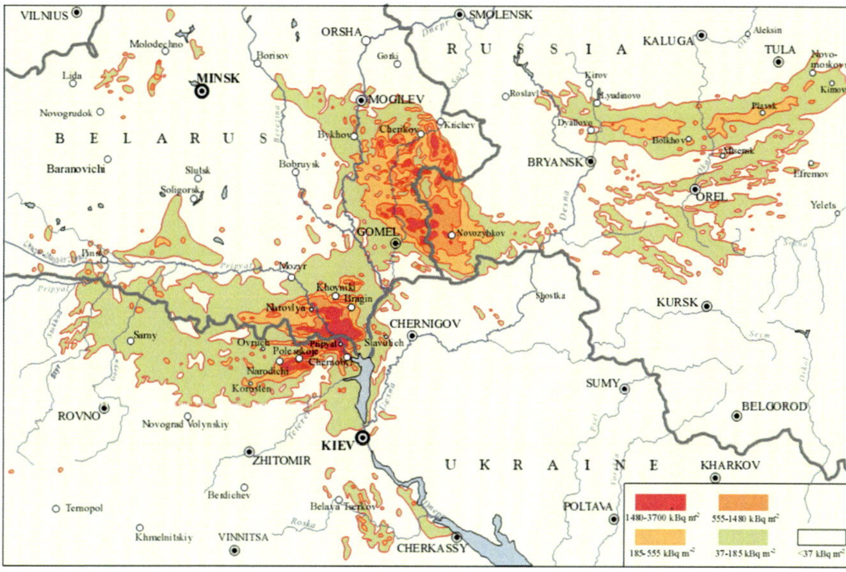

Fig. 1.9 Ground deposition of ^{137}Cs in the most contaminated territories of Ukraine, Belarus, and Russia (taken from UNSCEAR 2000)

meteorological parameters, as well as a number of landscape characteristics of the radioactively contaminated territory (alternation of watersheds, alternation of elevated open plowed areas with fragments of forested areas, etc.).

Some parts of the activity released during the first days of the accident migrated to Poland, entered over the Polish territory on 29 April. Later the contaminated air masses moved to Central Europe, as well as the Northern Mediterranean and the Balkans (NEA 2002). On 2 May, the radioactively contaminated air masses arrived at Great Britain. Due to wet deposition processes, some territories were contaminated in Austria, Germany, Norway, Romania, Finland, Sweden, and Great Britain (Wheeler 1988). In Finland, the highest obtained deposition values were 70 kBq m^{-2} for ^{137}Cs and 420 kBq m^{-2} for ^{131}I and about 30 kBq m^{-2} for ^{95}Zr (Paatero et al. 2011).

From 30 April, the wind direction in the region of the Chernobyl NPP changed to the south, and the radioactive contamination of the Central and Southern Ukraine begun, including Kyiv. During the first days of May, the wind direction continued rotated clockwise, and up to 5 May, it made almost a full turn of 360°, turning to the west again. As a result, almost all territory of Ukraine was contaminated in this period.

The Chernobyl releases were detected in the Northern Hemisphere as far away as Japan—57 mBq m^{-3} of ^{137}Cs, 230 mBq m^{-3} of ^{131}I on 6 May 1986 (Aoyama et al. 1986) and North America—10 mBq m^{-3} of ^{137}Cs and 20 mBq m^{-3} of ^{131}I in New York on 11 May 1986 (Larsen et al. 1989). But on the whole, countries outside Europe received very little deposition of radionuclides from the accident. No

Table 1.2 Activity concentration in depositions in the near zone of the Chernobyl NPP relative to ^{137}Cs concentration (on 26 April 1986) (based on Izrael et al. 1990; Muck et al. 2002; IAEA 2006)

Nuclide	Western trace (near zone)	Northern trace (near zone)	Southern trace (near zone)	In cesium hot spots (far zone)
^{90}Sr	0.5	0.13	1.5	0.014
^{95}Zr	5	3	10	0.06
^{99}Mo	8	3	25	0.11
^{103}Ru	4	2.7	12	1.9
^{132}Te	15	17	13	13
^{131}I	18	17	30	10
^{137}Cs	1.0	1.0	1.0	1.0
^{140}Ba	7	3	20	0.7
^{144}Ce	3	2.3	6	0.07
^{239}Np	25	7	140	0.6
^{239}Pu	0.0015	0.0015	–	–

deposition was detected in the Southern hemisphere by the surveillance networks of environmental radiation (UNSCEAR 2008).

It was found that the radionuclide composition of depositions varied greatly in different contaminated areas. Radionuclide ratios significantly changed in the release due to the processes occurring in the reactor (in particular, the increase in the fuel temperature in the early days of May, which led to an increase in the relative fraction of refractory elements in the release). Therefore, radioactive fallout, formed due to emissions in different periods (northern, western, and southern traces), differs in composition. Further, radionuclide ratios vary as a function of the distance from the source due to their different physicochemical properties, which determine the dry deposition velocity of nuclides on the underlying surface. Finally, the composition of deposition in the wet deposition spots differs significantly from spots due to dry deposition. In particular, the ratio of $^{131}I/^{137}Cs$ in depositions depends on the type of fallout: wet or dry (Zvonova et al. 2010). The spots of high contamination with ^{137}Cs were found after wet deposition. But ratio $^{131}I/^{137}Cs$ inversely depends on rain amount (Kruk et al. 2004). Radionuclide activities in the deposits in the near (100 km) zone of the ChNPP relative to ^{137}Cs are shown in Table 1.2. The activity ratios for the western and the northern trace are very similar. For the southern trace, generally larger ratios than for the other traces are observed. Muck et al. (2002) analyzed obtained measurement results in detail and proposed the best estimate of the ratios of the various radionuclides in ground depositions as functions of the distance from the Chernobyl NPP. All activity ratios, with the exception of $^{132}Te/^{137}Cs$, show a decrease in activity with increasing distance to the ChNPP. Such a decrease is less significant for ^{95}Zr and ^{144}Ce (about three times) than for ^{99}Mo and ^{140}Ba (two orders of magnitude) or ^{90}Sr and ^{103}Ru (one order). The $^{131}I/^{137}Cs$ ratio decreases about four times at a distance of more than 1000 kilometers. In the radius of the first 200 km, there was actually no change in the ratio (IAEA 2006).

The contamination of the territory outside the Chernobyl exclusion zone with ^{90}Sr is more local, compared to ^{137}Cs. In Belarus, the levels of soil contamination with strontium above 5.5 kBq m^{-2} are found in an area of 21.1 thousand km^2, which is 10% of the territory of the republic. The highest activity of ^{90}Sr in the soil in the far zone was found at a distance of 250 km from the ChNPP: in the Cherikovsky district of the Mogilev region, 29 kBq m^{-2}; and in the northern part of the Gomel region, in the Vetka district, 137 kBq m^{-2} (BNR 2011). In Russia, the maximum levels of contamination with ^{90}Sr are in the western part of the Bryansk region (Novozybkovo–Zlynka area)—about 18 kBq m^{-2}.

1.4.3 Hot Particles in the Chernobyl Release

Contamination of air, soil, and other environmental components after the Chernobyl accident was determined by radionuclides released in gaseous and particulate forms. Particulate matter consisted of fine radioactive aerosol particles with a size of a tenth of μm to several μm. These components of the release could be transported to large distances, and they formed a field of radioactive depositions around the world. From the point of view of the atmospheric transport modeling problem, this part of the Chernobyl release could be treated as a weightless pollutant, and calculation of air parcel trajectories originating from the Chernobyl NPP during the accident could be regarded as a good enough method for initial estimations of spatial distribution of contamination. Soon after the Chernobyl accident, however, small particles of relatively high specific activity with a wide range of chemical compositions were found on the ground surface at various distances from the ChNPP. They could cause local, nonstochastic damage in their closest location in a human body and may introduce possible long-term effects. These fuel and monoelemental particles (see Sect. 1.2.4) found at different distances from the Chernobyl NPP are often referred to as hot particles. Pöllänen (2002) reviewed in detail problems concerning with ambiguity of definition for the concept of a "hot particle," the same as different threshold values in detecting the particles and different analysis methods. Nevertheless, the analysis of individual particles of different types complements conventional bulk analysis and proves to be useful for estimation of radioactive contamination of the environment caused by the Chernobyl accident (Sandalls et al. 1993; Pöllänen 1997; Pöllänen et al. 1997; Devell 1988; Paatero et al. 2010; IAEA 2011).

The number of hot particles per area, their activity, and size are proved to be dependent on the distance from the Chernobyl NPP. Large fuel particles have been identified as a major contributor to the total activity deposited within 60 km of the plant. Of the particles identified within 10 km, more than 95% were attributed to fuel and only less than 3% to condensed particles. Owing to the high temperature involved, fuel particles were largely depleted in Cs isotopes and to a lesser extent in ^{90}Sr. However, larger fuel particles were less depleted in Cs isotopes than smaller particles. Near the plant, fuel particles up to hundreds of micrometers in diameter and of activity above 1 MBq were found (Salbu et al. 1994).

Hot particles from Chernobyl have also been identified in distant areas, for example, in Poland, Greece, Bulgaria, Hungary, as well as the Nordic countries (Güntay et al. 1996). In Norway, small-sized fuel particles containing U and mixed fission products depleted in Cs isotopes have been identified (Salbu et al. 1994). A typical activity of fuel particles found in Finland was 100 Bq (Saari et al. 1989).

Nearly monoelemental ruthenium/rhodium/molybdenum particles were also found (Pöllänen et al. 1997). Some of the Ru particles were more than 10 μm in diameter. The most active particle found in Sweden was 49 kBq (Kerekes et al. 1991), in Poland 308 kBq (Broda 1987) and in Greece 43 kBq (Kritidis et al. 1988). Persson et al. (1987) reported that at several sites in eastern Svealand (eastern central Sweden) and in southern Finland, relatively large particles with a diameter of about 5 μm or more were found during the period 27–30 April 1986. The most radioactive particles contained only molybdenum and ruthenium isotopes. Particle samples from Stockholm and Uppsala contained up to 10 kBq ^{99}Mo, 30 kBq ^{103}Ru, and 5 kBq ^{106}Ru (as of April 26).

Most particles found at a large distance (>500 km) from the plant were less than 1 μm in diameter although coarse particles up to tens of micrometers were also found; inactive material was sometimes present in these particles (Pöllänen et al. 1997).

Pöllänen (2002) reviewed the values of the hot particle number concentration in air. It was found from 0.0011 to 0.08 m^{-3} in Finland in the period 27–30 April 1986, from 0.0002 to 0.068 m^{-3} in Germany, and 5×10^{-5} m^{-3} in Hungary in May 1986. The wide range of values is because of difficulties in locating the particles, different threshold values in detecting the particles, different analysis methods, and different timing of the estimates.

Osuch et al. (1989) analyzed 206 hot particles found on the ground in early autumn 1986 in northeastern Poland. The threshold for detection of particles was about 50 Bq. The activities of Ru hot particles were found to be greater on the whole. Distribution of their total activity has a maximum at the activity measurement interval 20–50 kBq, and the whole range of found particle activity was 0.25–500 kBq. The range of detected U particles was from about 0.07 to 4 kBq with maxima in the activity measurement interval 0.5–1 kBq. However, the authors noted that the left side of the distribution for this type is likely to be disturbed by the detection threshold; thus, the measured averaged activity of U particles (1.1 $\pm$ 0.1 kBq) overestimated the real value. On the contrary, the measured averaged activity of Ru particles (19 $\pm$ 3 kBq) could be expected close to their real value because of their considerable higher activities. The obtained conclusion that the uranium particles increased in radioactivity with size but the smaller ruthenium particles had the highest activity was confirmed by measurement data in other countries (e.g., Raunemaa et al. 1987).

It was found that fuel (uranium) particles were depleted in ^{134}Cs, ^{137}Cs, ^{103}Ru, and ^{106}Ru (Osuch et al. 1989). For particles collected in Poland, the values of their average activity ratios to the activity of ^{95}Zr were found to be 71%, 45%, 77%, and 80% of inventory, respectively. These data for hot particles differed markedly from isotopic composition in the general fallout in Poland (Pienkowski et al. 1987), where excess of the ruthenium isotope ratio to ^{95}Zr was obtained larger up to seven times compared with their values in the fuel of the accidental Unit 4, and of the cesium isotopes—above 80 times. The reason for these differences is the different degrees of radionuclide volatility during the accidental release. The reduction of Cs activity in hot particles is stronger than of Ru isotopes due to large volatility of Cs compared with less-volatile Ru. The same reason results in adverse effects during deposition field formation at large distances from the Chernobyl NPP. The varying activity ratios of nonvolatile elements refer to fuel with different burnups.

The distribution of Ru particles over the territory of Poland was found to differ from that of U particles due to these particles getting to the atmosphere during different stages of the accidental release or different properties of their atmospheric transport and deposition.

Unfortunately, the sizes of hot particles were measured in only a few cases. So, Pöllänen et al. (1997) estimated the value of the particle aerodynamic diameter for available data of hot particle activity measurements made in 15 European countries. For U particles found near the ChNPP and within 30-km zone, the particle aerodynamic diameter was estimated to be in the range of 100–1000 μm. In Finland and Sweden, most of the found U particles had a diameter in the range of 15–40 μm (their total activity was several hundreds of Bq). The similar estimations were obtained for particles detected in Eastern and Southern Europe, while in Central Europe (Germany and Switzerland), the typical aerodynamic diameter of found particles was about 10 μm (the measured total activity of a particle is about 50 Bq). The aerodynamic diameter of Ru particles was estimated to be about 5 μm in Finland (for particles with activity about 100–200 Bq), and up to 37 μm in Sweden (activity 49 kBq). The largest value of the Ru particle diameter 58 μm was estimated for the particle with activity 308 kBq found in Poland.

The gravitational sedimentation velocity of particles with an aerodynamic diameter of 37 μm and 58 μm equals 0.041 and 0.101 m s^{-1}, respectively. Hence, long-distance transport of both condensed and fuel particles released during a nuclear accident has been more extensive than previously anticipated (Salbu et al. 1994).

Pöllänen et al. (1997) calculated the length of the particles' trajectories as a function of a particle size and an effective release height with the use of the TRADOS model. They obtained that particle with an aerodynamic diameter of 20 μm (gravitational sedimentation velocity is 0.012 m s^{-1}), which released in the first period of the accident, could not reach Sweden even if the effective release height was 2000 m. So, they assumed that either the maximum effective release height must have been considerably higher than previously reported (>2000 m) or convective cells with air updrafts, developing during the day of April 26 in Belarus and Baltic states, lifted the particles to higher altitudes. The same explanation—convective uplift in cumulus clouds—was made for large particle transport to northeastern Poland (500–700 km from the ChNPP).

The similar estimations were obtained in Persson et al. (1987). They calculated that particles with such diameter might have been transported to Sweden on April 28, provided that the initial plume rise was 500–1000 m and there were upward air motions over Belarus up to heights of 2500–5000 m.

The Norwegian Meteorological Institute simulated the transport of the large particles in the Chernobyl accident using the atmospheric transport model SNAP (Bartnicki et al. 2001). The simulation results showed the deposition of radionuclides was dependent mainly on particle size, and particles with a diameter of 10 μm could travel long distances before being deposited in Norway during the Chernobyl accident.

The most important conclusion as a result of made studies (Pöllänen et al. 1997) was that large particles could be transported to other areas differently than small particles and gaseous species owing to sedimentation and different wind conditions over height during transport. So, calculations of air parcel trajectories are not sufficient to identify the fallout areas of radioactive material. In the case of large particle release, the size distribution and number of particles released in each size class should be available as input parameters for following atmospheric transport modeling in the case of severe emergency. To estimate the radiological consequences of the accident, the simulations must be performed separately for each size class.

1.5 Atmospheric Transport Modeling

1.5.1 The Experience of Using the Radionuclide Atmospheric Transport Modeling to Assess the Consequences of the Chernobyl Accident

The accident at the Chernobyl nuclear power plant in 1986 was an example of a major radiation emergency that impacted the population far beyond the NPP responsibility area. Radiological protection of the population of radioactively contaminated territories in the acute phase of the accident was not fully ensured: (1) due to the absence of on-line computer decision support systems (DSSs) that would provide operational assessment and forecast of the development of the radiation situation on the national and global scale; (2) as a result, the existing radiation monitoring facilities in the former USSR were not optimally used for the environmental monitoring in radioactively contaminated areas, and the general pattern of radioactive contamination on the regional and national scales was clarified only a few months after the accident; and (3) this, in turn, led to the loss of the opportunity to obtain factual information on radioactive contamination of environmental objects by short-lived and middle-lived radionuclides, in particular, radioactive iodine (in particular, in the territory of Ukraine).

The Chernobyl disaster has shown that in a severe nuclear accident it is necessary to obtain information about the radioactive contamination of air, soil, and vegetation simultaneously at different spatial scales (from hundreds of meters to thousands of kilometers) under conditions of substantial uncertainty of the release source term and lack of direct measurement data. In this situation, methods of mathematical modeling of radioactive emissions into the atmosphere and deposition on an underlying surface are an important tool for assessing the consequences of a radiological emergency. They solve the following main practical tasks:

1. Operational forecast of emission dispersion and assessment of the radioactive contamination of the environment for the organization of radiological monitoring during the early phase of the accident;
2. Reconstruction of the source term using measurement data by solving the inverse task of radioactivity atmospheric transport; and
3. Reconstruction of the fields of fallout and dynamics of the nuclide activity concentration in the air in the initial period of the accident, especially short-lived ones (which is important for the retrospective dosimetry of the population in the affected regions).

The experience of the Chernobyl accident indicated two main directions for solving the problem of estimating and predicting the transport of emergency radioactive releases during the acute period of the accident: (1) the necessity of developing and improving the regional and long-range models of radionuclide atmospheric transport; (2) the development and implementation of computer support systems for the operative assessment of radiation accident consequences and for decision-making on population protection.

At the time of the Chernobyl accident, the agencies responsible for the assessment of the consequences of accidental emissions from nuclear energy facilities, not only in the former USSR but throughout the world, practically did not use regional and long-distance atmospheric transport models in operational mode. This can be explained not only by the complexity of the physical description of radionuclides dispersion processes and its interaction with the underlying surface at these scales, and by computer capacities for this period that was not sufficient for the operational use of such models in real time. The one more reason was that most specialists in radiation safety did not see a special need for their practical operation since the probability of a radiation accident that would lead to radioactive contamination of the territories at a distance of hundreds of kilometers from the source was considered virtually negligible. The Chernobyl accident in 1986 showed the need to radically revise the view on this problem.

Studies on modeling atmospheric transport of Chernobyl emissions can be classified according to the following criteria:

1. *Scales of the atmospheric transport problem considered*: (a) dispersion of radionuclides over the territory of Europe and global dispersion; (b) transport of radioactive release over the territory of the former USSR (regional scale); (c) formation of radioactive contamination within the near zone of the Chernobyl NPP zone (local scale). For each of these scales, specific mathematical models could be needed, as well as various sets of input information on meteorological conditions and parameters of the emission source.
2. *Nuclides considered.* Most of the work is devoted to the simulation of volatile ^{137}Cs and ^{131}I, with the exception of calculations for the near zone, as well as

Table 1.3 List of some atmospheric transport models used for modeling of the Chernobyl accident

Source	Model type	Space scale	Nuclide	Results presented
Gudiksen et al. (1989)	Lagrangian particle-in-cell	From local (50×50 km) area to the Northern Hemisphere	^{137}Cs, ^{131}I	Source term was estimated, dose due to external exposure and inhalation
ApSimon and Wilson 1987	Lagrangian puff	Europe	^{137}Cs, ^{131}I	^{137}Cs deposition, collective dose to European population
Albergel et al. (1988)	Lagrangian segmented plume	Europe	^{137}Cs	Air concentration and deposition
Haas et al. (1990)	Eulerian	Europe	^{137}Cs, ^{131}I	Air concentration and deposition
Brandt et al. (2002)	Combined Lagrangian and Eulerian	Europe	^{137}Cs, ^{134}Cs, ^{131}I	Deposition
Langner et al. (1998)	Eulerian	Europe	^{137}Cs	Air concentration and deposition
Terada and Chino (2005)	Lagrangian particle-in-cell	Europe	^{137}Cs	Air concentration and deposition
Sedunov et al. (1989)	Eulerian	Regional scale	^{137}Cs, ^{131}I	Air concentration and deposition
Borzilov and Klepikova (1993)	Eulerian	Regional (European territory of the former USSR) and local scales	^{137}Cs, ^{90}Sr, $^{239,240}Pu$	Air concentration and deposition
Izrael et al. (1987)	Lagrangian	Up to 100 km from the source	Polydisperse aerosol particles	Exposure dose rate
Vakulovskii et al. (1993)		Domain 140×140 km	^{137}Cs	Deposition

estimates of the transport of hot particles over long distances (Pöllänen et al. 1997; Bartnicki et al. 2001).

3. *Approach to determining the source parameters in modeling*. It is possible to outline the use of initially specified emission scenarios and own reconstruction on the basis of the solution of the inverse problem of atmospheric transport using data of the air concentration and deposition measurements.

In Table 1.3, the main features of several models which were used for modeling of the Chernobyl radionuclide atmospheric transport are summarized. More detailed information is available in Sects. 1.5.2–1.5.4.

1.5.2 Long-Range Transport

Concerning the atmospheric transport modeling, the post-Chernobyl period can be conditionally subdivided into three periods: (1) very few "quasi-online" calculations of the Chernobyl release transport made with using simple models in first days and weeks after the accident; (2) later period of more thorough analysis of the radionuclide transport aiming to reconstruct a time and space variability of the contamination fields as much as possible; and (3) the validation of developed models on the Chernobyl measurement data.

Immediately after the accident, three groups used the atmospheric transport models to estimate the Chernobyl release dispersion and possible radioactive contamination over Europe and the Northern Hemisphere. The staff at the Atmospheric Release Advisory Center (ARAC), the Lawrence Livermore National Laboratory, USA, utilized a three-dimensional atmospheric dispersion model PATRIC to derive the spatial and temporal evolution of the radioactivity distribution over local to hemispheric spatial scales with first results presented by 7 May 1986 (Knox and Dickerson 1986). Using obtained atmospheric transport modeling results and radiological measurements, Gudiksen et al. (1989) estimated the ^{137}Cs release as 89 PBq and ^{131}I as 1300 PBq. The simulation results showed that the cloud became segmented on 26 April, with the lower section heading toward Scandinavia and the upper part heading in a southeasterly direction with subsequent transport across Asia to Japan, the North Pacific, and the west coast of North America. The model reasonably predicted the time of arrival of the radionuclides at several locations over the Northern Hemisphere.

The second group from Imperial College, London, presented first results of calculations of the Chernobyl release transport on 21 May 1986 in the Nuclear Installations Inspectorate, UK (ApSimon and Wilson 1987). They used the Lagrangian trajectory model MESOS and calculated the trajectories and dilution of a sequence of puffs. The model used meteorological data routinely reported by synoptic weather stations throughout Western Europe. It was stated that the calculated map of the ^{137}Cs deposition over Europe indicates the large-scaled regions of higher deposition only, but there were a lot of relatively small areas where local heavy rains resulted in deposition spots. This problem was especially distinct in the UK, so more detailed studies were undertaken here. ApSimon et al. (1988, 1989) used data of the five UK weather radars, which enabled to estimate precipitation intensity with high spatial (2 km) and temporal (5 min) resolution. They included this information in modeling of the radioactivity passage over England and Wales with use of models RAINPATCH (ApSimon et al. 1988) and MESOS (ApSimon et al. 1989). The remarkably good agreement was obtained on the calculated and measured patterns of the highest areas of contamination within England and Wales even when the simple two-dimensional trajectory model has been used. ApSimon et al. (1989) made a more thorough analysis of the meteorological situation over the UK between 2 and 4 May, when the Chernobyl release passed over. This time some deposition spots were formed due to precipitation from convective systems associated with heat lows advected in from France and a cold front. In this type of front, the deposition spots are formed relatively close to the frontal zone, where the air from the warm sector rises steeply, giving heavy precipitation. However, they concluded that the situation might be more complicated in other forms of a frontal system, in which radionuclides could be transported aloft much longer distances before being brought down in precipitation. It may result in situations when, although concentrations of radionuclides are low near the ground, there are significant concentrations at higher altitudes.

The third group from Recherche en Prevision Numerique, Canada, made a simplified simulation of the radionuclide transport from the site of the Chernobyl NPP supposing a hypothetical source of a passive tracer with constant emission and duration of 5 days (Pudykiewicz 1988). The tracer simulation was performed between the 29 April and 15 May 1986, in emergency mode, and was used solely as an indicator of the horizontal radionuclide transport. Despite the extreme simplicity of the model, the results produced were quite significant for estimating which regions would be affected by radioactive debris and when.

In the past few years after the Chernobyl accident, several papers (Albergel et al. 1988; de Leeuw et al. 1988; Smith and Clark 1988; Haas et al. 1990) with results of the release modeling were published aimed to reconstruct the features of the radioactive contamination formation in Europe the same as the source-term parameters.

Albergel et al. (1988) presented the results of calculations carried out using the segmented plume Lagrangian model. They computed three-dimensional trajectories of radionuclide transport using the synoptic wind field and vertical velocities obtained from the analyses of the European Centre for Medium-term Weather Forecasting (ECMWF). It was shown that the consideration of large-scale vertical movements in the release transport layer leads to the radionuclide trajectories rising to a height of more than 5000 m at a distance of several thousand kilometers from the source.

The results of a similar study on the radionuclide transport over Europe, performed using the EURAD numerical model, are presented in Haas et al. (1990). The model contained two major modules, a mesoscale meteorological model MM4 and a chemistry-transport model CTM/RADM. In contrast to the abovementioned trajectory and Lagrangian models, the EURAD is a Eulerian model with a horizontal resolution of predicted meteorological variables and a calculated activity concentration of 80 km. The time dependencies of the ^{137}Cs and ^{131}I activity concentration in the air for a number of cities in Europe have been calculated. The obtained values are satisfactorily consistent with the measurement data concerning the description of the activity time dynamics in the initial period of the accident, although the difference in absolute values for some measuring points could reach two or more orders.

Simulation of the Chernobyl release on a global scale was made in Pudykiewicz (1988). He used a simple isobaric advection Eulerian model with the parameterization of scavenging and neglecting turbulent fluxes and vertical motions. A resolution of the horizontal grid of the model was 150 km, and computations were performed only for two atmospheric layers: 0–2000 and 2000–4000 m. The bottom layer winds were taken as the 850 hPa, and the 700 hPa winds were used in the upper layer. An isobaric advection on these levels was sufficient to explain the time of the arrival and even the observed values of the activities at several receptor points in Sweden and Canada.

In the next period, which is continued up to now, available and newly developed long-range transport models were validated on the Chernobyl data. Modeling prediction results are compared with measurements of the ^{137}Cs and ^{131}I activity in the air during April to May 1986 in Western Europe (REM databank mainly) and with the ^{137}Cs deposition maps over Europe created for the "Atlas of cesium deposition on Europe after the Chernobyl accident" (De Cort et al. 1998).

Several models have been validated in the ATMES (Atmospheric Model Evaluation Study) project, which was initiated as far back as in 1987 by the European Commission, the International Atomic Energy Agency (IAEA), and the World Meteorological Organisation (WMO). ATMES assembled an extensive quality-controlled database of environmental measurements from many institutes and arranged access to archived meteorological data at the ECMWF. The best estimate of the source term was established. The results of the evaluation of 22 participated models (Klug et al. 1992) were encouraging for the following development of models, although the comparison of the results of the models was proved to be rather complicated and ambiguous.

During this period, a wide range of long-distance models was validated, including simple trajectory (Galmarini et al. 1992), Lagrangian puff trajectory (Bonelli et al. 1992), Lagrangian particle (Desiato 1992; Ishikawa 1995; Terada and Chino 2005; Sofiev et al. 2007; Suh et al. 2009), Eulerian (Langner et al. 1998; Piedelièvre et al. 1990; Quélo et al. 2007; Evangeliou et al. 2013), and combined Lagrangian and Eulerian (Brandt et al. 2002) models. The main purpose of these studies was to assess the sensitivity of models to the choice of parameterizations of the main processes described in the model, their spatial and temporal resolution, the emission source parameters, and the input meteorological information used.

Despite the differences in the models, the used release scenario and used meteorological data almost all published simulation results for long-range atmospheric transport of the Chernobyl release over Europe showed a satisfactory agreement with measurement data obtained, including the arrival time and the radionuclide activity concentration in the air. It means than even relatively simple long-range models were able to reproduce main features of the Chernobyl release dispersion over Europe on the whole.

All modelers noted that one of the main sources of possible discrepancies between calculation results and radionuclide concentration measurements in air is a rather large uncertainty of the source parameters, first the dynamics of ^{137}Cs and ^{131}I daily emissions. The common result of most works is the high sensitivity of the simulation results to the parameterization of the initial distribution of the emission intensity from the source over height, the information about which is also quite contradictory. Therefore, significant errors in the source-term parameters did not allow making certain conclusions about the preference of using a particular model for calculating the Chernobyl release transport, and it was confirmed by the results of the ATMES project.

In so doing, the typical horizontal resolution of the calculated fields is in the range of dozens of kilometers (80 km for the EURAD model, for example). It is obvious that the results of computations averaged over such a coarse grid can be compared with the measurement data only at large distances from the source where the radioactive contamination fields are smoothed. In the near zone with high spatial gradients of the activity fields in the air and in depositions, the results of calculations using such models can significantly differ from those observed.

Satisfactory results of modeling of most of the long-range transport models were largely determined by the quality use of input meteorological information. The use of the results of objective analyses of meteorological variables made by national meteorological centers (ECMWF on grid points $1.5^\circ \times 1.5^\circ$ in Albergel et al. (1988)) makes it possible to model the Chernobyl release transport over Europe with satisfactory quality. Incorporation of the mesoscale weather prediction model in the radionuclide modeling system, for example, MM4 in EURAD (Haas et al. 1990), improves the quality of input meteorological data for the following transport modeling.

It should be noted that in general a reasonable agreement of modeling and measured data on the air activity concentration is not only a consequence of the high level of development of the models used but also to a certain extent related to the specific meteorological situation of the Chernobyl release dispersion. As a rule, modeling results are better for central European countries and worse for Scandinavia and the UK. The greatest problems in modeling arose for those regions where the radioactive clouds from the Chernobyl nuclear power plant fell into complex meteorological conditions, particularly in the area of atmospheric fronts (in England, for example). Due to the presence of a frontal system, the vertical exchange could be much more intense than it was considered by many models. Another problem for some models was the situation over the Baltic Sea on 27 April when the radioactive plume of the first release from the ChNPP divided: at lower altitude, radionuclides moved toward southern Sweden, while at higher altitudes, the radioactive release continued northward, passing over southern Finland (Persson et al. 1987). In fact, similar conclusions about the significant sensibility of the results not only of a specific model but also of the entire set of validated models were obtained under simulating two releases in the course of the ETEX project of the atmospheric model validation (Nodop 1997). The second release of ETEX-2 was characterized by a more complex meteorological situation due to a cold front passage during the experiment. In the result, for all the models, which participated in the model intercomparison, the differences between the model results and the measured concentration were obtained significantly larger than for the ETEX-1 experiment.

Among the internal long-range model parameters, it has been shown that the accuracy of the parameterizations of the vertical exchange, the horizontal diffusion, and the boundary-layer height has a great impact on the accuracy of the model results.

For the simulations of the deposition pattern, the key issues are parameterizations of dry and wet deposition processes. The formation of a number of local spots of radioactive fallout at various distances from the Chernobyl nuclear power plant occurred largely due to the washing out of radionuclides by rains. The possibility of their reproducing during modeling was determined, first of all, by the treatment of the cloud scavenging by wet deposition in a model, and above everything, by the resolution of input data about precipitation (measured or obtained with using weather forecast models). The low resolution of precipitation data, and as a result, calculated deposition field, makes a model unsuited for the prediction of local scale precipitation. So it is important to parameterize wet scavenging taking into account the subgrid nature of convective precipitation.

1.5.3 Simulation on a Regional Scale

Most of the works on modeling the Chernobyl emission transport at the regional and local scale were made with a view to reconstructing the parameters of the emission source and the characteristics of the radioactive contamination of the territory of the former USSR.

By using the Eulerian model, Sedunov et al. (1989) made one of the first attempts to simulate the ^{131}I deposition field on a regional scale. The main feature of the model was the space moving computational Eulerian grid, the horizontal dimensions of which are determined by the radioactive cloud size. Due to this, a good spatial resolution of the model was achieved from the release moment to the deposition on the underlying surface. Three main traces of deposition at the regional scale were obtained, and the total ^{131}I activity in the release was estimated as 480 PBq (without decay correction).

Izrael et al. (1989) calculated the field of the exposure dose rate from depositions at distances up to 150 km from the Chernobyl NPP by means of a regional atmospheric transport model. The model consists of three units: calculation of trajectories of air particle movement, calculation of the vertical flow of polydisperse contaminant on the underlying surface, and computation of integral depositions. The transport was calculated along the air flow direction at the levels of 1000, 925, and 850 hPa, depending on the height of the source. The main objective of this approach was to find the parameters of a given a priori lognormal distribution of released particles on size comparing the calculated and measured spatial field of the exposure dose rate. The values of the distribution parameters were found to be 25 μm for the median diameter and 0.4 for the standard deviation of the distribution.

Borzilov and Klepikova (1993) calculated the fields of ^{131}I and ^{137}Cs deposition for the European territory of the former USSR using the regional model of atmospheric transport REGION and their own scenario of emissions from the Chernobyl NPP. Detailed analysis of meteorological conditions of the radionuclide transport over this territory was made, and the largest deposition spots were obtained in modeling—central one around the ChNPP, in Belarus and in Russia. Despite some discrepancies in the location of Belarus and Russian spots, these modeling results obtained in 1986 were of large importance, because they led to an intensification of soil sampling program in this contaminated area (especially in Orel, Tula, and Kaluga regions in Russia).

An atmospheric transport model of the decision support system RECASS (Pitkevich et al. 1994) was used for the reconstruction of the population external exposure doses in the contaminated areas. The atmospheric transport and deposition of ^{137}Cs, ^{131}I, and ^{144}Ce were calculated, and the contribution of other nuclides to the dose rate was estimated by correlation ratios. The scenario of the release from the Chernobyl NPP was built considering the hypothesis that there were two periods of substantial additional emissions of radioactivity after May 5, 1986. Using modeling results, the time dynamics of the activity concentration in the air was reconstructed for a number of settlements of the Bryansk and Kaluga regions of Russia.

Besides abovementioned long-range transport modeling by using the WSPEEDI system, Terada and Chino (2005) additionally made similar calculations for a region of 1700 km $\times$ 1700 km with 10-km horizontal resolution to obtain a more detailed picture of the ^{137}Cs deposition field near the release source. Lately, they improved the model GEARN due to simultaneous atmospheric dispersion calculations of two nested domains, local and regional areas, with exchanging information between the domains (Terada and Chino 2008).

The regional model of atmospheric pollutant transport LEDI was used for the reconstruction of the dynamics of radioactive contamination in Ukraine during an initial phase of the Chernobyl accident (Talerko 2005a). The agreement of the calculated ^{137}Cs deposition density field with soil contamination measurement data is rather good. The formation of main deposition spots in the territory of Ukraine has been reconstructed under modeling, including western and northeastern arms of the deposition field just as the ^{137}Cs spots in central and southern parts of Ukraine (Fig. 1.9). The analysis of rain measurement data and the calculation results showed that the radioactive deposition field in the central part of Ukraine formed mainly due to dry deposition on the underlying surface. The contribution of wet deposition in this region was negligibly low. Nevertheless, according to soil activity measurement data, there are several spots of relatively large (over 37 kBq m^{-2} of ^{137}Cs) deposition in the Cherkasy region and the south part of the Kyiv region. According to the simulation results, the main reason of formation of such large-scale radioactive deposition spot at the distances 200–400 km from the source was an influence of changed meteorological conditions of release transport (due to daily variations of parameters of the atmospheric boundary layer where the main part of release moved). At night, in the conditions of the shallow stable boundary layer, the turbulent exchange is low, so the deposition of pollutants released from an elevated source was small in the near zone of the source. Therefore, for the release from elevated source with the maximal height that exceeds the nocturnal boundary layer height (with typical values about 100–300 m), the main part of released material rose over the layer of turbulent mixing. In this situation, the released material was transported by horizontal wind without considerable vertical mixing. So the high-level releases during several hours at night could result in the pollutant transport to the distances up to several dozens or even hundreds of kilometers (depending on the wind speed at the levels of transport). As a consequence of daily variations of turbulent parameters of the boundary layer, its height and the turbulence intensity increase in the morning and in the daytime due to a warming up of the surface. As a result of increasing planetary boundary layer height (it could reach the value 1000–2000 m), in some moments, the boundary layer "captures" the elevated release, and therefore the pollutants are involved in the intensive vertical turbulent exchange. Sometime later, the pollutant reaches the ground resulting in the formation of a deposition spot far away from the source.

This mechanism of the deposition spot formation at the mesoscale distances due to the combination of diurnal variations of the transport conditions and the high-elevated prolonged release source is quite obvious in principle. But just Chernobyl release with a large maximal rise height apparently became the first and for the present the only case in reality. On the other hand, any possible heavy radiation emergency at night could result in analogous consequences. In this situation, the radiological measurements in the near zone of the source could not ensure control in faraway regions.

1.5.4 Modeling of Radioactive Contamination in the Near Zone of Chernobyl NPP

All studies devoted to modeling in the near zone of the ChNPP were aimed at the reconstruction of the source parameters, including not only the estimation of the emission dynamics but also the determination of the size of aerosol particles in the emission. The direct measurement data used in this case was the deposition fields of individual radionuclides, as well as the field of the exposure dose rate in the near zone.

Izrael et al. (1987) calculated with the help of a kinematic model the fallout field of large polydisperse radioactive aerosol particles at a distance of up to 100 km from the ChNPP. The model takes into account the gravitational sedimentation of aerosol particles, and the horizontal dimensions of the radioactive cloud were parameterized by a simple method as the linear function of the distance from the emission source. As for the regional problem, the parameters of the logarithmically normal distribution of aerosol particle sizes in the emission were estimated from the comparison of the calculated and actual values of the area of radioactive fallout in the near zone. It was obtained that the value of the median diameter of distribution was 50 μm (at a density of 2.5 g cm^{-3}) and the standard deviation was 0.25.

Vakulovskii et al. (1993) presented the results of modeling of ^{137}Cs fallout around the ChNPP in the domain of 140 × 140 km using the Lagrangian particle model of atmospheric transport of the RECASS decision support system. Calculations were made for fixed values of the aerosol particle size and the initial release height (they are not given in the paper). By comparing the results of calculations of the cesium deposition field with the measurement data, reconstruction of the dynamics of cesium emission was performed.

Borzilov and Klepikova (1993) presented the ^{137}Cs, ^{90}Sr, and $^{239,240}Pu$ deposition fields calculated as early as in September 1986 using their own release scenario, the same as for the regional modeling task (Sect. 1.5.4). They used the Eulerian model MESO with the variable horizontal step increased from 1 to 5 km depending on the distance from the source. The distributions of the nuclide activities on the particle size in the release were used in modeling taking into account the presence of finely condensed and large fuel particles.

Despite the different estimation of total release, its temporal dynamics, and particle sizes in the release, all mentioned studies were able to obtain the main features of the deposition field in the near zone in accordance with the data measured fallout available for that period. Nevertheless, the problem of modeling the air contamination and deposition fields at a small scale was found to be rather complicated due to the greater sensitivity of the simulation results to the values of the emission parameters, primarily the initial emission height and the particle sizes, in comparison with long-range transport. In addition, all authors pointed out possible significant errors in the obtained results due to the scare data of meteorological observations in the near zone.

1.5.5 Reconstruction of the Source Parameters on the Basis of Inverse Modeling of Radionuclide Atmospheric Transport

As it was mentioned above, one of the ways to estimate the activity amount emitted from the destroyed reactor of the Chernobyl NPP was to reconstruct the Chernobyl source term on the basis of atmospheric transport simulation using the values of the activity in the air, deposition, and the gamma-dose rate, measured during different periods of the accident at different distances from the source. The results were obtained to be quite contradictory, since different sets of measurement data, models of atmospheric transport of various complexities, and various input meteorological information were used to solve this problem. An additional challenge was the choice of effective mathematical numerical methods for estimating the source characteristics. Over the past several decades, this problem has been intensively studied, and up to now several methods have been developed and brought to practical use.

Surveys of a variety of methods of source-term estimates developed in last years were made in Rao (2007), Redwood (2011), Bieringer et al. (2017), and Hutchinson et al. (2017). The effectiveness of these methods for estimating emissions under accidents at nuclear facilities was demonstrated for reconstruction of the release from Fukushima NPP in 2011 (Terada et al. 2012; Stohl et al. 2012; Saunier et al. 2013; Winiarek et al. 2014; Katata et al. 2015). In the period of the reconstruction of the Chernobyl release, only simple methods of the source-term estimation were available. So the results of studies in this period presented below are of mainly historical interest. However, they have contributed to the research of the characteristics of the Chernobyl release and an assessment of the consequences of further atmospheric transport of radionuclides.

The solution of the inverse tracer transport problem has been most frequently obtained by calculating back trajectories from the receptor, which are an easy tool for assessing possible source locations. Robertson (2004) (using a Eulerian model MATCH) and Sofiev et al. (2007) (using a Lagrangian Monte-Carlo random-walk model) presented results of the initial period of emergency response on the Chernobyl accident when the source of the accidental release was not determined. They made inverse modeling of the radioactivity transport with measurement data of several European stations to calculate the probability distribution for the location of the release source and obtained the Chernobyl NPP site as a release point with very good accuracy. However, they noted that the results obtained to be largely dependent on used data of few stations near the ChNPP (primarily, of Baryshivka, Kyiv region).

Careful evaluation of the Chernobyl source term has been made by Gudiksen et al. (1989) by using a three-dimensional Lagrangian particle model PATRIC for calculation of the Chernobyl release on various computational meshes ranged from the 50-km scale surrounding the reactor to hemispheric scale. An estimation of the source term (including 89 PBq of ^{137}Cs) was obtained by optimizing the agreement between the forward modeling results with the measurement data of airborne radioactivity over Europe, Japan, and the USA.

Several works of this period searched iteratively for an optimal set of the Chernobyl source parameters using different methods of nonlinear optimization in conjunction with forward atmospheric transport simulations. An optimal set of source parameters (and in some studies the meteorological input parameters) were iteratively adjusted by minimizing the multidimensional cost function. The cost function constructed as a sum of squares of deviations of calculated and measured values in several points can be quite complex and contain multiple local minima (Bieringer et al. 2017). Various methods were used to find the global cost minimum (Sedunov et al. 1989; Pitkevich et al. 1993; Borzilov and Klepikova 1993). Golubenkov et al. (1996) developed the method of conjugated gradients for solving of the cost function minimization problem, and used it for estimation of the Chernobyl release using the data of deposition measurements made in Russia only (on a regional scale, without the near zone and transboundary transport). The release was estimated as 7.2 PBq of ^{137}Cs, 61 PBq of ^{131}I, 39 PBq of ^{140}Ba, 24 PBq of ^{95}Zr, 17 PBq of ^{106}Ru, and 24 PBq of ^{144}Ce, which is less than 10% of UNSCEAR (2008) estimations for volatile nuclides and from 16% to 49% for refractory ones. Such low estimations were obtained due to the too-restricted area for which measurement data have been used.

A similar effect was obtained in Talerko (2010) for the inverse atmospheric transport problem solution using measurements of gamma doze rate made in 31 points in Pripyat town (2–4 km from the ChNPP) during the period from 03 h 26 April to 23 h 29 April 1986. A minimax method (Klepikova et al. 2010) was used

Table 1.4 Different scenarios of ^{131}I daily release (in PBq) from the Chernobyl nuclear power plant

Date	GKIAE (1986)	Izrael et al. (1990)	Haas et al. (1990)	Borzilov and Klepikova (1993)	UNSCEAR (2000)	Talerko (2005b)
26 April	192	33	149	241	704	339
27 April	56	111	46	70	204	227
28 April	41	152	42	32	150	130
29 April	28	54	33	23	102	113
30 April	19	11	20	26	69	23
1 May	17	11	20		62	45
2 May	28	15	42		102	29
3 May	30	21	136		107	45
4 May	35	31	64		130	16
5 May	36	44	120		130	9
Total release	480	482	670	392 (during first 5 days only)	1760	975

for reconstruction of the release intensity and the effective source height, and an adjustment of wind direction during a release. The restored values of daily ^{137}Cs and ^{131}I emissions were obtained from 47% (for April 26) of UNSCEAR (2008) results to only 13–19% for the remaining three days. This result can be an indication that the reconstruction of the emission parameters only on the basis of measurements made at distances of 2–4 km from the source does not fully reflect the patterns of the emission dispersion in the atmosphere. In the case of a large emergency, some of the activity due to thermal or dynamic impact can rise to a height of more than 200–300 m. In so doing, it becomes practically "invisible" for measuring devices located in the nearest zone of the nuclear power plant.

Borzilov and Klepikova (1993) made a reconstruction of the Chernobyl source parameters on the basis of the atmospheric transport modeling for the near and far zones. They adjusted not only the temporal dynamics of radionuclide release (^{137}Cs, ^{131}I, ^{90}Sr, and $^{239+240}$Pu were considered) but the activity distribution on aerosol particle size also. The ^{137}Cs and ^{90}Sr activity distributions were obtained as bimodal ones. In terms of the gravitational sedimentation velocity used in the paper, the maxima were found at 0 and 8 cm s^{-1} for ^{137}Cs, and 1 and 10 cm s^{-1} for ^{90}Sr (the particle size can be estimated to be 0 and 33 μm, and 12 and 37 μm, respectively). According to these data, the share of finely dispersed particles in the ^{137}Cs released activity can be estimated to be about 70% and for ^{90}Sr about 25%. For plutonium, the distribution function was found as unimodal with a maximum at 10 cm s^{-1}. The ^{131}I release was assumed in modeling in the gaseous fraction only. According to simulation results for the near and far zones, the ^{137}Cs emission estimate was 63 PBq during the first six days, iodine—392 PBq for the first 5 days (see Table 1.4).

In a later period, Davoine and Bocquet (2007) reconstructed the values of the release intensity and the effective height of the Chernobyl source as functions of time using data of the activity concentrations in air measured in Europe provided by the REM database. The estimated total released activity was obtained as 1620 PBq of

^{131}I (close to UNSCEAR (2008)), 136 PBq of ^{137}Cs (larger in 1.6 times), and 35 PBq of ^{134}Cs. As in other reconstructions, the dependence of the release intensity on time includes three stages, with intensive releases on 26 and 27 April, followed by a weak release period (28 April to 1 May) and intensification of release during 2–5 May. The release estimations for the third release period were obtained lower compared with the majority of previous works. It is interesting to note that a very similar result was previously obtained in Talerko (2005a, b) under the solution of the inverse problem of atmospheric transport using an entirely different set of measurement data of the ^{137}Cs deposition density in the territory of Ukraine. Also, in these studies, similar conclusions were obtained about the variability of the initial emission height, which are different from most of the previously proposed scenarios (see Sect. 1.2.6). Davoine and Bocquet (2007) have obtained the release altitude from the ground to the top of the calculation domain (over 5000 m) during the first stage and up to 3500 m during the third stage. For the regional task considered in Talerko (2005a), the maximal height has been limited to 1200 m, and the reconstructed maximal release height obtained was equal to this value not only in the first day (as supposed in the most modeling studies) but during all release periods, including the last third stage.

Besides the results of direct atmospheric transport modeling, the source parameters reconstructed in inverse modeling are found to be very sensitive to used input data, including radiological measurement data and meteorological information, especially precipitation fields. Another important question still remained—sensitivity of the reconstruction results to the physical process parameterization in the used atmospheric transport model. Bocquet (2012) studied the possibility of including the optimization of some model parameters into the solution of the joint inverse modeling problem. He applied the developed fast four-dimensional variational scheme (4D-Var) for the reconstruction of the source term jointly with the model parameters using the activity concentration measurements during the Chernobyl accident from the REM database. Based on this method, reestimated values of the total release were obtained as 110 PBq of ^{137}Cs (between estimations of UNSCEAR (2008) and Davoine and Bocquet (2007)), 57 PBq of ^{134}Cs (close to UNSCEAR (2008)), and 1160 PBq of ^{131}I (significantly lower than UNSCEAR (2008) and close to Talerko (2005b)). Several numerical experiments on joint optimization of the source parameters and the model parameters (the dry deposition velocity, the wet scavenging factor, and horizontal and vertical diffusivity) were made, and it was shown that the 4D-Var optimal values match the exact optimal values.

The results of the reconstruction of the total release of ^{137}Cs and ^{131}I by the method of the inverse problem of atmospheric transport are quite well coincident with each other, despite the differences in the models and algorithms of data assimilation. They coincide with the results of similar assessments conducted in other ways. This suggests that integral estimates of radionuclide emissions with sufficient accuracy can be obtained even with the help of rather simple atmospheric transport models. Greater discrepancies occur when restoring emission dynamics. This is due to the fact that it is much more difficult to calculate the spatial characteristics of the radioactive contamination field formation in separate periods of the initial stage of the accident. Therefore, for the reconstruction of the dynamics of the nuclide activity on a given territory, the quality of the model itself, that is, the detailed description of diffusion and deposition processes, as well as the completeness and reliability of the input radiological and meteorological information play a decisive role.

1.6 Reconstruction of Iodine Contamination Using Measurement Data and Atmospheric Transport Modeling Results

1.6.1 Available Measurement Data

During the Chernobyl accident, ^{131}I and its short-lived isotopes were among the most dangerous emissions. Radioactive iodine (^{131}I mainly), which enters the human body during the passage of the radioactive cloud by inhalation and then with food, was the main source of the thyroid doses of the population living in the contaminated territories. It has been responsible for the increase in children's thyroid cancer observed in Ukraine, Belarus, and Russia since 1986. So the problem of reconstruction of thyroid doses as a result of the Chernobyl accident is of particular importance (Likhtarov et al. 2014).

Unfortunately, due to the emergency situation and short half-life of ^{131}I (8.04 days), only a small number of reliable measurements of the spatial distribution of radioactive iodine in food, air, soil, and other parts of the environment were made, especially in the territory of Ukraine. The last measurements of ^{131}I contamination in soil with values above the detection limit were made at the beginning of July 1986. So, data of direct measurements of iodine contamination in the environment do not provide enough information for the reconstruction of the population thyroid doses after the Chernobyl accident. Moreover, even if the data on time-integrated values of radioiodine activity in the components of the environment in the initial period of the accident are available, the results of thyroid dose reconstruction are significantly dependent on the temporal dynamics of iodine intake into the human body and, in particular, on the beginning and the end of the period of intensive radionuclide contamination due to atmospheric transport into this area, which are different for each locality.

On the territory of the former USSR, there are two main arrays of iodine measurement data: (1) the ^{131}I daily depositions on collectors and the daily-averaged particulate ^{131}I activity concentration in meteorological stations of the USSR Goskomgidromet (see Sect. 1.4.2); and (2) results of measurements of ^{131}I content in the soil of 534 settlements of the Gomel and Mogilev regions of Belarus (Drozdovitch et al. 2013; Khrushchinskii et al. 2014) obtained in May to July 1986. Unfortunately, in the contaminated territories of Ukraine and Russia, there were no measurements of iodine content in soil in similar volumes, except for single measurements.

Outside the former USSR, the data of the ^{131}I activity concentration in the air and the daily depositions were obtained in several counties during the initial stage of the Chernobyl accident and were included in the REM Data Bank (see Sect. 1.4.2).

Besides a short half-life of iodine isotopes, another important feature of radioactive pollution of the environment by iodine isotopes is that they were released from the accidental reactor, transported in the atmosphere, and deposited on the underlying surface in various physicochemical forms, including aerosol particles and gaseous elemental (I_2) and organic (CH_3I) forms. It created additional problems for measurements of the iodine activity concentration because the great part of radioactivity sampling in the former USSR during the initial period of the accident was made using gauze collectors and measuring the aerosol part of the release only. A few measurements of the gaseous component, which required the use of special filters for its catching, were carried out in the initial period of the Chernobyl accident in 1986, and, mainly, at great distances beyond the borders of the former USSR territory. There were rare measurements of the ratio of various radioiodine forms in the release and in depositions within the near zone of the source.

Further, during the atmospheric transport, the transformation between these physicochemical forms occurred due to reaction with other atmospheric chemicals and air moisture, and due to different rates of their dry and wet deposition. Due to the slow rate of photodissociation of ^{131}I in organic and particulate forms, the main processes of iodine chemistry during atmospheric transport are transformation of its elemental form into organic and particulate ones (Nair et al. 2000). Elemental iodine released from the reactor interacted with the condensation nucleus of the atmospheric air and adsorbed on them. The rate of condensation nuclei conversion to submicron size droplets depends on relative humidity in the atmosphere and determines considerably a transformation intensity of iodine chemical forms. The ^{131}I behavior in the air and in depositions on soil and vegetation may change depending on current weather conditions, type of local underlying surface, type of vegetation, and so on. In particular, sublimation of deposited iodine occurred that complicated the interpretation of measurement results and decreased the reliability of obtained results. So, the dynamic equilibrium between the aerosol and gaseous forms of ^{131}I, which is set in the atmosphere and in depositions, is maintained only for space and time-averaged values.

Analysis of the state of the emergency reactor in the period of active emissions showed that the ratio between the activities of these forms in the release was changing in time. It varied not only depending on the state of the nuclear fuel in the reactor but also on the physicochemical processes that took place in it, in particular, the fuel temperature variability and formation of graphite combustion products.

During the explosion on 26 April, a large amount of ^{131}I and other short-living iodine isotopes was released in the aerosol form. Later, in the periods of the smothering of the accidental unit with inert materials, the graphite stack burning and, later on, the part of ^{131}I in the gaseous forms were seen to be increasing

(Makhon'ko et al. 1996). The individual measurements made at a distance of about 1000 km north of the ChNPP at 2 km height gave the proportion of the aerosol fraction of ^{131}I as 45–55% on 14–15 May, 24% on 17 May, and 10% on 19 May (Borisov et al. 1990). According to Ogorodnikov et al. (2008), the proportion of aerosol ^{131}I decreased with time from 70% to 9%. In the opinion of the authors, the most likely reason for the increase in the radioiodine fraction in the gas phase was that the gradual decay of the graphite burning led not only to a general decrease in the release of radioactive products but also to a change in the physicochemical characteristics of the iodine release composition.

The only measurements of iodine forms in the territory of the former USSR were made by Styro et al. (1992) in Vilnius (Lithuania). The highest concentration of ^{131}I in the air was measured on 30 April and was found to be 55 Bq m^{-3}. During this day, the ratio of the aerosol, elemental, and organic forms was obtained to be 24%:22%:54%, and on average from 30 April to 10 May, it was 35%:14%:51%. The variability of the part in the aerosol form during a set of daily measurements was in the range from 24% to 50% on 7 May (differed by a factor of 2), the elementary form from 7% (7 and 10 May) to 22% (by a factor of 3), and the organic form from 43% (7 May) to 58% (10 May); that is, the organic form was most stable.

According to data of measurements made in many countries after the Chernobyl accident, a part of particle form of ^{131}I was obtained to be 20–25% in Sweden (Devell et al. 1986), 25% in UK (Cambray et al. 1987), about 45% in Poland (Jaworowski and Kownacka 1988), 15–23% in Germany (Reineking et al. 1987), $\leq$40% in USA (Bondietti and Brantley 1986), and 30–35% in Japan (Aoyama et al. 1986). Winkelmann et al. (1987) reported that in the later stages of the accident, 40% of the iodine was associated with aerosol particles, 35% was gaseous elemental iodine, and 25% was organic iodide. With time, the organic iodide fraction increased. Noguchi and Murata (1988) measured four physicochemical forms of airborne ^{131}I, released during the Chernobyl accident, in Japan during the period 6–19 May 1986. During this period, activity concentration decreased from 0.14 to 0.022 Bq m^{-3}. The proportions of ^{131}I species identified during that period were 19% particulate iodine, 5% I_2, 70% organic iodides, and additionally 6% hypoiodous acid (HOI).

On the basis of reviewed ^{131}I measurement data, Styro et al. (1992) concluded that similarity of these results over the world is evidence that reactions of gaseous iodine released during the Chernobyl accident with atmospheric aerosols (i.e., adsorption and desorption) reached an equilibrium state sufficiently fast during their atmospheric transport. They noted that the mass concentration of accidental ^{131}I and other iodine isotopes released from the accidental ChNPP were many orders of magnitude less than the mass concentration of natural iodine isotopes (^{127}I mainly). Therefore, the presence of natural stable iodine may have an effect on the transformation of the isotopes of emergency origin.

An additional argument in support of this hypothesis is that the particulate iodine fractions measured after the Chernobyl accident agree not only with those for stable iodine in the environment but also for the radioiodine measured at both the TMI-2 accident and in boiling water reactors (BWRs) during normal operation. Ramsdell

Jr. et al. (1994) concluded that these data of iodine partitioning into different forms in the plume following the Chernobyl release are consistent both with one another and with the results of previous experimental studies at Hanford made by JD Ludwick in 1963 and 1966. There the ratio of particulate, organic, and elemental forms was measured to be 30%: 36%: 34% at a distance of about 3 km from the source of elemental iodine and 15%: 43%: 42%, respectively, at a distance of five miles during another experiment.

Similar conclusions were made by Georgi et al. (1988). They found that for measurements in the air in Germany and Austria after the Chernobyl accident, the ^{131}I particle size distribution was significantly different from other volatile nuclides, shifting to smaller particle sizes (see Sect. 1.2.4). They explained this difference to the fact that the nuclides ^{137}Cs, ^{132}Te, and ^{103}Ru were transported only in the particulate form, but ^{131}I was released mainly in the gas phase and was adsorbed continuously on local aerosols. Therefore, local processes were found to be dominant for ^{131}I particle formation. As evidence of this statement, they showed that the ^{131}I particle size distribution is in good agreement with the local urban aerosol distribution (for bromine and lead).

1.6.2 Methods of Reconstruction of Iodine Contamination

There are several possible ways to solve the problem of reconstruction of both the integral values of iodine fallout density in areas contaminated by the Chernobyl accident and temporal dynamics of radioactive iodine in the air and soil, depending on the quantity and quality of available information:

1. using measurements of ^{137}Cs activity in soil, together with the assumption that the $^{131}I/^{137}Cs$ deposition ratio at the time of the accident was constant and known.
2. space interpolation of data of direct iodine deposition measurements made in the initial period of the Chernobyl accident (because of the sparseness of available information, this approach has been complemented with the results of the previous one).
3. using measurements of the long-lived (half-life 1.57×10^7 years) isotope ^{129}I in soils, together with a determination of the $^{129}I/^{131}I$ ratio at the time of the accident.
4. calculation of the iodine deposition density field on the basis of mathematical modeling of atmospheric transport and deposition on the underlying surface using a scenario of the Chernobyl release.

A brief description of each of these methods and the results obtained with its help is given below.

1.6.2.1 Regression Models for ^{131}I and ^{137}Cs Soil Contamination

The distribution of ^{137}Cs activity in soil was well studied soon after the Chernobyl accident over all contaminated territories, so cesium could be treated as a reference nuclide for retrospective ^{131}I dosimetry. In different studies a regression model in the form

$$D_I = A \cdot (\Delta D_{Cs})^B$$

has been proposed for reconstruction of ^{131}I activity in soil using measurement data on ^{137}Cs soil contamination, where D_I is the estimated ^{131}I deposition density (kBq m^{-2}) as of 10 May 1986; $\Delta D_{Cs} = D_{Cs} - D_{Csb}$ is the difference between the measured ^{137}Cs deposition density D_{Cs} (kBq m^{-2}) and the ^{137}Cs global background contamination level D_{Csb}, which is equal to 2.2 kBq m^{-2}.

Pitkevich et al. (1993) analyzed data of ^{131}I and ^{137}Cs contamination in 122 soil samples collected in Ukraine, Belarus, and Russia in 1986 and proposed a linear regression model ($B = 1$) for reconstruction of ^{131}I deposition field for the territory of Russia (northeastern trace of the Chernobyl depositions). The regression coefficient A was estimated as a function of a distance from the ChNPP and obtained to be decreasing from about 6 for near 30-km zone to 3.34 for distances 400–600 km from the source. In particular, the last value was obtained for the relatively very high wet deposition area (^{137}Cs deposition level is up to 500 kBq m^{-2}) in Plavsk district, Tula region (Russia).

Mironov et al. (2002) analyzed the results of measurements of 24 soil samples from the Belarus part of the 30 km zone of Chernobyl NPP. The derived activity ratios at the time of the accident are estimated as $^{131}I/^{137}Cs = 10.0 \pm 3.1$ for ^{137}Cs depositions in the range from 0.04 to 9 MBq m^{-2} on 26 April 1986, so they obtained parameters of the regression model were equal to $A = 2.97$ (decay-corrected on 10 May 1986) and $B = 1$. Besides it, they provide measurement results of the $^{131}I/^{137}Cs$ activity ratios in air at various distances in the former USSR, Western Europe, and Japan during the first 12 days after the Chernobyl accident. The mean value (12.7 Bq/Bq) is very close to that estimated in the reactor at the time of the accident (10.4). All of them made it possible for the authors to suggest that there was little fractionation between iodine and cesium during atmospheric transport and in the deposition process, at least in Belarus, and that ^{137}Cs can provide reasonably good estimates ($\pm$50%) for ^{131}I.

Makhon'ko et al. (1996) proposed the values $A = 23.2$ and $B = 1.14$ for territories with cesium contamination in the range 0.37–37 kBq m^{-2}. This relation was obtained using data of radionuclide deposition measurements at a network of meteorological stations in the former USSR in April–May 1986. Additionally, they found a high correlation of ^{131}I and ^{137}Cs contamination density in the soil samples (about 100) collected in May–June 1986 (for cesium values in the range 0.37–7400 kBq m^{-2}) and proposed alternative regression model with values $A = 5.81$ and $B = 0.847$ for estimation of iodine contamination. The power value in both relations is close to unity, and the results of all regression models, including the linear one, differed up to 2–4 times among them depending on ^{137}Cs deposition density. The former relation of Makhon'ko and the relation of Pitkevich et al. (1993) could be regarded as upper estimation of the ^{131}I contamination density for the ranges $0.37 < \Delta D_{Cs} < 37$ kBq m^{-2} and $0.37 < \Delta D_{Cs} < 3700$ kBq m^{-2}, respectively.

Knatko and Dorozhok (2002) reconstructed ^{131}I soil contamination in settlements in Belarus using two regression dependencies obtained separately: $A = 33.8$, $B = 0.63$ for the eastern part (between Gomel and Mogilev cities, including the large deposition spot formed by wet removal of radionuclides) and $A = 69.4$, $B = 0.59$ for the southern part (Gomel region, contaminated territory near to the ChNPP formed due to dry depositions mainly). For ^{137}Cs deposition density of 37 kBq m^{-2}, the estimated ^{131}I/^{137}Cs ratio is 8.9 for wet depositions in the east of Belarus, and 15.8 for dry depositions in the south. For ^{137}Cs deposition density of 370 kBq m^{-2}, the used regressions result in this ratio diminishes to 3.8 for the eastern area and to 6.1 for the southern one. According to measurement data used by Knatko and Dorozhok (2002), the median value of ^{137}Cs contamination in the settlements of the eastern area is about 35% larger than in the case of the southern area. But due to different mechanisms of deposition formation for ^{131}I contamination, a situation is obtained to be opposite: the median value of ^{131}I deposition in the settlements in the southern area is about one third larger than for the eastern area. For 90% settlements, the estimated ^{131}I deposition density is about 500–2300 and 700–3500 kBq m^{-2} for the eastern and the southern area, respectively. The ratio of median values ^{131}I/^{137}Cs for settlements in the eastern area is estimated to be about 1.8 times smaller than for the southern area. These results agreed with data that in the territories where Chernobyl depositions were formed due to wet removal by rains, the ^{131}I/^{137}Cs ratio in soil is lower than that in the territories where dry depositions only took place (Kruk et al. 2004).

These relations between ^{131}I and ^{137}Cs soil contamination were used in the program EMRAS of validation of models of the iodine radioisotope transport in the environment (Zvonova et al. 2010) for evaluation of the Chernobyl ^{131}I deposition in Tula region, Russia. Predictions of the iodine concentration in soil of the various models were obtained to be within a factor of three relative to the observations.

While widely used, there are still some questions regarding the reliability of this assumption. The reported ratios of ^{131}I/^{137}Cs vary substantially over space (Izrael et al. 1990) due to several reasons. During the course of the accident, the composition of the nuclides, including the ratio of iodine and cesium, might have been changed in the release, and it resulted in their different relative activities in deposition traces (southern trace vs. northern and western ones). Additionally, this approach does not take into account the regional features of deposition (such as region-specific deposition time, deposition duration, and relative input of wet and dry deposition). Due to the large variability of the ratio of ^{131}I and ^{137}Cs in the Chernobyl fallouts, this approach can lead to large errors in the reconstructed values for ^{131}I, and, as a result, in the reconstructed doses. Some measurements for soils in Belarus and the Baltic Republics show ratios as large as 40–60 and in Lithuania and Poland as high as 92 (Orlov et al. 1992, 1996).

In the abovementioned study of Jaworowski and Kownacka (1988), the value of the ^{131}I/^{137}Cs activity concentration ratio (corrected with the filter efficiency factor) at a day of sampling (29 April to 20 May 1986) was calculated for near-ground and high-altitude measurements. The low values, ranging from about 3 to 11, were observed at all altitudes up to 15 km on 29 and 30 April and 1 May and then on 11 May. During the period from 2 to 8 May, this ratio was found to increase ranging from 36 to 70 (with maximal values at heights 6–9 km). The obtained values could be explained by an increase in the iodine relative activity in the release during the first days of May 1986, and in subsequent days by the influence of the radioactive decay of ^{131}I released into the atmosphere.

Straume et al. (1996) reported a poor correlation between ^{137}Cs and iodine contamination in soil in Belarus (iodine-to-^{137}Cs ratios vary by more than a factor of 20) and stated that ^{137}Cs deposition density does not appear to be an adequate surrogate for iodine deposition density.

1.6.2.2 Reconstruction on the Basis of Data of Direct ^{131}I Deposition Measurements

Direct iodine deposition measurements made in the initial period of the Chernobyl accident were used by different authors for reconstruction of spatial distribution of the ^{131}I deposition field. Makhon'ko et al. (1996) have built a map of ^{131}I soil contamination in the European part of the former USSR territory by 15 May 1986 based on the data of ^{131}I daily deposition measurement at 71 meteorological stations (Sect. 1.4.2). Compared with the previous method, this approach makes it possible to restore not only the value of the cumulative iodine fallout but also the dynamics of daily air contamination and depositions. For the map creation, available data of measurements at stations have been area interpolated. For the eastern part of this territory (in Russia mainly) where ^{131}I depositions were not measured, they were estimated using results of an aerial spectrometric survey of ^{137}Cs contamination density and two alternative empirical relations between ^{137}Cs and ^{131}I soil contamination (see Sect. 1.6.2.1). Because of a relatively sparse network of meteorological stations, the obtained ^{131}I deposition map could be considered rather general and smoothed. A fairly good general agreement between both reconstruction maps was found, in spite of the fact that the latter approach is more detailed (the map resolution is estimated to be about 10 km). It was stated that for construction of a map with better detailing, in May–June 1986, a comprehensive aircraft gamma-spectrometric survey should have been arranged with small distances between routes, but this was not done in a timely manner.

The direct measurements of soil contamination with iodine-131, made in Belarus, provided the foundation for building a map of reconstructed ^{131}I fallout density field in this country (Germenchuk et al. 1996). The reconstruction for the most contaminated Gomel and Mogilev regions was made using the data of gamma-spectroscopy measurement of ^{131}I soil contamination obtained in May–July 1986 (about 1000 measurements in total). For the rest of Belarus, the gamma rate measurement data were used, and the input of iodine components into this value was estimated for different regions depending on the isotopic composition of depositions.

Additionally, the iodine soil contamination was estimated using the measurement data of iodine contamination of milk made in Belarus in May–July 1986 and a model of ^{131}I transfer in the "soil–grass–milk" chain. The $^{131}I/^{137}Cs$ ratio in depositions (decay-corrected on 10 May 1986) was obtained to be from 2 to 9 for two regions close to the source (distances up to 200 km), and it reached up to 60 for distant Vitebsk, Grodno, and Minsk regions (up to 400 km). The largest iodine deposition values in Belarus were obtained in the 30-km Chernobyl zone: around 37,000 kBq m^{-2}, and within the Gomel–Mogilev spot in the Gomel region, they were up to 20,000 kBq m^{-2}. For these highly contaminated territories, the reconstruction results are obtained to be close to the results of Makhon'ko et al. (1996). But there are substantial distinctions for the rest of Belarus where cesium depositions were smaller.

Unfortunately, this method of ^{131}I soil deposition reconstruction is inapplicable for the territory of Ukraine because of very few direct measurements of ^{131}I contamination in soil and other environmental components were made in 1986. The map of the ^{131}I deposition field created by Makhon'ko et al. (1996) is rather rough in the most contaminated regions on account of the nearest to the Chernobyl NPP measurement sites were located at distances about 100 km in Kyiv and Baryshivka (Ukraine) and Gomel (Belarus).

1.6.2.3 Reconstruction of ^{131}I Deposition Density Using Data of ^{129}I Activity Measurements in Soil

The next method is an evaluation of ^{131}I deposition density, using data of direct measurements of the long-lived ^{129}I content in soil, known isotopic ratios of these isotopes, and evaluation of preaccident ^{129}I background. Isotope ^{129}I was also produced by fission in the reactor and released into the atmosphere together with ^{131}I during the accident. Measurements of ^{129}I in soil were performed either by the neutron activation analysis or the accelerator mass spectrometry. Although ^{129}I measurements are more complicated than those for ^{137}Cs, the fact that one is dealing with chemically identical isotopes means that this approach could enable a potentially more reliable estimation of the ^{131}I dispersion pattern. Simultaneously, as a rule, the correlation between iodine isotope content and ^{137}Cs activity in soil was investigated in such works.

Straume et al. (1996) calculated the $^{129}I/^{137}Cs$ and $^{129}I/^{131}I$ ratios in soil samples taken at eight settlements in five regions of Belarus in May 1993. There were three settlements in the deposition spot near to the Chernobyl NPP at distances 30–80 km ("central spot"), and five settlements were located within the "Gomel–Mogilev" spot. The central spot was formed on 26–27 April 1986, and the Gomel–Mogilev spot was formed mainly due to wet depositions two days later. Substantial differences were observed in the $^{129}I/^{137}Cs$ ratios for the various settlements evaluated, indicating a poor correlation between the deposition of ^{129}I and ^{137}Cs in general. However, for settlements within the Gomel–Mogilev spot, the atomic ratio $^{129}I/^{137}Cs$ varied in the range 0.05–0.12, while for settlements in the central spot, it was within the range 0.35–1.1. The same variability in the atomic ratio of $^{129}I/^{137}Cs$ from 0.38 to 1.1 for soil samples taken within the 30 km zone has been obtained by Sahoo et al. (2009). So, ratios of $^{129}I/^{137}Cs$ within the deposition spots formed by different deposition mechanisms differ by a factor of about three. The value of the decay-corrected (for 26 April 1986) atomic ratio for $^{129}I/^{131}I$ was estimated to be 12 $\pm$ 3 using the results of direct measurements of one archived soil sample collected on 1 May 1986 north of Gomel. Taking into account the results of other authors, Straume et al. (1996) concluded that the atomic ratio of $^{129}I/^{131}I$ released from the Chernobyl reactor during the 1986 accident was in the 10–20 range. These results agree with the data of Mironov et al. (2002) (the value of $^{129}I/^{131}I = 15.2 \pm 4.7$ for 24 soil samples taken within the Chernobyl 30-km zone) and to a lesser extent of Paul et al. (1987) (19 $\pm$ 5, 35 $\pm$ 9 and 9.4 based on three rain samples). So, the $^{129}I/^{131}I$ ratio could provide a reliable means to scale from contemporary ^{129}I measurements to the initial levels of ^{131}I.

Pietrzak-Flis et al. (2003) estimated ^{131}I deposition densities using measurements of soil samples collected in 16 locations in Poland during 1997–2000. The measured deposition of ^{129}I in chosen locations ranged from about 3 to 34 mBq m^{-2}, whereas the range of ^{137}Cs deposition was from 1.6 to 118 kBq m^{-2}. The highest ^{137}Cs depositions in three settlements were explained by wet deposition, whereas in the remaining locations, only a dry deposition took place. The $^{129}I/^{137}Cs$ atomic ratios in depositions varied considerably (0.13 $\pm$ 0.01–5.31 $\pm$ 0.59) for the locations studied. The lowest values are close to the $^{129}I/^{137}Cs$ ratio in the reactor core 0.14 at the time of the accident, and they were obtained for locations where wet deposition took place. For locations without rains during the Chernobyl release transport, a fractionation between iodine and cesium took place due to differences in the deposition of iodine and cesium isotopes. Corresponding $^{129}I/^{137}Cs$ activity ratios in depositions varied for these samples in the range (2.5–100) 10^{-7}. For comparison, the $^{129}I/^{137}Cs$ activity ratio calculated from the inventory data (UNSCEAR 2000) is 3.1×10^{-7}. The weighed mean atomic ratio of $^{129}I/^{131}I$ at the time of deposition in Poland was calculated to be 32.8 using the value of this ratio in the core inventory of 22.8 (Kirchner and Noack 1988). The corresponding activity ratio of $^{129}I/^{131}I$ is obtained to be 4.7×10^{-8}. As a result, an estimated ^{131}I deposition density varied for the chosen location in a wide range from about 63 to 729 kBq m^{-2}. Activity ratios of $^{131}I/^{137}Cs$ varied from 5.41 to 214. Usually, this ratio was lower in the locations where rainfall occurred.

The ^{129}I levels were determined in eight soil samples collected in the 30-km zone around the Chernobyl NPP in 1994/1995 (Sahoo et al. 2009). The $^{129}I/^{131}I$ ratio value in the reactor at the Chernobyl accident was used to be 11–15 according to Ermilov and Ziborov (1993). The $^{129}I/^{131}I$ activity ratio in the fallout at the time of the Chernobyl accident was estimated to be 2.6×10^{-8} (UNSCEAR 2000). For these samples collected in the central (at 6 km from the ChNPP) and southern (at 26–28 km from the ChNPP) parts of the Chernobyl zone, the $^{129}I/^{137}Cs$ atomic ratios were found to be in the range 0.38–1.06, which are close to data of Straume et al. (1996) for the territory of Belarus to the north of the ChNPP. Corresponding $^{129}I/^{137}Cs$ activity ratio is from 7.3×10^{-7} to 20.2×10^{-7}. Using measurement results of ^{129}I levels in soil samples (74.5–1100 mBq m^{-2}) and the average pre-Chernobyl level 26 mBq m^{-2} (Mironov et al. 2002), Sahoo et al. (2009) estimated the range of the ^{131}I deposition at the accident time in this area to be 2.6–57 MBq m^{-2}.

On contrary to the abovementioned works, Hou et al. (2003) made a more optimistic conclusion concerning the possibility of using ^{129}I/^{137}Cs ratios and ^{137}Cs levels to reconstruct the ^{131}I content in soil. In this work, samples of undisturbed soil collected from 11 locations in Russia, Belarus, and Sweden were analyzed for ^{129}I and ^{137}Cs. In Belarus and Russia, the samples were taken in two settlements located in the Chernobyl "central" deposition spot (at distances 12 and 110 km from the ChNPP) and in two settlements in the Gomel–Mogilev spot (about 150 and 180 km away). Despite the different nature of the spot formation and different distance from the source, it was obtained that the atomic ratio of ^{129}I/^{137}Cs ranges between 0.10 and 0.30, with an average of 0.18, and no clear geographic variation is observed. So they concluded that the ^{137}Cs concentration could, in combination with the measured ratios of ^{129}I/^{137}Cs and ^{129}I/^{131}I, be used to evaluate and reconstruct the Chernobyl ^{131}I contamination level in areas strongly contaminated by the Chernobyl accident. But all the same, even for this territory, as the results of many studies show, the iodine to cesium ratio is too variable to use the ^{137}Cs concentration as a basis for ^{131}I reconstruction. A much higher ratio of ^{129}I/^{137}Cs was observed in less Chernobyl-contaminated soil in Nora, Sweden. Authors attributed this to the contribution of ^{129}I releases from European reprocessing plants at La Hague (France) and Sellafield (UK) and implied that ^{129}I concentrations or ^{129}I/^{137}Cs ratios could not be used to reconstruct the Chernobyl-derived ^{131}I in the less Chernobyl-contaminated areas of Western Europe.

For the territory of Ukraine, the measurements of ^{129}I soil contamination have been made by the group from the Center for Radiation Protection and Radioecology of the University Hannover (Germany) and the State University of Agriculture and Ecology (Zhytomyr, Ukraine) (Michel et al. 2005). They investigated 42 soil samples taken in the Zhytomyr region. Eleven samples were taken in a low contaminated area, 24 samples in four moderately contaminated villages of Korosten district (about 120 km southwest of the ChNPP), and seven samples in three highly contaminated villages of Narodichi district (about 70 km southwest from the ChNPP). According to the Ukrainian laws, these settlements in Narodichi district belong to the second contamination zone (including territory with ^{137}Cs deposition densities in the range between 555 and 1480 kBq m^{-2}), and chosen settlements in Korosten district belong to the third contamination zone (including territory with ^{137}Cs deposition densities in the range between 185 and 555 kBq m^{-2}).

Total ^{129}I inventory (mBq m^{-2}) measured equals 38 for low contaminated villages (averaged over all settlements), from 67 to 130 for villages of Korosten district, and from 602 to 941 for villages of Narodichi district. They reported the best estimated value for the ^{131}I/^{129}I ratio to be $5.3 \pm 0.3 \times 10^7$ in terms of an activity ratio, or 13.6 ± 2.8 in terms of an atomic ratio. Using these values, the value of total ^{131}I activity in soil of chosen settlements was estimated. As a result, Michel et al. (2005) concluded that, despite uncertainties of this method (in particular, impact of iodine isotope migration in soils after deposition and choice of the initial $^{131}I/^{129}I$ activity ratio), retrospective dosimetry via ^{129}I can yield valuable information for the thyroid dose estimates due to the Chernobyl accident in areas with high ^{129}I contamination, which statistically significantly exceed the preaccident depositions from nuclear atmospheric weapons tests and from the emissions from European reprocessing plants.

1.6.2.4 Reconstruction of ^{131}I Radioactive Contamination in Ukraine and Belarus Using Atmospheric Transport Modeling

The method of mathematical modeling of the radionuclide atmospheric transport and their deposition on the underlying surface can be useful instrument for the reconstruction of dynamics of ^{131}I air activity concentration and fallout formation of in the initial period of the Chernobyl accident, especially for the territory of Ukraine where a few data of direct iodine measurements were available. This method enables taking into account the peculiarities of the formation of the radioactive contamination fields, including their variability in time and space.

The main aim of this approach was to reconstruct the formation features of time-dependent airborne and deposited ^{131}I fields in the territory of Ukraine during the initial stage of the Chernobyl accident (26 April to 7 May 1986) for the purpose of the thyroid dose reconstruction. The retrospection in Ukraine was conducted in order to include the dynamics of territories' contamination into the system of retrospective dosimetry of thyroid exposure. In the initial period of the accident, the direct thyroid activity measurements were made for some parts of the population that lived in the most contaminated Kyiv, Zhytomyr, and Chernihiv regions only. The data of the thyroid dose reconstruction for the rest of people of these regions and another territory of Ukraine may be obtained now on the basis of the atmospheric transport modeling. The results of the activity transport simulation enable us to use some ecological models for the thyroid dose estimation taking into account that the dose estimations are considerably dependent on time-variable airborne concentration values, deposition values, and especially on the activity arrival time and the finishing of fallout.

The method is based on the use of a mesoscale atmospheric transport model, LEDI (Lagrangian–Eulerian DIffusion model) (Talerko and Garger 2005). The first step of this work was modeling of ^{137}Cs transport and deposition in Ukraine and the comparison of obtained results with available data of soil cesium contamination measurements (Talerko 2005a). It enables us to build the release scenario from the accidental unit of the Chernobyl nuclear power station (taking into account the time variability of its parameters) and to assess the large- and small-scale features of the radioactivity deposition field in the territory of Ukraine and Belarus.

Scenario of ^{131}I Release Dynamics

Several scenarios of the ^{131}I release dynamics from the fourth unit of Chernobyl power plant during the first days after the accident are shown in Table 1.4 including the first estimation made by Soviet scientists after the accident (GKIAE 1986), the UNSCEAR (2000) scenario, and four scenarios used in atmospheric transport modeling works (Izrael et al. 1990; Haas et al. 1990; Borzilov and Klepikova 1993; Talerko 2005b).

The total ^{131}I release according to Talerko (2005b) was estimated to be 975 PBq or 31% of the core inventory. The corresponding value decay-corrected to 26 April is 1170 PBq (37%). It is necessary to note that this value is lower than estimations of total ^{131}I release made in UNSCEAR (2000), which gives that the time-averaged ratio $^{131}I/^{137}Cs$ equals 1760 PBq/85 PBq = 20.7. So it is possible that this ratio varied to an even greater degree during the release, and during some periods of the release, it was greater than the values used in this work. Accordingly, it could result in a larger value of this ratio in depositions in comparison with obtained values.

The time dependence of the iodine release scenario of Talerko (2005b) is similar to most scenarios, but the version of Izrael et al. (1990) suggested the release maximum on 28 April. For the rest, the largest release was on 26 April and then decreased, and the second maximum was assessed on 3–5 May. It should be noted that all scenarios based on the comparison of activity atmospheric transport and deposition modeling suggest much less values of iodine release in the comparison with the UNSCEAR scenario, which could be considered the most reliable one. This difference is the largest for assessments of the release during the first day. The possible reason is that the greater part of modeling works uses the available data of activity measurements in soil and air for the reconstruction of source-term parameters, including daily release estimation. However, the radioactive contamination on a local or regional scale has been formed not due to total radioactive release but only its part, which dispersed in the lower part of the atmosphere and could deposit on the underlying surface during the first days of the accident. On the other hand, another part of activity from the accidental unit could rise to much higher levels, resulting in radioactive contamination of heights about several kilometers (Jaworowski and Kownacka 1988) and did not affect the radioactive contamination of ground level air near the source. So, this difference between total release estimation in UNSCEAR (2000) and other estimations (including ours) may be regarded as the part of activity injected in the troposphere and lower stratosphere and dispersed at large distances from the Chernobyl nuclear power plant.

As for the cesium modeling, the Chernobyl source is treated as a volume one due to the formation of a convective plume over the destroyed unit and the release of a part of radioactive materials through its lateral surface. The parameters of this volume source (maximal height and initial vertical distribution of release intensity) were considered under the modeling of radionuclide atmospheric transport as functions of time during the period of intensive releases (Talerko 2005a). The radioactivity release intensity was set in four grids: at heights 200, 500, 800, and 1200 m.

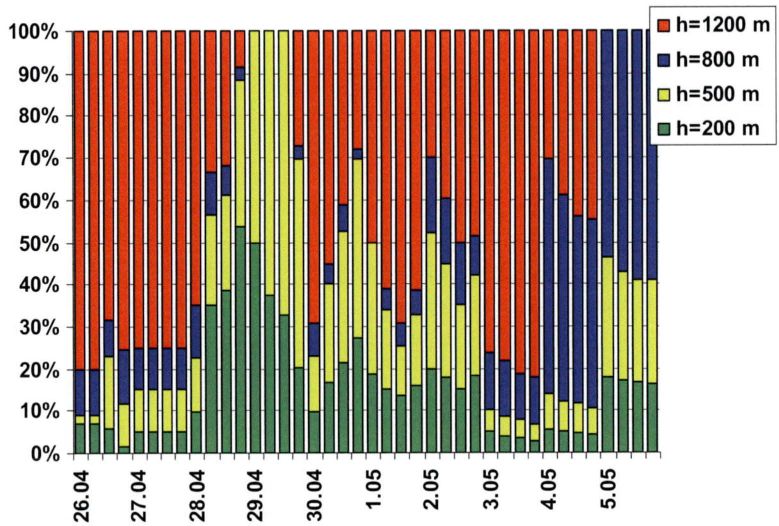

Fig. 1.10 ^{131}I release relative vertical distribution over heights 200, 500, 800, and 1200 m

The relative ^{131}I vertical distribution release intensity used in calculations (averaged over 6-h intervals) is shown in Fig. 1.10.

Iodine-131 Dry and Wet Deposition Parameterization

As mentioned above, the ratios between different iodine forms in the initial release from the Chernobyl NPP and the rates of their transformations during atmospheric transport were poorly investigated. So, the iodine isotopes were considered under atmospheric transport modeling as a uniform substance. From the point of modeling of atmospheric transport, this assumption is quite acceptable because the typical size of transported aerosol particles does not exceed several micrometers, and the gravitational sedimentation of aerosol particles may be neglected. But another great problem is the great difference between the values of dry deposition of the iodine forms. The results of laboratory and natural measurements of radioiodine aerosol deposition velocity give the values varying in the range from 0.1 cm s^{-1} to 1 cm s^{-1} (Ilyin et al. 1972; Nicholson 1988) depending on the particle size, the underlying surface characteristics, and meteorological conditions. The dry deposition of ^{131}I on the underlying surface in the organic form is estimated to be 0.01 cm s^{-1}, whereas the dry deposition velocity of elemental iodine is estimated to be 2 cm s^{-1} on dry grass and 3 cm s^{-1} on wet grass (Makhon'ko et al. 1996).

A comparison of works on atmospheric transport modeling of the Chernobyl release reveals the tendency to decreasing of the dry deposition velocity used in models when increasing the space scale of the modeling region. Izrael et al. (1987) obtained parameters of size distribution of the Chernobyl release aerosol particles from the comparison of the results of atmospheric transport modeling with the data of gamma-dose measurements. Assuming the lognormal distribution, they calculated the values of particle median diameter ξ and the logarithm of the geometric standard deviation σ. It was obtained that $\xi = 50$ μm and $\sigma = 0.25$ for mesoscale distances (up to 100 km), $\xi = 20$ μm and $\sigma = 0.4$ for regional scale (up to 400 km), and $\xi = 5$ μm and $\sigma = 0.25$ for the problem of long-distance transport modeling over Europe. The dry deposition velocity for these particle median diameters equals 19, 3, and 0.2 cm s^{-1} accordingly. The modeling works on the reconstruction of ^{131}I and ^{137}Cs deposition on the territory of the former USSR (Pitkevich et al. 1993; Borzilov et al. 1988) used the value of $V_g = 1$ cm s^{-1}. From another side, the typical values of ^{131}I dry deposition $V_g = 0.1$–0.3 cm s^{-1} have been used in most modeling works concerning the long-range transport of the Chernobyl release in Western Europe (Albergel et al. 1988; Izrael et al. 1990). The last values seem well grounded for the problem of the Chernobyl release deposition on the distances over 1000 km from the source. However, it is necessary to note that presumably only thinly dispersed part of the initial release was transported as far as such large distances. Furthermore, the part of iodine release in the elemental form was small at large distances as deposited earlier. So, the situation may be quite different in the close region near the Chernobyl power plant the same way as on meso- and regional scales (up to 200–500 km from the source). Both the elemental form contribution and the relatively large part of aerosol size distribution can result in considerably larger deposition values.

The results obtained for the Chernobyl release suggest the same tendency of decreasing of dry deposition velocity with increasing of the distance from the source. A lot of measurements of the air activity and deposition during the first days after the accident were made in the territory of Western Europe. It yields a deposition velocity of about 0.04–0.2 cm s^{-1} for ^{137}Cs deposition to grass (Nicholson 1988). The radionuclide dry deposition velocity in the closest region of the Chernobyl power plant was calculated by Izrael et al. (1990) using data of daily measurements of volume activity and deposition values obtained in the initial period of radioactive releases at the meteorological stations in Baryshivka (Kyiv region) and in Minsk. These dry deposition values for main radionuclides (including aerosol part of ^{131}I) are in the range from 0.4 cm s^{-1} to 1.8 cm s^{-1} that are essentially large than ones obtained from the measurements in Western Europe.

Makhon'ko et al. (1996) suggested the value of the ^{131}I deposition velocity $V_g = 0.8$ cm s^{-1} based on the data of radioactive iodine measured in Obninsk (Russia) on 30.04–01.05.1986. As it was outlined, this value is very close to one obtained for the global radioactive deposition as $V_g = 0.7$ cm s^{-1}. The dry deposition velocity was estimated using the abovementioned ratio between different physico-chemical forms measured in Vilnius and dry deposition values for each form. The value $V_g = 0.6$ cm s^{-1} was obtained for deposition on dry grass and $V_g = 0.9$ cm s^{-1} for deposition on wet grass. Taking these results into account, we consider the transport of ^{131}I accidental release over the territory of Ukraine as a uniform substance without division between different forms and use the averaged ^{131}I dry deposition velocity $V_g = 0.7$ cm s^{-1}.

For calculation of wet deposition of radioactivity on the underlying surface in the model, the washout coefficient Λ (s^{-1}) varied with rainfall rate J (mm h^{-1}) according to $\Lambda = \beta_0 \times J$, where β_0 (h mm^{-1} s^{-1}) is the constant depending on physicochemical form of radionuclides. According to Makhon'ko (1990), the value of β_0 is estimated to be 2.6×10^{-5} h mm^{-1} s^{-1} for particulate ^{137}Cs and 1.6×10^{-5} h mm^{-1} s^{-1} for ^{131}I (taking into account the percentage of the aerosol, molecular, and organic forms of iodine in the Chernobyl release).

Numerical Simulation for Ukraine and Belarus

The model of atmospheric transport LEDI was used for the reconstruction of dynamics of ^{131}I air and underlying surface contamination in Ukraine and in the three most contaminated (Gomel, Mogilev, and Brest) regions of Belarus during an initial phase of the Chernobyl accident. The data of radiosounding measurements of the atmosphere performed by a network of aerological stations in the former USSR were used as the input meteorological information. To describe the input of ^{131}I wet deposition under modeling, the data of rainfall measurements at meteorological stations in Ukraine and Belarus were used.

The procedure of ^{131}I transport and deposition modeling was similar to the one used for ^{137}Cs (Talerko 2005a). The prolonged release in the period from April 26 to May 5 in the calculations has been simulated by a sequence of instant puffs with a periodicity of 1 hour. The amount of activity in each such puff was selected so that the summarized ^{131}I activity in all puffs for every day corresponded to the chosen release scenario (Table 1.4). The summarized activity of the release within every hour was arranged between separate puffs with a different initial release height (200, 500, 800, and 1200 m) according to the chosen form of the initial distribution of release over height (Fig. 1.10). So, the total fields of ^{131}I deposition and volume concentration were calculated as a sum of the contributions of all 960 puffs with different release times and different initial heights.

One of the main tasks of this work was to create the database of calculated ^{131}I activity in air and ground during the period from 26 April to 7 May 1986 in every settlement in Ukraine to provide with necessary information the retrospective dosimetry studies of the reconstruction of thyroid doses of the Ukrainian population caused by the Chernobyl accident. So, values of daily deposition and time-averaged air volume concentration were calculated in 12,715 settlements of Ukraine, the same as in 8327 settlements of Belarus.

The atmospheric transport model LEDI and used meteorological data enable to obtain the relatively large-scale (with a scale of about several dozens of kilometers) features of radioactive contamination fields. Though, as the results of radioactive contamination measurements show, there is a considerable heterogeneity of deposition field with a scale of about 1 km. This variability may be explained by different reasons: impact of local features of underlying surface that was not taken into account under simulation (horizontal resolution of used set of topography and landuse data was 10 km), different characteristics of deposited material (e.g., variability of relative input of different physical and chemical forms of radioiodine), a possible input of wet deposition caused by local rains that has not been detected with the rainfall gauge network, and so on. So, to describe the local variability of radioactive deposition field, we used the approach similar to the one used by Pitkevich et al. (1994) and named by them as "the method of local effective depositions." According to it, the correction factor is introduced for every settlement to come into agreement the modeling results with measured values. This factor may be calculated as the ratio between ^{137}Cs deposition density D_{calc} calculated using atmospheric transport modeling and the median value of local measurements of the total ^{137}Cs deposition density D_{meas} taken from the database of radioactive contamination of soil in settlements of Ukraine and Belarus. As stated in Talerko (2005a), this value does not differ more than 1.5 times for 61.5% of the total number of settlements, whereas for 1.5% settlements, this ratio is greater than 5. This value may be treated as a measure of the difference between real local depositions and smoothed calculated deposition field. It was supposed that these local differences are the same as for ^{137}Cs as for ^{131}I deposition fields. So, the correction factor of "local effective depositions" is equal to $R = D_{meas}^{Cs\text{-}137}/D_{calc}^{Cs\text{-}137}$. It was calculated for every settlement and applied to the results of ^{131}I deposition modeling. The final value of iodine deposition density is equal to $D^{I\text{-}131} = R \times D_{calc}^{I\text{-}131}$. This correction is applied to both daily deposition and cumulative deposition density values.

Simulation Results

The general pattern of ^{131}I air and ground contamination field formation on the regional scale includes the following stages:

1. Formation of the western trace of contamination field during 26 and 27 April. According to the simulation results, during the first hours of April 26, the release transport in the near zone of the ChNPP occurred in the western and northwestern directions. Further, the activity moved to Belarus, and then to the direction of Poland, the Baltic Sea, and Scandinavia. During this period, the formation of radioactive contamination of the southern and western parts of Belarus, as well as the northwestern part of Ukraine took place.
2. Changing the main transport direction to the north and northeast to Belarus and Russia. On the evening of April 26 and during the first half of April 27, the prevailing direction of radioactive releases in the area of the ChNPP location gradually changed clockwise and radioactive contamination in the central part of Belarus began. On the night of April 28, the atmospheric circulation changed over the territory of Belarus, which led to the turn of the predominant direction of radioactivity transfer in the east direction. In the morning and afternoon of April 28, radioactive releases reached the Mogilev and northern parts of the Gomel region. Under the influence of heavy rainfalls (up to 15–20 mm per day) on April 28 and 29 in the eastern parts of the Gomel and Mogilev regions, large deposition spots in the territory of Belarus were formed. Further, radioactivity moved to the

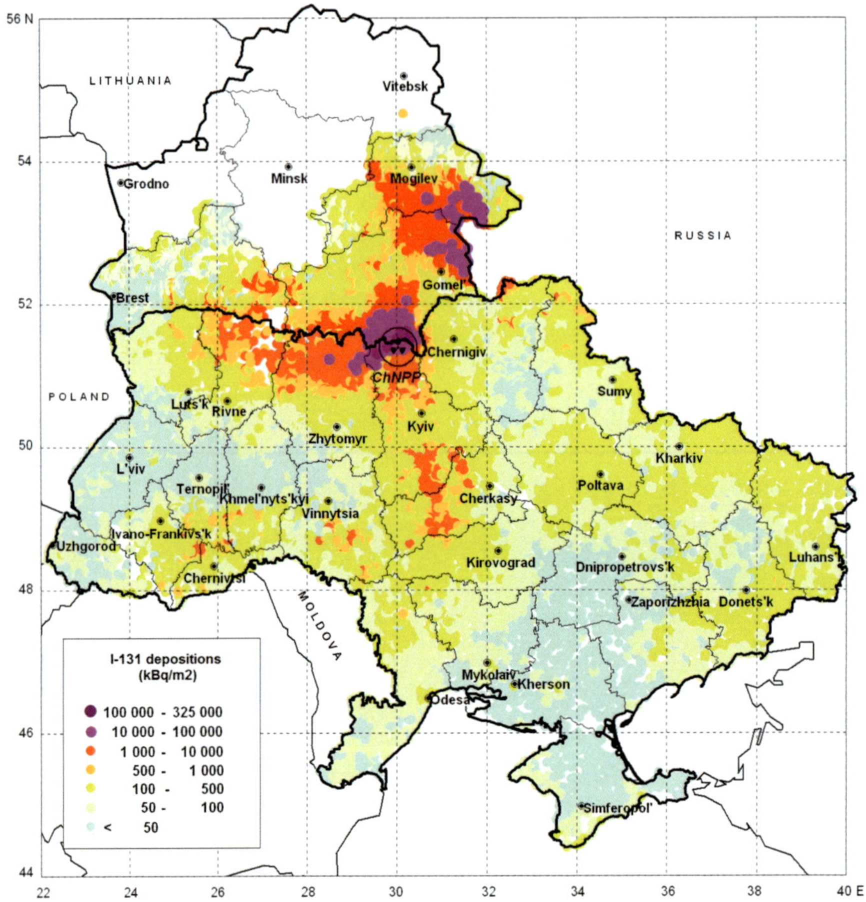

Fig. 1.11 Calculated field of cumulative ^{131}I surface ground deposition in Ukraine and Belarus (kBq m^{-2}) due to the Chernobyl accident. The inset shows deposition values for each settlement

east, resulting in radioactive contamination of the northeastern part of Ukraine and several regions of Russia.

3. Clockwise transport direction rotation resulted in the beginning of radioactive contamination of central (including Kyiv), southwestern, and eastern parts of Ukraine by 30 April and 1 May.
4. Following the formation of southern part of radioactive contamination field in Ukraine during the period 2–6 May 1986.
5. Finishing of radioactive deposition formation in the territory of Ukraine by 7 May.

The cumulative ^{131}I deposition density values (a simple sum of daily deposition values without decay correction) obtained for settlements in Ukraine and Belarus with the help of the LEDI model are shown in Fig. 1.11.

The largest calculated values of integral ^{131}I deposition (decay-corrected to 26 April 1986) are 127 MBq m^{-2} in village Varovichi (40 km west from the ChNPP) and 126 MBq m^{-2} in village Tolsty Les (22 km west from the ChNPP). The integral ^{131}I deposition in town Pripyat (2 km northwest from the ChNPP) was assessed to be 102 MBq m^{-2}. All these settlements are within the exclusion zone of the ChNPP. The largest values of integral ^{131}I deposition outside the exclusion zone are assessed to be 54.6 MBq m^{-2} in Yasen (Kyiv region, 52 km west from the power plant) and 53.1 MBq m^{-2} in Martynovichi (Kyiv region, 65 km west).

In Kyiv, the obtained integral ^{131}I deposition value is 737 kBq m^{-2} and cumulative ^{131}I one is 471 kBq m^{-2}. According to modeling in Kyiv, the maximal daily-averaged air volume ^{131}I concentration was 355 Bq m^{-3} on 1 May, and the air volume concentration averaged over the entire period of intensive fallout in Kyiv (from 30 April to 5 May) was 130 Bq m^{-3}.

Unlike Belarus, in Ukraine, the deposition field is formed due to dry deposition mainly (Talerko 2005a). The contribution of wet deposition was negligibly low there (with the exception of several small local spots in different regions of Ukraine). According to the simulation results, the formation of large-scale deposition spots in the southern part of Kyiv region (near Bila Tserkov) and in Cherkasy region (spot near Cherkasy and Kaniv towns) could be explained by the transport and deposition of the activity released from the high-elevated prolonged source taking into account of diurnal variability of turbulent parameters of atmospheric planetary layer. The transport of activity released during the nighttime of 1 May under conditions of large (more than 10 m s^{-1}) wind velocity resulted in the formation of these radioactive deposition spots at large (up 200–400 km from the source) distances in this territory.

The calculated ratio of integral (decay-corrected to 26 April 1986) ^{131}I to ^{137}Cs deposition densities in Ukraine decreases as a function of the distance from the Chernobyl NPP and varies from about 28 at the south trace of radioactive contamination field (near Kyiv) to less than 5 in Lugans'k region on the east of Ukraine. According to the simulation results, in Belarus, this ratio is obtained mostly in the range of 10–25 for areas where there were no rains during the first days of the accident or they were small, However, for the eastern part of Belarus (the north of Gomel and the south of the Mogilev regions), the typical calculated ratios of $^{131}I/^{137}Cs$ activity in fallout are from 6 to 8 due to significant wet removal of radionuclides from the air by rains. Such model results are confirmed by the available measurement data on the territory of Belarus (Kruk et al. 2004; Drozdovitch et al. 2013; Khrushchinskii et al. 2014).

The total value of ^{131}I deposition in the territory of Ukraine during the first 12 days after the accident is assessed as 185 PBq (without decay correction). It means 19% of our assessment of total ^{131}I release (975 PBq), or about 11% of the release estimation made in UNSCEAR (2000), or about 6% of the ^{131}I inventory (UNSCEAR 2000).

1.7 Conclusions

The accident at the Chernobyl nuclear power plant has become the largest one in the history of nuclear power and has led to the radioactive contamination of huge areas in almost the whole of Europe. One of its main lessons was that the response to the accident in its acute phase in order to protect the population from the effects of radiation took place in conditions of *total deficit: deficit of time, measuring resources for radiation monitoring, information, and scientific knowledge*. The conclusions made after the Chernobyl accident resulted in a significant modernization of emergency preparedness and emergency response systems in countries developing nuclear power.

The *deficit of time* for emergency response is due to the fact that the effectiveness of preventing the radiation dose of the population in radioactively contaminated areas is reduced in time. For example, according to Prister (2007), the reduction in thyroid dose is 10–12 times when taking stable iodine preparations in 1 hour after absorption of ^{131}I, 4 times in 2 h, and only 1.2 times if stable iodine is taken in 15 h. Thus, atmospheric transport models could be an important tool for obtaining a forecast for the development of the radiation situation. The efficiency and operationability of their use are provided in the framework of computer decision support systems (DSSs) for protecting the population in the event of a radiation accident. After the Chernobyl accident, a number of decision support systems have been developed for the first assessments and the operational forecast of the radiation situation in the event of a nuclear accident, including NARAC (USA), ARGOS (Sweden), RODOS (EU), RECASS (Russia), and WSPEEDI (Japan) (Sugiyama et al. 2014; Hoe et al. 2009; Ehrhardt 1997; Shershakov and Trakhtengerts 1996; Nakanishi et al. 2011).

The absence of the results of forecasts made with such online systems in the initial period of the Chernobyl accident led to the existing radiation monitoring facilities that had not been optimally used to control the state of the environment in radioactively contaminated areas. It took several months to get an objective picture of the radioactive contamination of the territories, and the refinement of the radiation situation took several years. Thus, as a consequence of the need for radiation monitoring in large areas, there was *a deficit of measuring resources for monitoring* in the initial period of the accident. The Chernobyl experience led to the need to organize an information exchange between predictive models of atmospheric transport and radiation monitoring systems, which allows us to correctly choose priorities in time and space for the adoption of protective countermeasures.

However, the experience of the reconstruction of radioactive contamination due to the Chernobyl accident has shown that even the use of the most advanced atmospheric transport models leads to uncertainties in estimates of radionuclide concentrations in the environment and the associated radiation doses of the population that can reach several orders of magnitude. This is due to the *deficit of information*, which is typical in the acute period of the accident. First of all, this is the uncertainty of estimates of the source parameters: the emission intensity, the radionuclide composition, and the effective emission height. After the Chernobyl accident, methods for construction of the source parameters using radiological measurement data are intensively developing, including those integrated into the DSS for online use. An important problem remains to provide computational models with reliable meteorological information with a high resolution both in time and in spatial coordinates. First of all, it is necessary to have information about precipitation, determining the formation of deposition spots, including small area ones. The existing network of meteorological stations cannot always provide such detailed information. Therefore, the use of information for mesoscale weather forecast models, as well as observational data from meteorological radars, can solve this problem.

Another important lesson of the Chernobyl accident was that for the prediction and assessment of the consequences of a radiation accident, simulation of radionuclide atmospheric transport and their deposition onto the underlying surface should be carried out not separately from calculations of other models of radionuclide migration in the environment, but in interaction with them. Nonconsideration of the radionuclide migration properties in soils of different types led to an initial underestimation of the Chernobyl accident consequences for the population living in the northwestern part of Ukraine (Ukrainian Polissya with its organic soils). Under relatively low deposition densities (about 37 kBq m^{-2} of ^{137}Cs), the doses of population exposure due to the consumption of local food products were higher than those living at distances of 40–50 km from the Chernobyl NPP in areas with mineral soils.

Finally, to this day, there is a *deficit of scientific knowledge* about the features of the dispersion of radioactive release as a result of a major radiation accident at a nuclear power plant. The variability of estimates of the dry deposition velocity of finely dispersed aerosol particles in model calculations of ^{137}Cs deposition as a result of the Chernobyl accident reached one order of magnitude. The problems of parameterization of the relative contribution of various physicochemical forms of radioactive iodine in the modeling of atmospheric transport were not completely solved. In addition, considerable difficulties arose in many studies in modeling the features of the radioactive release transport during its intersection of atmospheric fronts, and the related processes of changing radionuclide trajectories and their wet removal out of the atmosphere. The development of atmospheric transport models and perfect parameterization of these processes can solve these problems.

References

Albergel A, Martin D, Strauss B, Gross JM (1988) The Chernobyl accident: modelling of dispersion over Europe of the radioactive plume and comparison with activity measurements. Atmos Environ 22:2431–2444

Andriesse CD, Tanke RH (1984) Dominant factor in the release of fission products from overheated urania. Nucl Technol 65:415–421

Aoyama M, Hirose K, Suzuki Y, Inoue H, Sugimura Y (1986) High level radioactive nuclides in Japan in May. Nature 321:819–820

ApSimon HM, Wilson JJN (1987) Modelling atmospheric dispersal of the Chernobyl release across Europe. Bound-Lay Meterol 41:123–133

ApSimon HM, Simms KL, Collier CG (1988) The use of weather radar in assessing deposition of radioactivity from Chernobyl across England and Wales. Atmos Environ 22(9):1895–1900

ApSimon HM, Wilson JJN, Simms KL (1989) Analysis of the dispersal and deposition of radionuclides from Chernobyl across Europe. Proc R Soc Lond Ser A 425:365–405

Balonov MI, Bruk GY, Golikov VY, Erkin VG, Zvonova IA, Parkhomenko VI, Shutov VN (1996) Exposure of the population in the Russian Federation as a result of the Chernobyl accident. Radiat Risk 7:8–48

Bartnicki J, Salbu B, Saltbones J, Foss A, Lind OC (2001) Gravitational settling of particles in dispersion model simulations using the Chernobyl accident as a test case. DNMI research report no. 131. Norwegian Meteorological Institute (DNMI), Oslo

Begichev SN, Borovoi AA, Burlakov EV et al (1990) Fuel of unit 4 reactor at the Chernobyl NPP (Short reference book). Kurchatov Institute of Atomic Energy, Moscow. Preprint 5268/3

Bieringer PE, Young GS, Rodriguez LM, Annunzio AJ, Vandenberghe F, Haupt SE (2017) Paradigms and commonalities in atmospheric source term estimation methods. Atmos Environ. https://doi.org/10.1016/j.atmosenv.2017.02.011

BNR (Belarus National Report) (2001) 15 Years after the Chernobyl accident: consequences in the Republic of Belarus and their overcoming. BNR, Minsk. (in Russian)

BNR (Belarus National Report) (2011) A quarter of a century after the Chernobyl disaster: results and prospects for overcoming. BNR, Minsk. (in Russian)

Bocquet M (2012) Parameter field estimation for atmospheric dispersion: application to the Chernobyl accident using 4D-Var. Q J Roy Meteorol Soc 138:664–681

Bondietti EA, Brantley JN (1986) Characteristics of Chernobyl radioactivity in Tennessee. Nature 322:313–314

Bonelli P, Calori G, Finzi G (1992) A fast long-range transport model for operational use in episode simulation. Application to the Chernobyl accident. Atmos Environ 26A(14):2523–2535

Borisov NB, Ogorodnikov BI, Skitovich VI et al (1990) Composition and concentration of radioiodine in the atmosphere during the Chernobyl accident and a four-year post-emergency period. In: Proceedings of the 2nd all-union meeting following the liquidation of the accident, Chernobyl, 2–11. (in Russian)

Borovoi A (1992) Characteristics of the nuclear fuel of power unit No.4 of Chernobyl NPP. In: Kryshev II (ed) Radioecological consequences of the Chernobyl accident. Nuclear Society International, Moscow, pp 9–20

Borovoi AA, Gagarinskii AY (2001) Emission of radionuclides from the destroyed unit of the Chernobyl nuclear power plant. Atom Energy 90:153–161

Borzilov VA, Klepikova NV (1993) Effect of meteorological conditions and release composition on radionuclide deposition after the Chernobyl accident. In: Mervin SE, Balonov M (eds) The Chernobyl papers, Doses to the Soviet population and early health effects studies, vol 1. REPS, Washington, DC, pp 47–70

Borzilov VA, Klepikova NV, Kostrikov AA, Trojanova NI, Khvalensky YF (1988) Meteorological conditions of long-distance transport of radionuclides released into the atmosphere due to the Chernobyl accident. In: Radiation aspects of the Chernobyl accident: proceedings of the 1st All-Union conference, Obninsk, June 1988. Vol. 1. Radioactive contamination of environment. Hydrometeoizdat, St.-Petersburg, pp 87–92. (in Russian)

Brandt J, Christensen JH, Frohn LM (2002) Modelling transport and deposition of caesium and iodine from the Chernobyl accident using the DREAM model. Atmos Chem Phys 2:397–417

Broda R (1987) Gamma spectroscopy analysis of hot particles from the Chernobyl fallout. Acta Phys Polonica B18:935–950

Buzulukov YP, Dobrynin YL (1993) Release of radionuclides during the Chernobyl accident. In: Mervin SE, Balonov M (eds) The Chernobyl papers. Vol.1. Doses to the Soviet population and early health effects studies. REPS, Washington, DC, pp 3–21

Cambray RS, Cawse PA, Garland JA, Gibson JAB, Johnson P, Lewis GNJ, Newton D, Salmon L, Wade BO (1987) Observations on radioactivity from the Chernobyl accident. Nucl Energy 26:77–101

Davoine X, Bocquet M (2007) Inverse modelling-based reconstruction of the Chernobyl source term available for long-range transport. Atmos Chem Phys 7:1549–1564

De Cort M, Dubois G, Fridman SD, Germenchuk MG, Izrael YA, Janssens A et al (1998) Atlas of cesium deposition on Europe after the Chernobyl accident. EUR Report Nr. 16733. Office for Official Publications of the European Communities, Brussels-Luxemburg

Desiato F (1992) A long-range dispersion model evaluation study with Chernobyl data. Atmos Environ 26A(15):2805–2820

Devell L (1988) Nuclide composition of Chernobyl hot particles. In: Von Philipsborn H, Steinhausler F (eds) Hot particles from the Chernobyl fallout. Proceedings of an international workshop held in Theuern, 28–29 Oct 1987. Bergbau- und Industriemuseum, Ostbauern, Band 16, Theuern, pp 23–34

Devell L, Güntay S, Powers DA (1996) The Chernobyl reactor accident source term. Development of a consensus view. NEA/CSNI/R(95)24

Devell L, Tovedal H, Bergström U, Appelgren A, Chyssler J, Andersson L (1986) Initial observations of fallout from the reactor accident at Chernobyl. Nature 321:192–193

Dobrynin YL, Khramtsov PB (1993) Data verification methodology and new data for Chernobyl source term. Radiat Prot Dosimetry 50:307–310

Drozdovitch V, Zhukova O, Germenchuk M, Khrutchinsky A, Kukhta T, Luckyanov N, Minenko V, Podgaiskaya M, Savkin M, Vakulovsky S, Voillequé P, Bouville A (2013) Database of meteorological and radiation measurements made in Belarus during the first three months following the Chernobyl accident. J Environ Radioact 116:84–92

EGE (2005) Environmental consequences of the Chernobyl accident and their remediation: twenty years of experience. Report of the UN Chernobyl Forum Expert Group "environment". IAEA, Vienna

Ehrhardt J (1997) The RODOS system: decision support for off-site emergency management in Europe. Nucl Technol Publ 1–4:35–40

Ermilov A, Ziborov A (1993) Radionuclide relations in the fuel component of the radioactive fallout in the near Chernobyl NPP. Radiat Risk 3:134–138

Evangeliou N, Balkanski Y, Cozic A, Møller AP (2013) Simulations of the transport and deposition of ^{137}Cs over Europe after the Chernobyl Nuclear Power Plant accident: influence of varying emission-altitude and model horizontal and vertical resolution. Atmos Chem Phys 13:7183–7198

Evangeliou N, Hamburger T, Talerko N, Zibtsev S, Bondar Y, Stohl A, Balkanski Y, Mousseau T, Møller A (2016) Reconstructing the Chernobyl Nuclear Power Plant (CNPP) accident 30 years after. A unique database of air concentration and deposition measurements over Europe. Environ Pollut 216:408–418

Galmarini S, Graziani G, Tassone C (1992) The atmospheric long range transport model LORAN and its application to Chernobyl release. Environ Softw 7:143–154

Gavrilin Y, Khrouch V, Shinkarev S et al (2004) Case-control study of Chernobyl-related thyroid cancer among children of Belarus. Part I: estimation of individual thyroid doses resulting from intakes of ^{131}I, short-lived radioiodines (^{132}I, ^{133}I, ^{135}I), and short-lived radiotelluriums (^{131m}Te and ^{132}Te). Health Phys 86:565–585

Gavrilin Y (2001) Consequences of two scenarios for the development of the accident at the Chernobyl NPP. Bull Atom Energ 8:20–28. (in Russian)

Georgi B, Helmeke H-J, Hietel B, Tschiersch J (1988) Particle size distribution measurements after the Chernobyl accident. In: von Philipsborn HV, Steinhausler F (eds) Hot particles from the

Chernobyl fallout. Proceedings of an international workshop held in Theuem, 28–29 Oct 1987. Bergbau- und Industriemuseum, Ostbauern, Band 16, Theuern, pp 39–52

Germenchuk MG, Zhukova OM, Shagalova ED, Matweenko II (1996) Methodical approaches to the reconstruction of iodine-131 deposition and the features of its distribution over the Belorussian territory after the Chernobyl accident. Med Biol Aspects Chernobyl Accident 4:72–78. (in Russian)

GKIAE (Gosudarstvennyj Komitet po Ispol'zovaniyu Atomnoj Energii) SSSR, Moscow (1986) The accident at the Chernobyl nuclear power plant and its consequences (INIS-mf–10523). International Atomic Energy Agency, Vienna

Golubenkov AV, Borodin RV, Soifer A (1996) RECASS source term estimation sub-system and its application for reconstruction of the source rate of the chernobyl accident. Radiat Prot Dosimetry 64(1-2):49–55

Gudiksen PH, Harvey TF, Lange R (1989) Chernobyl source term, atmospheric dispersion, and dose estimation. Health Phys 57(5):697–706

Güntay S, Powers DA, Devell L (1996) The Chernobyl reactor accident source term: development of a consensus view. One decade after Chernobyl: summing up the consequences of the accident. In: IAEA-TECDOC-964, vol 2. IAEA, Vienna, pp 183–193

Haas H, Memmesheimer M, Geiss H, Jakobs HJ, Laube M, Ebel A (1990) Simulation of the Chernobyl radioactive cloud over Europe using the EURAD model. Atmos Environ 24A:673–692

Hoe S, McGinnity P, Charnock T et al (2009) ARGOS decision support system for emergency management. Technical University of Denmark, Kongens Lyngby. http://orbit.dtu.dk/files/3924948/Hoe_paper.pdf

Hou XL, Fogh CL, Kucera J, Andersson KG, Dahlgaard H, Nielsen SP (2003) Iodine-129 and caesium-137 in Chernobyl contaminated soil and their chemical fractionation. Sci Total Environ 308:97–109

Hutchinson M, Oh H, Chen W-H (2017) A review of source term estimation methods for atmospheric dispersion events using static or mobile sensors. Inform Fusion 36:130–148

IAEA (International Atomic Energy Agency) (1986) Summary report on the post-accident review meeting on the Chernobyl accident, safety series no. 75-INSAG-1. IAEA, Vienna

IAEA (International Atomic Energy Agency) (1992) The chernobyl accident: updating of INSAG-1. A report by the International Nuclear Safety Advisory Group, safety series no. 75-INSAG-7. IAEA, Vienna

IAEA (International Atomic Energy Agency) (2006) Environmental consequences of the chernobyl accident and their remediation: twenty years of experience. Report of the UN Chernobyl Forum Expert Group "environment". IAEA, Vienna

IAEA (International Atomic Energy Agency) (2011) Radioactive particles in the environment: sources, particle characterization and analytical techniques. TECDOC-1663. IAEA, Vienna

Ilyin LA, Arkhangelskaya GV, Konstantinov YO, Likhtarev IA (1972) Radioactive iodine in the radiation safety problem. Atomizdat, Moscow. (in Russian)

Ishikawa H (1995) Evaluation of the effect of horizontal diffusion on the long-range atmospheric transport simulation with Chernobyl data. J Appl Meteorol 34:1653–1665

Ivanov Y, Kashparov V, Sandalls J, Laptev G, Victorova N, Kruglov A, Salbu B, Oughton D, Arkhipov N (1996) Fuel component of ChNPP release fallout: properties and behaviour in the environment. In: The radiological consequences of the Chernobyl accident. IAEA, Vienna, pp 173–177

Izrael YA, Petrov VN, Severov DA (1987) Modelling of radioactive deposition in the near zone of the Chernobyl nuclear power station. Meteorol Gidrol 7:5–12. (in Russian)

Izrael YA, Petrov VN, Severov DA (1989) Regional model of radionuclide transport and deposition after the accident at the Chernobyl NPP. Meteorol Gidrol 6:5–14. (in Russian)

Izrael YuA, Vakulovskii SM, Vetrov VA, Petrov VN, Rovinsky FYA, Stukin ED (1990) Chernobyl: radioactive contamination of the environment. Gidrometeoizdat, Leningrad. (in Russian)

Jaworowski Z, Kownacka L (1988) Tropospheric and stratospheric distributions of radioactive iodine and cesium after the Chernobyl accident. J Environ Radioact 6:145–150

Jost DT, Gaggeler HW, Baltensperger U, Zinder B, Haller P (1986) Chernobyl fallout in size-fractionated aerosol. Nature 321:22

JRC (2015) Radioactivity environmental monitoring (REM). European Union – Joint Research Centre, Ispra. http://rem.jrc.ec.europa.eu/RemWeb/Index.aspx#

Kashparov VA (2001) Formation and dynamics of radioactive contamination of the environment during the accident at the Chernobyl NPP and in the post-accidental period. In: Cheornobyl. The exclusion zone. Naukova Dumka, Kyiv, pp 11–46. (in Ukrainian)

Kashparov VA (2016) Chernobyl: 30 years of radioactive contamination legacy report. Ukrainian Institute of Agricultural Radiology, Kyiv

Kashparov VA, Ahamdach N, Zvarich SI, Yoschenko VI, Maloshtan IM, Dewiere L (2004) Kinetics of dissolution of Chernobyl fuel particles in soil in natural conditions. J Environ Radioact 72:335–353

Kashparov VA, Ivanov YA, Zvarich SI, Protsak VP, Khomutinin YV, Kurepin AD, Pazukhin EM (1996) Formation of hot particles during the Chernobyl nuclear power plant accident. Nucl Technol 114:246–253

Kashparov VA, Kalinina GV, Ivliev AI, Cherkisyan VO (1995) Hot particles in soil from Chernobyl AES region. Radiat Meas 25(1-4):413–414

Kashparov VA, Lundin SM, Khomutinin YV, Kaminsky SP, Levtchuk SE, Protsak VP, Kadygrib AM, Zvarich SI, Yoschenko VI, Tschiersch J (2001) Soil contamination with ^{90}Sr in the Chernobyl accident near-field. J Environ Radioact 56(3):285–298

Kashparov VA, Lundin SM, Zvarich SI, Yoschenko VI, Levtchuk SE, Khomutinin YV, Maloshtan IN, Protsak VP (2003) Territory contamination with the radionuclides representing the fuel component of Chernobyl fallout. Sci Total Environ 317(1–3):105–119

Kashparov VA, Oughton DH, Zvarich SI, Protsak VP, Levchuk SE (1999) Kinetics of fuel particle weathering and ^{90}Sr mobility in the Chernobyl 30-km exclusion zone. Health Phys 76 (3):251–259

Katata G, Chino M, Kobayashi T, Terada H, Ota M, Nagai H, Kajino M, Draxler R, Hort MC, Malo A, Torii T, Sanada Y (2015) Detailed source term estimation of the atmospheric release for the Fukushima Daiichi Nuclear Power Station accident by coupling simulations of an atmospheric dispersion model with an improved deposition scheme and oceanic dispersion model. Atmos Chem Phys 15:1029–1070

Kauppinen EI, Hillamo RE, Aaltonen SH, Sinkko KTS (1986) Radioactivity size distributions of ambient aerosols in Helsinki, Finland, during May 1986 after Chernobyl accident: preliminary report. Environ Sci Technol 20(12):1257–1259

Kerekes A, Falk R, Suomela J (1991) Analysis of hot particles collected in Sweden after the Chernobyl accident. SSI-rapport 91-02. Statens Stralskyddinstitut, Stockholm

Khrushchinskii AA, Kuten' SA, Minenko VF, Zhukova OM, Podgaiskaya AA, Germenchuk MG, Kukhta TA, Vakulovskii SM, Drozdovitch VV (2014) Radionuclide ratios in precipitation on the territory of Belarus after the Chernobyl accident: calculation from gamma-spectrometric measurements on soil in May–July 1986. Atom Energy 117(2):143–148

Kirchner G, Noack CC (1988) Core history and nuclide inventory of the Chernobyl core at the time of accident. Nucl Safety 29:1–5

Klepikova NV, Freinmunt GN, Ladeikin YA, Kamaev DA (2010) Method for estimation of a source parameters using measurements in the near zone. In: Problems of hydrometeorology and monitoring of environment, vol 3. Obninsk, SPA Typhoon, pp 164–176. (in Russian)

Klug W, Graziani G, Pierce D, Tassone C (1992) Evaluation of long range atmospheric models using environmental radioactivity data from the Chernobyl accident. Elsevier Science Publishers, Barking, England, ATMES Report

Knatko VA, Dorozhok IN (2002) Estimation of thyroid doses from inhalation of ^{131}I for population of contaminated regions of Belarus. In: Imanaka T (ed) Recent research activities about the

Chernobyl NPP accident in Belarus, Ukraine and Russia. Research Reactor Institute, Kyoto University, Kyoto, pp 160–167

Knox JB, Dickerson MB (1986) ARAC (Atmospheric Release Advisory Capability) preliminary dose estimates for Chernobyl reactor accident. Lawrence Livermore National Laboratory, Livermore, CA

Konoplev AV, Borzilov VA, Bobovnikova CI et al (1988) Distribution of radionuclides deposited after the accident at the Chernobyl NPP in the "soil-water" system. Meteorol Gidrol 12:63–74. (in Russian)

Konoplev A (2020) Mobility and bioavailability of the Chernobyl-derived radionuclides in soil-water environment: review. In: Konoplev A, Kato K, Kalmykov SN (eds) Behavior of radionuclides in the environment II: Chernobyl. Springer Nature, Singapore, pp 157–193

Kritidis P, Catsaros N, Probonas M (1988) Hot particles in Greece after the Chernobyl accident, estimations on inhalation probability. In: von Philipsborn HV, Steinhausler F (eds) Hot particles from the Chernobyl fallout. Proceedings of an international workshop held in Theuern, 28–29 Oct 1987. Bergbau- und Industriemuseum, Ostbauern, Band 16, Theuern, pp 115–120

Kruk JE, Pröhl G, Kenigsberg JI (2004) A radioecological model for thyroid dose reconstruction of the Belarus population following the Chernobyl accident. Radiat Environ Biophys 43:101–110

Kuriny VD, Ivanov YA, Kashparov VA, Loschilov NA, Protsak VP, Yudin EB, Zhurba MA, Parshakov AE (1993) Particle-associated Chernobyl fall-out in the local and intermediate zones. Ann Nucl Energy 20:415–420

Kutkov VA, Arefieva ZS, Muravev YB et al (1995) Unique form of airborne radioactivity: nuclear fuel "hot particles" released during the Chernobyl accident. In: Environmental impact of radioactive releases. Proceedings of a symposium, Vienna, 8–12 May 1995. IAEA, Vienna, pp 625–630

Langner J, Robertson L, Persson C, Ullerstig A (1998) Validation of the operational emergency response model at the Swedish Meteorological and Hydrological Institute using data from ETEX and the Chernobyl accident. Atmos Environ 32:4325–4333

Larsen RJ, Haagenson PL, Reiss NM (1989) Transport processes associated with the initial elevated concentrated of Chernobyl radioactivity in surface air in the United States. J Environ Radioact 10:1–18

de Leeuw FAAM, van Aalst RM, van Dop H (1988) Modelling of transport and deposition over Europe of radionuclides from the Chernobyl accident. In: Air pollution modeling and its application VIII: proceedings of the 16th NATO/CCMS international technology meeting, Lindau, 6–10 Apr 1987, New York. NATO/CCMS, London, pp 499–507

Likhtarov I, Kovgan L, Masiuk S, Talerko M, Chepurny M, Ivanova O, Gerasymenko V, Boyko Z, Voillequé P, Drozdovitch V, Bouville A (2014) Thyroid cancer study among Ukrainian children exposed to radiation after the Chernobyl accident: improved estimates of the thyroid doses to the cohort members. Health Phys 106(3):370–396

Linnik VG, Sokolov AV, Sokolov PV (2016) Multi scales of Cs-137 contamination levels of the Bryansk region landscapes (according to aerial gamma survey data). In: Shershakov VM (ed) Proceedings of the international scientific and practical conference "Radioactivity after nuclear explosions and accidents: consequences and ways of overcoming" Obninsk, 19–21 Apr. Postoyannyj Komitet Soyuznogo Gosudarstva, Moscow, pp 267–297. (in Russian)

Makhon'ko KP (ed) (1990) Guide on organization of environmental monitoring in the area of a nuclear power plant location. Gidrometeoizdat, Leningrad. (in Russian)

Makhon'ko KP, Kozlova EG, Volokitin AA (1996) Radioiodine accumulation on soil and reconstruction of doses from iodine exposure on the territory contaminated after the Chernobyl accident. Radiat Risk 7:90–129

Michel R, Handl J, Ernst T, Botsch W et al (2005) Iodine-129 in soils from Northern Ukraine and the retrospective dosimetry of the iodine-131 exposure after the Chernobyl accident. Sci Total Environ 340:35–55

Mironov V, Kudrjashov V, Yiou F, Raisbeck GM (2002) Use of ^{129}I and ^{137}Cs in soils for the estimation of ^{131}I deposition in Belarus as a result of the Chernobyl accident. J Environ Radioact 59:293–307

Muck K, Prohl G, Likhtarev I, Kovgan L, Meckbach R, Golikov V (2002) A consistent radionuclide vector after the Chernobyl accident. Health Phys 82:141–156

Nair SK, Apostoaei AI, Hoffman FO (2000) A radioiodine speciation, deposition, and dispersion model with uncertainty propagation for the Oak Ridge dose reconstruction. Health Phys 78:394–413

Nakanishi C, Sato S, Furuno A, Terada H, Nagai H, Muto S (2011) WSPEEDI-II system user's manual for a nuclear or radiological emergency. JAEA Technol 2011-005:141

NEA (2002) Chernobyl. Assessment of radiological and health impacts. 2002 update of chernobyl: ten years on. OECD Nuclear Energy Agency, Paris

Netterville D (1990) Plume rise, entrainment and dispersion in turbulent winds. Atmos Environ 24:1061–1081

Nicholson KW (1988) The dry deposition of small particles: a review of experimental measurements. Atmos Environ 22:2653–2666

Nodop K (ed) (1997) Proceedings ETEX symposium on long-range atmospheric transport, model verification and emergency response, Vienna, 13–16 May, 1997. Office for Official Publications of the European Communities, Luxembourg

Noguchi H, Murata M (1988) Physicochemical speciation of airborne ^{131}I in Japan from Chernobyl. J Environ Radioact 7:65–74

Nosovsky AV, Vasilchenko VN, Kluchnikov AA, Prister BS (2006) Accident at the Chernobyl nuclear power plant: experience of overcoming, lessons learned. Tehnika, Kyiv. (in Russian)

Ogorodnikov BI, Pazukhin EM, Kluchnikov AA (2008) Radioactive aerosols of the "Shelter" object. Institute for Safety Problems of NPPs, Chernobyl. (in Russian)

Orlov M, Snykov V, Yu K, Volokitin A (1996) Contamination of the soil of the European part of the territory of the USSR with ^{131}I after the Chernobyl nuclear accident. Atom Energy 80 (6):439–444

Orlov MY, Snykov VP, Khvalenskii YA, Teslenko VP, Korenev AI (1992) Radioactive contamination of the territory of Belorussia and Russia after the Chernobyl Nuclear Power Plant disaster. Atom Energy 72:334–339

Osuch S, Dabrowska M, Jaracz P, Kaczanowski J, Van Khoi L, Mirowski S, Piasecki E, Szeflińska G, Szefliński Z, Tropiło J, Wilhelmi Z (1989) Isotopic composition of high-activity particles released in the Chernobyl accident. Health Phys 57(5):707–716

Paatero J, Hameri K, Jaakkola T, Jantunen M, Koivukoski J, Saxen R (2010) Airborne and deposited radioactivity from the Chernobyl accident – a review of investigations in Finland. Boreal Environ Res 15:19–33

Paatero J, Hämeri K, Jantunen M, Hari P, Persson C, Kulmala M, Mattsson R, Hansson H-C, Raunemaa T (2011) Chernobyl: observations in Finland and Sweden. In: Ensor DS (ed) Aerosol Science and technology: history and reviews. RTI Press, Research Triangle Park, NC, pp 339–366

Paul M, Fink D, Hollos G, Kaufman A, Kutschera W, Magaritz M (1987) Measurement of iodine-129 concentrations in the environment after the Chernobyl reactor accident. Nucl Instrum Meth B B29:341–345

Persson C, Rodhe H, De Geer L-E (1987) The Chernobyl accident – a meteorological analysis of how radionuclides reached and were deposited in Sweden. Ambio 16(1):20–31

Piedelièvre JP, Musson-Genon L, Bompay F (1990) MEDIA – an Eulerian model of atmospheric dispersion: first validation on the Chernobyl release. J Appl Meteorol 29:1205–1220

Pienkowski L, Jastrzebski J, Tys J (1987) Isotopic composition of the radioactive fallout in Eastern Poland after the Chernobyl accident. J Radioanal Nucl Chem 117(6):379–409

Pietruszewski A, Jagielak J, Kozub M, Wołoszyn Z, Sosińska A (1990) High activity and hot particles data for 1986 year samples measured in CLRP after Chernobyl accident. In:

International symposium on post-Chernobyl environmental radioactivity studies in East European Countries, Kazimierz Dolny, Poland, 17–19 Sep, pp 127–164
Pietrzak-Flis Z, Krajewski P, Radwan I, Muramatsu Y (2003) Retrospective evaluation of ^{131}I deposition density and thyroid dose in Poland after the Chernobyl accident. Health Phys 84 (6):698–708
Pitkevich VA, Duba VV, Ivanov VK, Shershakov VM, Golubenkov AV, Borodin RV, Kosykh VS (1994) Methodology for reconstruction of absorbed external radiation doses for population living on the territory of Russia contaminated due to the ChNPP accident. Radiat Risk 4:95–112
Pitkevich VA, Shershakov VM, Duba VV et al (1993) Reconstruction of composition of the Chernobyl radionuclide fallout in the territories of Russia. Radiat Risk 3:62–93
Pöllänen R (1997) Highly radioactive ruthenium particles released from Chernobyl accident: particle characteristics and radiological hazard. Radiat Prot Dosimetry 71(1):23–32
Pöllänen R (2002) Nuclear fuel particles in the environment – characteristics, atmospheric transport and skin doses. Dissertation. STUK – Radiation and Nuclear Safety Authority, Department of Physics, University of Helsinki, Helsinki
Pöllänen R, Valkama I, Toivonen H (1997) Transport of radioactive particles from the Chernobyl accident. Atmos Environ 31:3575–3590
Prister BS (2007) Chernobyl disaster: the effectiveness of population protection measures, the experience of international cooperation. Ukrainian Nuclear Society, Kyiv. (in Russian)
Pudykiewicz J (1988) Numerical simulation of the transport of radioactive cloud from the Chernobyl nuclear accident. Tellus 408:241–259
Puhakka T, Jylhä K, Saarikivi P, Koistinen J (1990) Meteorological factors influencing the radioactive deposition in Finland after the Chernobyl accident. J Appl Meteorol 29:813–829
Quélo D, Krysta M, Bocquet M, Isnard O, Minier Y, Sportisse B (2007) Validation of the POLYPHEMUS platform on the ETEX, Chernobyl and Algeciras cases. Atmos Environ 41:5300–5315
Raes F, Graziani G, Stanners D, Girardi F (1990) Radioactivity measurements in air over Europe after the Chernobyl accident. Atmos Environ 24A:909–916
Ramsdell JV Jr, Simonen CA, Burk KW (1994) Regional Atmospheric Transport Code for Hanford Emission Tracking (RATCHET). Hanford Environmental Dose Reconstruction Project. PNWD-2224 HEDR; UC-000. Battelle Pacific Northwest Laboratories, Richland, WA
Rao KS (2007) Source estimation methods for atmospheric dispersion. Atmos Environ 41:6964–6973
Raunemaa T, Lehtinen S, Saari H, Kulmala M (1987) 2–10 μm sized hot particles in Chernobyl fallout to Finland. J Aerosol Sci 18(6):693–696
Redwood M (2011) Source term estimation and event reconstruction: a survey. ADMLC/2011/1 report. Atmospheric Dispersion Modelling Liaison Committee, London
Reineking A, Becker KH, Porstendörfer J, Wicke A (1987) Air activity concentrations and particle size distributions of the Chernobyl aerosol. Radiat Prot Dosimetry 19:159–163
Robertson L (2004) Extended back-trajectories by means of adjoint equations. Report RMK no. 105. Swedish Meteorological and Hydrological Institute, Norrköping
Saari H, Luokkanen S, Kulmala M, Lehtinen S, Raunemaa T (1989) Isolation and characterization of hot particles from Chernobyl fallout in Southwestern Finland. Health Phys 57:975–984
Sahoo SK, Muramatsu Y, Yoshida S, Matsuzaki H, Rühm W (2009) Determination of ^{129}I and ^{127}I concentration in soil samples from the Chernobyl 30-km zone by AMS and ICP-MS. J Radiat Res 50:325–332
Salbu B, Krekling T, Lind OC, Oughton DH, Drakopoulas M, Simionovichi A, Snigireva I, Snigirev A, Weitkamp T, Adams F, Janssens K, Kashparov V (2001) High energy X-ray microscopy for characterization of fuel particles. Nucl Instr Meth Phys Res A467–468:1249–1252
Salbu B, Krekling T, Oughton DH (1998) Characterization of radioactive particles in the environment. Analyst 123:843–849

Salbu B, Krekling T, Oughton DH, Østby G, Kashparov VA, Brand TL, Day JP (1994) Hot particles in accidental releases from Chernobyl and Windscale nuclear installations. Analyst 119:125–130

Sandalls FJ, Segal MG, Viktorova N (1993) Hot particles from Chernobyl: a review. J Environ Radioact 18:5–22

Saunier O, Mathieu A, Didier D, Tombette M, Quélo D, Winiarek V, Bocquet M (2013) An inverse modeling method to assess the source term of the Fukushima Nuclear Power Plant accident using gamma dose rate observations. Atmos Chem Phys 13:11403–11421

Savonenkov VG, Anderson EB, Smirnova EA, Shabalev SI (2009) Radiogeochemical study of fuel-containing new formations caused by Chernobyl NPP accident. Papers Khlopin Rad Inst 14:87–117. (in Russian)

Sedunov YS, Borzilov VA, Klepikova NV, Chernokozhin EV, Troyanova NI (1989) Physicomathematical modeling of the regional transport of radioactive pollutants in the atmosphere in consequence of the Chernobyl accident. Meterorol Gidrol 9:5–10

Shershakov VM, Trakhtengerts EA (1996) Development of the RODOS/RECASS system as a distributed, decision making support system in an emergency. Radiat Prot Dosimetry 64:143–147

Shestopalov VM (ed) (1996) Atlas of the Chernobyl exclusion zone. Kartographiya, Kyiv. (in Russian)

Sich AR (1994) The Chernobyl accident revisited: source term analysis and reconstruction of events during the active phase. Dissertation. Massachusetts Institute of Technology, Cambridge, MA

Sinkko K, Aaltonen H, Mustonen R, Taipale TK, Juutilainen J (1987) Airborne radioactivity in Finland after the Chernobyl accident in 1986. STUK – A56. Helsinki, Finnish Centre for Radiation and Nuclear Safety

Smith FB, Clark MJ (1988) The transport and deposition of airborne debris from the Chernobyl nuclear power plant accident with special emphasis on the consequences to the United Kingdom. HMSO, London

Sofiev M, Valkama I, Fortelius C, Siljamo P (2007) Forward and inverse modelling of radioactive pollutants dispersion after Chernobyl accident. In: Borrego C, Renner E (eds) Developments in environmental science, vol 6, pp 283–292

Steinhauser G, Brandl A, Johnson TE (2014) Comparison of the Chernobyl and Fukushima nuclear accidents: a review of the environmental impacts. Sci Total Environ 470–471:800–817

Stohl A, Seibert P, Wotawa G, Arnold D, Burkhart JF, Eckhardt S, Tapia C, Vargas A, Yasunari TJ (2012) Xenon-133 and caesium-137 releases into the atmosphere from the Fukushima Dai-ichi nuclear power plant: determination of the source term, atmospheric dispersion, and deposition. Atmos Chem Phys 12:2313–2343

Straume T, Marchetti AA, Anspaugh LR, Khrouch VT, Gavrilin YI, Shinkarev SM, Drozdovitch VV, Ulanovsky AV, Korneev SV, Brekeshev MK, Leonov ES, Voigt G, Panchenko SV, Minenko VF (1996) The feasibility of using ^{129}I to reconstruct ^{131}I deposition from the Chernobyl reactor accident. Health Phys 71(5):733–740

Styro BI, Nedvetskayte TN, Filistovich VI (1992) Iodine isotopes and radioactive safety. Gidrometeoizdat, St-Petersburg

Sugiyama G, Nasstrom J, Larsen S, Pobanz B, Simpson M (2014) National Atmospheric Release Advisory Center (NARAC) source estimation capabilities: R&D to operations. CTBTO Workshop, Stockholm, Sweden, 23–25 Sep 2014. Presentation, LLNL-PRES-660636

Suh K-S, Han M-H, Jung S-H, Lee C-W (2009) Numerical simulation for a long-range dispersion of a pollutant using Chernobyl data. Math Comput Model 49:337–343

Talerko N (1990) Calculating the lift of a radioactive pollutant from the Chernobyl NPP accidental unit. Meteorol Gydrol 10:39–46

Talerko N (2005a) Mesoscale modelling of radioactive contamination formation in Ukraine caused by the Chernobyl accident. J Environ Radioact 78(3):311–329

Talerko N (2005b) Reconstruction of ^{131}I radioactive contamination of Ukraine caused by the Chernobyl accident using atmospheric transport modeling. J Environ Radioact 84(3):343–362

Talerko MM (2010) Reconstruction of Chernobyl source parameters using gamma dose rate measurements in town Pripyat. Nucl Phys Energ 11(2):169–177. (in Russian)

Talerko NN, Garger EK (2005) Experience of atmospheric transport model LEDI testing using field experiments and Chernobyl data. Preprint 05-1. Institute for safety Problems of Nuclear Power Plants, Chernobyl, Kyiv. (in Russian)

Terada H, Chino M (2005) Improvement of worldwide version of system for prediction of environmental emergency dose information (WSPEEDI), (II) Evaluation of numerical models by ^{137}Cs deposition due to the Chernobyl nuclear accident. J Nucl Sci Technol 42:651–660

Terada H, Chino M (2008) Development of an atmospheric dispersion model for accidental discharge of radionuclides with the function of simultaneous prediction for multiple domains and its evaluation by application to the Chernobyl nuclear accident. J Nucl Sci Technol 45:920–931

Terada H, Katata G, Chino M, Nagai H (2012) Atmospheric discharge and dispersion of radionuclides during the Fukushima Dai-ichi Nuclear Power Plant accident. Part II: verification of the source term and analysis of regional-scale atmospheric dispersion. J Environ Radiat 112:141–154

UNR (Ukraine National Report) (2011) 25 years of the Chernobyl accident. Safety of future. KIM Publishers, Kyiv. (in Ukrainian)

UNSCEAR (2000) United Nations Scientific Committee on the effects of atomic radiation. Sources and effects of ionizing radiation. Report to General Assembly. Annex J. Exposures and effects of the Chernobyl accident. UN, New York, NY

UNSCEAR (2008) United Nations Scientific Committee on the effects of atomic radiation. Sources and effects of ionizing radiation. Report to General Assembly. Annex D. Health effects due to radiation from the Chernobyl accident. UN, New York, NY

Vakulovskii SM, Orlov MY, Snykov VP (1989) Activity of radionuclides as a source of information about the accident at the Chernobyl NPP. In: Chernobyl-88. Proceedings of the 1st all-union scientific meeting on the results of the liquidation of the Chernobyl accident consequences. Chernobyl 1:4–22. (in Russian)

Vakulovskii SM, Shershakov VM, Golubenkov AV, Baranov AY et al (1993) Computer-information support of analysis of radiation environment in the territories polluted as a result of the Chernobyl accident. Radiat Risk 3:39–61

Vakulovskii SM, Valetova NK, Nikitin AI (2012) Radioactive contamination of a site in Kiev oblast on April 30, 1986. Atom Energy 111:445–449

Velikhov EP, Ponomarev-Stepnoy NN, Asmolov VG et al (1991) Current understanding of occurrence and development of the accident at the Chernobyl NPP. In: Selected proceedings of the international conference on nuclear accidents and the future of energy. The Lessons of Chernobyl, Paris, pp 12–36

Wheeler DA (1988) Atmospheric dispersal and deposition of radioactive material from Chernobyl. Atmos Environ 22(5):853–863

Whitehead NE, Ballestra S, Holm E, Walton A (1988) Air radionuclide patterns observed at Monaco from the Chernobyl accident. J Environ Radioact 7:249–264

Winiarek V, Bocquet M, Duhanyan N, Roustan Y, Saunier O, Mathieu A (2014) Estimation of the caesium-137 source term from the Fukushima Daiichi nuclear power plant using a consistent joint assimilation of air concentration and deposition observations. Atmos Environ 82:268–279

Winkelmann I et al (1987) Radioactivity measurements in the Federal Republic of Germany after the Chernobyl accident, ISH-116. Institut für Strahlenhygiene, Neuherberg

Zhurba M, Kashparov V, Ahamdach N, Salbu B, Yoschenko V, Levchuk S (2009) The "Hot particles" data base. In: Oughton DH, Kashparov V (eds) Radioactive particles in the environment. Springer, Dordrecht, pp 187–196

Zvonova I, Krajewski P, Berkovsky V, Ammann M, Duffa C, Filistovic V, Homma T, Kanyar B, Nedveckaite T, Simon SL, Vlasov O, Webbe-Wood D (2010) Validation of ^{131}I ecological transfer models and thyroid dose assessments using Chernobyl fallout data from the Plavsk district, Russia. J Environ Radioact 101(1):8–15

Chapter 2
Re-entrainment of the Chernobyl-Derived Radionuclides in Air: Experimental Data and Modeling

Evgeny Garger and Mykola Talerko

Abstract The chapter presents extensive experimental results and theoretical estimations of radioactive aerosol re-entrainment into the air from radioactive depositions on the ground following the Chernobyl accident, carried out generally within a 30-km exclusion zone of the Chernobyl nuclear power plant since May 1986 up to the present day. In the investigations on field conditions, the main attention was given to derive the integral characteristics of the resuspension process of radioactive particles used in practice, in particular, the generalized data on the resuspension factor and resuspension rate. The special feature of the chapter is the generalization of the data on the radionuclide activity concentration distribution with respect to particle size under various conditions of aerosol resuspension in the atmospheric surface layer. Results of long-term measurements of the characteristics of the radioactive aerosols released from the "Shelter" object are given. Observational and experimental data of re-entrainment of radioactive aerosols in the atmosphere caused by wildland fires and extreme meteorological conditions in the radioactive-contaminated territory are presented, which are the same as the results of modeling works.

Keywords Resuspension · Aerosol · Chernobyl exclusion zone · Radioactive contamination · Modeling · Wildland fires

2.1 Sources of Secondary Contamination

Predominantly, this chapter presents extensive experimental results and theoretical estimations obtained under conditions of various operative works for eliminating the consequences of the Chernobyl accident, carried out generally within a 30-km

---

E. Garger (✉) · M. Talerko
Institute for Safety Problems of Nuclear Power Plants, National Academy of Sciences of Ukraine, Kyiv, Ukraine

A. Konoplev et al. (eds.), *Behavior of Radionuclides in the Environment II*,
https://doi.org/10.1007/978-981-15-3568-0_2

exclusion zone of the Chernobyl nuclear power plant (ChNPP) since May 1986 up to the present day.

The analysis of the obtained re-entrainment measurement results is a rather complicated task due to the complex features of the Chernobyl zone. First, the air activity concentration is determined from the land surface, which is the constant source of radioactive aerosols changing in intensity with time, depending on the season, a characteristic of the underlying surface and meteorological conditions. Second, the radionuclide deposition after the accident was formed by radioactive aerosols of different sizes, nuclide composition, and chemical properties (including "hot" particles). Third, the radionuclide air activity concentration within the zone was influenced not only by resuspension in a traditional sense but also by mechanical disturbances enhanced by intensive human activity (the constant emissions from radioactive substances into the atmosphere from the "Shelter" object above Unit 4 of the ChNPP). Additionally, the mesoscale meteorological phenomena such as the passage of cold fronts with squalls have been marked; during dry seasons, there were fires in the contaminated forests and grass territories. Though these phenomena are short term, they may cause a considerable transport of radioactive substances over long distances into relatively clean territories.

In the investigations on field conditions, the main attention was given to derive the integral characteristics of the resuspension process of radioactive particles used in practice, in particular, the generalized data on the resuspension factor and the resuspension rate. The special feature of the chapter is the generalization of data on the radionuclide activity concentration distribution with respect to particle size under various conditions of aerosol resuspension in the atmospheric surface layer. Measurement with impactors has allowed us to estimate the contribution of particles of various ranges to the total activity concentration value.

2.1.1 Parameters for Description of Redistribution of Radioactive Substances and Estimation of Inhalation Dose Assessment

Garger (2008) presented mainly the empirical data on the resuspension of radioactive particles in the field after the Chernobyl accident. Especially, many measurement data have been obtained in Ukraine, Russia, Belarus, and in other European countries in these 30 years after the Chernobyl accident. As a rule, they were analyzed and generalized in frameworks of integrated empirical models. Thus, the accents were made on a practical significance of determination of characteristics which are necessary for the description of redistribution of radioactive substances or the estimation of the inhalation dose.

The activity concentration is quite often estimated as the measured density of deposition multiplied by the resuspension factor taken based on the available models from the literature, assuming uniformity of the underlying surface and deposition densities. It should be noted that the resuspension factor enables us to evaluate the activity concentration of radionuclides at the point where it is necessary to assess the inhalation dose. The resuspension rate that characterizes the emission intensity of radioactive substance is needed to assess the activity concentration and deposition of radioactivity over space and time. This allows for evaluating both an inhalation dose and a peroral dose due to the consumption of contaminated food.

2.1.2 Density of ^{137}Cs Contamination Within the Chernobyl Exclusion Zone

The total radionuclide inventory in the territory of the exclusion zone was estimated to be 4100–5600 TBq of ^{137}Cs, 2600–3700 TBq of ^{90}Sr, and 30 TBq of $^{239+240}$Pu excluding the radioactive substances located at the ChNPP site and in waste burials (Kashparov et al. 2001). The 30-km exclusion zone is located in an agricultural region containing mixed (i.e., deciduous and coniferous) forests. The forests occupy 30–35% of the area within the zone, with most of the remainder comprising agricultural land that was used to grow grass and cereal crops (and since left fallow). The soil in the region primarily comprises sod-podzolic (94%), sod with low humus (5%), and swampy peat (1%) soils. The topography of the 30-km zone is generally flat. Throughout the 30-km zone there is extensive vegetation and the soil is generally moist.

Garger et al. (1990) and Garger (1987) reported a preliminary assessment of the effects of resuspension in the 30-km Chernobyl exclusion zone (ChEZ). In these studies, the atmospheric activity concentration values and the soil contamination levels were reported for the period August/September 1986 and the results are used to determine the possible levels of resuspension intensity and their variability depending on time and location. The results are applicable in general for the evaluation of inhalation doses and in the assessment of the potential for the transport of contamination due to resuspension. During the period of the reported measurements there were major decontamination activities in the vicinity of ChNPP, which could have significantly contributed to resuspension, and the effects of this technogenic (i.e., mechanical) resuspension were considered.

As a whole, the natural region Polesye, where the ChEZ is located, is characterized as a region with light winds, and, consequently, the processes of saltation and surface soil movement are rare; whereas, in the steppe and semiarid regions of Ukraine and Russia, the input of these processes into particles lifted up from soil increases sharply. In Polesye, a scale of wind-driven resuspension and movement of substance in the polluted territory is local mainly, and mesoscale is rare. However, a regional transportation of radioactive dust particles also can be observed from the contaminated regions of Ukraine, Belarus, and Russia by moderate east winds in the surface and the boundary layers of the atmosphere for a long period (7–10 days) toward the Central Europe. It may happen in spring (April–May), when a steady anticyclone sets in over Eastern Europe and the soil is not protected yet by a green cover. The resuspension and atmospheric transport of aerosols may last up to ten and more days, as it occurred in the spring of 1993.

2.1.3 Samplers, Installations, and Measurements

Soon after the Chernobyl accident, studies on the radioactive aerosol characteristics and wind-driven resuspensions in the 30-km exclusion zone of ChNPP were launched by the Hydrometeorological Committee of the former USSR. These studies included measurements of resuspension characteristics and the temporal behavior of the activity concentration in the atmospheric surface layer.

At the time of measurements, there was little knowledge on the spatial variability of the radionuclide deposition field in the 30-km zone. However, γ-radiation measurements demonstrated significant spatial variations. The part of the zone where intensive soil activity measurements were made is shown in Fig. 2.1. Additionally, the sites where aerosol measurements were performed since 1986 are shown. It was

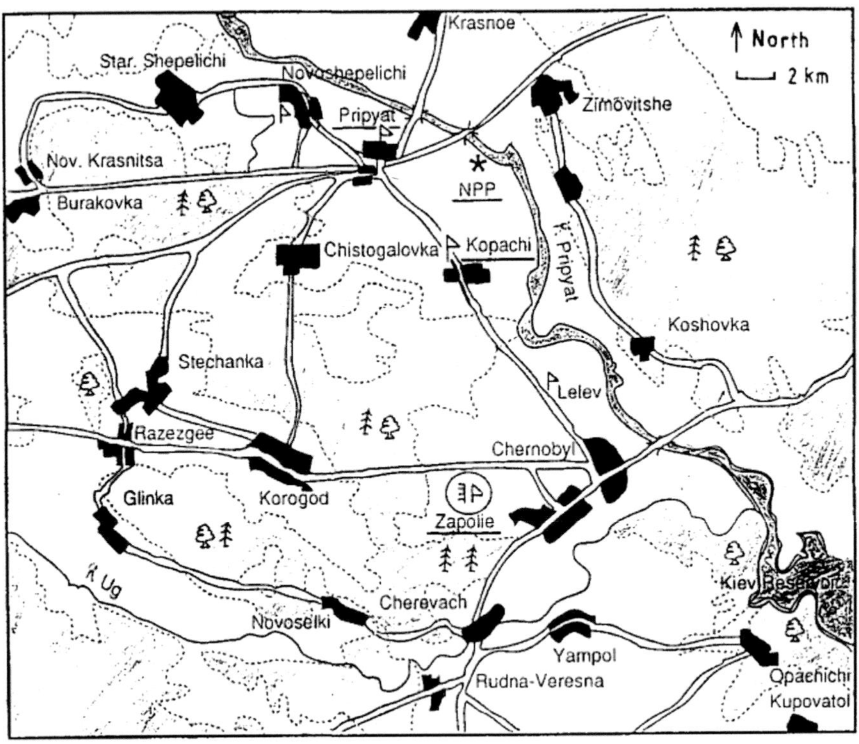

Fig. 2.1 Sampling locations in the 30-km exclusion zone (ЗР = main field measurement site; ⊳ = air activity concentration measurement sites; NPP = location of ChNPP)

decided to set up a permanent field site, at which various types of samplers could be used and where it would be possible to measure the air activity concentration depending on the height above the ground. This site is located close to the village of Zapolie, about 14 km south of ChNPP. The Zapolie site is located on the largest open place (a grass field of approximately 1800 m by 600 m). The Zapolie site was selected because of the relatively homogeneous surface contamination within 200 m of the sampling location. The site area is sufficient for the formation of the local surface boundary layer, in which the resuspended radioactive material is representative of the sampling location (Byzova et al. 1989). The activity variation (sigma/mean) of 20 soil samples taken at the site ranges from 22% (^{137}Cs) to 10% (^{103}Ru) (Garger 1987). This is a relatively high homogeneity for the 30-km exclusion zone. The surface roughness for this grass-covered site is approximately 0.1 m.

Permanent meteorological observations were carried out at the Chernobyl meteorological station (16 km from the NPP). Meteorological data collected in the 30-km zone indicate that the wind direction frequency varies between 9% and 16% over the eight primary wind directions, with the highest frequency observed for the southwest winds. The months with the highest wind velocities are February to April.

In June 1987, the Pripyat city (4 km from the NPP in the northwest direction) was selected as a second permanent site (Fig. 2.1). Besides Pripyat, after the accident, the high-volume sampler at the meteorological station in Chernobyl provided daily filter sampling. In both Chernobyl and Pripyat, high-volume samplers designed by SPA "Typhoon", Russia operated at a flow rate of 100,000 m^3 day^{-1} (1.16 m^3 s^{-1}) at a sampling height of about 1.5 m. The Russian-type Petryanov cloth FPP-15-1.5 with an area of 1.05 m^2 was used in the filter.

In the period from the end of May to December 1986, size-segregated aerosol measurements in the radioactive plume from Unit 4 of the ChNPP were made with the use of a helicopter at different distances and heights. The measurements enabled us to estimate the emission rate from Unit 4 and the size distribution of particles (Gaziev et al. 1993; Gaziev and Kabanov 1993; Skitovich et al. 1993).

Additionally, several similar high-volume sites were set up at other places in the exclusion zone (Garger et al. 1997a). Episodic measurements were carried out at the Pripyat-Beach and Kopachi sites between 1986 and 1993. Between August and September 1986, observations were made every 10 days at the following sites: Pripyat, Kopachi, Lelev, Zapolie (village), Zalesie, Korogod, Opachihi, Yampol, Zimovitshe, Chernobyl, and Novoselki (Garger 1994).

In 1992 and 1993, special aerosol measurements were conducted at Zapolie (field), Pripyat-Beach, and Kopachi. Pripyat-Beach is a plateau of artificial origin comprising hydraulic fill sand. The azimuth of the measurement site is 315°, and the distance from Unit 4 of the ChNPP is 3.8 km. Vegetation is sparse; only separate spots of wild grass and lichens are observed. The soil is pure friable sand. Kopachi is a grass field located about 3 km to the south of Unit 4. The surface roughness in this site is about 0.02 m. At a distance of 250–300 m from this measurement site, highly contaminated installations, armatures, and reinforced concrete are put together.

Airborne particle material was collected using passive "cone" air samplers. They consist of a gauze placed on a conical (19 cm base radius and approximately 80 cm

Table 2.1 Surface contamination in measurement sites

Place	Distance from source (km)	Surface contamination (MBq m^{-2})							
		^{144}Ce	^{141}Ce	^{103}Ru	^{106}Ru	^{137}Cs	^{134}Cs	^{95}Zr	^{95}Nb
Pripyat	4	55.5	4.4	4.9	14.4	5.2	2.0	28.7	53.7
Kopachi	4	17.4	1.5	2.2	4.4	1.6	0.7	8.7	12.6
Lelev	10	4.8	0.5	0.8	1.7	0.7	0.3	2.6	3.9
Zapolie (village)	14	2.8	0.2	0.4	1.2	0.3	0.2	1.3	2.0
Zapolie (site)	14	6.9	1.2	1.2	–	0.5	0.2	3.3	5.8
Zalesie	16	3.0	0.3	0.4	0.9	0.4	0.2	0.2	2.2
Korogod	14	0.5	<0.1	0.1	0.2	0.5	0.2	0.2	3.4
Opachichi	25	1.7	0.2	0.3	0.6	0.2	0.1	0.8	1.3
Yampol	21	2.1	0.2	0.4	0.9	0.3	0.1	1.3	1.9
Zimovitshe	7	70.3	9.7	6.8	7.5	2.4	0.3	–	–
Chernobyl	16	4.8	0.4	0.9	1.9	0.5	0.2	2.4	3.4
Novoselki	20	2.0	–	0.6	<0.1	0.3	0.1	1.6	–

long) wire frame. These frames rotate so that the open end faces the wind, and the particulate material is collected by filtration of the air passing through the gauze. The volume of aspirated air is estimated as the product of the base area of the cone, with the wind speed measured alongside the sampler. The efficiency of the cone samplers used has been evaluated on the basis of special experiments. The obtained efficiency factor of the cone samplers was compared with that of Belyaev et al. (1967). This factor is based on empirical data but does not take the particle size into consideration. According this work, under fine weather, the cone efficiency increases with increasing wind speed and decreases with increasing exposure time.

The work has been made to increase the efficiency of these cone samplers and to include the effects of particle size. They are, in any event, likely to be the most efficient at collecting large (> several μm in diameter) particles, and it is likely that the most resuspended material will fall within this particle diameter range since resuspended material is usually associated with host particles (Sehmel 1980). Air samples were collected at eight heights, between 0.25 and 15 m, in order to evaluate the vertical air concentration profiles. While, currently, there is only little validation data on the calibration of cone samplers, it is considered that the results gained by its use would be reasonable indicators of real air concentration.

In addition to air and soil samples, dry deposition samples have been collected at a height of 1 m. The samples were collected on planchettes, which are gauze-covered flat plates.

The estimation of the thermal stratification of the atmospheric surface layer was carried out on the basis of gradient measuring data of wind velocity and air temperature in a layer of 0.5–30 m using the "Sosna" gradient device (Garger 1994). At the remaining measurement points (see Table 2.1), gauze cones were used for aerosol sampling to determine the concentration of radionuclides in the air.

A gauze cone was mounted on a tripod at a height of 1 m paired with a horizontal planchette on which a Petryanov filter was exposed. A gauze cone was chosen in summer 1986 because of its availability and simplicity of manufacturing in large quantities, and its complete autonomy from power supplies. The cone was provided with a mechanical M-92 anemometer. This enabled us to measure the atmospheric activity concentration and its vertical profile with a large exposition time up to 2–3 days, with a high reliability of the spectrometer analysis data.

A contamination of the underlying surface was measured at various sites in the 30-km exclusion zone using a 15-cm diameter ring and the samples were taken to a depth of 5 cm. Twelve sites were chosen in the southern part of the zone (see Table 2.1 and Fig. 2.1) and a minimum of three samples were taken at each site. Samples were collected at various times during August and September 1986, and, for consistency, the results have been decay corrected to September 25, 1986.

It is evident that the density of surface contamination in the site territory for the measurement period of 2 months decreased with time. Apparently, this decrease was due to the natural decay of short-lived radionuclides ^{141}Ce, ^{103}Ru, ^{95}Nb, ^{95}Zr, and partly ^{144}Ce, ^{106}Ru.

For an estimation of the wind-driven resuspension rate, only tests with vertical profiles of the radionuclide activity concentration and the average wind velocity close to logarithmic were taken into account, that is, the measurement conditions approximately corresponded to a horizontally homogeneous pollutant source. Totally, 22 experiments with an exposition time of about 3 days were carried out during the period from August 1986 to September 1987.

In 1992, European Experimental collaboration project No. 1 (ECP1) was started, with the operations experimental period being completed in 1994 (ECP1 1996). The various joint field operations in the study of the radionuclide activity concentration flows and deposition were carried out. A principal advantage of field joint experiments was the simultaneous use of a large quantity of various measurement devices. Their inter-calibration was required to be performed to receive comparable results. There are no standardized aerosol samplers covering the complete particle size range. Only for the sampling of particle diameters up to 10 μm was there standard equipments available with well-defined collection characteristics (PM10). For quantitative sampling of large aerosol particles, which are of special interest in the case of resuspension, devices of special design were developed and deployed in the project. Most aerosol samplers which are widely in use in Ukraine, Russia, and Belarus have not been characterized in the western literature. Nevertheless, in order to use and to correctly interpret the results obtained from the experiments, it was necessary to compare the sampling instruments which were operated in the experiments.

In the field experiment of ECP1, size-integrating samplers as well as samplers which separate different size ranges have been used. All devices separate the particles by filtration. The designs of the inlet and the flow rate are different, and anisokinetic sampling errors are likely to exist.

In Table 2.2, all samplers used in the field experiments, which separate the particles according to their size, are specified. Andersen PM10 has an upper size limit but the deposition in the inlet is used to indicate the large particle fraction. Two

Table 2.2 Size-separated aerosol samplers used in the field experiments

Sampler name	Reference	Flow rate ($m^3\,h^{-1}$)	Size range cut point (μm)	Remarks
Andersen	Garger et al. (1997a)	67.8	7.2, 3.0,1.5,0.95, 0.49, <0.49	Slotted six-stage
PM10				Impactor (1992)
	Garger et al. (1997a)		4.9, 2.3, 1.4, 0.8, <0.8	Circular five-stage
				Impactor (1993/94)
Berner impactor	Berner and Lürzer (1980)	1.73	16, 8, 4, 2, 1, 0.5, 0.25, 0.13, 0.06	Mass size distribution
IBP impactor	Frank et al. (1996)	1.05	29, 14, 5.4, 1.6, 0.56, <0.56	Inside tractor cabin
IK impactor	ECP1 (1992b)	40	22, 10, 5, 3.2, 1.5,<1.5	
PK impactor	ECP1 (1992b)	190	20, 12, 7, 4, 2, <2	
RAI	Wagenpfeil et al. (1994)	8.8, 35, 18	10, 20, 28	Parallel impaction
UP impactor	ECP1 (1992b)	400	13, 4.5, 2.0, 0.65,<0.65	
Wind tunnel	ECP1 (1992b)	100, 47, 18	15, 24, 35	Parallel impaction
WRAC	Hollander et al. (1989)	100	9.2, 20.4, 48.6, 60.2	Parallel impaction

different cascade impactors were fitted to the PM10 in the field campaigns. The Russian IBP (Institute of Biophysics, Moscow) cascade impactor was characterized through laboratory experiments (Frank et al. 1996). It was used for aerosol sampling inside a tractor cabin. The rotating arm impactor (RAI) (Wagenpfeil et al. 1994) and the wind tunnel sampler collect particles larger than a certain size, that is, each size fraction with a different flow rate (Frank et al. 1996). The WRAC (wide-ranging aerosol classifier) parallel impactors (Hollander et al. 1989) collect particles smaller than a certain size on a filter. For details, see the quoted literature. The impactors IK, UP, and PK were widely used in the observations within the 30-km zone, especially the IK impactor in Zapolie from September 1986 and the UP impactor in Pripyat from September 1987. The flow rate of the well-characterized Berner impactor (Berner and Lürzer 1980) was too small to detect the nuclide size distribution, but the mass size distribution could be determined. The aerodynamical particle sizer (Blackford et al. 1987) was used to measure the particle number concentration with a high time resolution (1–20 min) in the size range of 0.6–30 μm.

2.1.4 Inter-comparison of Integrating and Size Differentiating Aerosol Samplers

In field experiments, which lasted typically for 3 weeks, ten institutes from eight countries participated. Resuspension measurements began under wind-driven conditions. Andersen PM10, IK, and PK (designed by SPA "Typhoon", Russia) impactors, which separate the particles according to their size, were the main measurement instruments. It was a general understanding that it was necessary to compare all methods of measurement of the atmospheric activity concentration, deposition, and the size distribution of radioactive particles at the same time.

During the ECP1 project implementation, several integrating aerosol samplers were used, including FOA (Swedish abbreviation of the operating institution - National Defence Research Institute, Sweden) (Vintersved 1994), Grad installation (SPA Typhoon, Russia) (ECP1 1992b), IPA (isokinetic sampler of aerosol, Ukraine) (Garger 2008), IPSN (ECP1 1992a), and "Typhoon" sampler (SPA "Typhoon", Russia) (ECP1 1992b).

The first comparison was made at the Pripyat-Beach and Zapolie sites in 1992. Devices were placed in the limited area of 10 m × 15 m at Pripyat-Beach and 30 m × 20 m at Zapolie. Care was taken to ensure that the devices did not influence each other. During a special period in 1993, observations of the activity concentration in air and the deposition density for inter-comparison were repeated because some divergence was observed, for example, for the sampler IPA. Good characterization of the performance of this device is very important because ten IPA samplers are being used for the measurement of the concentration at different distances from a line source of dust.

The installation Grad, which consists of four identical high-volume samplers with inlets at heights 1.0, 1.8, 2.5, and 3.5 m, was chosen as the reference instrument for the measurements without size resolution. The sampler had been operated since 1990 at the site and showed consistent results in comparison with the "Typhoon" sampler (at the same sampling height). The design of the FOA and the "Typhoon" samplers is similar to that of the Grad sampler, but the Grad sampler provides results at different sampling heights. In all cases, the Grad sampler of similar inlet height as the considered instrument for comparison was taken as reference for comparing the other instruments.

Sampler Andersen PM10, UP impactor, rotating impactor RAI, FOA sampler, IPSN sampler, and Grad sampler were all situated close to each other. The PK impactor, Grad sampler, and WRAC had their inlets at designed heights between 3.2 and 3.6 m, and were placed 20 m apart from each other. The experimental area at the Kopachi site was about the same scale. All together, 10 experiments were conducted in 1992 and 19 in 1993.

In 1992, a comparison was made between the ^{137}Cs air activity concentration values measured using samplers "Typhoon" and Grad and the integral activity concentration over all cascades of the impactor PK. The first four measurements were made for wind-driven resuspension and the the rest were made during agricultural work (imitation of cultivation). The PK concentration values are systematically

Table 2.3 Mean ratio of the atmospheric ^{137}Cs and ^{7}Be activity concentration values based on different samplers

Mean ratio	^{137}Cs		^{7}Be	
	Wind resuspension	Anthropogenic resuspension	Wind resuspension	Anthropogenic resuspension
FOA/Grad	0.73 ± 0.09	0.71 ± 0.16	0.92 ± 0.10	0.67 ± 0.10
IPSN/Grad	1.38 ± 0.06	1.29 ± 0.55	0.79 ± 0.04	0.56 ± 0.14
WRAC/Grad	–	0.60 ± 0.01	0.65 ± 0.02	0.61 ± 0.04
IPA/Grad	1.80 ± 0.21	0.50 ± 0.51	0.22 ± 0.02	0.40 ± 0.18
RAI/Grad	0.76	0.39 ± 0.15	–	–

less than the Grad values. The ratio between individual samples differs a little. The mean ratio equals 0.82±0.13. The total individual error is determined generally by the accuracy of the gamma-spectrometric analysis, which accounts to an error of not more than 10–15% of the measured activities. This value does not exceed the statistical variance of 16% of the total ensemble. The mean ratio of the samplers "Typhoon"/Grad is 0.91 ± 0.34; the scatter is larger, but it is never more than a factor of 2.

For comparison of the instruments, the results of the joint measurement campaign in May 1993 at Zapolie are chosen. During this period, all instruments operated close to each other, allowing a most complete comparison. The data from the isokinetic sampler IPA, the FOA sampler, the isokinetic sampler IPSN, the rotating impactor, and WRAC are compared with those from the Grad sampler (see Table 2.3).

The samplers were selected such that their heights were the same and the observation periods were equal. The FOA sampler construction is similar to the Grad design, and differs mainly in the kind of the filter material used. Grad and FOA samplers "cut off" the aerosol particles with sizes near 25–30 μm. The isokinetic sampler IPA aspirates apparently a wider spectrum of particle sizes than the Grad and FOA samplers but has less aerosol capacity, which can well be seen in periods of very high concentration during enhanced resuspension due to anthropogenic activity. The IPSN sampler is adjusted according to the wind velocity and wind direction in order to match the isokinetic conditions and has two separate filters (a vertical filter for small particles and a horizontal filter for large particles). The RAI samples only particles larger than 10 μm, and, therefore, the measured activity concentration must be less than the FOA or Grad data. The aerosol samples of the WRAC were taken isokinetically from a 60-cm diameter duct with an average flow of 2.4 m s^{-1}. The samplers were operated using the individual routine filter material. The ratios between the WRAC data and the Grad data were systematically less than unity. In Table 2.3, the mean values of the relative atmospheric activity concentration values for ^{137}Cs and ^{7}Be from different instruments are provided.

In Fig. 2.2, the activity concentration values of ^{137}Cs, as determined by the active (pump-operated) integral instruments, are given for the whole experimental period.

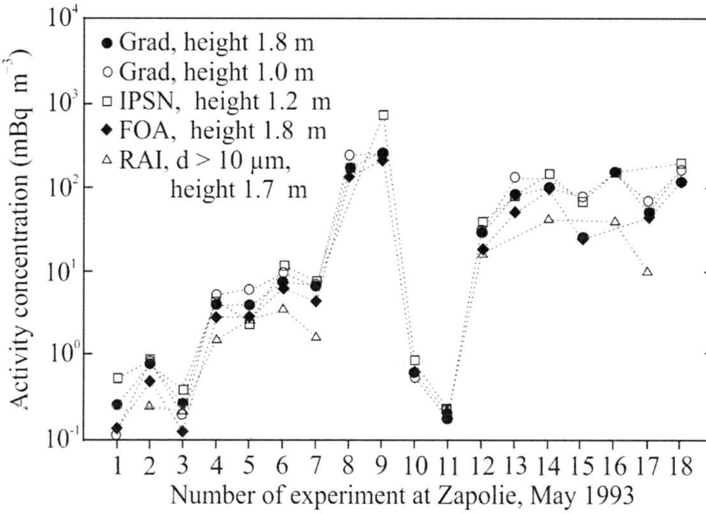

Fig. 2.2 A comparison of the ^{137}Cs activity concentration based on measurements using integrating samplers (Grad, IPSN, FOA) during wind resuspension (experiments 1, 3, 10, and 11) and different anthropogenic-enhanced resuspension (all other experiments) at the Zapolie site, May 1993. Sampler RAI, which collected only particles larger than 10 μm aerodynamic diameter, generally shows about 40% of the activity of the integrating instruments

Results of Grad samplers with two different inlet heights are shown together with samplers FOA and IPSN.

The results of the RAI sampler (sampling only particles larger 10 μm) are included in Fig. 2.2 for comparison. In general, the measured concentration values are in good agreement within a factor of 2. Of special interest are the experiments conducted on May 13 (experimentd 7 and 8) with a very high emission of large particles. Whereas for the Grad and FOA samplers, an upper cutoff of 25–30 μm is estimated, the isokinetic IPSN sampler collects larger particles effectively as well. Thus, this sampler detects higher concentration values during the period with a higher proportion of large particles.

The ^{7}Be atmospheric activity concentration (due to the stratospheric origin) is uniform at a given sampling site and is generally associated with smaller particles, which are unlikely to be prone to anisokinetic sampling errors. Several samplers are compared on the basis of this nuclide. Again, the agreement is, in general, within a factor of 2.

A comparison of the size distribution measurements of radionuclides by the PK, UP, and Andersen PM10 impactors for wind-driven conditions is given in Table 2.4. The results of the measurements of the ^{137}Cs activity concentration during 5 days in the absence of any other work at the observation site are shown. The measurements by the PK and PM10 impactors agree satisfactorily. The UP impactor data for the integral total activity were in the limit of a factor of 2 when compared with the PM10 results.

Table 2.4 ^{137}Cs activity concentration of different aerosol particle size fractions based on the Andersen PM10, PK, and UP impactors during simultaneous sampling

Andersen PM10 impactor (h = 1.8 m, V = 3832 m^3, sampling period 7185 min)		PK impactor (h = 3.2 m, V = 13,264 m^3, sampling period 7335 min)		UP impactor (h = 1.5 m, V = 11,845 m^3, sampling period 6855 min)	
Range (μm)	Concentration (mBq m^{-3})	Range (μm)	Concentration (mBq m^{-3})	Range (μm)	Concentration (mBq m^{-3})
>10	0.044	>12	0.036	>13	0.014
4.9–10.0	0.019	4.0–12.0	0.050	2.0–13.0	0.032
2.3–4.9	0.027	2.0–4.0	0.020		
0.8–2.3	0.045	<2.0	0.059	0.65–2.0	0.014
<0.8	0.011	–	–	<0.65	0.014
Total	0.146	Total	0.165	Total	0.074

Thus, a comparison of the integrating samplers shows that the samplers, in general, agree within a factor of 2 according to the ^{137}Cs and ^{7}Be activity concentration measurements. Larger discrepancies were found for the IPA sampler, and an additional investigation of its characteristics is demanded. The isokinetic sampler IPSN has shown that ordinary integrating samplers may have about 30% loss in the ^{137}Cs-activity associated with large particles due to anisokinetic sampling. This sampler is very useful for the investigation of the anthropogenic resuspension, for example, decontamination work, when a lot of radioactive dust with a wide size distribution is lifted up into the air. The mean ratio of the ^{137}Cs activity concentration between RAI (sampling only particles larger than 10 μm) and Grad is 0.39 ± 0.15 during enhanced resuspension due to anthropogenic activity, which supports the results of the IPSN sampler. Samplers with similar design as Typhoon, Grad, and FOA have found concentration values which are very close. These data can be interpreted jointly for the analysis of different tasks of monitoring of radioactivity and radiation protection.

2.2 Resuspension Factor and Resuspension Rate

Several different reviews have been devoted to the description of the general concepts, mechanisms, and models for obtaining a quantitative assessment of the resuspension of radioactive particulates from the surface layer of soil (Linsley 1978; Healy 1980; Sehmel 1984; Smith et al. 1982; Nicholson 1988; Makhonko 1992; Garger et al. 1990; Garland et al. 1992). One of the general results of these efforts is that the resuspension factor is used most frequently for environmental assessments of exposure to humans.

The resuspension factor K (m^{-1}) is defined as $K = q/D$, where q is the mean atmospheric activity concentration at the breathing height over a radioactive-contaminated surface (Bq m^{-3}) and D is the initial surface deposit (Bq m^{-2}). This factor is useful in localized operative situations for the characterization of the relationship between surface and airborne contamination, and its use implies an equilibrium between the resuspended and deposited aerosols (Sehmel 1984).

2.2.1 *Experimental Estimation of the Resuspension Factor After the Chernobyl Accident*

There is considerable variation in the levels of radionuclide deposition density in the 30-km ChEZ (see Table 2.1). There is also a rapid decline in deposition with increasing distance from ChNPP. Consequently, any interpretation of atmospheric activity concentration values which arise from resuspension should consider the possible effects of the advection from upwind surfaces.

The atmospheric activity concentration values of radionuclides were measured between September 14 and 17, 1986. During this period, the wind direction varied from the west to northwest and the advection of resuspended aerosols from highly contaminated areas to the measurement sites would have been small. The wind speed, however, was quite low, between 1.0 and 1.5 m s^{-1}, for the greater part of period. The air activity concentration values were generally higher at the most contaminated sites, Pripyat, Kopachi, and Zimovitshe. Calculations of the resuspension factor give values in the range of 6×10^{-9}–3×10^{-6} m^{-1} for the same period (see Table 2.5). The values of Pripyat town, a highly contaminated site, are similar to most other sites, indicating a relationship between the atmospheric activity concentration and the surface contamination of the underlying ground.

The high variability of the resuspension factor shown in Table 2.5 is likely to reflect a number of influences. First, the type of surface has an important effect on resuspension, with the presence of vegetation significantly reducing the occurrence

Table 2.5 Resuspension factor: September 14–17, 1986

	Resuspension factor × 10^{-8} (m^{-1})							
Place	^{144}Ce	^{141}Ce	^{103}Ru	^{106}Ru	^{137}Cs	^{134}Cs	^{95}Zr	^{95}Nb
Pripyat	9.7	10.0	11.0	7.6	7.7	0.9	8.7	8.5
Kopachi	85.0	72.0	60.0	59.0	59.0	58.0	64.0	88.0
Lelev	16.0	322.0	15.0	26.0	5.2	42.0	9.0	18.0
Zapolie	11.0	9.0	3.0	0.6	5.3	10.0	6.0	6.5
Zalesie	10.0	11.0	10.0	21.0	6.8	17.0	55.0	9.2
Korogod	10.0	23.0	10.0	28.0	4.2	11.0	11.0	9.90
Opachichi	9.0	7.0	5.6	21.0	13.0	21.0	4.0	6.7
Yampol	3.3	7.5	5.5	26.0	5.9	16.0	4.1	2.8
Zimovitshe	18.0	10.0	22.0	30.0	42.0	120.0	–	–
Chernobyl	15.0	18.0	19.0	10.0	20.0	15.0	17.0	23.0

of saltation (i.e., skipping of particles across a surface), which has been found to be important in the resuspension process. Second, the vertical migration of contaminants within the surface layer of soil will be important in determining the amount available for resuspension. Although, for consistency, the top 5 cm of the soil was taken for radionuclide analysis, it is unlikely that a material near the bottom of this sample would be normally available for resuspension (Nicholson 1988).

Third, there were the effects of mechanical actions which could greatly influence resuspension. This effect is likely to be important at Kopachi, where the resuspension factor was of the order of 10^{-7} m^{-1} for all nuclides, since intensive decontamination activities around ChNPP during the period of study would have resulted in resuspension and advection into the sampling area. The passage of vehicles has also been shown to have an effective influence on resuspension (Nicholson and Branson 1990), although it is difficult to be quantified for the measurement period.

The importance of advection from upwind sources is illustrated in Table 2.6 (Garger 1994). A comparison of the air activity concentration measurements during the north winds, when the wind was passing across the most contaminated part of the region in which decontamination activities were taking place, with the air activity concentration measurements during the south winds shows significantly lower values for the latter. These results relate to measurements made at an agricultural site and may be considered to be typical of the ChEZ. It is also shown in Table 2.6 that the atmospheric activity concentration values are significantly lower during rainfall and the occurrence of a moist surface.

Surface contamination and atmospheric activity concentration in the 30-km ChEZ varied significantly in space during 1986. The atmospheric activity concentration was dependent on resuspension, as well as the radionuclide advection was an important factor in many cases. The effects of decontamination activities around ChNPP were also found to be important in determining the downwind air activity concentration.

During 1986, the resuspension factor was found to be mostly in the range of 10^{-8}–10^{-7} m^{-1}, depending on the radionuclide.

In 1993, in Zapolie, six series of measurments were carried out using the PK impactor, with the purpose of estimating whether the ^{137}Cs resuspension factor was dependent on the aerosol particle size (Table 2.7).

Table 2.7 shows that the maximal value of the resuspension factor corresponds to the range of inhalational particles with an aerodynamic diameter less than 2 μm. It is seen that the resuspension factor for respiratory particles with diameters less than 12 μm is equal 3.6×10^{-10} m^{-1}, and exceeds the resuspension factor for particles >12 μm by more than two times.

In the period from July 23 to August 15, 1992, an investigation of the resuspension of radioactive particulates near Vertebu (Novozybkov district, Bryansk region, Russia) was conducted on a cultivated field with a total area of about 276 ha (Lukoyanov et al. 1994). In the surface layer of the atmosphere, a considerable concentration of dust was observed. In some days it exceeded the daily averaged threshold limit value (TLV) of a dust in air equal to 0.15 mg m^{-3}, adopted in Belarus

Table 2.6 Comparison of the air activity concentration values for north and south winds at the Zapolie site

Period of air sample	Surface conditions	Wind direction	Wind speed at 1 m (m s^{-1})	Air activity concentration (Bq m^{-3})					
				^{144}Ce	^{141}Ce	^{103}Ru	^{137}Cs	^{134}Cs	^{95}Zr
August 6–9, 1986	Dry	North	0.8	9400	1600	2000	700	250	7700
August 9–12, 1986	Wet	North	1.0	850	78	220	92	37	740
August 12–13, 1986	Dry	South	2.9	67	9.2	5.2	3.2	1.5	28
August 27–29, 86	Dry	South	1.7	100	3.7	4.8	5.9	2.6	6.7
September 2–3, 1986	Dry	South	1.0	25	7.8	2.6	9.6	7.0	8.5

Table 2.7 Resuspension factors obtained from the PK impactor data

Aerosol particle size range, μm	<2	2–4	4–7	7–12	12–20	>20
Resuspension factor, m^{-1}	1.5×10^{-10}	0.85×10^{-10}	0.51×10^{-10}	0.71×10^{-10}	0.71×10^{-10}	0.77×10^{-10}

Table 2.8 Resuspension factors for various fractions of soil particles

Dust fraction of soil (μm)	Number concentration of particles (l^{-1})	Calculated concentration of ^{137}Cs $\times 10^{-4}$ (Bq m^{-3})	K_1 $\times 10^{-8}$ (m^{-1})	K_2 $\times 10^{-8}$ (m^{-1})
0.5–1.0	11,631	1.22	110.0	6.25
1.0–5.0	1274	8.79	52.7	2.99
5.0–10.0	20	3.78	6.7	0.38
10.0–25.0	5	1.79	1.9	0.11

and Ukraine. The measurements of concentration have shown that the share of respiratory particles ($d < 10$ μm) was 46% of the mass of all particles and the share of inhalation particles (<1.0 μm) was on an average 14%. Thus, 87% of ^{137}Cs activity fell on respiratory particles (≤10 μm). The experimental data of that work have enabled us to estimate a relationship between the activity concentration of ^{137}Cs bound to the particles and their sizes, and to calculate the dependence of the resuspension factor on the particle size. In Table 2.8, values of the resuspension factor K_1 for the density of contamination in 1-cm layer and K_2 for the density of contamination corresponding to the inventory of ^{137}Cs in soil depending on the soil particle size are given (Lukoyanov et al. 1994).

In that work, the mass particle concentration values measured and calculated using data of the number concentration considering a measured soil density 2.6 g m^{-3} have been compared. It turned out that the average value of the experimental mass concentration exceeds the calculated one by 1.4 times. Lukoyanov et al. (1994) proposed the empirical dependence of the ^{137}Cs resuspension factor on particle size as

$$K = A \cdot d^{-1.61}, \tag{2.1}$$

where K is the resuspension factor (m^{-1}); $A = 6.2 \times 10^{-7}$ and 8.5×10^{-8} corresponding to the K_1 and K_2 values; and d is the mean diameter of particle size interval (μm). The dependence (2.1) was obtained for respiratory particles of soil at a wind velocity not exceeding 3 m s^{-1} at a level of 1 m under weak unstable conditions of thermal stratification of the atmosphere.

The generalized data on the resuspension factor estimations for ^{137}Cs and $^{239+240}Pu$ at six measuring sites in Belarus, Russia, and Ukraine under the ECP1 project are presented in Hollander and Garger (1996). The highest values of the resuspension factor have been obtained in Novozybkov, where the soil was the most eroded. Despite the different environmental conditions at the six sites, the differences between the minimum and maximal values of the resuspension factor do not exceed four times. Rather low values were obtained for the Pripyat-Beach site though it had the largest radionuclide deposition density. It is probably due to the characteristic of the beach surface, which consists of coarse hydraulic fill sand into which the radioactivity easily penetrates deeply during monsoon. It seems that the resuspension factor for $^{239+240}Pu$ is of the same order of magnitude as that for ^{137}Cs.

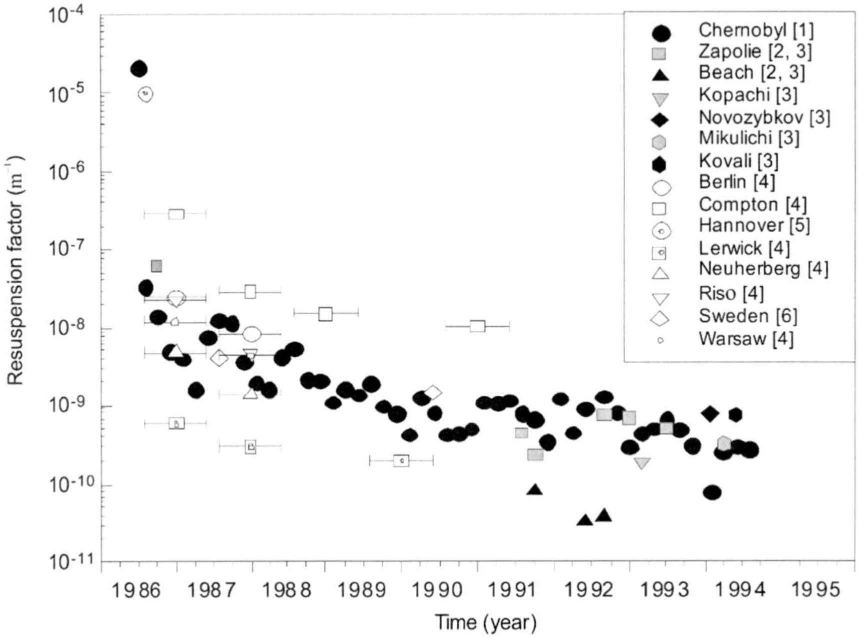

Fig. 2.3 Summary of the available data of measured resuspension factors in Europe. The bars show the sampling periods. [1]—Garger et al. (1997a), [2]—Garger et al. (1994), [3]—Hollander and Garger (1996), [4]—Garland and Pomeroy (1994), [5]—Hollander (1994), [6]—Vintersved et al. (1991)

Measurements of the atmospheric activity concentration of radioactivity after the Chernobyl accident were made in many countries. According to a Swedish data (Vintersved et al. 1991), there are at least three sources which contribute to the atmospheric activity. The first one is natural wind resuspension, that is, wind-blown dust of soil and plant origin from the ground in the areas surrounding the measurement sites. The second source is radioactivity transported over large distances from heavily contaminated areas like the Chernobyl zone, which could originate from resuspension too. The third source may be the local combustion of contaminated wood. The atmospheric activity concentration of ^{137}Cs declined until 1991 but has remained approximately constant after this. The resuspension factors have been derived for each site using the mean values of the atmospheric activity concentration values of ^{137}Cs from January to December 1991, divided by the total deposition of ^{137}Cs. It is interesting to note that the resuspension factor is similar in all the sites and shows little dependency on the value of Cs deposition due to the Chernobyl accident. A value of 5×10^{-10} m^{-1} was chosen as the resuspension factor value and was multiplied by the total deposition that provided a good approximation of the atmospheric ^{137}Cs activity concentration during 6–8 years after the initial deposition.

In Fig. 2.3, the relative monthly averaged values of the resuspension factor are presented according the measurement data in Chernobyl and Baryshevka (Garger et al. 1997a). Despite the difference in the distance of their settlements from ChNPP (18 and 150 km, respectively) and their different soil contamination values, the time dependence of their mean resuspension factors is similar. Thus, it is possible to make a conclusion that high-frequency oscillations of the resuspension factor and the radionuclide air activity concentration values are well smoothed by their monthly averaging, and it is evident that the resuspension process characteristics in the northern part of Ukraine are basically similar.

Garger et al. (1990) showed that for short periods after the accident (about 10 days), the resuspension factor exceeded a value of 10^{-5} m^{-1}; for the subsequent days, the value was about 10^{-6} m^{-1} in Chernobyl and after 3 months was about 10^{-8} m^{-1}. In this later period, the exceptionally high values of resuspension factor between 10^{-6} and 10^{-7} m^{-1} were obtained in places with high anthropogenic activities near ChNPP and main roads with intense traffic (Bondarenko et al. 1993). Eight years after the accident the resuspension factor value was about 10^{-10} m^{-1}.

Finally, the integral measurements of wind resuspension are used to determine the resuspension factor for comparison with the resuspension factors measured after the Chernobyl accident in different European countries. This general characteristic of the local process responsible for the activity concentration of ^{137}Cs in the atmospheric surface layer was determined for our measurements from 1986 to 1993 at the different sites in Ukraine, Belarus, and Russia. We have used the Garland and Pomeroy's (1994) paper as the database (Fig. 2.3). Additionally, we present the data of measurement that started after the Chernobyl accident in 1986: measurements at the sites Zapolie, Pripyat-Beach, and Kopachi (Ukraine) (Garger et al. 1990; Hollander and Garger 1996); Mikulichi and Kovali (Belarus); Novozybkov (Russia) (Hollander and Garger 1996); and information from Great Britain (Compton, Lerwick), Germany (Berlin, Neuherberg, Hannover), Poland (Warsaw) (Garland and Pomeroy 1994), and Sweden (Vintersved et al. 1991).

Despite the difference in their distance from ChNPP and their different contamination levels, one may notice a similarity in their time dependence of resuspension factors of these sites. The greatest differences were observed in Chernobyl till the end of anthropogenic activity in the town (Nicholson 1988).

The data from Compton and Lerwik have shown the range of variability. One may note that the resuspension factor depending on the underlying surface characteristic mainly varies in the range of two factors of magnitude. It shows a similar characteristic as wind resuspension over enough moist soil at moderate latitudes of Europe.

2.2.2 Empirical Models of Resuspension Factors and Comparison with Observations

The resuspension factor K is useful in operative situations for characterizing the relationship between surface and airborne contaminations; however, using it implies an equilibrium between the resuspended and deposited aerosols (Linsley 1978). In practice, the surface contamination is not homogenous and the airborne concentration is a sum of the local resuspended contamination and that advected from upwind sources of resuspension (Garger 1994).

Airborne concentration and surface deposition of radioactivity are interrelated functions of the co-ordinates, time, and soil properties. Therefore, K is an integral function with many variables. The following are the results of testing the widely applied simple models for K on the basis of data obtained in the vicinity of ChNPP.

One empirical predictive resuspension model describes the time dependency as

$$K(t) = K(0)\exp(-\lambda t), \tag{2.2}$$

where λ is the half-time life and $K(0)$ is a constant representing the initial value of the resuspension factor. Such a formulation appears to simulate, reasonably well, the available observations for time periods of up to several weeks after the deposition.

Another simple model is used to approximate the time dependency in the following manner (Anspaugh et al. 1975):

$$K(t) = K(0)\exp\left(-\lambda\sqrt{t}\right) + 10^{-9}, \left(\mathrm{m}^{-1}\right), \tag{2.3}$$

where t is the time in days. The second term of Eq. (2.3) was proposed on the basis of the 17 years of data from the observations of plutonium in the air and the initial value of $K(0)$ was equal to $10^{-4}\ \mathrm{m}^{-1}$, $\lambda = 0.15\ (\mathrm{days})^{-1/2}$.

USAEC (1974) and USAEC (1975) used the following K formulas:

$$K(t) = 10^{-5}\exp(-0.0139t) + 10^{-9}, \left(\mathrm{m}^{-1}\right), \tag{2.4}$$

$$K(t) = 10^{-5}\exp(-0.00185t) + 10^{-9}, \left(\mathrm{m}^{-1}\right), \tag{2.5}$$

respectively.

In addition, Linsley (1978) proposed the formula

$$K(t) = 10^{-6}\exp(-0.01t) + 10^{-9}, \left(\mathrm{m}^{-1}\right), \tag{2.6}$$

where the initial value of K was selected for the relatively damp climate of England.

Using this formula, the background value of K is achieved after 2 years. Thus, all these expressions are empirical interpolation formulas describing the transitional process from an initial high level of K to the background.

In contrast to Eqs. (2.2–2.6), another expression that does not have the low background level for K (Garland 1982; Garland et al. 1992) is described as follows:

$$K(t) = 1.2 \times 10^{-6} t^{-1}, \left(\mathrm{m}^{-1}\right). \tag{2.7}$$

The empirical equations that account for the changes in the ^{137}Cs activity concentration in the air with time have been obtained in Neuherberg (Germany) from mid-May 1986 to the end of 1988 and from June 1986 to the end of 1990. Normalizing these equations for the density of contamination of ^{137}Cs in the 0–1 cm soil layer, Hötzl et al. (1989) and Hötzl et al. (1992) give the following:

$$\begin{aligned} K(t) &= 3.4 \times 10^{-6} \exp(-0.152t) + 18.4 \times 10^{-9} \exp(-0.003t) \\ K(t) &= 2.67 \times 10^{-6} t^{-1.07}, \end{aligned} \tag{2.8}$$

respectively.

In Eqs. (2.2–2.6), the first general uncertainty is the selection of the initial value of the resuspension factor which can change from 10^{-6} to 10^{-4} m^{-1} during the first 10 days, producing a difference of two orders of magnitude among model estimates of K. The second general source of the uncertainty is connected to the rate of decrease of K with time; the time scale λ in Eqs. (2.2–2.6) may change by 2–3 orders of magnitude. The background term causes errors in the estimates of K from Eqs. (2.3–2.6) for a very long time after the initial starting point.

When comparing the model predictions with the empirical data, a question arises about the correspondence of the time- or space averaging of the calculated and measured magnitudes of K. Because the model values of K assume homogeneity of external conditions (soil, contamination, underlying surface) and stationary meteorological conditions for the measuring period, we opted to use the empirical data averaged for the whole year.

The annual averages of the resuspension factor, determined experimentally and calculated by models, are shown in Table 2.9 (Garger et al. 1995). Table 2.9 also shows the lower and upper limits of K from 1986 to 1991, and the ratio of calculated values of K to the experimental ones. The model of Hötzl et al. (1989), derived from the empirical data measured in Munich after the Chernobyl accident, agrees best with the data.

The model of Garland et al. (1992) agrees well, but it systematically underpredicts the experimental values. The calculated values of K from the Linsley model are conservative, especially in the first 2 years. If the calculated values are compared with the upper limits of the measured values, as is important in practice, the data of the Hötzl et al. (1989) and Hötzl et al. (1992) models are found to be lower than the upper limits of K by a factor of 2–3. It is important to note that the lower and upper limits were determinated for the monthly averaged resuspension factors. The real values of K for the daily data will be even larger.

It is also interesting to compare modeled values of concentration with the measured ones. For this purpose, concentration data were selected for 2 months

Table 2.9 Annual mean resuspension factor *K* for ^{137}Cs measured at the Chernobyl city in 1986 (after May 20, 1986) and during 1987–1991, compared with model calculations[a]

Year	Lower limit of *K* (10^{-7} m^{-1})	Mean value of *K* (10^{-7} m^{-1})	Upper limit of *K* (10^{-7} m^{-1})	K_{Gar} (10^{-7} m^{-1})	K_{Hoet} (10^{-7} m^{-1})	K_{Lin} (10^{-7} m^{-1})	$\frac{K_{Gar}}{K_{exp}}$	$\frac{K_{Hoet}}{K_{exp}}$	$\frac{K_{Lin}}{K_{exp}}$
1986	0.053	0.330	0.830	0.270	0.510	3.90	0.82	1.54	11.8
1987	0.013	0.082	0.170	0.030	0.042	0.27	0.36	0.51	3.30
1988	0.004	0.032	0.070	0.014	0.021	0.016	0.44	0.66	0.50
1989	0.002	0.014	0.034	0.010	0.014	0.010	0.71	1.00	0.71
1990	0.0008	0.006	0.015	0.008	0.010	0.010	1.33	1.66	1.66
1991	0.001	0.008	0.018	0.006	0.008	0.010	0.75	1.00	1.25

[a]K_{Gar} calculated according to the Garland et al. (1992) model; K_{Hoet} calculated according to the Hötzl et al. (1989) model; K_{Lin} calculated according to the Linsley (1978) model; K_{exp} derived from experimental measurements (Garger et al. 1995)

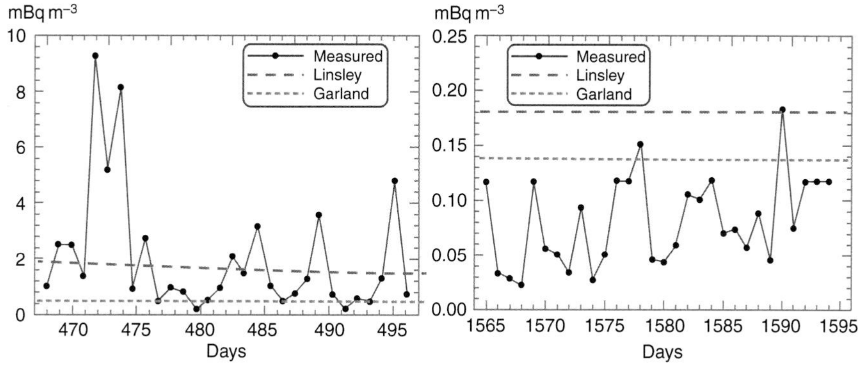

Fig. 2.4 Measured and calculated ^{137}Cs air activity concentration values in Chernobyl: air activity concentration (**a**) September 1987, (**b**) September 1990

(Fig. 2.4) (Garger et al. 1995). The concentration values calculated by the two models, Linsley (1978) and Garland et al. (1992), are presented for the daily concentration values of ^{137}Cs obtained in September 1987 and 1990. In 1987, these model estimations yielded insufficient values for many days, especially for the Hötzl et al. (1989) model. In 1990, the situation became opposite. A certain amount of overestimation is found in both the annual and daily concentration values. The inverse power law model of Garland (Eq. (2.7)) can be matched to the long-term trend of *K* dependence of time.

Finally, we compare the K models with the experimental data obtained at different points in the 30-km zone. Air concentration and surface deposition of eight radionuclides were measured at 10 points during the same period between September 14 and 17, 1986 (Garger 1994). The calculated resuspension factors range from 6×10^{-9} to 3×10^{-6} m^{-1}. The experimental data and the calculated results differ by up to more than one order of magnitude. The high range of the resuspension factors is due to different underlying surfaces, soils, and the effects of anthropogenic actions in the 30-km zone. Anthropogenic activities were especially important for places which were situated near the NPP or main roads. For specific locations, however, the *K* values may be higher than those estimated by the Garland and Hötzl models by 2–3 orders of magnitude.

The models for *K* were compared with experimental monthly data and with each other as the normalized resuspension factor $K(t)/K_0$ with respect to time (Fig. 2.5). Presentation of the model predictions as $K(t)$ normalized for K_0 enables a comparison of the dependence of *K* on time without the effect of differences in the initial value of *K*. The value K_0 is dependent on a substantial amount of subjective estimation.

The Chernobyl and Baryshevka data are shown, along with five equations and a new equation derived from the experimental data, $K(t)/K_0 = t^{-1.4}$. This equation was selected after study of the value of the power function for the entire time series of measurements at Chernobyl (Garger et al. 1997b). It provides the best fit to the

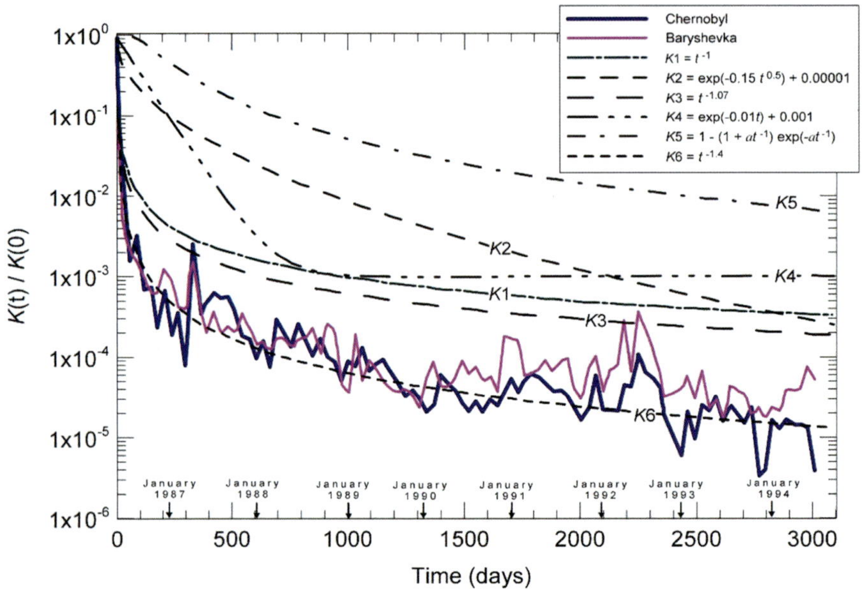

Fig. 2.5 Comparison of observed and predicted resuspension factors over time, normalized for the initial resuspension factor K_0. $K1$, model of Garland; $K2$, model of Anspaugh; $K3$, model of Hötzl; $K4$, model of Linsley; $K5$, model of Makhonko; $K6$, empirical dependence (Garger et al. 1997b). For all equations, t is the time in days

experimental data and shows that the process of stabilization is going more rapidly, with stabilization of the experiment occurring near 1200 days. The processes of migration in soil are probably appreciable by this time.

A similar characteristic of the inverse curves with powers from 1.0 to 1.4 is easily seen (Fig. 2.5). The results of Linsley (1978) and Anspaugh et al. (1975) models exceed the experimental data in the first 1000 days by 2–3 orders of magnitude. The curve by Makhonko (1992), developed for long-term resuspension after the deposition of radionuclides from the stratosphere for stationary atmospheric conditions in the midland of Russia, describes the resuspension factor controlled by the process of soil migration (Makhonko 1984): it differs sharply in value from the other curves in Fig. 2.5. In spite of the inverse dependence with a predicted power near 1.1 (Hall and Reed 1989), the decrease in the resuspension ("the decrease in the power of the spring and capacity of well") is more rapid than predicted. This is explained by the downward migration processes of radiocesium initially deposited on the soil surface.

2.2.3 Models of Resuspension Rate and Measurements

For the calculation of transport of radionuclides from a surface, it is necessary to know whether the emission ability is equal to a vertical turbulent flux of radionuclides at the ground surface J (Bq m^{-2} s^{-1}) normalized to a ground surface contamination density, in other words, the resuspension rate.

According to Healy (1974) and Healy (1980), the resuspension rate is defined as a share of contamination on ground, which rises in air in a unit of time caused by wind or a mechanical perturbation. According to this definition, the resuspension rate Λ (s^{-1}) is expressed as $\Lambda = J/D$, where D is the mean density of surface contamination by radionuclides (Bq m^{-2}).

The total resuspension rate as function depending on various mechanisms of particles lifted up in air may be represented as follows (Chepil 1951):

$$\Lambda = f(\Lambda_{air}, \Lambda_{salt}, \Lambda_{sc}, \Lambda_{w}, \Lambda_{m}), \tag{2.9}$$

where Λ_{air} is the resuspension rate of particles (<100 μm) moving actually as air suspension; Λ_{salt} is the resuspension rate for saltation particles (50–500 μm); Λ_{sc} is the resuspension rate of particles due to large particle (>1000 μm) motion on the surface by creep. The author considered that these three functions should be taken into account though one of them can dominate; Λ_{w}, the wind resuspension rate, should also be considered in the absence of the three mentioned types of transport observed during strong winds and soil erosion; Λ_{m} is the resuspension rate caused by the mechanical reasons of lift up.

At middle latitudes with a wet climate and low wind speed, a soil erosion is rare following which the dust particles blow-off from the surface layer of ground and other elements of surface polluted with radioactive materials or others toxics, and their lift up becomes the important mechanism of further transport over different distances. In this case, an estimation of Λ_{m} is necessary, because despite a little resuspension intensity under these conditions, the process of secondary lift up takes place within extensively polluted territories and for a long time, continuing to many years. On a local scale, various anthropogenous activities strengthen resuspension of substances and their redistribution in space sharply. In this connection, in practice, experimental estimations of Λ_{m} are necessary in calculations of contamination redistribution in space and time after an accident caused by a mechanical resuspension, atmospheric transport, and deposition of radioactive dust, and, at the end, estimation of possible dose loadings on the population.

Generally, for an area source with a nonuniform surface and a surface contamination density, the radionuclide resuspension intensity can depend on the physical characteristics of the surface and radionuclides, and also on the spatial coordinates and time. The knowledge of the resuspension rate as functions of coordinates and time enables us to describe a field of concentration in any point around the polluted territory with the help of the equation of eddy diffusion for an area source. It was illustrated in Healy (1974) for the territory polluted with plutonium. This approach was used widely for the estimation of fields of the ^{137}Cs activity concentration in the air in territorird extensively contaminated after the Chernobyl accident in the Kyiv region (Ukraine), and Gomel and Mogilev regions (Belarus) using the experimental estimation of the Λ value in 1986 (Garger et al. 1992; Garger 2008).

The uncertainties in the estimation of the resuspension rate are the same as that for the resuspension factor estimation. It is clear that the vertical turbulent flow of substances due to resuspension should vary in time caused by the daily and seasonal variabilities of meteorological parameters, and, by a depletion of ground surface contamination, wash-off, erosion, and penetration of contamination into ground. The estimation of the resuspension rate also depends on the air sampling method employed.

The estimation of Λ can be carried out by several methods:

1. Direct measurements of vertical turbulent flow of radionuclides above a flat source.
2. Gradient measurements of the mean wind speed, mean air temperature, and the mean activity concentration of radionuclides in the surface layer of the atmosphere that enable estimating the vertical turbulent flow of radionuclides from homogeneous and stationary sources.
3. Measurements of the radionuclide activity concentration in air at a fixed height outside of a finite source or directly above a source.

In Garger et al. (1990) and Garger et al. (1994), the resuspension rate by wind was determined with the help of gradient measuring of the mean radionuclide concentration $\overline{q}$, wind velocity, and air temperature, provided the conditions appropriate to the case of a flat homogeneous constant source are valid. The vertical flux of the deposited material J_{A} in the surface layer of the atmosphere is considered to be constant

$$J_{\mathrm{A}} = -k_z \frac{\mathrm{d}\overline{q}}{\mathrm{d}z} = \mathrm{const}, \qquad (2.10)$$

where k_z is the vertical coefficient of turbulent diffusion and there is no gravitational sedimentation of aerosol particles. According to the Monin and Obukhov theory (Zilinitkevich 1970), the similarity is observed for the vertical gradient of the radionuclide concentration in the surface layer of the atmosphere:

$$\frac{\mathrm{d}\overline{q}}{\mathrm{d}z} = \frac{J_{\mathrm{A}}}{\kappa u_* L} q_{\mathrm{a}}\left(\frac{z}{L}\right), \qquad (2.11)$$

where q_a (z/L) is the universal function of similarity in the Monin and Obukhov theory; L is the Monin–Obukhov length, and κ is the Karman constant. After the integration of Eq. (2.11) along z, we obtain

$$\overline{q}(z_2) - \overline{q}(z_1) = -\frac{J_{\mathrm{A}}}{\kappa u_*}\left[f_{\mathrm{a}}\left(\frac{z_2}{L}\right) - f_{\mathrm{a}}\left(\frac{z_1}{L}\right)\right], \qquad (2.12)$$

where f_{a} (z/L) is an antiderivative of q_{a} (z/L). For the case when $L \to 0$, that is, for conditions of neutral stratification, at small values of z/L we have approximately (Byzova et al. 1991)

$$\overline{q}(z_2) - \overline{q}(z_1) \cong -\frac{J_A}{\kappa u_*}\left(\ln\frac{z_2}{z_1} + \beta_a\frac{z_2 - z_1}{L}\right), \tag{2.13}$$

where β_α is an empirical constant. For simplicity, it is considered that the nature of the turbulent transport of substances does not differ from the nature of turbulent momentum transfer. Dividing Eq. (2.13) by a density of contamination (Garger et al. 1990), (Garger 1994) we have final expression for the radionuclides resuspension rate by wind at small values of *z/L*:

$$\Lambda = \frac{[\overline{q}(z_1) - \overline{q}(z_2)]\kappa u_*}{D\left(\ln\frac{z}{z_1} + \beta_a\frac{z_2 - z_1}{L}\right)}, \tag{2.14}$$

where $\beta_a = 1.45$ for $-0.16 < z/L \leq 0$ and $\beta_a = 9.9$ for $z/L > 0$ according to Byzova et al. (1991), D is the soil contamination density.

If Λ is determined for the period more 1 day, when thermal conditions at the average almost always could be considered as equilibrium, Eq. (2.14) becomes simpler

$$\Lambda = \frac{\kappa u_*[\overline{q}(z_1) - \overline{q}(z_2)]}{D\ln\frac{z}{z_1}}. \tag{2.15}$$

The gradient measurements of the activity concentration gives the information about a sign of the flux in the surface layer and allow us to determine the occurrence of radionuclide advection from more intense, distant surface contamination sources.

The vertical air activity concentration profiles of ^{137}Cs, ^{103}Ru, and ^{144}Ce are shown in Fig. 2.6, along with a measured wind speed profile, for an individual experiment between September 13 and 14, 1986 (Garger 2008). The decrease in the atmospheric concentration with height is consistent with upward diffusion and indicates a distant source of resuspension. So, data have to be carefully chosen so that the effects of upwind advection are minimized.

Table 2.10 illustrates some calculated values of the resuspension rate for various surface types. These results are appropriate for conditions when advection of radionuclides from upwind sources was considered negligible (i.e., by choosing results when the upwind territory was relatively uncontaminated). The experimental sites include an agricultural field, a forest, and an area of sand without vegetation which had been prepared for building work just prior to the Chernobyl accident. All the resuspension rate values relate to similar conditions (i.e., a dry surface with a moderate wind speed of about 2 m s ^{1} at 1 m above ground level at the site). It can be seen from Table 2.10 that similar resuspension rates were obtained for the forest and sandy sites, approximately 2×10^{-9} s^{-1}, while values several times lower were found for the agricultural site.

The systematic difference in the values of wind resuspension intensity between a forest and a field for all radionuclides could be seen well. These data have allowed us

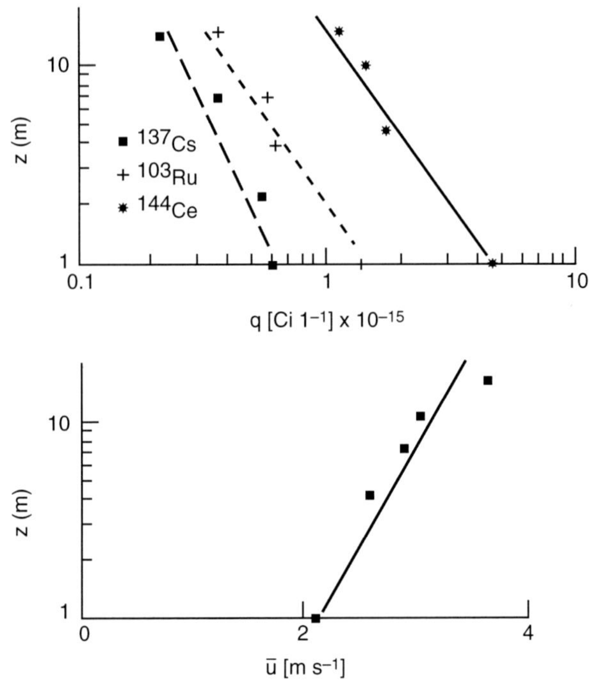

Fig. 2.6 Vertical air activity concentration profiles of ^{137}Cs, ^{103}Ru, and ^{144}Ce, and wind speed profiles measured between September 13 and 14, 1986 at Zapolie ($1\ Ci\ l^{-1} = 37\ TBq\ m^{-3}$)

Table 2.10 Resuspension rate for different surface types (Garger 1994)

Surface type	Resuspension rate $\times 10^{-9}$ (s^{-1})		
	^{144}Ce	^{137}Cs	$^{95}Zr + ^{95}Nb$
Field	0.3 ± 0.1	1.0 ± 0.7	0.4 ± 0.2
Forest	2.1 ± 0.9	2.1 ± 0.8	3.7 ± 0.9
Sand beach without vegetation	2.2	3.7	2.4

to estimate the dust-forming power of the underlying surface and the radioactivity transport from the 30-km zone to the surrounding regions in the first 2 years after the accident. It is observed that the transport of a radioactive dust within 1 year did not exceed several tenths of percent of the density of contamination.

Belyaev and Surnin (1991) measured the dust concentration at heights from 0.6 up to 34 m during June 1986 in the Chernobyl zone with the help of samplers with three-layer filters AFA-HA. The upper layer of the filter was imbued with NaOH solution and the third one was covered by a layer of activated carbon. The samplers were fastened to a rope anchored to a mast of high-voltage line at a height of 36 m. The vertical profiles of gamma-radiating radionuclide activity concentration values were obtained. The estimations of the resuspension rate were carried out only at the east wind during observations with the help of the formula by Garger et al. (1990). The estimations have shown that in June 1986, the resuspension rate was about $2 \times 10^{-8}\ s^{-1}$ for ^{144}Ce, ^{95}Zr, and ^{95}Nb, what was related by the authors to

Table 2.11 Resuspension rate ($year^{-1}$) for various underlying surfaces

Surface type	Place and time of measuring	Resuspension rate ($year^{-1}$)		
		^{144}Ce	^{137}Cs	$^{95}Zr + {}^{95}Nb$
Urban area (Belyaev and Surnin 1991)	Chernobyl, June 1986	6.3×10^{-1}	1.3×10^{0}	6.3×10^{-1}
Grass field (Garger 1994)	Zapolie (14 km), September 1986	9.4×10^{-3}	3.2×10^{-2}	1×10^{-2}
Forest (Garger 1994)	Forest (14–15 km), September 1986	6.6×10^{-2}	6.6×10^{-2}	1.2×10^{-1}
Sand beach (Garger 1994)	Pripiat town (2 km), Ukraine, September 1986	6.3×10^{-2}	1.2×10^{-1}	8×10^{-2}
Field, sand beach (Hollander and Garger 1996)	Pripiat town, 1993	–	2.5×10^{-3} 2×10^{-4}	–
Cultivated field (Naydenov and Lukoyanov 1994)	Novozybkov district, Briansk region, Russia, 1992	–	4.7×10^{-2}[a] 2.8×10^{-3}[b]	–
		^{90}Sr	^{137}Cs	$^{239+240}Pu$
Half rural area (Rosner and Winkler 2001)	Bavaria, Germany 1987 1997	2.5×10^{-4} 1.0×10^{-4}	4.3×10^{-3} 2.0×10^{-4}	1.8×10^{-4} 0.4×10^{-4}
Cultivated field (Shinn et al. 1982)	South Carolina, USA, old deposit	–	–	1.4×10^{-4}
Desert (Shinn et al. 1982)	Nevada, USA, old deposit	–	–	0.8×10^{-4}–1.5×10^{-2}
Forest-steppe (Makhonko and Robotnova 1982)	Southern Ural, Russia, polluted soils from 1957 up to 1960	1.4×10^{-4} – 0.4×10^{-1}	–	–

[a]Resuspension rate at ^{137}Cs density in 1 cm of soil layer
[b]At the inventory of ^{137}Cs in 20 cm arable soil layer

large-dispersed dust; 4×10^{-8} s^{-1} for ^{103}Ru and ^{137}Cs, what was related to low-dispersed dust; and about 4×10^{-6} $s^{-1}$for ^{131}I.

Naydenov and Lukoyanov (1994) experimentally studied the parameters of wind-driven resuspension radioactive dust above a ^{137}Cs contaminated cultivated field in the Novozybkov district of the Bryansk region (Russia). The measurement was carried out in July–August 1992 with the help of a gradient device by the method described by Naydenov and Lukoyanov (1994). As a result of measuring of the ^{137}Cs activity concentration vertical profile for 3 days under the absence of a noticeable advection flow of radioactivity, the values of resuspension rate were calculated in the range of 1.2×10^{-9}–1.5×10^{-9} s^{-1} for a surface density of contamination taken within 1 cm soil layer and in the range of 7×10^{-11}–9×10^{-11} s^{-1} for an inventory of ^{137}Cs within 20 cm of arable soil layer.

Table 2.11 summarizes the resuspension rate estimations for different underlying surfaces in the polluted territory as a result of the Chernobyl accident within the 30-km exclusion zone in Ukraine (in brackets the distance of the measurement sites from the fourth unit of the ChNPP is indicated) and in the Bryansk region (Russia).

Table 2.11 also contains data obtained in Bavaria (Rosner and Winkler 2001) using Slinn's assumption (Slinn 1978), which states that, on an average, in a year, radionuclide losses from an underlying surface are equal to the radionuclide deposition. Also, data on the study of radioactive plutonium resuspension at different places in USA, which differ sharply in the characteristics of the underlying surface, particularly the soil, are presented.

From Table 2.11, it follows that the weakening of resuspension process with time occurs in the surface layer of the atmosphere by 1–2 orders of magnitude during 4 months, based on the characteristics of the underlying surface. In 1986, the resuspension rate values differed noticeably depending on the distance from the source and the type of underlying surface. It is clear that during eight incomplete years after the accident, the values of resuspension rate in Zapolie on a grass field have decreased by one order of magnitude, and in Pripyat-Beach, consisting of coarse sand hydraulic fill before the accident, have decreased by three orders of magnitude, apparently, due to more intense radionuclide vertical migration into the soil. The data which were obtained in the Novozybkov district of the Bryansk region well illustrate the dependence of the resuspension rate values on the form of normalization of a vertical turbulent flux of radioactive aerosols. According to these data, the normalization into a total inventory of radionuclide in soil underestimated the resuspension rate value by an order of magnitude. Measurement data of old radionuclide deposits with age from 8 up to 20 years result in resuspension rates of the order of 10^{-4}–10^{-3} year^{-1}, regardless of the radionuclide type. In other words, only one thousandth or one ten-thousandth share of deposition in the surface layer of soil resuspends in the air within 1 year.

2.3 Statistical Characteristics of the ^{137}Cs Activity Concentration in Air

2.3.1 Long-Term Dynamics of the Atmospheric ^{137}Cs Activity Concentration

Daily filter samples of aerosols were taken with high-volume samplers in the towns of Pripyat and Chernobyl, which are located 18 km apart from each other within the 30-km zone of Chernobyl (Garger et al. 1994). There were used samplers, designed by SPA "Typhoon", Russia, with Petryanov-type filters. The exposed filters were pressed into discs and analyzed by γ-spectrometry. On 5–7% of the days of the year, it was not possible to obtain data because of various technical problems, but the series of measurements as a whole can still be used for the estimation of basic characteristics of the fluctuations in the activity concentration.

Statistical characteristics of the ^{144}Ce and ^{137}Cs activity concentration values in air are presented in Table 2.12 (for Pripyat) and Table 2.13 (for Chernobyl). In addition to the first to fourth statistical moments of the ^{137}Cs and ^{144}Ce activity

Table 2.12 Statistical characteristics of distribution functions of the daily average activity concentration values of ^{144}Ce and ^{137}Cs in Pripyat

	Nuclide	1987	1988	1989	1990	1991
Number of days with data	^{144}Ce	165	294	223	124	–
	^{137}Cs	184	366	365	365	365
Average activity	^{144}Ce	16,283	5841	1334	385	–
concentration (μBq m^{-3})	^{137}Cs	4145	2989	935	516	463
Median (μBq m^{-3})	^{144}Ce	4214	1324	495	188	–
	^{137}Cs	1616	1075	498	341	320
Standard deviation (μBq m^{-3})	^{144}Ce	34,829	16,265	2727	781	–
	^{137}Cs	8154	11,524	1623	762	741
Skewness	^{144}Ce	4.04	6.4	5.21	6.52	–
	^{137}Cs	4.75	13.3	6.38	4.91	6.84
Kurtosis	^{144}Ce	18.8	46.8	34.5	53.1	–
	^{137}Cs	27.7	209.1	54.9	30.4	60.7
Intensity of fluctuation	^{144}Ce	2.14	2.97	2.04	2.03	–
	^{137}Cs	1.97	3.86	1.74	1.48	1.60
Ratio of average activity	^{144}Ce	3.93	4.14	2.67	2.05	–
concentration to median	^{137}Cs	2.56	2.78	1.88	1.51	1.45

Table 2.13 Statistical characteristics of distribution functions of the daily average activity concentration values of ^{144}Ce and ^{137}Cs in Chernobyl

	Nuclide	1987	1988	1989	1990	1991
Number of days with data	^{144}Ce	163	306	240	104	–
	^{137}Cs	184	366	365	365	365
Average activity	^{144}Ce	5216	1233	307	85.6	–
concentration (μBq m^{-3})	^{137}Cs	4145	2989	248	117	137
Median (μBq m^{-3})	^{144}Ce	2738	629	191	60.6	–
	^{137}Cs	888	370	148	93	96
Standard deviation (μBq m^{-3})	^{144}Ce	7346	1989	464	96.0	–
	^{137}Cs	1638	620	357	150	184
Skewness	^{144}Ce	3.20	7.39	6.74	3.64	–
	^{137}Cs	2.27	2.53	7.00	6.77	8.14
Kurtosis	^{144}Ce	13.3	80.2	61.5	17.6	–
	^{137}Cs	6.09	8.59	69.1	60.6	91.5
Intensity of fluctuation	^{144}Ce	1.41	1.61	1.51	1.12	–
	^{137}Cs	1.10	1.08	1.44	1.28	1.34
Ratio of average activity	^{144}Ce	1.90	1.96	1.61	1.41	–
concentration to median	^{137}Cs	1.67	1.54	1.68	1.26	1.43

distributions, the intensity of the fluctuation values (determined as the ratio of the standard deviation to the average concentration) are given. During all years, the intensity of fluctuation values exceeded unity and was the largest in Pripyat during 1987–1988 when the technogenic activity, including decontamination and reconstruction, was the most intense. The strong asymmetry of the distribution of ^{137}Cs and ^{144}Ce activity concentration values is reflected by the large ratio of the average concentration to the median and the values of skewness. The monthly mean values for both the radionuclides, ^{137}Cs and ^{144}Ce, declined by 1–2 orders of magnitude in these 4 years. At that time, the majority of decontamination work in the 30-km zone was completed and the resuspension by technogenic (i.e., mechanical) activity became less important. In 1987 and 1988, the years of high efforts in the decontamination of large areas, there is some evidence of the contribution by technogenic resuspension to the atmospheric activity.

The decrease in the atmospheric activity concentration was observed to be higher than expected, assuming radioactive decay only. There must be further processes which are responsible for the reduction of the activity concentration in the air, mechanisms for instance that reduce the source term of resuspension, such as vertical migration in soil, runoff with rain or melting water, and snow cover.

The frequency distributions of the ^{137}Cs atmospheric activity concentration values in Pripyat and Chernobyl showed very wide distributions of the ^{137}Cs activity concentration in the second half of 1987 and 1989, but very narrow distributions in 1989. The description of these empirical distributions by such theoretical functions as the lognormal, exponential, or gamma-distribution was not successful for the ensemble of measurement series in the sense of the χ^2 criterion. The empirical distributions reflect more complex situations than described by the model distributions. Apparently, the empirical function of the activity concentration distribution results from the influence of several different processes: local resuspension by wind, transport of radioactive aerosols from distant, highly contaminated areas, and resuspension by technogenic (mechanical) activities. The superposition of these processes gives rise to a complex distribution function which changes with time as the different processes are also changing with time.

For specific periods, however, certain distributions may describe the empirical data well. For example, while no lognormal distribution could be fitted to the empirical distribution in 1987, it became possible between 1988 and 1990. The activity concentration values exceeding the average annual concentration (positive fluctuations) are discussed because of their significance on the dose rate and their potential health effects. For Pripyat and Chernobyl, Garger et al. (1994) gave the cumulative frequency distribution for the ratio of these positive fluctuations to the standard deviation $((q-\overline{q})/\sigma)$ in a probability plot. A lognormal distribution follows a straight line in this kind of a plot. The smaller fluctuations are rather well described by the lognormal distribution (in Pripyat better than in Chernobyl), whereas the very high fluctuations (amplitudes larger than 2–3 times the standard deviation) lie further from a straight line. From year to year, the cumulative distribution is shifted toward lower values. For example, in both Pripyat and Chernobyl, the frequency of the concentration with relative amplitudes less than or equal to 1 was 60–70% in 1987 but 92% in 1991. Conversely, the frequency of high concentration decreases with time.

Finally, we consider the influence of the averaging period on the estimation of the normalized standard deviation of the activity concentration $\sigma(T)/\overline{q}$, where $\overline{q}$ is the annual mean concentration; T is the averaging period; and $\sigma(T) = \left[\overline{(q(T) - \overline{q})^2}\right]^{½}$. The normalized standard deviation was calculated for $T \in \{2, 3, 4, 6, 10, 15, 20, 30, 60 \text{ days}\}$ for the years 1987–1991 with the measured air concentration values in Pripyat and Chernobyl. The dependence of $\sigma/\overline{q}$ on T can be expressed by a power function as

$$\frac{\sigma(T)}{\overline{q}} \approx T^{-m}, \tag{2.16}$$

with the empirically determined exponent m. The mean value is $m = 0.33 \pm 0.08$. The above relation can be used to compare the different averaging periods, T_1 and T_2:

$$\sigma(T_2) = \sigma(T_1)\left[\frac{T_2}{T_1}\right]^{-m}. \tag{2.17}$$

We now can estimate the effect of increasing the internal averaging time by sampling for 3 or 10 days instead of 1 day: the standard deviation will decrease to 69% and 47%, respectively, of the daily value. Therefore, the averaging period is of importance in the assessment of activity measurements and in the design of a field measurement program. Besides, the presented formula helps to restore values of the relative standard deviations. According to Garger et al. (1994), the value of power m equals to 0.30 ± 0.05 from the data for 1986–1994.

2.3.2 *Statistical Prediction of the ^{137}Cs Activity Concentration in the Surface Layer of the Atmosphere of the ChEZ*

The knowledge of the activity concentration of the dose-relevant radionuclides is required at first for the estimation of the inhalation dose and related analysis, for example, the horizontal redistribution of the radionuclides. Since ^{137}Cs is a major radionuclide released from the Chernobyl accident with a relatively long half-life, information on the soil contamination density and long-time series data of its atmospheric concentration have been accumulated for inside and around the exclusion zone. In Garger et al. (1997b), the prospects and limitations are shown of the approach to predict the activity concentration in the air by using the soil contamination density and the resuspension factor. In particular, Garger et al. (1994) have found that the models based on the resuspension factor do not take into account the fluctuations of the airborne activity concentration, and that the predicted values may, therefore, differ from the measured values by several orders of magnitude.

An alternative statistical approach presented here uses the measurement of the ^{137}Cs activity concentration in the air to construct a long-term predictive model. The

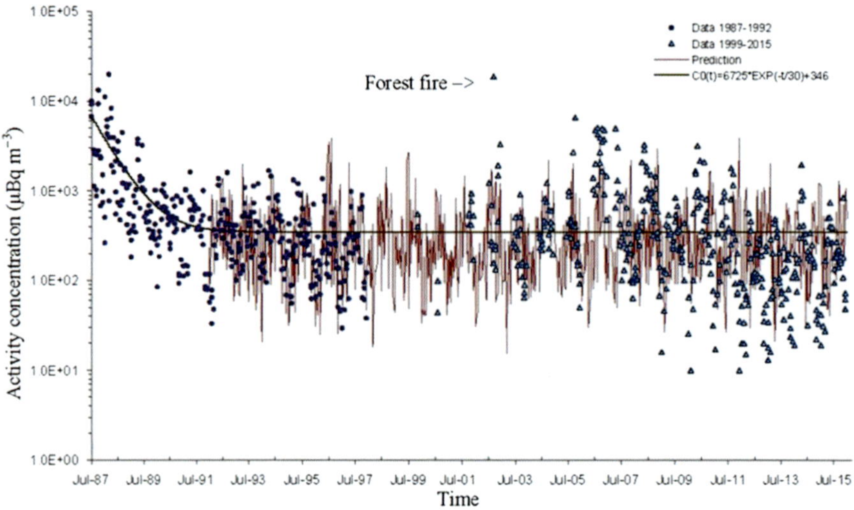

Fig. 2.7 Measured and predicted ^{137}Cs activity concentration values for Pripyat until 2015 (10-day average). A large value of the measured concentration (indicated with arrow) was due to forest fires in August–September 2012

model is developed on the data measured in the town of Pripyat in the period 1987–1991, which were validated using the measurement data in the same town in the later period 1991–1997. The Pripyat data demonstrates the significant temporal variation of the airborne ^{137}Cs activity concentration in the aftermath of the Chernobyl accident. Therefore, the time series is considered as the result of dynamic processes that include different components such as trend and low-frequency fluctuations and other stochastic fluctuations (Garger et al. 2012).

A first attempt to calculate a long-time prediction from the Pripyat air concentration data was made by Viswanathan et al. (2000). A technique specially adjusted for a high portion of missing data (about 7%) was applied for the calculations based on the least-square power spectrum analysis of Lomb (1976). Garger et al. (2012) suggested an improved approach of successive selecting the time trends of the ^{137}Cs activity concentration measurement data. It consists of the following three steps:

1. Identification and reduction of a descending trend.
2. Detection of a periodic component.
3. Calculation of the stochastic components for the prediction.

Further analysis is based on the ^{137}Cs air activity concentration time series collected in Pripyat from 1987 to 1991. In Fig. 2.7, the experimental data at Pripyat are shown for the measurement period from 1987 till 2015. The descending trend of the activity concentration in the air can be matched only by a model using a constant term. Finally, the predicted values of the airborne ^{137}Cs activity concentration $c^*(t)$ as shown in Fig. 2.7 were computed according to

$$c^{*}(t) = 10^{z(t)} \cdot [6725 \exp(-t/300) + 346], (\mu\text{Bq}\,\text{m}^{-3}), \tag{2.18}$$

where the latter factor is a descending exponent fit with the constant term, while $z(t) = u^{*}(t) + x(t)$ is a random cyclic component, which has the same low-frequency periods of fluctuation, variance, and mean value as $u(t) = \log[c(t)/c_0(t)]$, and t is the time in days. Here $c(t)$ is the measured air ^{137}Cs activity concentration, $c_0(t)$ is the estimated value of 10-day averaged concentration, and $x(t)$ is the calculated stochastic component with mean and standard deviation of a standard normal distribution (Garger et al. 2012).

The model predictions match the observed atmospheric activity concentration of ^{137}Cs pretty well. Two major deviations to the predicted values can be explained by unexpected phenomena: intense forest fires that occurred in 2002 and a significant release of radionuclides from the Chernobyl Shelter that occurred in 2005.

According to the sampling and measurements performed from 1987 to 1991, the prediction appeared satisfactory with respect to both the average values of the atmospheric activity concentration and the values of the amplitudes of fluctuations. It could be shown that the length of the sampling period was sufficient enough for setting up a reliable prediction model. In general, the magnitude and characteristics of fluctuations of the predicted activity concentration in the air are similar to those that were measured in the Pripyat area from 1991 to 1997 (validation period).

2.4 Activity Size Distribution of the Atmospheric Aerosol Particles

After the Chernobyl accident, numerous measurements were made to characterize the primary radioactive atmospheric aerosols in terms of nuclide composition and particle size at several locations in Europe (Jost et al. 1986; Tschiersch and Georgi 1987; Reineking et al. 1987; Horn et al. 1987). Within the 30-km ChEZ, a few measurements were made for the characterization of the primary aerosols (Garger 1987; Gaziev et al. 1993; Gaziev and Kabanov 1993). One was launched by the Hydrometeorological Committee of the USSR and the other (ECP1 1996) by the European Commission. This part includes an analysis of the size-resolved airborne activity concentration values of wind-driven resuspension which have been measured since 1986.

In Western and Central Europe, observations of radioactive aerosols began immediately after the accident (Bunzl et al. 1995; Besnus et al. 1997; Livens and Baxter 1988). In July 1986, a permanent field site near the Zapolie village (14 km to the south of ChNPP) was set up, in which various types of samplers could be deployed. The size distributions of radionuclides were measured at this site with the multicascade impactors IK and PK since 1991 (see Table 2.2). An additional multicascade impactor UP was set up in the Pripyat town in the autumn of 1987. The exposed filters were analyzed by γ-spectrometry. The detection limit was assumed to be twice the background value, which is 0.45 Bq/sample for ^{137}Cs and 0.03 Bq/sample for ^{144}Ce (compressed aerosol filter sample). More details of the measurement sites, instruments, and experiments are given in Garger et al. (1997a). Measurements of the physicochemical characteristics of the resuspension of radioactive aerosols, their dispersion in space and time, secondary contamination of the atmospheric surface layer due to various types of sources, including object "Shelter", were carried out.

Aerosol measurements were made in the surface layer of the atmosphere within the 30-km zone. They were conducted periodically from September 1986 till June 1993 at two sites, Zapolie and Pripyat. The main task was to estimate the time evolution of the particle size distribution of the radioactive aerosols.

2.4.1 Typical Form of the Activity Distribution of Particle Size in the Air During Wind-Driven Resuspension

Radionuclides deposited in particulate form on the ground do not remain unchanged with time in concentration and physicochemical form in the top soil layer, and are, therefore, accessible for resuspension. The first measurements at the experimental sites were made during August and September 1986 (Garger 1994). Surface contamination was measured using samples 15 cm in diameter and depth of 5 cm. Detailed data are presented for 12 locations in the southern part of the 30-km zone with distances to the NPP from 4 to 25 km. At the main experimental site Zapolie, the following mean values were measured: 0.5 MBq m^{-2} of ^{137}Cs contamination and 6.9 MBq m^{-2} of ^{144}Ce contamination. The activity ratio $^{144}Ce/^{137}Cs = 13.8$ is in good agreement with the data of Begichev et al. (1990) and Bogatov et al. (1990), confirming that this site is representative for the 30-km zone.

According to the measured residence half-times in the top soil layer for ^{137}Cs (Bunzl et al. 1995), two cases can be distinguished. In the far-field site in Bavaria, Germany, an increase in the residence time from about 0.85 years (2 years after deposition) to about 1.65 years (7 years after deposition) was observed, whereas in the near-field site at a distance of 6 km from ChNPP, a decrease in the residence time was observed from about 6.8 years (2 years after deposition) to about 3.2 years (7 years after deposition). The different behavior was attributed to the different matrix in which ^{137}Cs was fixed. In the near-field site, the major part of ^{137}Cs was deposited in the form of rather insoluble fuel particles, which obviously need several years for disintegration; in the far-field site, ^{137}Cs was transported mainly as a condensation aerosol in a more soluble form. Seven years after the deposition, about 17% of the deposited activity was still in the top soil layer (0–1 cm) available for resuspension in both the cases.

During the resuspension experiments conducted in Zapolie in 1992/1993, soil samples were taken and the activity concentration profiles were determined (Besnus et al. 1997). In the top soil layer, a relatively high variability of the activity concentration was found. About 20% of the deposited activity was still available in the top soil layer (0–1 cm), and the remaining 80% of the total activity was found in the 0–5 cm soil layer. There was no significant difference in the activity depth profiles observed for the several types of nuclides.

At Zapolie, the size distribution of soil particles and the specific activity distributions were measured (Besnus et al. 1997). The Zapolie soil texture consists mainly of silts and fine sand with 33% of particles ranging from 50 to 200 μm and 42% ranging from 2 to 50 μm. The fine volume fraction of soil (<2 μm) is about 8%. For particle sizes larger than 20 μm, a linear relationship was found between specific ^{137}Cs activity and size. The high increase in the specific activity measured for the smallest particle sizes seems not to depend just on their external surface. Clay minerals with structures have larger surfaces than their approximation by spheres; fuel particles are also in them.

For the retrospective inhalation dose assessment for workers who participated in the Chernobyl accident (liquidators), in general, the assumption was made that the size distribution of the radioactive aerosols has the form of a lognormal distribution. In order to check this assumption, the first two statistical moments of the measured activity size distributions were calculated.

Garger et al. (1998b) calculated the total activity concentration values of ^{137}Cs and ^{144}Ce and the main parameters of the activity size distributions for the measurement period 1986–1993: the mean weighed diameter d, the median diameter d_m, the geometric diameter d_g, and the standard deviation σ_g.

In Garger et al. (1998b), the total airborne activity concentration values, the ratios between the mean and median diameters, and the ratios between geometric mean and median diameters are documented. In the case of a lognormal distribution, these ratios must be unity. When the experimental distribution had two modes with a well-defined gap in between, it was possible to estimate two median diameters. Analysis shows that the measured distributions are generally very wide and mostly differ from a lognormal distribution. The statistical parameters of the activity size distributions are more uniform at Pripyat compared to the values at Zapolie.

A detailed analysis of the radioactivity size distribution has shown five typical shapes of distributions. Type (i) represents a lognormal distribution with a median diameter between 2 and 4 μm. Types (ii), (iii), and (v) have two modes and are together the most frequent distributions with 91% of the 88 cases at Zapolie. The medians of these bimodal distributions are in the following diameter ranges: about 0.5 μm and between 10 and 20 μm (type ii), between 1–2 μm and 10–20 μm (type iii), about 4 μm and between 20–30 μm (type v). Type (ii) is a very distinct bimodal distribution with a very large part of inhalable particles. Type (iv) has a broad maximum in the large particle size range (median at about 10 μm). The main parameters of the size distribution of type (iv) were reported to be representative of fuel particles inside the 10-km zone of ChNPP by other authors as well (Garger 1987; Begichev et al. 1990; Bogatov et al. 1990). However, the size range of particles, and the mean and median diameters depend on the transport direction of the primary radioactive clouds. In general, the interval of the median diameter for type (iv) ranges from 5 to 55 μm, with the size of the particles not exceeding 100 μm.

Table 2.14 Frequency of size distribution types in (%) at Zapolie and Pripyat for different ^{137}Cs activity concentration q (mBq m^{-3}) ranges

Distribution type	(i)	(ii)	(iii)	(iv)	(v)
Zapolie, all	3	19	16	6	56
Zapolie, $q \leq 0.12$ mBq m^{-3}	0	29	43	0	29
Zapolie, 0.12 mBq m^{-3} $< q <$ 1 mBq m^{-3}	6	0	6	6	82
Zapolie, $q \geq 1$ mBq m^{-3}	0	50	12.5	12.5	25
Pripyat, all	23	13	3	55	0

The relative frequencies of the distribution types measured at Zapolie and Pripyat are given in Table 2.14. Out of a total of 88 examined cases, the lognormal size distribution of type (i) was observed only three times at Zapolie (3%). Type (v) was observed in 56% of all cases, type (ii) in 19%, type (iii) in 16% cases, and type (iv) in 6% of all cases. In other words, at Zapolie, mainly a bimodal radioactive aerosol distribution was observed with a considerable part of respirable and inhalable particles and an important fraction of large particles. Taking this into account, the unimodal distribution which we expected from the soil activity size distribution will not give a correct estimation of the inhalation dose.

At Zapolie, the different types of distributions were split up into three groups according to the total ^{137}Cs activity concentration values: a background case, when the total concentration was equal or less than the mean annual value for Chernobyl in 1991 (0.12 mBq m^{-3}), a case with high airborne concentration (equal or more than 1.0 mBq m^{-3}), and an intermediate case with airborne concentration values between these values. For the intermediate case, it was found that 82% of the distributions were of type (v). For the background conditions, 29% of distributions were of type (v). In periods with high airborne concentration values, 50% of distributions were of type (ii), with a considerable quantity of fine particles.

In contrast to Zapolie, the distribution type (v) was not observed at all at Pripyat. The distributions of type (iv) were measured in 55%, type (i) in 23%, type (ii) in 13%, and type (iii) in 3% of all the 31 cases. In two other cases, the activity could be detected in only one or two cascades. A unimodal type of activity size distribution is predominant at Pripyat, which is situated close to the very highly contaminated area. For ^{137}Cs, the maximum of the mean aerodynamic diameter ranges from 8 to 12 μm, and that of the median aerodynamic diameter from 5 to 10 μm. For ^{144}Ce, the maxima are between 8–12.5 μm and 6–10 μm, respectively.

A special case of the radionuclide size distribution was found during two periods of forest fire (23.07–30.07.1992, 05.08–05.09.1992) (Fig. 2.8). The main part of radioactivity was connected to submicron particles with median diameters of 0.28 and 0.5 μm, respectively. These data are consistent with the measurements of the radionuclide size distribution during forest fires (Hollander and Garger 1996).

Thus, the analysis of 88 samples has shown that at Zapolie a bimodal distribution of ^{137}Cs was observed in 91% of all cases, which was formed by two processes: the local resuspension and the advective transport of radioactive aerosols from highly contaminated territories, for example, from the ChNPP.

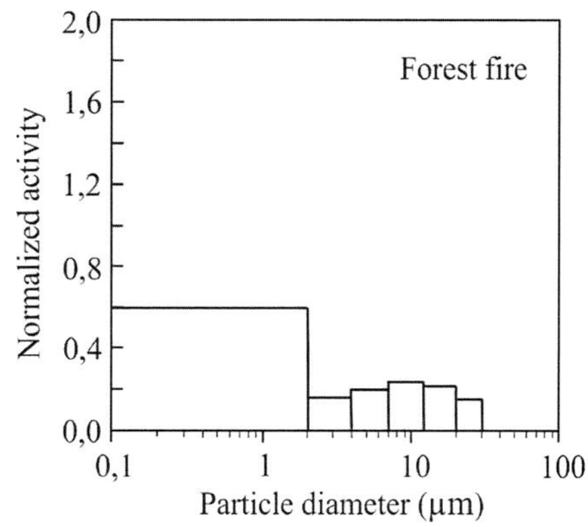

Fig. 2.8 ^{137}Cs activity size distribution measured at Zapolie during a forest fire at about 17 km distance from the burning area

However, at Pripyat, which is situated within a highly contaminated area, the shapes of size distributions were representative of local resuspension with only a weak transformation.

The observed variability of the radionuclide size distributions in the air of the 30-km zone makes it difficult to calculate resuspension parameters taking into consideration only local processes (e.g., the estimation of the resuspension factor or the estimation of the airborne activity concentration using the radioactive loading of soil particles). During the measurement period the lognormal radionuclide size distribution was observed in only 3% of all cases in Zapolie. Because of the heterogeneously contaminated territory of the 30-km zone, the use of the lognormal size distribution for airborne radionuclides does not seem to be correct for the calculations of the inhalation dose. An estimation of the error of dose calculation assuming a lognormal distribution is strongly suggested. At Zapolie, the mean air activity concentration values of ^{137}Cs discriminated in four size ranges showed an increasing part of inhalable particles with time since the accident. In 1993 the inhalable fraction was about 48% of the total concentration.

2.4.2 Anthropogenic Conditions

Anthropogenic activities, such as agricultural soil management (e.g., harrowing and ploughing) or vehicle traffic (especially on unpaved roads), can significantly enhance the atmospheric soil dust concentration. Radionuclides which are already deposited will get resuspended together with the soil dust. The enhanced resuspension of radionuclides will affect the inhalation dose and the spread of contamination, at least on a local scale. In particular, people who are involved in the activities or live downwind close to roads and agricultural land warrant concern.

To investigate the influence of the different types of anthropogenic activities on the resuspension process and on the secondary air contamination, measurements were performed within the 30-km ChEZ. In this chapter, the results are presented of measurements performed in 1993 during enhanced resuspension due to various anthropogenic activities.

Here, the emphasis is on describing the resuspended radioactive particles in terms of the number concentration and their size distribution for the estimation of human inhalation dose. The experimental information on the particle size distribution can be used directly, avoiding any general assumptions. Moreover, the shares of the fine and coarse particles can be estimated for modeling the processes of transport and redistribution of contamination. For different anthropogenic activities, this may result in different effective scales for potential effects. The redistribution estimation requires knowledge of the resuspension and deposition rates. Therefore, the dry deposition velocity was determined experimentally as well in measurements described below.

The resuspension measurements were carried out at the Zapolie site which is located on a large open grass field. Resuspension enhanced by anthropogenic activities was measured during different simulated agricultural activities and operations of different trucks. For that purpose, two different soil surfaces free of vegetation were prepared on which several tractor types were driven simulating soil management such as harrowing. The soil surface strips represented fixed line sources. In a typical experiment, the tractor started at one end of the strip, passed the sampler at a certain fixed distance, drove to the other end of the prepared surface, and returned. For a certain experiment, the strip was chosen for which the wind trajectories pass the sampling equipment after crossing the prepared surface. Detailed information about the orientation of the strip sources and the positions of the different samplers are given in ECP1 (1996) and Garger et al. (1998a).

The size distribution of radioactive particles was measured by a cascade impactor PK. The mass concentration for the different particle size ranges was determined with a Berner impactor (Berner and Lürzer 1980). The number concentration of airborne particles in the size range 0.6–30 μm was measured by a size-resolved aerodynamic particle sizer (APS) (Blackford et al. 1987). The samplers operated close to each other at a sampling height of 3.0–3.5 m and at a distance of approximately 45 m from the northeastern dust strip and approximately 120 m from the southwestern strip. The vertical profiles of the atmospheric nuclide activity concentration values were determined by the installation Grad, with 4 identical filter samplers operating at heights 1.0, 1.8, 2.5, and 3.5 m (Garger et al. 1997a). In addition to air samples, the deposited material was collected. The samples were collected on planchettes, which were gauze-covered flat plates. The planchettes were placed near each other at heights 0.5, 1.0, 1.5, and 2.0 m.

The results of the measurements show that the total airborne radionuclide activity concentration values and deposition rates increased considerably during the anthropogenic-enhanced resuspension. Depending on the experimental conditions, the increase was by a factor of several thousand in comparison to the concentration values occurring during wind resuspension at distance of 20–30 m from the dust source and a factor 10–100 at distances of 100 m or more (ECP1 1996).

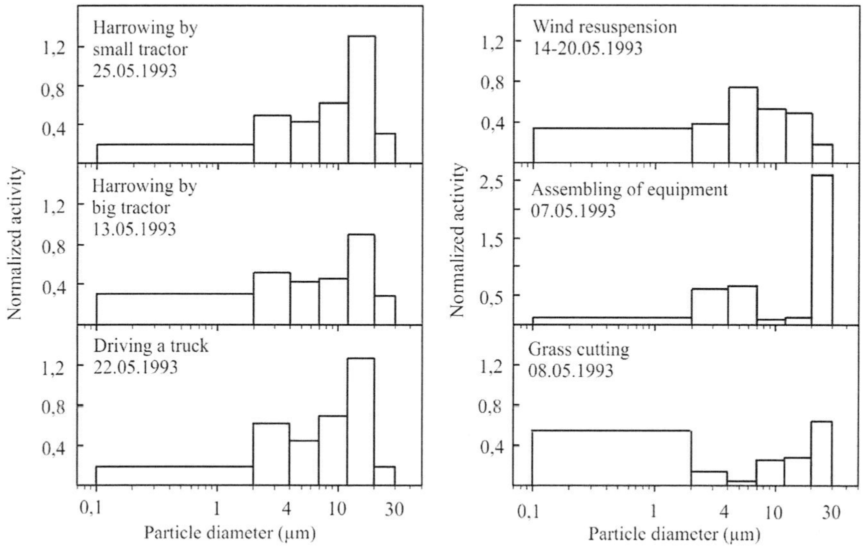

Fig. 2.9 ^{137}Cs activity size distributions in the normalized coordinates $\frac{\Delta c}{c \cdot \Delta \log d}$ and log *d*, where *c* is the air activity concentration of ^{137}Cs and *d* is the aerodynamic particle diameter during resuspension experiments in May 1993 at Zapolie. The first three distributions are typical for simulated soil management and traffic on unpaved roads. The other distributions are results of individual experiments

The measurements of the number of concentration values of hot particles showed an increase by three orders of magnitude, reaching 0.7–1.0 hot particles/m^3, with a maximum activity of 1.5–2.0 Bq/particle.

For the ten experiments conducted, the mean and median diameters were 7.1 ± 1.1 and 4.4 ± 1.5 μm, respectively. During assembling of equipments with dust resuspension caused by people walking close to the samplers, a mean diameter of 13.8 μm and a median diameter 11.0 μm were observed. In the experiment with only wind resuspension from the bare soil of the prepared dust strip, the mean diameter was 5.6 μm and the median diameter 3.0 μm.

Certain typical shapes of the measured size distributions can be distinguished. In Fig. 2.9, the ^{137}Cs size distributions are given in the normalized coordinates $\frac{\Delta c}{c \cdot \Delta \log d}$ and log *d*, where *c* is the air activity concentration of ^{137}Cs and *d* is the aerodynamic particle diameter. In the experiments with the tractors simulating harrowing and the trucks simulating traffic on unpaved road, very similar distributions were observed. The two maxima, the first in the range 2–4 μm (which is only weakly developed) and the second in the range 12–20 μm (which is more pronounced) are characteristic. A considerable part of the activity was measured in the fine particle range (0.1–2.0 μm): 33% ± 6% of the total activity. These size distributions are very similar to a certain type of distribution measured at Zapolie under wind-driven conditions (Garger et al. 1998b). We may explain these distributions under wind-driven conditions by the large-scale anthropogenic decontamination works close to the ChNPP at a distance of approximately 12–14 km from the sampling site.

Very different shapes of ^{137}Cs activity size distributions were measured during the other experiments. Wind resuspension across the bare soil of the prepared dust strip resulted in a unimodal distribution with the maximum in the size fraction 4–7 μm. During the assembling of equipment, a bimodal distribution was measured with a very high maximum in the range 20–30 μm and a secondary maximum in the range 4–7 μm. Grass cutting around the site produced a very high proportion of fine particles (0.1–2 μm), but a second maximum as well in the 20–30 μm range. A comparison of these distributions again with the wind-driven cases (Garger et al. 1998b) shows a striking similarity between the "assembling of equipment" distribution and the distribution type measured most frequently at Zapolie during wind resuspension. Local small-scale anthropogenic activities seem to significantly influence the particle size distribution. The enhanced radioactive loading of airborne particles is quantified by the enhancement factor, which is the ratio of the mass specific air concentration and the mass specific soil concentration.

2.5 Nuclear Fuel Particles from the Chernobyl Reactor in the Air

2.5.1 Fuel Particles Collected During Natural Wind Resuspension in 1987

After the Chernobyl accident, daily filter samples of aerosols were taken using high-volume "Typhoon" samplers (air flow rate 4500–5000 $m^3\ h^{-1}$) in the Pripyat town, located ca. 4 km to the west of the reactor building. All filters were divided into two parts: the first part was measured in 1987 and the second part was kept for future investigations, sealed individually in a rigid box made of polystyrene (PS), and stored in a dry place at room temperature in order to preserve the sampled nuclear material.

Two filters that collected during October 12 and 13, 1987 and November 22 and 23, 1987 were selected for investigations. The mean airborne radionuclide activity concentration during the first sampling period was 20.8 mBq m^{-3} for ^{137}Cs, 89.3 mBq m^{-3} for ^{144}Ce, and 30.5 mBq m^{-3} for ^{106}Ru. During the second period the concentration was 12.1 mBq m^{-3} for ^{137}Cs, 34.9 mBq m^{-3} for ^{144}Ce, and 24.3 mBq m^{-3} for ^{106}Ru.

At the same time, the particle size distribution of radioactivity was measured by the impactor UP (five cascades, flow rate ca. 400 $m^3\ h^{-1}$). More detailed information on the technical characteristics of the installations used can be found in Garger et al. (1997a). These measurements showed the differences in the distribution of activities for these two periods. For the first period, the distribution was approximately symmetric. For the second, the distribution was skewed toward larger particles.

For the first period, the median diameter was 3.0 μm and log $\sigma_g = 0.88$; for the second, the median diameter was 7.5 μm and log $\sigma_g = 0.44$. Obviously, the distribution characteristics of the particles for these two observation periods were not identical.

The surface of the filters was preliminary scanned by α-β-radiometer for revealing parts with higher radioactivity. Consequently, in total seven filter fragments with diameters of about 6.5 cm (the size of the filter holder diameter is about 8 cm) were cut out and labeled. The quantity of hot particles on each filter fragment was determined by autoradiography.

Measurements on the solubility of the fuel particles deposited on the filters were made using static in vitro test systems. A review and critical analysis of *in vitro* methods designed to estimate dissolution rates in the respiratory tract and a discussion on the factors affecting the in vitro dissolution rates were published in Ansoborlo et al. (1999). Based on these considerations, a static in vitro test system as described in Eidson and Griffith Jr (1984), Miglio et al. (1977), and Gamble (1967) was applied in this study. The filter holders were made from fluorinated polymer (PTFE). Fragments of the Petryanov filters were confined within two membrane filters with a pore size of 0.14 μm and of the MFA type (Dubna, Russia) as a filter sandwich. For the solvent, Gamble's solution was chosen (Gamble 1967). Several lung fluid simulates had been proposed which were designed to approximate the composition of extracellular fluid and are generally called serum ultrafiltrate simulate (SUF), serum lung fluid (SLF), and Gamble's or Ringer's solution. They differed slightly in their composition and were widely used by different investigators (Eidson and Mewhinney 1983; Dennis et al. 1982; Ansoborlo et al. 1998). The findings of a comparison of different solvent media (Ansoborlo et al. 1998) justify the choice of Gamble's solution as the lung fluid simulate. The filter holders were kept in a horizontal position for 7–28 days in Gamble's solution and the solution was analyzed at defined time intervals for the radionuclides ^{90}Sr, ^{137}Cs, ^{238}Pu, $^{239+240}$Pu, ^{241}Am, and ^{244}Cm, the most relevant nuclides for dosimetry.

For the measurement of ^{137}Cs on filters and in the solutions, a γ-spectrometer based on the semiconductor detector GMX-30190 (effectiveness 32.5%, energy resolution 1.89 keV on the line 1.33 MeV) and the multichannel buffer 919 Spectrum Master (Ortec Inc.) was used.

Nuclides ^{238}Pu, $^{239+240}$Pu, ^{241}Am, and ^{244}Cm were separated and concentrated without carriers from aerosol filters and leaching solutions according to the method described by Ageev et al. (2003).

The total activities of the significant radionuclides in the seven individual aerosol filter fragments are presented in Table 2.15. It can be seen that all principal long-lived transuranium radionuclides are present in the aerosol particles. From the composition, it can be concluded that the particles originated from the irradiated fuel, and that the radionuclides are fixed in the matrix of uranium dioxide. Usually, for the description of radionuclide characteristics of different radioactive products of the Chernobyl accident, a fractionating factor of the radionuclides with respect to the refractory ^{144}Ce is used (Ermilov and Ziborov 1993; Dubasov et al. 1996; Bogatov et al. 1990; Kutkov and Kamarinskaya 1996; Wagenpfeil and Tschiersch 2001).

Table 2.15 Radionuclide activities in the aerosol filter fragments before the solution experiments (in 2000)

Number of fragment	Sampling date	^{137}Cs Bq/sample	^{90}Sr Bq/sample	$^{239+240}Pu$ Bq/sample	^{238}Pu Bq/sample	^{241}Am Bq/sample	^{244}Cu Bq/sample
1	12–13.10.1987	3.6 ± 0.4	2.7 ± 0.2	0.035 ± 0.005	0.016 ± 0.005	0.042 ± 0.008	[a]
2	12–13.10.1987	2.9 ± 0.5	3.2 ± 0.3	0.065 ± 0.007	0.025 ± 0.004	0.073 ± 0.013	[a]
3	12–13.10.1987	6.1 ± 0.5	9.6 ± 0.6	0.18 ± 0.01	0.084 ± 0.009	0.21 ± 0.02	0.008 ± 0.002
4	12–13.10.1987	8.0 ± 0.6	9.7 ± 0.5	0.20 ± 0.02	0.075 ± 0.008	0.23 ± 0.02	[a]
5	12–13.10.1987	28 ± 1	31 ± 1	0.61 ± 0.02	0.38 ± 0.05	0.81 ± 0.05	0.045 ± 0.006
6	22–23.11.1987	6.1 ± 0.4	6.1 ± 0.2	0.11 ± 0.01	0.07 ± 0.01	0.14 ± 0.02	[a]
7	22–23.11.1987	17 ± 1	21 ± 1	0.43 ± 0.02	0.29 ± 0.04	0.56 ± 0.03	0.034 ± 0.004

[a]Below detection limit

In Garger et al. (2004c), nuclides ^{239}Pu + ^{240}Pu were taken as the reference radionuclides, as these plutonium isotopes were products of the ^{238}U(n,γ) ^{239}U$^{-\beta} \rightarrow$ ^{39}Np$^{-\beta} \rightarrow$ ^{239}Pu(n,γ) ^{240}Pu reaction chain and were more solidly connected with the uranium matrix of the hot particles. In Table 2.15, the activity ratio related to that reference is shown. The values are in good agreement with the calculated values for the reactor fuel just before the Chernobyl accident (Ageev et al. 1998), which confirms that the investigated aerosol particles represented irradiated fuel at different grades of dispersion.

The mean density ρ of hot particles from samples of the zone close to the ChNPP was found to be equal to 10 ± 0.8 g cm^{-3} (Kutkov and Kamarinskaya 1996). The mean specific activity of $^{239+240}$Pu in the irradiated fuel was 1.3 × 10^{7} Bq g^{-1} (U_3O_8) or 1.15 × 10^{7} Bq g^{-1} (UO_2) (Ageev et al. 1998). The assumption used was that particles have a spherical form and the estimated size of the particle equivalent diameter d_0 is given by the formula

$$d_0 = 2\left(\frac{3A}{4\pi\rho_p q}\right)^{1/3}, \tag{2.19}$$

where A is the activity (Bq) of the particle and q is the specific activity (Bq g^{-1}) of the particle.

On the seven individual filter fragments, some hot particles and an unknown quantity of small submicron particles of low specific activity were deposited. In the calculations made in this chapter, the activity values of $^{239+240}$Pu from Table 2.15 were used. In the radiographic films, only one single large spot could be found on fragment number 7. An equivalent diameter of hot particles was calculated according to Eq. (2.19) for each filter fragment. This made it possible to estimate the dissolution rate constant from the next equation (Mercer 1967) $m = m_0\left(1 - \frac{\alpha_s kt}{3\alpha_v \rho_p d_0}\right)^3$, where m_0 is the initial mass of the particle and d_0 is its initial diameter; t is time; s is the particle surface area; $\alpha_s = s/d^2$ is the surface shape factor; $\alpha_v = m/\rho_p d^3$ is the volume shape factor, ρ_p is the particle density; and k is the dissolution rate constant.

In Table 2.16, the activity of the radionuclides in the simulated lung fluid solution (SLF) is shown, together with the derived equivalent diameter of the particles. The much lower solubility of activity in fragment 7 (particularly for ^{137}Cs and ^{90}Sr) compared with the other fragments reflects that the activity in fragment 7 is concentrated in only one hot particle. The large scattering in the activities found in the solution, especially for $^{239+240}$Pu, may be related to the sampling period of the filters of only 1 day in 1987. There were many different secondary sources of radioactive particles in the zone near to the "Shelter" object causing large airborne concentration fluctuations. The leaching of the radionuclides from the hot particles into the SLF decreases in the order ^{137}Cs > ^{90}Sr ≫ ^{241}Am > $^{239+240}$Pu.

Table 2.17 presents the shape-modified dissolution rate constant $k_s = k\alpha_s/\alpha_v$ as derived from the measured activities of the radionuclides in the solution (Table 2.16). The dissolution rate was assumed to be constant in time, and the

Table 2.16 Activity of radionuclides in the solution (% of the total activity)

Number of fragment	Exposition in solution (days)	Equivalent diameter of particle d_0 (μm)	^{137}Cs	^{90}Sr	$^{239+240}Pu$	^{241}Am
1	28	8.4	27	9.2	1.8	2.2
2	28	10.3	23	7.0	0.52	1.2
3	28	14.4	14	2.2	0.25	1.0
4	28	14.9	10	3.2	0.09	0.07
5	25	21.6	4.2	0.8	0.18	0.06
6	21	12.2	5.9	4.4	0.88	0.40
7	21	19.3	0.1	0.41	0.10	0.16
Mean		14.4 ± 4.7	12 ± 9.9	3.9 ± 3.2	0.54 ± 0.62	0.73 ± 0.79

Table 2.17 Dissolution characteristics of hot particles

Number of fragment	$^{239+240}Pu$ activity in the solution (%)	Dissolution rate constant $k_s \times 10^{-6}$ (g cm^{-2} day^{-1})	Time for complete dissolution of the particle (years)
1	1.8	5.4	13
2	0.52	1.9	45
3	0.25	1.3	91
4	0.09	0.72	170
5	0.18	1.6	110
6	0.88	5.1	20
7	0.10	0.92	170
Mean		2.4 ± 2.0	

time t_∞ when the particle will be completely dissolved was estimated according to (Mercer 1967)

$$t_\infty = 3\alpha_\nu \rho_p d_0 / \alpha_s k. \tag{2.20}$$

From these data, it can be seen that the solution process for $^{239+240}Pu$ will need many years for the dissolution of fuel particles in the size of the order of 10 μm. It was found in Mercer (1967), however, that the particles that dissolve completely as calculated from Eq. (2.20) of Garger et al. (2004c) is an overestimation.

The solubility rate constant can be expected to increase rapidly as the particle size diminishes below about 0.01 μm. From Table 2.17, it follows that the dissolution rate constants of fragments 1 and 6 are about 3–7 times higher than the others, and that the complete dissolution time is much less as well. The mean equivalent diameter for these particles is 10.3 μm and $k_s = 5.2 \times 10^{-6}$ g cm^{-2} day^{-1}; for the other fragments, the mean equivalent diameter is higher (16.1 μm) and the dissolution rate is proportionately lower, $k_s = (1.29 \pm 0.48) \times 10^{-6}$ g cm^{-2} day^{-1}.

Experimental estimations of k are discussed in Mercer (1967) and it is concluded that the *in vivo* solubility rate constants for compounds including U_3O_8 and UO_2

dusts are in the order of approximately 10^{-7} g cm^{-2} day^{-1}, which is in agreement with the data of Table 2.17. In Bogatov et al. (1990), the dependence of leaching agents and the destruction of matrix-bond radionuclides were discussed and were found to range from 5.7×10^{-6} to 1.6×10^{-4} g cm^{-2} day^{-1} for ^{137}Cs, ^{134}Cs, and ^{90}Sr.

The leaching constants (k_1, day^{-1}) for ^{137}Cs, ^{90}Sr, $^{239+240}$Pu, and ^{241}Am were calculated according to Mercer (1967), $dA(t)/dt = -k_1 A(t)$, where $A(t)$ is the activity at time t and k_1 is the leaching rate constant (Table 2.18).

From these data, it can be seen that the leaching constants for ^{137}Cs and ^{90}Sr are 4–100 times higher than that for $^{239+240}$Pu, including that for fragment 7, where the leaching constant for cesium is surprisingly small, and the range of the values is very high. In Gudihy et al. (1989), a leaching constant of about 0.004 day^{-1} for all the analyzed γ-emitting radionuclides was found. In Kutkov and Kamarinskaya (1996), the leaching constant for the total of all α-emitting nuclides was determined to be 0.005 ± 0.002 day^{-1} and for ^{137}Cs to be 0.002 ± 0.001 day^{-1}. For ^{137}Cs and ^{90}Sr, these data are in good agreement with the estimations here, but the leaching constants for $^{239+240}$Pu and ^{241}Am were 1–2 orders magnitude lower than those for the radionuclides presented in Kutkov and Kamarinskaya (1996).

2.5.2 *Inhalation Dose Coefficients Applying Measured Solubility Parameters*

In Garger et al. (2004c), the inhalation dose coefficients for the radionuclides ^{137}Cs, ^{90}Sr, ^{239}Pu, and ^{241}Am were calculated by Roth, applying the integrated modules for bioassay analysis (IMBA) computer code developed by NRPB (Birchall et al. 2003). The program implements the dosimetric model for the human respiratory tract (ICRP 1994), the systemic biokinetic models as described in ICRP (1990, 1993), and uses the ICRP (1983) nuclear decay data to calculate internal doses. The dissolution and absorption of the material deposited in the respiratory tract into the blood is described in the ICRP model by two components: a rapid component (f_r) with a dissolution rate (s_r) of 100 day^{-1} and a slow component ($1 - f_r$) with dissolution rates (s_s) of 5×10^{-3} day^{-1} (Type M) and 1×10^{-4} day^{-1} (Type S). The fraction that dissolved rapidly (f_r) is assumed to be 1 for Type F, 0.1 for Type M, and 0.001 for Type S material.

For the particles investigated here, individual inhalation dose coefficients were calculated, applying the particle characteristics given in Tables 2.16 and 2.18. In Table 2.18, the values of the inhalation dose coefficients averaged over all 7 fragments are shown in bold.

A comparison of the experimentally determined dose coefficients with ICRP values (see Table 2.18) indicates that nuclear fuel particles closely resemble the material characteristics of type M for ^{137}Cs and ^{90}Sr and those of type S for ^{239}Pu and ^{241}Am. The discrepancies between the dose coefficients as derived from the experimental parameters and the default ICRP values predominantly arise from the differences in solubility, whereas different particle sizes play a minor role. The

Table 2.18 Slow solubility rate s_s (assumed to be represented by the measured leaching constants k_1) and inhalation dose coefficients for the radionuclides investigated

	^{137}Cs		^{90}Sr		$^{239+240}Pu$		^{241}Am	
Number of the fragment	$s_s \times 10^{-4}$ (day^{-1})	Dose coefficient (Sv Bq^{-1})	$s_s \times 10^{-4}$ (day^{-1})	Dose coefficient (Sv Bq^{-1})	$s_s \times 10^{-4}$ (day^{-1})	Dose coefficient (Sv Bq^{-1})	$s_s \times 10^{-4}$ (day^{-1})	Dose coefficient (Sv Bq^{-1})
1	112	3.9×10^{-9}	34	1.7×10^{-8}	6.5	7.5×10^{-6}	7.9	7.2×10^{-6}
2	93	3.3×10^{-9}	26	1.6×10^{-8}	1.9	6.0×10^{-6}	4.3	4.9×10^{-6}
3	54	3.0×10^{-9}	7.9	1.5×10^{-8}	0.89	4.0×10^{-6}	3.6	2.9×10^{-6}
4	38	3.2×10^{-9}	12	1.3×10^{-8}	0.32	3.8×10^{-6}	0.25	4.1×10^{-6}
5	17	2.7×10^{-9}	2.9	1.2×10^{-8}	0.64	2.7×10^{-6}	0.24	2.9×10^{-6}
6	22	4.6×10^{-9}	16	1.5×10^{-8}	3.2	3.7×10^{-6}	1.4	5.1×10^{-6}
7	0.36	6.1×10^{-9}	1.5	1.4×10^{-8}	0.36	3.0×10^{-6}	0.57	3.2×10^{-6}
Median	38	$\mathbf{3.3 \times 10^{-9}}$	12	$\mathbf{1.3 \times 10^{-8}}$	0.89	$\mathbf{4.0 \times 10^{-6}}$	1.4	$\mathbf{4.3 \times 10^{-6}}$
ICRP Default 5 μm AMAD	Type F	6.7×10^{-9}	Type M	2.4×10^{-8}	Type M	3.3×10^{-5}	Type M	2.7×10^{-5}
ICRP 14.4 μm AMAD	Type M	3.1×10^{-9}	Type M	1.0×10^{-8}	Type S	4.2×10^{-6}	Type S	4.3×10^{-6}

experimental investigations clearly demonstrate the effect of decreasing solubility with increasing particle size as predicted from the theoretical considerations.

2.5.3 Radioactive Aerosols Released from the "Shelter"

At any time after the nuclear accident in Chernobyl, former Unit 4 of the ChNPP was a source of radionuclide emissions into the environment. Soon after the fire in the reactor was stopped in 1986, a containment structure ("Shelter") was constructed to prevent any further release of radioactive material. The "Shelter" was erected very quickly and had many small and some large openings (up to a few tens of square meters) through which radionuclides could be swept away by wind and rain. Investigations of the processes occurring in the "Shelter" (e.g., resuspension and deposition of the radioactive indoor dust) and of the characteristics of the released radioactive aerosols (Bogatov et al. 1990; Bogatov 2000; Borovoy and Gagarinskiy 2001) began immediately after the construction of the "Shelter"was completed. Two types of "Shelter" aerosol particles could be distinguished: aggregates of fine particles from the irradiated fuel (which are usually low in levels of volatile radionuclides) and atmospheric aerosol particulates with attached radionuclides (Bogatov 2000). The contaminated aerosols under the roof space of the "Shelter" are formed by roughly dispersed fuel particles, with an activity median aerodynamic diameter (AMAD) close to 5 μm (Bogatov 2000). Particles redistributed from the surfaces inside the "Shelter" are in the range of 6–9 μm AMAD. The total mass of the fuel dust is close to 5000 kg, from which 100 kg is potentially available for resuspension into the environment. Therefore, the radioactive aerosols still represent one of the potential radiological risks originating from the ChNPP. In particular, the aerosols were one of the major factors of radioactive exposure to the personnel working in the industrial site and in the local zone of the "Shelter".

In regard to the building of a new "Shelter-2", the aim of the study by Garger et al. (2006) was to specify the actual radioactive releases into the surface layer of the atmosphere and to assess the inhalation dose for workers in that place. For these purposes, the atmospheric activity concentration of the relevant radionuclides and their activity size distribution were measured at the gaps of the "Shelter" (Fig. 2.10) and in the near environment of the "Shelter" (Fig. 2.11). To estimate inhalation doses, the results were used to determine the individual input parameters for the applied standard dose assessment model.

The places of allocation of sampling sites for measurements of the airborne radionuclide activity concentration at the "Shelter" site from 1997 until 2003 were chosen. Such monitoring efforts required continuous representative measurements of the concentration field of the radioactive aerosol. The following aerodynamic considerations were made to identify the appropriate location for the samplers.

Obstacles cause reverse (backward) flows near the surface. Measurements of stream lines of different building models performed in aerodynamic tunnels (Slade 1968) and under natural conditions (Garger et al. 1988) have shown that this

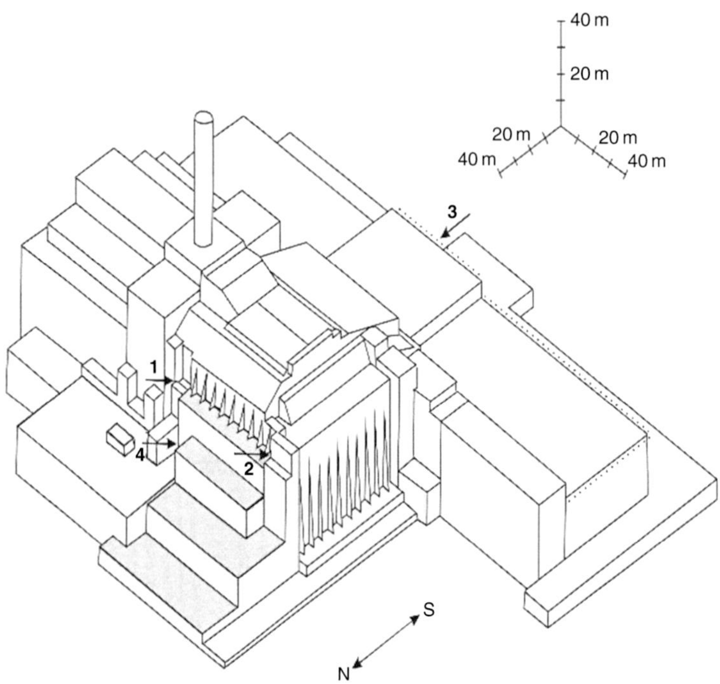

Fig. 2.10 Containment ("Shelter") of the former Unit 4 of the ChNPP with the main gaps 1–4 where aerosol measurements were made

slipstream area depends on the height and width of buildings. The horizontal dimension of that zone is larger than two heights of the obstacle (Garger et al. 1988).

At a distance of two building heights, the height of the slipstream zone is less than that of the obstacle height. Thus, with the given maximum "Shelter" height of 68 m, for different directions of the flow, shadow lengths of 190–270 m were estimated (Bogatov 2000). From the sketch of the "Shelter" site (Fig. 2.11) it can be seen that the lee part of the industrial site is always in the aerodynamic shadow of the "Shelter" building. The zone around the "Shelter" is characterized by a high turbulence and reverse flow, which means that the air activity concentration values do not vary much within this zone. Three samplers were therefore placed 50–100 m to the north, northwest, and south of the "Shelter", respectively. The fourth sampler was positioned at the southwest side near a personnel road (Fig. 2.11). The filter samples were taken with high-volume samplers Typhoon, Grad, and SES for the determination of the activity concentration. For size-segregated sampling, impactors IBP and Andersen PM10 were used. More details of the experimental setup can be found in Garger and Kashpur (2000). Several large gaps with areas between 10 and 90 m^2 (about 200 m^2 in total) were monitored between May 1996 and December 2000. Within the gaps, the samplers were installed as centrally as possible, depending on the availability of fixing possibilities at the building (Garger et al. 2002). The total release through a certain gap was estimated using the following equation (Garger

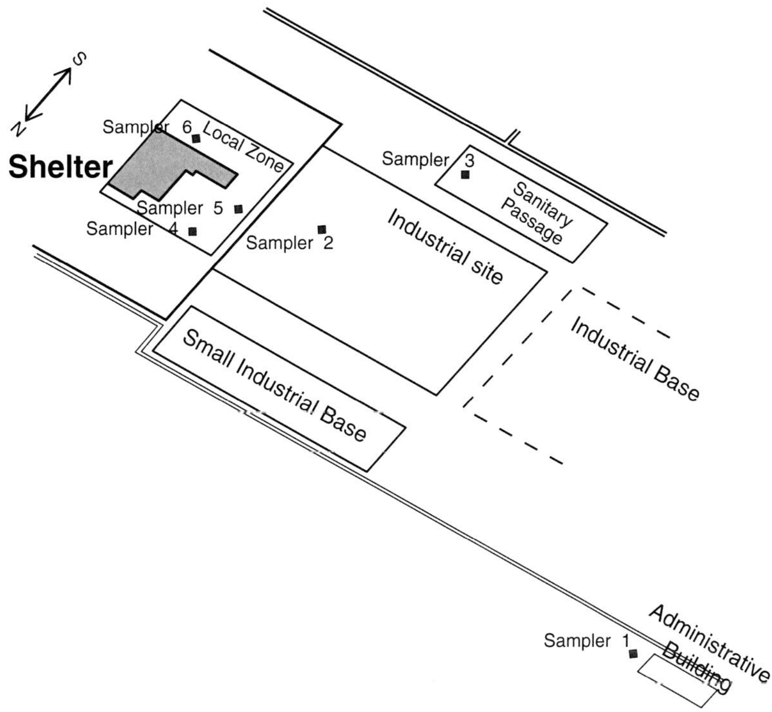

Fig. 2.11 Sketch of the immediate environment of the "Shelter" ("Shelter" site) including the location of the aerosol samplers. The following samplers were deployed: Typhoon (sampler 1), Grad (samplers 2, 4, 5, 6), and SES (sampler 3)

et al. 2004a), $Q_j = q_j \sum_{i=1}^{n} A_i v_i$, where Q_j (Bq s^{-1}) is the total mean activity flow rate through a gap j, which was divided into n rectangles of area A_i (m^2); v_i (m s^{-1}) is the wind velocity at the center of the rectangles, and q_j (Bq m^{-3}) is the mean activity concentration in the air. While the mean activity concentration q_j was measured typically for a period of 2–4 days, the flow parameters were determined by short-time measurements, on a daily basis. More detailed information on the experimental equipments is given in Garger et al. (2004a, 2006).

Radiochemical analyses were performed only for those size-segregated samples that were high in ^{137}Cs. For those low in ^{137}Cs, $^{239+240}$Pu and ^{241}Am activity concentration values were below the detection limit. For these nuclides, the mean activity concentration values must therefore be considered as upper limits. Details of the activity measurements of ^{137}Cs, ^{7}Be, $^{239+240}$Pu, and ^{241}Am on the filters and in the leaching solutions are given in Garger et al. (2004a).

The total flow rate of activity through the gaps of the "Shelter" was 274.1 Bq s^{-1}, corresponding to 8.64×10^9 Bq year^{-1}. Of the release, 78.5% was due to ^{137}Cs, 21.1% due to ^{90}Sr, and 0.4% due to $^{239+240}$Pu.

The activity size distributions of the particles released at the largest gaps were determined by impactor measurements. On an average (i.e., weighted by the number of measurements), the measured activity concentration values and their standard deviations were 243 ± 228 mBq m^{-3} for ^{137}Cs, 119 ± 60.3 mBq m^{-3} for ^{90}Sr, 1.83 ± 0.70 m^3 for $^{239+240}Pu$, and 2.04 ± 1.73 mBq m^{-3} for ^{241}Am. During the period 2001–2003, significantly lower activity concentration values were determined. The corresponding mean activity concentration values and their standard deviations were 14.0 ± 10.0 mBq m^{-3} for ^{137}Cs, 3.12 ± 2.57 mBq m^{-3} for ^{90}Sr, 0.095 ± 0.025 mBq m^{-3} for $^{239+240}Pu$, and 0.845 ± 0.515 mBq m^{-3} for ^{241}Am.

The activity size distributions of ^{137}Cs, ^{90}Sr, and $^{239+240}Pu$ were measured in the outgoing flow of gap 1 at the northern side of the "Shelter" (Fig. 2.11). The distributions are wide and show two maxima, one in the fine-particle range (at about 2 μm) and the other in the coarse-particle range (above 10 μm). In comparison, the activity in the submicrometer particle range is low for $^{239+240}Pu$, medium for ^{90}Sr, and high for ^{137}Cs. Size-segregated measurements performed with the Andersen PM10 sampler at locations 3, 4, and 6 are presented in Table 2.19. The mean ^{137}Cs activity concentration was 1.74 ± 2.68 mBq m^{-3}. Again, the highest ^{137}Cs activity concentration values were measured in the southern side of the "Shelter" (sampler 6).

The AMAD values ranged from 0.6 to 3.6 μm. Their mean geometric standard deviations ranged from 1.3 to 4.7. During several experiments with increased activity concentration, the measured AMADs were in the submicrometer range.

This has to be considered for inhalation dose assessment. The measurements in Pripyat may be taken as background concentration values for the "Shelter" site, as Pripyat is situated 4 km northwest of the "Shelter". Nine series of measurements were performed there using the Andersen PM10 sampler in the 1996–2000 period. The mean activity concentration values were found to be in the range of the lowest concentration values at the "Shelter" site. It is interesting to compare the present mean parameters of the particle size distributions of ^{137}Cs with those measured at Pripyat in 1987–1992 (Garger et al. 1994). Probably due to previous decontamination, less anthropogenic activities, and weathering in general, the atmospheric activity concentration was reduced by one order of magnitude and the AMAD decreased from 5.4 to 1.2 μm (Table 2.19), compared to the earlier values (Garger et al. 1994).

Effective dose rates were calculated for all sampling periods as discussed in Garger et al. (2004c). The experimental data on the Chernobyl aerosols (e.g., on the solubility) were utilized instead of model default values, if available. The resulting individual effective dose rates were summarized.

At the "Shelter" gaps, the mean total effective dose rate was calculated to be of 97.5 nSv h^{-1} for the period 1996–1999. The main dose contribution resulted from ^{241}Am, while that from ^{137}Cs was about 3% of the total inhalation dose. The reduced emissions observed in the period 2000–2003 resulted in a dose rate decrease by more than one order of magnitude, for most radionuclides. Because dose rate from ^{241}Am decreased only by a factor of 5, the mean total effective dose rate was determined to be 15.6 nSv h^{-1} in the 2000–2003 period (Table 2.20).

Table 2.19 Results of size-resolved aerosol measurements performed in the immediate environment of the "Shelter": ^{137}Cs activity concentration, standard deviation σ of N samples, AMAD, geometric standard deviation σ_g, in the years 1997–2003, and the related effective dose rate

^{137}Cs measurement period	Activity conc. $\pm\sigma$ (N) (mBq m^{-3})	AMAD (μm)	σ_g	Eff. dose rate (pSv h^{-1})	Sampler location
10.07–31.07.1997	2.2 ± 1.2 (7)	2.2 ± 1.1	4.3 ± 2.4	25.1	Sampler 4, 50 m north
05.11–15.11.2002	9.95 ± 8.64 (3)	0.7 ± 0.5	3.1 ± 1.7	139	Sampler 6, 50 m south
12.08–30.09.2003	0.97 ± 0.42 (7)	3.6 ± 0.8	4.3 ± 1.3	9.50	Sampler 6, 50 m south
30.09–10.12.2003	0.33 ± 0.17 (10)	2.1 ± 0.3	2.2 ± 0.4	2.57	Sampler 6, 50 m south
30.07–07.12.2002	0.87 ± 0.23 (4)	1.2 ± 0.1	4.7 ± 0.9	12.1	Sampler 3, 200 m southwest
07.08–16.10.2003	0.79 ± 0.20 (7)	2.1 ± 0.6	2.1 ± 0.9	11.3	Act. conc. <1.0 mBq m^{-3}
	3.36 ± 2.26 (3)	0.57 ± 0.06	1.3 ± 0.2	30.4	Act. conc. >1.0 mBq m^{-3}
1987–1992	**2.44 ± 2.19 (17)**	**5.4 ± 3.2**	**1.7 ± 1.2**	**21.9**	**Pripyat, 4 km northwest**
1996–2000	**0.34 ± 0.18 (9)**	**1.2 ± 0.1**	**4.7 ± 1.0**	**4.57**	**Pripyat, 4 km northwest**

Values measured in the abandoned city of Pripyat are given for comparison (in bold in the last two lines)

Table 2.20 Weighted mean of individually calculated effective dose rates (nSv h^{-1}) from inhalation at the "Shelter" gaps

Period	^{137}Cs	^{90}Sr	$^{239+240}Pu$	^{241}Am	Total
1996–1999	2.70	5.46	16.7	72.6	97.5
2000–2003	0.149	0.196	1.60	13.7	15.6

In the vicinity of the "Shelter", it was only possible to analyze ^{137}Cs distribution characteristics because the activities were generally much lower there compared with those at the "Shelter" gaps. The resulting calculated dose rates were on an average about two orders of magnitude lower than those directly at the gaps, in the early period. The reduced emissions during 2002–2003 led to a minor dose rate reduction and to a ^{137}Cs dose rate which was smaller by about a factor of 7 than that directly at the gaps, for that period. The integral measurements performed in the "Shelter" vicinity may be used for the dose assessment of the rest of the radionuclides considered. A literature data for Pripyat and the "Shelter" premises were used for a very conservative (i.e., an upper limit for the effective dose rate from inhalation) estimation for the single period of high airborne activity concentration values in the

Table 2.21 Weighted mean of individually calculated effective dose rates (nSv h^{-1}) from inhalation in the close environment of the "Shelter"

Period	^{137}Cs	^{90}Sr	$^{239+240}$Pu	^{241}Am	Total
1997	0.0251				
2002–2003	0.0214				
14.11–23.11.2001					
$d = 0.5$ μm, $\sigma_g = 2.46$	1.87	10.3	69.4	25.6	107
$d = 1$ μm, $\sigma_g = 2.47$	1.87	9.4	61.3	23.0	95.6
$d = 5$ μm, $\sigma_g = 2.50$	1.33	5.55	43.0	13.2	63.1

lee of the "Shelter" (sampler 6). For this estimate, dose rate calculations were performed for three different realistic particle size distributions. As a result, total effective dose rates of up to 107 nSv h^{-1} were calculated, which are comparable to those calculated at the gaps (Table 2.21).

The main contribution to inhalation dose results from the nuclide $^{239+240}$Pu. However, the data suggest the mean total effective inhalation dose rate, if averaged over time, to be two orders of magnitude lower, confirming the result found for ^{137}Cs. Effective dose rates due to inhalation in the range of 100 nSv h^{-1} as determined at the gaps or during some periods in the "Shelter" vicinity would result in additional doses which are not negligible for any construction workers of "Shelter-2". Assuming an annual working time of 1.710 h (standard working time in the European Union and Ukraine), the inhalation dose rate would sum up to about 0.2 mSv year^{-1}. This dose rate does not conflict with the legal regulations of radiation exposure at work places, but must be taken into consideration if assessing the total radiation exposure in the case of any potential external exposure. This additional dose rate must be considered for the dose assessment of construction workers who are deployed at the Chernobyl site to build "Shelter-2".

Measurements at the containment building of the destroyed Chernobyl reactor and in the immediate environment have demonstrated that the reactor remains are a source of airborne radionuclides, even many years after the accident. The mean activity concentration in the aerosols measured directly at the gaps was about 240 mBq m^{-3}, with an AMAD of 2.4 μm for ^{137}Cs, 120 mBq m^{-3} with an AMAD in the range of 3.1–13 μm for ^{90}Sr, 1.8 mBq m^{-3} with an AMAD in the range of 3.5–11 μm for $^{239+240}$Pu, and 2.0 mBq m^{-3} with an AMAD of 1.5 μm for ^{241}Am. In the near environment, the mean ^{137}Cs activity concentration in the aerosol form was 2.2 mBq m^{-3} with an AMAD of 2.2 μm. The activity concentration values at the "Shelter" gaps and within the "Shelter" site were measured to be significantly higher than those in the abandoned city of Pripyat, at a distance of 4 km from the site of the "Shelter".

2.6 Re-entrainment of Radioactive Aerosols in the Atmosphere Caused by Wildland Fires and Extreme Meteorological Conditions

2.6.1 Radionuclide Emission Due to Wildland Fires at Radioactive-Contaminated Territories (Observations and Experiments)

The highly contaminated territories of Ukraine, Belarus, and Russia after the Chernobyl accident still remain potential sources of radioactive contamination due to forest fires. The radionuclides in the vegetation and soil would be emitted into the atmosphere during vegetation fires, which could lead to serious environmental damage. Radioactive emissions from contaminated wildfires could represent a risk for firefighters. In addition, populations are affected by radioactive smoke particles transported over long distances.

The main source is the ChEZ with an area of 2600 km^2. The forests covered 53% of the area before the disaster. After 1986, economic activities stopped and the forest area extended. Now, about 38% of the territory is Scots pine forests, 30% broadleaf forests, and the other 32% is deforested and former agriculture areas. However, there was no active forest and fire management after the accident. Scots pine stands became overstocked and stressed, predisposing them to attack by insects and diseases. The lack of fire management allowed the vegetation to overgrow, creating conditions favorable for fires to ignite and spread (Hao et al. 2009). Forest inventory data shows that 15,300 ha of forests in ChEZ are damaged, including 5300 ha damaged by pests that are now very fire prone (Hohl et al. 2012).

Radionuclides ^{137}Cs, ^{90}Sr, ^{238}Pu, and $^{239+240}$Pu are concentrated mostly in the top layer of soil in the forests and grasslands in ChEZ (Yoschenko et al. 2006a). The radioactivity in litter is higher than that in live foliage, bark, or live grasses.

2.6.1.1 Wildland Fires Within the ChEZ

According to Goldammer et al. (2015), from 1993 to 2013, more than 1250 wildland fires of different type and intensity have occurred in the ChEZ. Approximately 55% of the fires occurred in former agricultural lands. An additional 33% occurred in forested land. Although these fires consumed only 3300 ha of vegetation, larger fires have occurred in the region. The strongest fire on a total area of 17,000 ha of agricultural and forest land (including crown fire over an area of 5000 ha) occurred during July–August 1992 over a 2-week period, and hundreds of firefighters were involved in extinguishing it. Separate combustion sources were developed in different parts of ChEZ, including the most polluted central part. Additionally, 1200 ha of forest were burnt in the Belarusian part of the ChEZ (Dusha-Gudym 2005).

After 1992, on the most territories contaminated by radionuclides, the ratio of forest and nonforest lands burnt by fires changed appreciably. During the period 1993–2001, a total of 770 wildfires in the ChEZ of Ukraine affected 2482 ha (4.4% forest, >95% abandoned lands and agricultural estates). During the period 1993–2000, 186 wildfires occurred in the closed zone of Belarus and affected an area of 3136 ha, including 1458 ha of forest (46.5%) (Dusha-Gudym 2005).

Garger et al. (2004b) presented the measurement results of the air radioactive contamination at the industrial site of ChNPP for two periods: (1) August 2001 and (2) August–September 2002. They reported that in the period of forest fires of different scales in 2001–2002, the activity concentration of ^{137}Cs increased by 200–700 times in the ChNPP area. In the surface layer of the atmosphere, the particles of smoke with AMAD less than 1 μm predominated.

On August 16, 2001 there was a fire of forest area located at a distance of 2 km from ChNPP in a westerly direction on the territory of a sufficiently high level of radioactive contamination of the soil (10^6–10^7 Bq kg^{-1}). The fire has covered an area of no more than 10 ha. The values of the activity concentration of aerosols in the atmospheric boundary layer during the active phase of a fire (burning of litter, shrubs, dead trees, abandoned buildings) amounted to 69 mBq m^{-3} on August 16 at a distance of 500 m from the source, and 293 mBq m^{-3} on 17 August at a distance of 50 m from the source. On the stage of smoldering on 21 August, 2001 the aerosols activity concentration was obtained to be about 25 mBq m^{-3} at a distance of 5 m from the source. Even during the smoldering phase the activity concentration in the air at ChNPP industrial site has increased by two orders of magnitude compared with the background values equal to 0.1–0.2 mBq m^{-3}.

In all phases of fire during August 16–23, 2001, the maximum number of particles with aerodynamic diameters in the range of 0.5–1.0 μm was observed. The activity size distribution of ^{137}Cs particles during the active phase of the fire between 16 and 17 August, 2001 differ slightly in shape from each other. The median diameters and the standard deviations for a lognormal distribution were found to be $d_m = 1.4$ μm, $\sigma_g = 1.4$ and $d_m = 1.5$ μm, $\sigma_g = 1.1$, respectively. In contrast, during the phase of smoldering, the median diameter of radioactive particles was found to be 0.37 μm and the standard deviation 2.49. So, most of them belonged to the sub-micron size particles and the distribution width was considerably wider than that in the days of active forest burning. Similar characteristics were observed in 1992 in Zapolie at a distance of 19 km from the fire near Burakovka village in the 30-km zone (Garger et al. 1998b), where the median diameter of aerosol particles in the period of the fire 23.07–30.07.1992 was found to be 0.28 μm.

In the period from August 29 to September 3, 2001, as a result of severe forest fires outside of the ChEZ (in Belarus and in the northeastern part of Zhytomyr region of Ukraine), there was heavy smokiness in the vicinity of ChNPP. The maximum value of the activity concentration of aerosols in the air of 187 mBq m^{-3} was observed between September 2 and 3, 2002. During this period, the number of fine smoke aerosol particles increased by 12–20 times, which can be attributed to the transfer of smoke particles from Belarus. The distributions of counting aerosol concentration on the particle size during the fire in 2001 and the heavy smokiness in

2002 have differences caused partly by different distances (from 50 m to tens of kilometers) between the aerosol sampling points and the fire places. So, in 2001, a share of particulate matter with an aerodynamic diameter in the range of $0.5 < d \leq 1.0$ μm was 77.9%,and in 2002 was 92.7%; that is, in 2002, aerosol particles with large diameters deposited from the atmosphere on the ground before reaching the territory of the ChNPP industrial site.

According to the measurement results, the activity concentration of aerosols in the atmospheric boundary layer during the wildfires was determined mainly by ^{137}Cs (in August 2001—98.2% of the total activity, in September 2002—97.2%).

Kashparov et al. (2015) presented the results of measurements during a wildland fire in April 2015. The grass fire covered 6250 ha of meadows, with the forest ground and crown fires covering 2737 and 1140 ha, respectively. The maximum density of contamination in the areas of the ground forest fire in Lubyanka Forestry reached by ^{137}Cs was 1040, ^{90}Sr was 368, $^{238+240}$Pu was 11.4, and ^{241}Am was 14.4 kBq m^{-2}. Within a smoke zone, the atmospheric activity concentration was 0.77 mBq m^{-3} of ^{137}Cs and 0.35 mBq m^{-3} of ^{90}Sr on April 27 in Stechanka village; 7.6 mBq m^{-3} of ^{137}Cs and 10 mBq m^{-3} of ^{90}Sr on April 28 in Stara Krasnica village; and 1.4 mBq m^{-3} of ^{137}Cs and 0.71 mBq m^{-3} of ^{90}Sr on April 29 in Lubyanka village. When extinguishing meadow and forest fires, the total committed effective dose of firefighters from Chernobyl radionuclides does not exceed 42 μSv. During forest fires, more than half of the expected effective dose of internal radiation may be caused by the inhalation of ^{90}Sr. During grassland fires, the inhalation intake of ^{90}Sr, $^{238-241}$Pu, and ^{241}Am can give approximately equal contributions in the expected effective dose of internal radiation of firefighters. The contribution of beta-emitting ^{241}Pu in the formation of the expected effective dose of internal exposure of personnel is commensurate with the contribution of each of the alpha-emitting radionuclides of plutonium (^{238,239,240}Pu).

2.6.1.2 Wildland Fires Outside the ChEZ

Besides forest fires within the ChEZ, some large contaminated regions in Ukraine, Belarus, and Russia during some hot and dry years were substantial sources of secondary air contamination both in these territories and within the ChEZ itself.

In 1992, fires covered 870 ha of forest lands in Gomel and Mogilev regions (Belarus), outside the exclusion zone. In the Bryansk region (Russia), the area of radioactive fires was less than 200 ha. As a result of the fires ^{137}Cs radionuclides were lifted and transported by the smoke to the territory of Russia in May and August 1992 (Dusha-Gudym 2005). In 2002, a number of regions of the Russian Federation were characterized by a hard situation. The forest land area burnt by fires made up to 1986 ha (Dusha-Gudym 2005).

In 2001–2002, forest fires of different scales influenced a radiation situation in the ChEZ (Garger et al. 2004b).

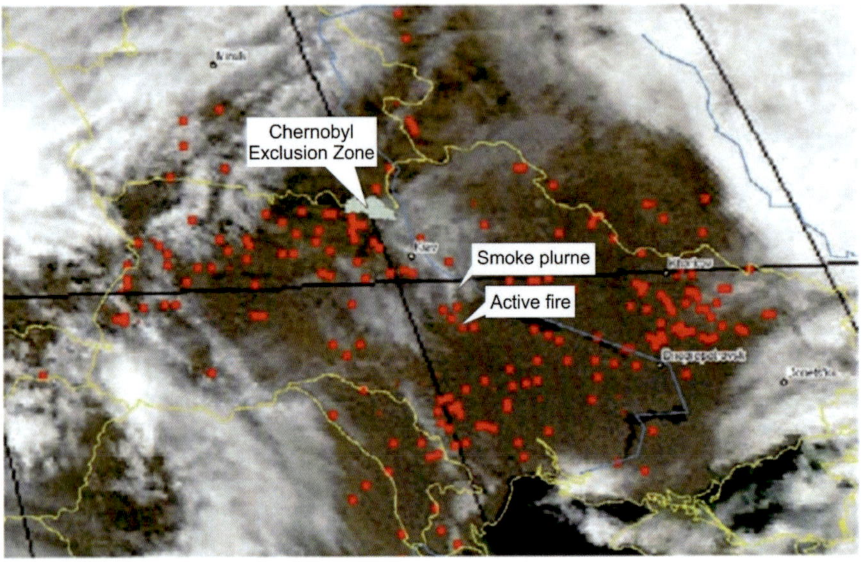

Fig. 2.12 MODIS satellite image of fire locations (red dots) and smoke in Ukraine and its neighboring countries, April 16, 2006 (from Goldammer et al. 2015)

Figure 2.12 shows the spatial distribution of fires and smoke in Ukraine and neighboring countries on April 16, 2006. Smoke plumes originating from active fires south of Kyiv traveled north to about 100 km.

The long-range transport of Chernobyl's smoke plumes were not only observed by satellites but also were detected by the ground radioisotope monitoring stations in Sweden. During the 1972–1985 period, the monthly ^{137}Cs levels peaked in May, which were caused by the nuclear test fallout in the stratosphere. However, during the 1987–2000 period, the ^{137}Cs levels peaked in April and October, which coincided with the periods of most intense burning in the ChEZ. The resuspension of mineral dust in Chernobyl may also have contributed to the elevated ^{137}Cs levels (Kulan 2006).

Lujaniene et al. (2006) presented the results of the air activity measurements in Vilnius, Lithuania made after the Chernobyl accident. A considerable ^{137}Cs activity concentration in the atmosphere was registered between August 31 and September 1, 1992 (up to 300 μBq m^{-3}) and between September 6 and 7 (190 μBq m^{-3}) during extensive forest fires in the 30-km exclusion zone. In 1996, the maximum activity concentration of ^{137}Cs was observed in April. The mean monthly activity concentration was 25.9 μBq m^{-3}, but in 2-day old samples, it increased up to 253 μBq m^{-3} (April 9, 1996). Another event of transport of products of forest fires was registered during September 6–9, 2002 after extensive fire in the forests and peat bogs of Belarus (up to 200 μBq m^{-3}). The measurements carried out on aerosol samples collected in Vilnius during 1995–2003 indicated the presence of alpha-emitting radionuclides. The activity concentration values of $^{239+240}Pu$ and ^{241}Am ranged from 1 to 500 and 0.3 to 500 nBq m^{-3}, respectively.

Several events in 2002 redistributed and transported ^{137}Cs over Europe (Evangeliou et al. 2015). These fires were so serious that peat bogs also burned, followed by forests, causing massive smoke emissions that persisted for a period of nearly 2 months, sometimes reducing visibility to below 60 m. These fires resulted in significant exposures of smoke to major population centers such as Kyiv and Moscow. The highest release occurred at the end of July, with ^{137}Cs fallout reaching Sweden, Finland, and Central Europe.

Fires in 2008 were of lower intensity and frequency. However, some took place inside highly contaminated regions, injecting a significant plume into the atmosphere. In March 2008, ^{137}Cs covered Belarus and a large part of Russia, and another event in August distributed ^{137}Cs more locally (Evangeliou et al. 2015).

Paatero et al. (2009) reported about two biomass burning events in 2006 in the western Russia. The deposition of ^{137}Cs in these areas has been estimated to be mostly 2–10 kBq m^{-2}, with limited areas of higher values, 10–40 kBq m^{-2}. During the spring smoke episode, the ^{137}Cs activity concentration in the air in Helsinki increased by a factor of 10 (up to 13 μBq m^{-3}) when compared with values just before the episode, due to a long-range atmospheric transport of the radionuclides. Simultaneously there was an increase of the same order of magnitude in the concentration values of PM10. The increase in the PM10 concentration is related to soot emissions from the fires, in other words, the incomplete combustion of organic material. The summer episode was less severe even though there were wildfires close to the Finnish-Russian border. During the summer smoke episode, the ^{137}Cs activity concentration in the air increased by a factor of 5 compared with values just before the episode. Based on the measurement data obtained, they concluded that from the radiological point of view, the exposure to the increased radionuclide activity concentration was insignificant when compared with the health hazards due to the increased concentration of aerosol particles and their chemical components.

2.6.1.3 Wildland Fire Experiments in Radioactive-Contaminated Territories

Kashparov et al. (2000) reported the results of measurements of radioactive aerosol characteristics (emission parameters, distribution within the experimental site, dispersal composition, and deposition parameters) during the controlled burning of forest and meadow ecosystems on two experimental sites: (1) the site of 100×200 m size vegetated with pine trees and contaminated mainly by ^{137}Cs at 1.1 ± 0.2 MBq m^{-2} and (2) the site of 50×140 m with dry grass from previous year, with a ^{137}Cs inventory of grass per unit area of 550 ± 30 Bq m^{-2}. At these sites, the ^{137}Cs background airborne activity concentration was $0.5-1.5$ and 3 mBq m^{-3} respectively.

In the active phase of the forest fire (duration 2 h), the airborne activity concentration of ^{137}Cs in the lower (1 m) layer of the atmosphere varied within the range of 10−100 mBq m^{-3} at 15–270 m distances downwind the source, the largest values being obtained at the maximal distance of measurements, 270 m. After the active phase, during the period of smoldering (duration ~21 h), the activity concentration was 5−20 mBq m^{-3}, and, during the postfire period, 1 day later, 2−5 mBq m^{-3}.

During the active and smoldering phases of the experiment, the airborne concentration of aerosol particles larger than 10 μm doubled at 100 m from the source, and that of particles smaller than 10 μm increased by 20−700 times compared with the natural background levels. In the postaccident period, the airborne concentration of fine particles (smaller than 10 μm) exceeded background values by 3−30 times. The number of large particles decreased with time from the beginning of the fire and with increasing distance from the source. At a distance 15 m from the source, the share of large aerosol particles with size >8.5 μm in the ^{137}Cs total activity concentration in the lower air layer was found to be 46% and small particles with size <2.8 μm 37%. At a distance of 150 m, the shares were obtained to be 20% and 69%, respectively.

The additional terrestrial contamination due to forest fires can be estimated as a value in the range of 10^{-4}–10^{-5} of its background value; therefore, even during the most unfavorable conditions, radionuclide emission during forest fires will not provide a significant contribution to terrestrial contamination.

During grassland fires, the mean value of airborne ^{137}Cs aerosol activity concentration in the surface layer of the atmosphere was 19.2 Bq m^{-3}.

In Yoschenko et al. (2006a)**,** the focus of the investigation was on the potential impact of the radionuclides ^{90}Sr, ^{238}Pu, and $^{239+240}Pu$, besides ^{137}Cs. Controlled burning of experimental plots of forests or grasslands in the ChEZ has been carried out in order to estimate the parameters of radionuclide emission, transport, and deposition during forest and grassland fires and to evaluate the working conditions of firemen.

The active experiments were performed on two grassland sites (3600 and 5400 m^2 area) and one forest site (8770 m^2 area). Both grassland fires were carried out within the same experimental area near the former village Chistogalovka, at a distance of approximately 3 km west of the ChNPP site. The forest site is located 5 km of the ChNPP close to the former village Novoshepelichi. It is a cultivated pine forest surrounded by grasslands. At the three chosen sites, the background values of the airborne activity concentration was 2.1−4.6 mBq m^{-3} for ^{137}Cs, 0.8−4.4 mBq m^{-3} for ^{90}Sr, 1.5−8.3 μBq m^{-3} for ^{238}Pu, and 2.9−22 μBq m^{-3} for $^{239+240}Pu$.

During grassland fires, the airborne activity concentration values of ^{137}Cs and ^{90}Sr reached values of several Bq m^{-3} near the source of release and decreased with the distance. The activity concentration of plutonium was generally three orders of magnitude lower and an increase with the distance was observed in site #1. In site #2, rather constant values were measured at greater distances after an initial decrease.

During the forest fire experiment, an increase of several orders of magnitude of the airborne radionuclide activity concentration was observed in the territory near the fire area. But, in general, one order of magnitude lower activity concentration values for ^{137}Cs and ^{90}Sr were measured in comparison to the grassland fires. After an initial decrease in concentration over long distances a rather constant airborne concentration was observed for all nuclides. At the furthest sampling site the concentration increased again.

The additional contamination density of the grass after the fire would be about 1% ^{90}Sr, 1% ^{137}Cs, 7% ^{238}Pu, and 7% $^{239+240}Pu$ of the contamination density before the fire if all newly deposited material is fixed on the grass. Thus, the wildland fires in the ChEZ do not lead to a significant redistribution of contamination on the scale of the whole zone at scales of fires used in experiments. However, if the new vegetation was grown next to a site with burning biomass on uncontaminated soil, the vegetation could be contaminated with about 1 kBq kg^{-1} ^{90}Sr, 500 Bq $kg^{-1}$$^{137}Cs$, 0.1 Bq kg^{-1} ^{238}Pu, and 0.2 Bq $kg^{-1}$$^{239+240}Pu$ in close vicinity of the fire territory. So, for decontaminated areas inside the zone, secondary contamination by fire events might be a problem for areas at a distance up to several hundred meters from the burned territory.

According to a data obtained by Yoschenko et al. (2006a), the dose of the external irradiation from soil and vegetation can provide a significant contribution to the total dose by the inhalation intake of the plutonium radioisotopes and their progenies. Under some conditions, this contribution can reach half of the total dose. At the same time, taking into account the sharp decrease in the airborne radionuclide activity concentration with the distance from the source of release, they stated that the inhalation component of the total dose (as well as the external irradiation from radionuclides in air) is not important for the personnel of the exclusion zone not involved in the fire fighting.

Formation and transformation of aerosol particles carrying the radionuclides were studied during the forest fire experiment in a highly contaminated Briansk region, Russia in August 1993 (pine forest burning on 0.2 ha area) (Lujaniene et al. 1997). According to the obtained data, forest fires were responsible for the generation of a great amount of submicron (0.2 μm) aerosol carriers of ^{137}Cs. In aerosol samples collected during this active fire experiment, about 80% of submicron aerosol particles of ^{137}Cs of 0.1–0.4 μm contributed to the total aerosol concentration (Lujaniene et al. 2009). The formation mechanism of submicron ^{137}Cs particles can be attributed to evaporation, condensation on nuclei available in the atmosphere, and their further coagulation. These results agree well with the results of the experiments carried out in different laboratories. It was shown that about 95% of fine particles formed during the combustion of contaminated biomass had a size of 0.1–0.4 μm (Grebenkov et al. 2002). Natural experiments with burning of fir needles indicated the formation of aerosol particles of ^{137}Cs in the range of 0.62–0.72 μm below 600–700 °C (Ogorodnikov 2002). Simultaneously, the fraction of large particles with AMAD of 10–14 μm was formed due to underburning of needles.

2.6.2 *Modeling of the Air Secondary Radioactive Contamination Due to Forest Fires*

As can be seen from the previous section, wildland fires may result in additional contamination of the air both in the immediate vicinity of the fire territory and over long distances, including long-range atmospheric transport of fine aerosols. Therefore, different types of models of emission, following atmospheric transport and deposition of fire products, are developed for the local area around the fire (up to 10 km) and mesoscale and long-range atmospheric transport. The main features of the radionuclide atmospheric transport models for different spatial and temporal scales are determined primarily by the nature of the diffusion processes of pollutants in the atmosphere, and they are similar for both the models of emission from point sources (e.g., from a NPP stack) and for the resuspension models during forest fires. The main problem for the models of forest fires is the correct description of the radioactive aerosol source parameters (duration and area of the fire, emission intensity for the different phases of the fire, the initial raising height of smoke plume, activity size distribution of radioactive particles). Therefore, in general, the model of atmospheric transport of smoke particles must be a part of a more general model of wildland fire evolution (Grishin 1992; Mandel et al. 2011). Using a number of simplifying assumptions, a subtask of the assessment of radioactive contamination of the atmosphere by combustion products can be singled out of the general picture describing the development of the fire and its consequences. Then, this problem can be solved by relatively simple methods. The price for such simplification is the introduction into used models a number of constants whose values have to be determined empirically.

One of the main problems of radionuclide atmospheric transport models due to forest fires is the parameterization of the emission fraction of radionuclides after a fire event, namely, the fraction that will be emitted compared withwhat is present on the ground or in the biomass. Evidence from laboratory experiments and field studies are quite contradictory (Table 2.22).

2.6.2.1 Near-Range Atmospheric Transport Modeling of Radionuclides During Wildland Fires in the ChEZ

A model of aerosol emission, following atmospheric transport and deposition during forest fires, developed by Talerko has been applied by Kashparov et al. (2000) to clarify the measurement results obtained in the course of the controlled fire experiments. The model is based on the approach proposed in Talerko (1990). The model describes two possible scenarios of wildfires: formation of a convective plume over a fire area and a nonconvective diffusion regime of aerosol particle dispersion in the atmosphere. It was concluded that the process of admixture transport in the atmosphere, in this specific case, could be described by a superposition of the abovementioned processes. The main contribution to the field of the aerosol activity

Table 2.22 Radionuclide emission fraction during wildland fires

Nuclide	Emission fraction	Source	Comments
I, Cl	80–90%	Amiro et al. (1996)	Straw and wood burning experiments in field and laboratory conditions. Biomass were combusted with temperatures varying from 160 to 1000 °C
Cs	40–70%		
^{137}Cs	10–40% depending on the temperature (12% after a cool burning and 39% after a hot burning)	Horrill et al. (1995)	Laboratory experiments for heather burning (50 kg of dried material)
^{137}Cs	1% of totally available within the burnt region, or ~10% of available in the biomass	Wotawa et al. (2006)	Estimation for modeling of inter- and intracontinental transport of cesium released by boreal forest fires
^{137}Cs	40% of deposited	Evangeliou et al. (2014)	Conservative estimation for modeling of redistribution over Europe
^{137}Cs, ^{90}Sr (labile radionuclides)	20%	Evangeliou et al. (2016)	Conservative estimation for modeling
^{238}Pu, ^{239}Pu, ^{240}Pu, ^{241}Am (refractory ones)	10%		
^{137}Cs	~3–5% of deposited	Hollander and Garger (1996)	Estimation value has been obtained after the natural forest fire in the ChEZ in July 1992
^{137}Cs	4% of deposited	Yoschenko et al. (2006a)	Fire experiments in the ChEZ (plot length of about 50–100 m)
^{137}Cs and ^{90}Sr	Up to some ‰	Yoschenko et al. (2006b)	Grassland fires (comparison of modeling and experimental data)
Pu	Up to 1‰		
^{137}Cs and ^{90}Sr	Up to 3–4%		Forest fires (comparison of modeling and experimental data)
Pu	Up to 1%		

concentration values at a distance of 10 m is provided by nonconvective transport. In intermediate zones (10–110 m), the nonconvective transport still gives a comparable contribution, while at larger distances, the contribution of convective transport is much higher (Fig. 2.13).

Estimates of the additional terrestrial contamination by activity redistribution caused by fire have demonstrated that it is very small at all distances from the source. The maximum deposition, according to the above calculations, is observed at a distance of 1500–2000 m and is approximately 110 Bq m^{-2}. The additional terrestrial contamination due to forest fires can be estimated to be in the range 10^{-4}–10^{-5} of its background value. At 20 km from the fire source, the deposition

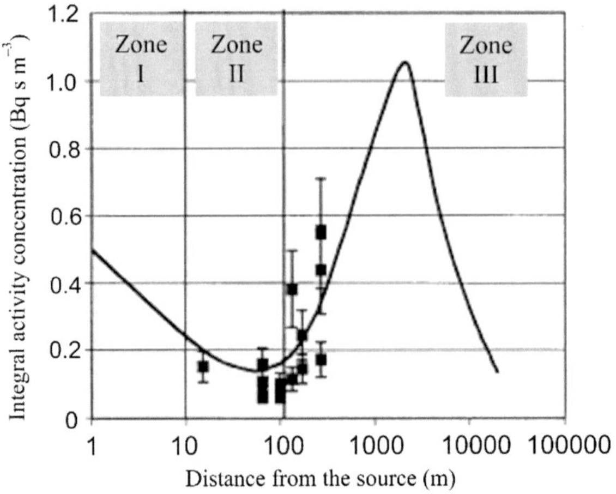

Fig. 2.13 Integral airborne ^{137}Cs activity concentration (kBq s m^{-3}) on the downwind axis (active burning phase) (Kashparov et al. 2000). The solid line represents the relationship obtained in the calculations by superposition of the convective and nonconvective modes of admixture transport in the atmosphere. Three zones of deposition are indicated: I—nonconvection flux deposition zone, 0–10 m; II—intermediate zone, 10–110 m; and III—convection flux deposition zone, >110 m. Solid line is the calculated result; observed values combined with their uncertainty range

is several tenths Bq m^{-2} and can be neglected even on territories contaminated only with a global fallout.

Yoschenko et al. (2006b) used the experimental results obtained in Yoschenko et al. (2006a) for verification of the model presented in Kashparov et al. (2000). They complete the model with equations for the calculation of an initial air temperature in a plume above a fire area and an initial plume radius. By comparing the experimental data and model calculation of the airborne activity concentration and the deposition density, they concluded that the developed model satisfactorily describes the studied process (see estimations of the radionuclide emission fraction value in Table 2.22).

2.6.2.2 Modeling of the Consequences of Mesoscale and Long-Range Atmospheric Transport of Radionuclides During Forest Fires

Evangeliou et al. (2015) simulated fires in Ukraine and Belarus during 2002, 2008, and 2010 using the Moderate Resolution Imaging Spectroradiometer (MODIS) satellite data. The satellite imagery recorded 185 fires in 2002, 50 in 2008, and 54 in 2010 that burned 473, 125, and 137 ha, respectively. The coupled model LMDZORINCA was used in the simulations consisting of the aerosol module INCA, the general circulation model LMDz, and the global vegetation model ORCHIDEE. The fire incidents were simulated using the zoom version for Europe, achieving a horizontal resolution of $0.51° \times 0.45°$ for 19 vertical levels. According to the modeling results, the highest exposure to the radioactive cloud from the 2002,

2008, and 2010 fires occurred over Central Europe, with the annual average air activity concentration of ^{137}Cs ranging from 17 to 120 mBq m^{-3}. These values are of the same order of magnitude as that of the ^{137}Cs air activity concentration measured in the experiments by Kashparov et al. (2000) directly near the fire territory and time-averaged for release periods only. Thus obtained modeling estimations of the ^{137}Cs concentration in the air, the secondary cesium deposition (up to 32.2 kBq m^{-2}), and the total amount of ^{137}Cs redeposited over Europe as a result of these fires (8% of total cesium depositions after the Chernobyl accident) should to be regarded as an overestimated, first of all, because of too conservative value of the radionuclide emission fraction of 40% (Table 2.22) used in the calculations.

For fires in ChEZ during spring and August 2015 (the estimated total area of fires was 10,882 and 5867 ha, respectively) Evangeliou et al. (2016) carried out simulation of the atmospheric transport and deposition of the radioactive plume released by the forest fires. They showed that in spring 2015, the release from the continuous burning in the ChEZ was transported to Western Russia, then to the north and reached southern Finland and Sweden and subsequently Latvia and Lithuania. In parallel, radionuclides were also transported to the east, and smaller amounts to the south, that is, toward the Black Sea. The aforementioned circulation only describes the more labile radionuclides ^{137}Cs and ^{90}Sr. The refractory transuranium elements were deposited closer to the source due to their larger particle size.

According to the results of modeling of radionuclide transport in August 2015, the plume was immediately transported to the east during the first 4 days. Thereafter, the fallout burden was divided into three parts: one part to Russia and Belarus, and another part of the plume was transported across Poland, Northern Germany, and Denmark and ended in a big anticyclone that covered whole Scandinavia. Finally, a third part of smaller intensity moved southwards affecting the Balkan region, the Aegean Sea and Turkey. The simulated activity concentration values were of a similar magnitude as for the spring fires for ^{137}Cs and ^{90}Sr, but ^{238}Pu, ^{239}Pu, ^{240}Pu and ^{241}Am concentrations were substantially lower.

Several researches made calculations of possible consequences of forest fires in the ChEZ and neighboring contaminated territories under different scenarios, including worst case scenarios.

Hohl et al. (2012) assessed the health implications of potential catastrophic wildfires in the Ukrainian portion of the ChEZ on populations living and working beyond the ChEZ. As a worst case scenario, it is assumed that a fire would completely consume the biomass of pine forests and former agricultural lands and release approximately 4×10^{14} Bq of radioactive material into the atmosphere. The wind was assumed to blow toward Kyiv throughout the event of a fire. The estimated activity concentration values of ^{137}Cs in the near-ground air were obtained to be 12 mBq m^{-3} at a distance of 30 km from a ChEZ center and 1.9 mBq m^{-3} at a distance of 100 km. An additional ^{137}Cs deposition density due to a fire at these distances was estimated as 240 and 38 kBq m^{-2}, respectively. The estimated exposure of populations at 30 or more kilometers from the source of the fire was 22 mSv, and is below the critical thresholds that would require evacuations (according to the Ukrainian safety norms NRBU-97, people should be evacuated if the prevented dose in the first 2 weeks exceeds 50 mSv). Since the estimated total ingestion doses to children (1 year) and adults were found to the exceed acceptable levels, a restriction on the intake of contaminated vegetation, meat, and milk indefinitely would be recommended. At 100 km (i.e., the approximate distance to Kyiv), the adult total exposure during the first year after the event was obtained to be 9.4 mSv, including 5.9 mSv through ingestion. For children, the equivalent values are 7.1 and 5.5 mSv.

Evangeliou et al. (2014) used three different scenarios of forest fires: (a) minor fires that cover 10% of the forests in Ukraine, Belarus, and Russia, which can be extinguished by humans, (b) intermediate fires that affect 50% of the same forests, and (c) a worst case scenario, in which the entire area is affected by fires. The range of the ^{137}Cs activity concentration values in atmospheric aerosols in the first scenario is only significant mostly for the local areas around the contaminated forests, when concentrations are in the range of few tens of mBq m^{-3}. This is not the case in the most intense scenarios, where the plumes reach North Europe and Central Russia with maximum concentrations up to three orders of magnitude higher. The activity concentration values of ^{137}Cs during a fire in the two last scenarios could reach to an order of 10 Bq m^{-3} in Belarus (to the north of the ChEZ). The estimates showed that 1–90 kBq m^{-2} can be redeposited over Europe in 1 fire year affecting an area of several million inhabitants. It was concluded that large fire events in contaminated forests can be classified as "accident with local or wider consequences" or even as "serious accidents". On the basis of the obtained results of atmospheric transport modeling, the number of excess lifetime cancer incidents was estimated to be between 20 and 240 for the global population, of which 10–170 may be fatal. These numbers are far lower than the obtained cancer cases after Chernobyl, but somehow comparable to those obtained from the deposition exposure to the Chernobyl-remaining ^{137}Cs over Europe and the total exposure after Fukushima. As in the previously mentioned work of these authors, it is necessary to note that these values have been obtained under assumptions which seem to be very conservative, so they can be regarded as overestimated.

Hence, as a result of forest fires in the ChEZ in the past years, a short-term (during few days) increase in the ^{137}Cs activity concentration up to three orders of magnitude was observed compared with the values in the other periods. During the last large fire in 2015, the daily average activity concentration of ^{137}Cs reached short-term (for 1 day), values up to 8 mBq m^{-3}, which do not exceed the value of the ^{137}Cs reference level of the activity concentration in the air for the ChEZ (10 mBq m^{-3}). Thus, even within the ChEZ, an increase in air contamination due to the forest fires does not result in the formation of the personnel doses exceeding the established standards. At the same time, estimates indicate that the within the fire zone and in the vicinity of burning area the radionuclide activity concentration in the air can be much greater. Therefore, special attention should be paid to control doses to personnel, involved in fire extinguishing, due to the scope for the potential inhalation of radioactive aerosols resulting from burning trees and grass.

During forest fires in the ChEZ, an increase in the activity concentration in the air outside of the ChEZ is much smaller and does not lead to a significant increase in the dose of the population of Ukraine and Belarus. The same conclusion can be drawn about the forest fires that may occur outside of the ChEZ near settlements, because the quantities of radioactive contamination of forest areas is much less compared to the ChEZ.

However, the problem of assessing the forest fires danger in the contaminated territories in Ukraine and Belarus remains relevant, since up to now forest fires are a source of additional increase in the activity concentration of radionuclides in the air. Each event of a forest fire in ChEZ attracts the attention of the media and the population living near the ChEZ, including the residents of Kyiv and other major cities. In the case of particularly large and prolonged fires, radioactive aerosols can spread over long distances outside Ukraine, including the countries of Western Europe. It is therefore necessary to have a reliable means for rapid assessment and prediction of atmospheric transport of radioactive aerosols rising into the air as a result of forest fires. To date, the main problem that requires further research is the development of methods for operational determining the parameters of the forest fire territory as an area nonstationary source of radioactive aerosols. Among them are the following:

1. Rapid assessment and prediction of the territory size covered by a forest fire, including the temporal dynamics of the spread of the fire front. In general, this is necessary to use a special surface fire spread models for the calculation of fire spread over the area taking into account the properties of the area and weather conditions. However, these models require detailed information on the characteristics of the forest, which is not always available. A feature of forest fires in the ChEZ is their relatively small areas (up to several thousand hectares) and duration (up to several days) compared to large-scale fires in Siberia (Russia) or California (USA). Therefore, for practical calculations of air pollution due to the forest fire it's advisable to consider the fire area as a short-term constant area source of radionuclides emission in the atmosphere. These values can be set according to the observations and by expert estimates.
2. Estimations of the value of the radionuclide emission fraction during forest fires are differed considerably: from a few percent to 70% according reviewed experimental and modeling works. The most valid are the results of field experiments and model estimates of the radionuclide emission fraction, carried out for the fires in the ChEZ, which give values 3–5% of ^{137}Cs and ^{90}Sr and up to 1% of the Pu isotopes. Conditions of experiments for biomass burning, for which estimates are obtained from 20% and above, as a rule, do not fully correspond to the real conditions of wood burning in a forest fire. In addition, it is necessary to note that in the second year after deposition, the bulk of radionuclides migrates to the forest litter and soil. As a result, the proportion of radionuclides that could potentially be raised to the atmosphere from forest fires, in their total stock in the contaminated area, is steadily decreasing with time. Thus, the model estimates the effects of forest fires in the territories of Ukraine and neighboring countries, obtained using the emission fraction value of 10% or more, can be considered as very conservative.

3. Effective height of the smoke plume lifting over a forest fire. Its value is determined primarily by the ratio between the kinetic energy of the rising air flow (determined by the intensity of heat release in the fire zone) and the wind kinetic energy (Byram 1959).
4. The distribution of particulate matter formed during burning in size. The large particles (more than 20 μm) that are formed in a fire, deposited near the burning territory up to 100 m). Ignoring this factor in the simulation could lead to an overestimation of the long-range transport of radionuclide's, and simultaneously to an underestimation of the value of the additional deposition in the near (up to 100–1000 m) fire zone.

Clarification of these parameters values and methods for their reliable estimate will enable to improve models of atmospheric transport of radioactive aerosols rising into the air as a result of forest fires, for a variety of spatial scales.

2.6.3 Radionuclide Resuspension Due to Extreme Meteorological Conditions

Secondary pollution of the atmosphere due to the rise of radionuclides from the earth's surface can occur not only due to the resuspension at low and moderate wind speeds, but also in extreme weather events, which include dust storms, tornadoes, squalls. Despite the relative short duration of such phenomena, their contribution to the overall balance of the flux of dust and related radionuclides from the earth's surface in the atmosphere can be considerable.

2.6.3.1 Atmospheric Dispersion of Radionuclides Resuspended During Dust Storms

Dust storms are a natural phenomenon under conditions of (1) strong wind, (2) cyclone movement, and (3) dry, loose, and bare sandy soila. They occur frequently in deserts and surrounding areas and in many arid and semiarid agricultural regions. Based on their reduction of the visibility in air, dust storms are conventionally classified as dust haze, blowing dust, dust storm, and dust devil, with corresponding visibilities of about 10, 1–10, 0.5–1, and <0.5, respectively (Wang et al. 2006).

Dust storms in Ukraine and Belarus can lead in a short time to the territorial redistribution of radioactive materials deposited on the ground after the Chernobyl accident.

According to Lipinskiy et al. (2006), dust storms were observed in the territories of the Kyiv region, although their frequencies are much less in the southern part of Ukraine. In the first years after the Chernobyl accident, a sharp increase in aerosol concentrations in the 30-km ChEZ was observed several times due to strong winds.

Table 2.23 ^{137}Cs activity concentration at 3 monitoring sites in the ChEZ during the dust storm between September 5 and 7, 1992

Sampling period	^{137}Cs activity concentration (mBq m^{-3})			Wind speed (m s^{-1})
	ORU-750 (1 km south of the ChNPP)	BNS (3 km east of the ChNPP)	OVD-2 (in the Pripyat city)	
Before the storm ("background")	5–10	5–10	0.5–1	
6 Sept (14^{00}–17^{00})	200			10–12
6 Sept (12^{00}–16^{20})			100	10–12
6 Sept (16^{20}–19^{30})			7	5–7
17^{00} 6 Sept to 14^{00} 7 Sept	5			5–9
14^{20} 6 Sept to 13^{20} 7 Sept		17		5–9
11^{00}–12^{40} 7 Sept			2	8–9

During a dust storm at the end of April 1987 the ^{137}Cs activity concentration increases sharply. It resulted in increasing of the monthly averaged ^{137}Cs activity concentration according measurement at meteorological station in Chernobyl: 0.3, 9.4 and 1.4 mBq m^{-3} in March, April and May 1987 accordingly.

Three months later (24 July 1987) there was a dust storm in the exclusion zone (Meshalkin et al. 1992). It was noted that the visibility dropped to 7–10 m, the concentration of radioactive aerosols at the monitoring site ORU-750 located half a kilometer to the south of the "Shelter" object has increased by two orders of magnitude.

The dust storm in the exclusion zone observed during July 28–29, 1988 (Izrael 1990). Under an average wind speed of 12–15 m s^{-1} the daily averaged activity concentration of $^{239,\,240}Pu$ in the Pripyat city reached 1.5×10^{-1} Bq m^{-3}, that is, 10–15 times higher than the year averaged value for 1988.

Table 2.23 presents measurement results of the activity concentration values of radioactive aerosols in the air during the dust storm between September 5 and 7, 1992 on the territory of Ukraine and Belarus polluted after the Chernobyl accident (Ogorodnikov 2011). The emergence and development of dust storms was due to the passage of the cyclone near the exclusion zone, which originated on September 4 over the Balkans. During the dust storm, the activity concentration values of radioactive aerosols in the ChEZ increased in by 10–100 times.

Outside of the ChEZ, the data on radioactive aerosols in August and September 1992 were obtained at several meteorological stations in Ukraine (Oster, Shchors), Belarus (Gomel, Minsk, Mogilev, Mozyr), and Lithuania (Vilnius) (Ogorodnikov 2011). During the dust storm that occurred between September 5 and 7, 1992, the daily averaged activity concentration of the mixture of β-emitting nuclides in Oster and Shchors (Chernihiv region, Ukraine) increased up to 1.2 mBq m^{-3} and 0.9 mBq m^{-3} (about five times above the "background" values). In Minsk and

Mozyr, the "peak" values were 6–7 times higher than the average for August. In Mogilev and Gomel, the activity concentration values of radioactive substances in the atmosphere between September 5 and 7 were increased in comparison with the average for August by 1.5–3 times, that is, less than that in Minsk and Mozyr. However, the increase in the mass concentration of aerosols was observed clearly. Especially, dustiness of the atmosphere increased sharply in Gomel: from 0.02–0.03 to 0.4 mg m^{-3}. The last value is almost three times greater than the daily averaged threshold limit value of dust in the air adopted in Belarus. In August and September 1992 in Vilnius (Lithuania) two ^{137}Cs concentration peaks were registered (Lujaniene et al. 2009). The former was on 31 August to 1 September and reached 300 μBq m^{-3}, and the latter—on 6–7 September with a maximum of 190 μBq m^{-3}, which are 2–3 orders of magnitude higher than background values this period.

2.6.3.2 Radionuclides Re-entrainment Due to Tornado Over Nuclear Installations, Radioactive Facilities, and Contaminated Territories in the ChEZ

Tornado belongs to the most dangerous natural phenomenon that causes a huge catastrophic destruction. The radiological consequences of air pollution as a result of the passage of the tornado over the objects that contain radioactive materials or radioactive-contaminated territories have not yet been fixed. However, the possibility of such situations is not excluded. The model estimates of the possible effects of the tornado on individual radiation hazardous objects in ChEZ were conducted, including the object "Shelter", the Vektor radioactive waste storage facility and the cooling pond of the ChNPP of the additional air pollution due to the rise of radioactive dust during the passage of a tornado in the ChEZ.

According to Bryuhan et al. (1989), the ChEZ is located in an area of high tornado risk. During the period from 1969 to 2012 seven tornadoes were recorded in the Kyiv region. Taking into account recent observations, a probability of a tornado in the Kyiv region within 1000 km^2 area can be estimated to be 1.5×10^{-2} $year^{-1}$, and within the ChEZ—6×10^{-3} $year^{-1}$. The most intense among observed in the Kyiv region was a tornado in 18 August 1969, which was assigned to be F3 category according the Fujita scale.

After the Chernobyl accident several calculations of possible secondary air pollution as a result of the removal of radioactive dust from the "Shelter" object with a tornado was carried out. They are differs by approaches to the description of the tornado, as well as methods of modeling the further diffusion of radioactive dust after the tornado dissipation.

Beskorovajnyj et al. (1994) found that in the area of the tornado dissipation the ground deposition density can exceed 185 MBq m^{-2} under estimation of the total release from the "Shelter" equals to 4.8×10^{16} Bq. It was concluded that due to a tornado propagation to the territory of Belarus a contamination of nearby areas may exceed pollution during the Chernobyl accident in 1986.

In SIP (2004), it was assumed that all fuel dust (500 kg) was lifted to the tornado base in a mesocyclone parent cloud at an altitude of about 1000 m for 1 min and further moved in a compact form within the cloud with a mesocyclone. The resulting radioactive cloud moves over the atmospheric boundary layer. Ninety percentage of the dust in the parent cloud is washed out by rainfall after the tornado and mesocyclone collapse, and the rest of the fuel dust rises over the troposphere and independently disperses after the collapse of the parent cloud. Under the adopted assumptions the radioactive fallout zone area is 16 km^2, the trace length is 10 km and a trace width is about 1.5 km. The central part of the radioactive fallout zone is located at a distance of 15–45 km from the "Shelter". The mass deposition density has been estimated to be 28 mg m^{-2}.

Bogorad et al. (2006) for the same scenario of the "Shelter" destruction using the atmospheric transport models CALPUFF obtained that the maximum value near the source of contamination could be 2.5 mg m^{-2} of fuel dust, and the estimated density of the fuel dust deposition at a distance of 100 km from the "Shelter" is about 0.001 mg m^{-2}. For numerical simulation the weather conditions related to a tornado formation on 11 June 2001 in the Zhytomyr region, Ukraine were chosen.

Rybalka et al. (2013) analyzed consequences of tornado passing over the Vector facility site for long-term storage and disposal of radioactive waste in the ChEZ. They estimated that about 0.2% of the radioactive waste total activity may be drawn into the funnel of tornado. For disposal of radioactive waste with estimated activity of 2.5×10^{16} Bq at the Vector site the dose of potential exposure near the site have been calculated. At a distance of 12 km from site it was estimated as 6.9 mSv per year exceeding the criteria of admissible dose of 1 mSv per year.

In the framework of the Ukrainian National Program of ChNPP decommissioning a partial drainage of the cooling pond (CP) of the ChNPP has been carried out, that is, bringing the territory occupied by the water body to its natural state, determined by the water level in the Pripyat River. Talerko and Garger (2013) evaluated the effects of a possible rise of radioactive aerosols in the air from the dried part of the cooling pond bottom caused by a tornado passing over it, including transboundary atmospheric transport of radionuclides to the territory of neighboring countries (Belarus and Russia). They performed calculations using a set of models, including the conceptual model of a tornado of Weber and Hunter (1996), the mesoscale atmospheric diffusion model of pollutant transport LEDI and dosimetric models. The activity raised by a tornado vortex of intensity F3 from the CP drained part was assessed as 13.9 TBq ^{137}Cs, 2.47 TBq ^{90}Sr, 0.0248 TBq Pu. Depending on the chosen meteorological scenario an maximal additional density deposition on the territory of Belarus (in Polesye State Radiation Ecological Reserve) was assessed as 35 kBq m^{-2} ^{137}Cs, 6 kBq m^{-2} ^{90}Sr and 60 Bq m^{-2} Pu, whereas in the nearest settlements of Belarus and Russia—5–9 kBq m^{-2} ^{137}Cs, 1.0–1.7 kBq m^{-2} ^{90}Sr and 10–17 Bq m^{-2} Pu, that is, it isn't exceed 1–2% of the actual one. The radiation dose to the people of Belarus will not exceed 1 μSv, Russia—10^{-2} μSv. Thus, in the event of a tornado over the ChNPP cooling pond a conservative estimation of total dose for residents of nearby towns of Belarus three orders of magnitude, and Russia five orders of magnitude less than the limit of effective dose for the population of 1 mSv per year established by the standards of radiation safety in these countries. The people of nearby settlements of Belarus and Russia can obtain the same radiation dose in the case of a F3 category tornado passage over the territory of the ChEZ with an average density of soil contamination equals to 1700 kBq m^{-2} ^{137}Cs, 300 kBq m^{-2} ^{90}Sr and 3 kBq m^{-2} Pu. Actual contamination of large part of the ChEZ in Ukraine and Polesye reserve in Belarus considerably exceeds these values.

2.7 Conclusions

This chapter does not claim to cover the problem completely. Predominantly, it presents extensive experimental results and theoretical estimations obtained under operational conditions associated with the elimination of consequences of the Chernobyl accident, carried out generally within the 30-km exclusion zone of the ChNPP since May 1986 up to now.

The analysis of the results of re-entrainment measurements is a rather complicated task due to the features of the Chernobyl zone. First, the measured air activity concentration is determined from the land surface, which is a constant source of radioactive aerosols with time-dependent intensity depending on the season, a characteristic of the underlying surface, and meteorological conditions. Second, radionuclide deposition after the accident consisted of radioactive aerosols of different sizes, nuclide composition, and chemical properties (including "hot" particles). Third, the activity concentration values of radionuclides in air within the zone were influenced not only by resuspension in a traditional sense but also by the antropogenic activities in the Chernobyl zone (including construction of the New Safe Confinement above Unit 4 of the ChNPP).

It should be noted that the resuspension factor enables the evaluation of the activity concentration of radionuclides at the point where it is necessary to assess the inhalation dose. The resuspension rate, which characterizes the emission intensity of radioactive substances, is needed to assess the activity concentration and deposition of radioactivity over space and time. This allows evaluating both an inhalation dose and an ingestion dose due to the consumption of contaminated food.

The mesoscale meteorological phenomena, such as passage of cold fronts with squalls, have been important; during dry seasons, there were also fires in contaminated forests and grasslands. Though these phenomena are short-term, they may cause considerable transport of radioactive substances over long distances above relatively clean territories.

In the investigations under field conditions, primary attention was given to derivation of the integral characteristics of the re-entrainment process of radioactive particles used in practice, in particular, the generalized data on the resuspension factor and the resuspension rate.

The special feature of this chapter is the generalization of data on the distribution of the radionuclide activity concentration with respect to particle size under various conditions of aerosol re-entrainment in the atmospheric surface layer.

Measurement with impactors has allowed estimation of the contribution of particles of various size ranges to the total activity concentration value.

Results of measurements obtained near ChNPP in the postaccidental period are presented for the number concentration of aerosol "hot" particles, the beta-activity distribution on "hot" particles sizes, and their contribution to the total beta-activity concentration. The estimation of the leaching rate and the size-dependent dissolution rate of "hot" particles in a simulator of lung fluid are also given.

The main approaches to the modeling of radionuclide re-entrainment and consequent redistribution in space and over time are reviewed. These approaches are necessary for estimating human inhalation doses and including resuspension factor models, a mass-loading approach, and mathematical models of radionuclide atmospheric transport and dispersion on the basis of the semiempirical equation of turbulent diffusion.

The verification of several models based on the resuspension factor was made on the basis of measurement results within the ChEZ. Besides wind resuspension, other processes in the environment that result in the emission of radionuclides over contaminated territories are briefly reviewed, including forest fires, dust storms, and tornadoes.

References

Ageev VA, Vyrichek SL, Klyuchnikov AA, Lashko AP, Levshin EB, Odintsov AA, Lashko TN (1998) Estimate of ^{242m}Am content in fuel from the no. 4 unit of the Chernobyl nuclear power plant. At Energy 84(4):278–282

Ageev VA, Odintsov OO, Sajeniouk AD (2003) Routine radiochemical method determination of ^{90}Sr, ^{238}Pu, $^{239+240}$Pu, ^{241}Am and ^{244}Cm in the different environmental samples. In: Book of abstracts, sixth international conference on methods and applications of radioanalytical chemistry – MARC-VI, 7–11 Apr 2003, Kailua-Kona, HI, pp 64–65

Amiro BD, Sheppard SC, Johnston FL, Evenden WG, Harris DR (1996) Burning radionuclide question: what happens to iodine, cesium and chlorine in biomass fires. Sci Total Environ 187:93–103

Ansoborlo E, Guilmette RA, Hoover MD, Chazel V, Houpert P, Henge-Napoli MH (1998) Application of in vitro dissolutions testes to different uranium compounds and comparison with in vivo data. Radiat Prot Dosimetry 79:33–37

Ansoborlo E, Henge-Napoli MH, Chazel V, Gibert R, Guilmette RA (1999) Review and critical analysis of available in vitro tests. Health Phys 77(6):638–645

Anspaugh LR, Shinn JH, Phelps PL, Kennedy NC (1975) Resuspension and redistribution of plutonium in soils. Health Phys 29:571–582

Begichev SN, Borovoy AA, Burlakov EV, Gavrilov SL, Dovbenko AA, Levina LA, Markushev VM, Marchenko AE, Stroganov AA, Tataurov AL (1990) Fuel of 4 unit reactor of the Chernobyl NPP, IAE-5268/3. Kurchatov Institute of Atomic Energy, Moscow

Belyaev SP, Makhonko KP, Mashkov ST (1967) Gauze cone for mass measurements of the radioactive dust concentration in the atmosphere. Trudy IPG 8. (in Russian)

Belyaev SP, Surnin VA (1991) Estimates of radioactive dust resuspension in Chernobyl in June 1986. Trudy IEM 20(153):133–140. (in Russian)

Berner A, Lürzer C (1980) Mass size distributions of traffic aerosol at Vienna. J Phys Chem 84:2079–2083

Beskorovajnyj VP, Kotovich VV, Molodykh VG, Skurat VV, Stankevich LA, Sharovarov GA (1994) Radiation consequences of collapse of structural elements of the sarcophagus. In: Sarcophagus safety '94, the state of the Chernobyl Nuclear Power Plant Unit 4, proceedings of international symposium Zeleny Mys, Chernobyl, Ukraine, 14–18 Mar 1994 (in Russian)

Besnus F, Peres JM, Guillou P, Kashparov V, Gordeev S, Mironov V, Knatko V, Bondar J, Kudrjashov V, Sokolik G, Leonova S, Aragon A, Espinosa A (1997) Contamination characteristics of podzols from districts of Ukraine, Belarus and Russia strongly affected by the Chernobyl accident. Report EUR 16912 EN. Office for Official Publications of the European Communities, Luxembourg

Birchall A, Puncher M, James AC, Marsh JW, Jarvis NS, Peace MS, Davis K, King DJ (2003) IMBA-EXPERT: internal dosimetry made simple. Radiat Prot Dosimetry 105:421–424

Blackford D, Quant F, Sem G (1987) An improved aerodynamic particle size analyzer. In: Proceedings of Fine Particle Society annual meeting, Boston, MA, Jul 1987

Bogatov SA (2000) Estimation of inventory and identification of the dust pollution properties in the roof space of the Shelter. National Academy of Sciences of Ukraine, ISTC Ukrytiye, 00–2, Chernobyl. (in Russian)

Bogatov SA, Borovoy AA, Dubasov UV, Lomonosov VV (1990) Form and characteristics of the fuel emission for the accident of Chernobyl NPP. At Energy 69:36–40

Bogorad VI, Zheleznyak MI, Kovalets IV, Kushchan A, Litvinskaya TV, Slepchenko AY (2006) Quantitative assessment of radioactive fallout caused by the potential destruction of the New Safe Confinement under the influence tornado class F 30. Yadernaya i Radiacionnaya Bezopasnost 1:28–34. (in Russian)

Bondarenko OA, Garger EK, Giriy VA et al (1993) The spatial and temporal course of concentration and deposition of radionuclides in the 30 km zone of the Chernobyl nuclear power plant. In: Izrael YUA (ed) Radiation aspects of the Chernobyl accident, proceedings of the 1st all-union conference, Obninsk, June 1988, vol 1. Hydrometeoizdat, Moscow, p 1. (in Russian)

Borovoy AA, Gagarinskiy AY (2001) Emission of radionuclides from the destroyed unit of the Chernobyl Nuclear Power Plant. At Energy 90:153–161

Bryuhan FF, Lyakhov ME, Pogrebnyak VN (1989) Tornado dangerous areas in the USSR and placement of nuclear power plants. Izvest Akad Nauk Seriya Geograficheskaya 1:40–48. (in Russian)

Byram GM (1959) Combustion of forest fuels. In: Davis KP (ed) Forest fire: control and use. McGraw-Hill, New York, NY, pp 61–89

Bunzl K, Schimmack W, Krouglov SV, Alexakhin RM (1995) Changes with time in the migration of radiocesium in the soil, as observed near Chernobyl and in Germany, 1986–1994. Sci Total Environ 175:49–56

Byzova NL, Ivanov VN, Garger EK (1989) Turbulence in the boundary layer of the atmosphere. Gidrometeoizdat, Leningrad. (in Russian)

Byzova NL, Garger EK, Ivanov VN (1991) Experimental investigations of atmospheric diffusion and calculations of pollution dispersion. Gidrometeoizdat, Leningrad. (in Russian)

Chepil WS (1951) Properties of soil which influence wind erosion. 4. State of dry aggregate structure. Soil Sci 72:387–401

Dennis NA, Blauer HM, Kent JE (1982) Dissolution fractions and half-times of single source yellowcake in simulated lung fluids. Health Phys 42:469–477

Dubasov YV, Savonenkov VD, Smirnova EA (1996) Ordering of radioactive products of the Chernobyl NPP disaster. Radiochemistry 38(2):101–116. (in Russian)

Dusha-Gudym SI (2005) Transport of radioactive materials by wildland fires in the Chernobyl accident zone: how to address the problem. Int Forest Fire News 32:119–125

ECP1 (1992a) Technical report 1991/92. IPSN report

ECP1 (1992b) Technical report 1991/92. UAAS report

ECP1 (1996) Contamination of surfaces by resuspended material. Final Report EUR 16527

Eidson AF, Griffith WC Jr (1984) Techniques for yellow cake dissolution studies in vitro and their use in bioassay interpretation. Health Phys 46(1):151–163

Eidson AF, Mewhinney JA (1983) In vitro dissolution of respirable aerosols of industrial uranium and plutonium mixed-oxide nuclear fuels. Health Phys 45(6):1023–1037

Ermilov AP, Ziborov AM (1993) Radionuclide ratios in the fuel component of fallouts in the near Chernobyl NPP zone. Radiat Risk 3:134–138

Evangeliou N, Balkanski Y, Cozic A, Hao WM, Møller AP (2014) Wildfires in Chernobyl-contaminated forests and risks to the population and the environment: a new nuclear disaster about to happen. Environ Int 73:346–358

Evangeliou N, Balkanski Y, Cozic A, Hao WM, Mouillot F, Thonicke K, Paugam R, Zibtsev S, Mousseau TA, Wang R, Poulter B, Petkov A, Yue C, Cadule P, Koffi B, Kaiser JW, Møller AP (2015) Fire evolution in the radioactive forests of Ukraine and Belarus: future risks for the population and the environment. Ecol Monogr 85(1):49–72

Evangeliou N, Zibtsev S, Myroniuk V, Zhurba M, Hamburger T, Stohl A, Balkanski Y, Paugam R, Mousseau TA, Møller AP, Kireev SI (2016) Resuspension and atmospheric transport of radionuclides due to wildfires near the Chernobyl Nuclear Power Plant in 2015: an impact assessment. Sci Rep 6:26062. https://doi.org/10.1038/srep26062

Frank G, Kashparov V, Protsak V, Tschiersch J (1996) Comparison measurements of a Russian standard aerosol impactor wirh several Western standard aerosol insruments. J Aerosol Sci 27:477–486

Gamble JL (1967) Chemical anatomy, physiology and pathology of extracellular fluid. Harvard University Press, Boston, MA

Garger EK (ed) (1987) Resuspension and diffusion of radionuclides in the area of the Chernobyl accident. Report. IEM (Institute of Experimental Meteorology Typhoon), Obninsk. (in Russian)

Garger EK (1994) Air concentrations of radionuclides in the vicinity of Chernobyl and effects of resuspension. J Aerosol Sci 25:745–753

Garger EK (2008) Reentrainment of radioactive aerosol in the surface layer of the atmosphere. Institute for the Safety Problems of Nuclear Power Plants, Chornobyl. (in Russian)

Garger EK, Zhukov GP, Lukoyanov NF (1988) Study of pollutant dispersion from low sources in the presence of a single obstacle. Trudy IEM 46(136):106–114. (in Russian)

Garger EK, Zhukov GP, Sedunov YS (1990) Estimation of radionuclides resuspension parameters in the Chernobyl nuclear power plant area. Meteorol Hydrol 1:5–10. (in Russian)

Garger EK, Gavrilov VP, Zhukov GP (1992) Estimation of the secondary contamination by resuspension within the 30 km zone of the Chernobyl Nuclear Power Plant and its comparison with measured data. In: Schwartz SE, Slinn WGN (eds) Precipitation Scavenging and Atmospheric-Surface Exchange, Proceedings of the Fifth International Conference, Richland, WA, 15–19 July 1991, vol 3. Hemisphere Publishing Corp., Washington, DC, pp 1592–1603

Garger EK, Kashpur VA, Gurgula BI, Paretzke HG, Tschiersch J (1994) Statistical characteristics of the activity concentration in the surface layer of the atmosphere in the 30-km zone of Chernobyl. J Aerosol Sci 25(5):767–777

Garger EK, Anspaugh RLR, Shinn JH, Hoffman FO (1995) A test of resuspension-factor models against Chernobyl data. In: Proceedings of an international symposium on environmental impact of radioactive releases. IAEA, Vienna, pp 369–376

Garger EK, Kashpur VA, Belov G, Demchuk V, Tschiersch J, Wagenpfeil F, Paretzke HG, Besnus F, Hollander W, Martinez-Serrano J, Vintersved I (1997a) Measurement of resuspended aerosol in the Chernobyl area. Part I: discussion of instrumentation and estimation of measurement uncertainty. Radiat Environ Biophys 36:139–148

Garger EK, Hoffman FO, Thiessen KM (1997b) Uncertainty of the long-term resuspension factor. Atmos Environ 31:1647–1656

Garger EK, Kashpur VA, Paretzke HG, Tschiersch J (1998a) Measurement of resuspended aerosol in the Chernobyl area. Part II: size distribution of radioactive particles. Radiat Environ Biophys 36:275–283

Garger EK, Paretzke HG, Tschiersch J (1998b) Measurement of resuspended aerosol in the Chernobyl area. Part III: size distribution and dry deposition velocity of radioactive particles during anthropogenic enhanced resuspension. Radiat Environ Biophys 37:201–208

Garger EK, Kashpur VA (2000) Investigation of the object Ukrytie as a radioactive aerosol source in the surface layer of the atmosphere. Regulation of the control of radioactive aerosol releases from the object Ukrytie, report of IAB UAAS, Vols. 1–3 (in Russian)
Garger EK, Kashpur VA, Korneev AA, Kurochkin AA (2002) Results of studies of radioactive aerosol release from the 'Shelter' object. Probl Chernobyl 10(2):60–71. (in Russian)
Garger EK, Kashpur VA, Sazhenyuk AD, Skoryak GG, Gora AD, Kurochkin AA (2004a) Aerosol characteristics of disorganized releases from the shelter object. Problems of Nuclear Power Plants' Safety and of Chernobyl 1:125–135. (in Russian)
Garger EK, Kashpur VA, Skoryak GG, Gora AD, Kurochkin AA, Lisnichenko V (2004b) Aerosol radioactivity and disperse structure at the Chernobyl NPP during the period of forest fires. Agroecol J 3:6–12. (in Russian)
Garger EK, Sazhenyuk AD, Odintzov AA, Paretzke HG, Roth P, Tschiersch J (2004c) Solubility of airborn radioactive fuel particles from the Chernobyl reactor and implication to dose. Radiat Environ Biophys 43:43–49
Garger EK, Kashpur VA, Li WB, Tschiersch J (2006) Radioactive aerosols released from the Chernobyl Shelter into the immediate environment. Radiat Environ Biophys 45:105–114
Garger EK, Kuzmenko YI, Sickinger S, Tschiersch J (2012) Prediction of the ^{137}Cs activity concentration in the atmospheric surface layer of the Chernobyl exclusion zone. J Environ Radioact 110:53–58
Garland JA (1982) Resuspension of particulate material from grass. Experimental programme 1979–1980. HMSO, AERE-R10106, London
Garland JA, Pattenden NJ, Playford K (1992) Resuspension following Chernobyl. In: Modelling of resuspension, seasonality and losses during food processing. First report of the VAMP Terrestrial Working Group, International Atomic Energy Agency, IAEA-TECDOC-647. IAEA, Vienna
Garland JA, Pomeroy IR (1994) Resuspension of fall-out material following the Chernobyl accident. J Aerosol Sci 25(5):793–806
Gaziev JI, Kabanov YI (1993) Study of contamination of the surface layer of the atmosphere by hot particles and inhalation fraction of the aerosol component of products of the accident on the Chernobyl NPP. In: Izrael YA (ed) Radiation aspects of the chernobyl accident, proceedings of the 1st all-union conference, Obninsk, June 1988, vol 1. Hydrometeoizdat, Moscow, pp 104–107. (in Russian)
Gaziev JI, Nazarov LE, Lachikhin AV, Valetova NK (1993) Investigation of physics characteristics of the radioactive gas-aerosol products of the accident of Chernobyl NPP and estimations of power of technogenic emission of these products in the atmosphere. In: Izrael YA (ed) Radiation aspects of the chernobyl accident, proceedings of the 1st all-union conference, Obninsk, June 1988, vol 1. Hydrometeoizdat, Moscow, pp 98–103. (in Russian)
Goldammer JG, Kashparov V, Zibtsev S, Robinson S (2015) Best practices to combat wildfires in contaminated areas and recommendations on firemen safety under fires on the radioactive contaminated territories. Global Fire Monitoring Center (GFMC), Freiburg. (in Russian)
Grebenkov A, Baxter LL, Fogh CL, Pleshchenkov IG, Roed J, Solovjev VN (2002) Formation and release of radioactive aerosols during combustion of contaminated biomass. In: Proceedings of international conference on radioactivity in the environment, Monaco, 1–5 Sep 2002, 99 493–496
Grishin AM (1992) Mathematical modeling of forest fires and new ways of fire-fighting. Nauka, Novosibirsk
Gudihy RG, Finch GL, Newton GJ, Hahn FF, Mewhinney JJ, Rothenberg SJ, Powers DA (1989) Characteristics of radioactive particles released from the Chernobyl nuclear reactor. Environ Sci Technol 23:89–95
Hall D, Reed J (1989) The time dependence of the resuspension of particles. J Aerosol Sci 20:839–842
Hao WM, Bondarenko OO, Zibtsev S, Hutton D (2009) Vegetation fires, smoke emissions, and dispersion of radionuclides in the Chernobyl exclusion zone. Dev Environ Sci 8:265–275

Healy JW (1974) A proposed interium standard for plutonium in soils, USAEC report LA-5483-MS. Los Alamos Scientific Laboratory, NTIS, Los Alamos, NM
Healy JW (1980) Review of resuspension models. In: Hanson WC (ed) Transuranic elements in the environment, DOE/TIC-22800. US Department of Energy, Washington, DC, pp 209–235
Hohl A, Niccolai A, Oliver C, Melnychuk D, Zibtsev S, Goldammer JG, Petrenko M, Gulidov V (2012) The human health effects of radioactive smoke from a catastrophic wildfire in the Chernobyl exclusion zone: a worst case scenario. J Earth Bioresour Life Qual 1:1–34
Hollander W (1994) Resuspension factors of ^{137}Cs in Hannover after the Chernobyl accident. J Aerosol Sci 25:789–792
Hollander W, Dunkhorst W, Pohlmann G (1989) A sampler for total suspended particulates with size resolution and high sampling efficiency for large particles. Part Syst Charact 6:74–80
Hollander W, Garger E (eds) (1996) Contamination of surfaces by resuspended material. Experimental collaboration project No 1, Final report, EUR 16527. Office for Official Publications of the European Communities, Luxembourg
Horn H-G, Bonka H, Maqua M (1987) Measured particle bound activity size distribution, deposition velocity, and activity concentration in rainwater after the Chernobyl accident. J Aerosol Sci 18:681–684
Horrill AD, Kennedy VH, Paterson IS, McGowan GM (1995) The effect of heather burning on the transfer of radiocaesium to smoke and the solubility of radiocaesium associated with different types of heather ash. J Environ Radioact 29:1–10
Hötzl H, Rosner G, Wincler R (1989) Long-term behaviour of Chernobyl fallout in air and precipitation. J Environ Radioact 10:157–171
Hötzl H, Rosner G, Wincler R (1992) Sources of present Chernobyl derived caesium concentrations in surface air and deposition samples. Sci Total Environ 119:231–242
ICRP (1983) Radionuclide transformations – energy and intensity of emissions. Ann ICRP 11–13. No. 38. ICRP Publication, London
ICRP (1990) Age-dependent doses to members of the public from intake of radionuclides part 1. Ann ICRP 20 (2). No. 56. ICRP Publication, London
ICRP (1993) Age-dependent doses to members of the public from intake of radionuclides part 2 ingestion dose coefficients. Ann ICRP 23 (3–4). No. 67. ICRP Publication, London
ICRP (1994) Human respiratory tract model for radiological protection. Ann ICRP 24 (1–3), no. 66. ICRP Publication, London
Izrael YA (ed) (1990) Chernobyl: radioactive pollution of the environment. Hydrometeoizdat, Leningrad
Jost DT, Gäggeler HW, Baltensperger U, Zinder B, Haller P (1986) Chernobyl fallout in size-fractionated aerosol. Nature 324:22–23
Kashparov VA, Lundin SM, Kadygrib AM, Protsak VP, Levtchuk SE, Yoschenko VI, Kashpur VA, Talerko NN (2000) Forest fires in the territory contaminated as a result of the Chernobyl accident: radioactive aerosol resuspension and exposure of fire-fighters. J Environ Radioact 51:281–298
Kashparov VA, Lundin SM, Khomutinin YV, Kaminsky SP, Levchuk SE, Protsak VP, Kadygrib AM, Zvarich SI, Yoschenko VI, Tschiersch J (2001) Soil contamination with ^{90}Sr in the near zone of the Chernobyl accident. J Environ Radioact 56:285–298
Kashparov VA, Zhurba MA, Kireev SI, Zibtsev SV, Myronyuk VV (2015) Evaluation of expected exposure doses for fire-fighting participants in the Chernobyl exclusion zone in April 2015. Yaderna Fizyka ta Energetyka 16:399–407. (in Russian)
Kulan A (2006) Seasonal ^{7}Be and ^{137}Cs activities in surface air before and after the Chernobyl event. J Environ Radioact 90:140–150
Kutkov VA, Kamarinskaya OI (1996) In vitro solubility of Chernobyl nuclear fuel aerosol with respect to collective behaviour of its radionuclides. In: IRPA 9 international congress on radiation protection, Vol 2, pp 445–447
Linsley GS (1978) Resuspension of the transuranium elements: a review of existing data. National Radiological Protection Board, Harwell

Lipinskiy VM, Osadchiy VI, Babichenko VM (eds) (2006) Natural meteorological phenomena in Ukraine over the past twenty years (1986–2005). Nika-Center, Kyiv. (in Ukrainian)
Livens FR, Baxter MS (1988) Particle size and radionuclide levels in some West Cumbrian soils. Sci Total Environ 70:1–17
Lomb NR (1976) Least-squary frequency analysis of unequally spaced data. Astrophys Space Sci 39:447–462
Lujaniene G, Ogorodnikov B, Budyka A, Skitovich V, Lujanas V (1997) An investigation of changes in radionuclide carrier properties. J Environ Radioact 35(1):71–90
Lujaniene G, Šapolaite J, Remeikis V, Lujanas V, Jermolajev A, Aninkevicius V (2006) Cesium, americium and plutonium isotopes in ground level air of Vilnius. Czech J Phys 56(D):D55–D61
Lujaniene G, Aninkevicius V, Lujanas V (2009) Artificial radionuclides in the atmosphere over Lithuania. J Environ Radioact 100:108–119
Lukoyanov NF, Naydenov AB, Meshkova VG (1994) Assessment of ^{137}Cs contamination of near-surface atmosphere and its distribution in the soil and the atmosphere above a cultivated field. Trudy IEM 57(159):37–50. (in Russian)
Makhonko KP (1984) Effective rate of wind pickup of dust from the ground. Meteorol Gidrol 2:105–107
Makhonko KP (1992) Wind uplift of radioactive dust from the ground. At Energy 72(5):465–472
Makhonko KP, Robotnova FA (1982) Resuspension and radioactive fallout from soil surface and particulate contamination of vegetative cover. Pure Appl Geophys 120:54–66
Mandel J, Beezley JD, Kochanski AK (2011) Coupled atmosphere-wildland fire modeling with WRF 3.3 and SFIRE 2011. Geosci Model Dev 4:591–610
Mercer TT (1967) On the role of particle size in the dissolution of lung burdens. Health Phys 13:1211–1221
Meshalkin GS, Arkhipov NP, Arkhipov AN, Afonin SB, Bakin RI, Vasilchenko DL, Ermakov AI, Ivanov YuP, Kovalev AV, Maguskin BB, Sukhoruchkin AK (1992) Water and wind migration of radionuclides in the territory of the Chernobyl exclusion zone. In: Proceedings of 3rd all-union scientific and technical conference on the basis of liquidation of the Chernobyl accident consequences, Vol 2, Chernobyl, pp 225–235 (in Russian)
Miglio JJ, Muggenburg BA, Brooks AL (1977) A rapid method for determining the relative solubility of plutonium aerosols. Health Phys 33:449–457
Naydenov AV, Lukoyanov NF (1994) Experimental evaluation of vertical flow and radioactive dust resuspension intensity over contaminated cultivated field. Trudy IEM 57(159):17–27. (in Russian)
Nicholson KW (1988) A review of particle resuspension. Atmos Environ 12:2639–2651
Nicholson KW, Branson JR (1990) Factors affecting resuspension by road traffic. Sci Total Environ 93:349–358
Ogorodnikov BI (2002) Influence of high temperatures on behaviour of ^{137}Cs in the environment. In: Proceedings of international science workshop on radioecology of Chernobyl zone, 18–19 Sep 2002, Slavutych, pp 45–46 (in Russian)
Ogorodnikov BI (2011) Dust storm in the territory of Ukraine and Belarus contaminated with radionuclides after the Chernobyl accident. Meteorol Hydrol 9:64–77. (In Russian)
Paatero J, Vesterbacka K, Makkonen U, Kyllonen K, Hellen H, Hatakka J, Anttila P (2009) Resuspension of radionuclides into the atmosphere due to forest fires. J Radioanal Nucl Chem 282:473–476
Reineking A, Becker KH, Porstendörfer J, Wicke A (1987) Air activity concentrations and particle size distributions of the Chernobyl aerosol. Radiat Prot Dosimetry 19:159–163
Rosner G, Winkler R (2001) Long-term variation (1986–1998) of post-Chernobyl ^{90}Sr, ^{137}Cs, ^{238}Pu and $^{239,240}Pu$ concentrations in air, depositions to ground, resuspension factors and resuspension rates in south Germany. Sci Total Environ 273:11–25
Rybalka N, Mykolaichuk O, Alekseeva Z, Kondratiev S, Nikolaev E (2013) Scenarios of radiological impacts in the long-term safety analysis of radioactive waste disposal at the Vector Site located in the Chernobyl exclusion zone. In: EUROSAFE forum, 4–5 Nov 2013, Cologne

Sehmel GA (1980) Particle resuspension: a review. Atmos Environ 22:2639–2651
Sehmel GA (1984) Deposition and resuspension: atmospheric science and power production. DOE/TIC-27601 (DE 84005177) 12:533–572
Shinn JH, Homan DN, Gay DD (1982) Plutonium aerosol fluxes and pulmonary exposure rates during resuspension from bare soils near a chemical separation facility, V/2 proceeding of the fourth international conference, Santa Monica, CA, 29 Nov to 3 Dec 1982, pp 1131–1143
SIP (Shelter Implementation Plan) (2004) Assessment of the radiological consequences of the loss of integrity. Research report. SIP-I-SC-21-310-SAR-005-01 (in Russian)
Skitovich VI, Budyka AK, Ogorodnikov VI (1993) Results of two-year observations of radioactive particles dimensions within the ChNPP 30-km zone. In: Izrael YuA (ed) Radiation aspects of the Chernobyl accident, vol 1. Proceedings of the 1st all-union conference radiation aspects of the Chernobyl accident, Obninsk, Jun 1988, , pp 115−121 (in Russian)
Slade DH (1968) Meteorology and atomic energy. TID-24190, US Atomic Energy Commission, Division of Technical Information, Oak Ridge, TN
Slinn WGN (1978) Parametrizations for resuspension and for wet and dry deposition of particles and gases for use in radiation dose calculations. Nucl Safety 19(2):205–219
Smith WJ, Whicker FW, Meyer HR (1982) Review and categorization of saltation, suspension, and resuspension models. Nucl Safety 23:685–699
Talerko NN (1990) Calculation of radioactive admixture ascent from Chernobyl NPP accidental unit. Meteorol Hydrol 10:39–46
Talerko NN, Garger EK (2013) Prognostic assessment of radionuclides transboundary transport due to a tornado over the Chernobyl NPP cooling pond. Problems of nuclear power plants' safety and of Chernobyl. Probl Chernobyl 20:85–93. (in Russian)
Tschiersch J, Georgi B (1987) Chernobyl fallout size distribution in urban areas. J Aerosol Sci 18:689–692
USAEC (1974) Reactor safety study – an assessment of accident risks in US commercial nuclear power plants; draft report WASH-1400, Appendix VI. USAEC, Washington, DC
USAEC (1975) Reactor safety study – an assessment of accident risks in US commercial nuclear power plants; final report WASH-1400, Appendix VI. USAEC, Washington, DC
Vintersved I, Arntsing R, Bjurman B, de Geer L-E, Jakobsson S (1991) Resuspension of radioactive caesium from Chernobyl accident. In: Moberg L (ed) The Chernobyl fallout in Sweden. Swedish Radiation Protection Institute, Stockholm, pp 85–106
Vintersved I (1994) Intercomparison of large stationary air samplers. In: Dahlgaard H (ed) Nordic radioecology. The transfer of radionuclides through Nordic ecosystems to man. Elsevier, Amsterdam
Viswanathan GM, Buldyrev SV, Garger EK, Kashpur VA, Lucena LS, Shlyakhter A, Stanley HE, Tschiersch J (2000) Quantifying nonstationary radioactivity concentration fluctuations near Chernobyl: a complete statistical description. Phys Rev E 62(3):4389–4392
Wagenpfeil F, Härtl T, Tschiersch J (1994) Size-fractionating sampler for giant particles. J Aerosol Sci 25:111–112
Wagenpfeil F, Tschiersch J (2001) Resuspension of coarse fuel particles in the Chernobyl area. J Environ Radioact 52:5–16
Wang X, Oenema O, Hoogmoed WB, Perdok UD, Cai D (2006) Dust storm erosion and its impact on soil carbon and nitrogen losses in northern China. Catena 66:221–227
Weber AH, Hunter CH (1996) Estimating dispersion from a tornado vortex and mesocyclone. WSRC-TR-94-0386, Westinghouse Savannah River Company, Aiken, SC
Wotawa G, De Geer L-E, Becker A, D'Amours R, Jean M, Servranckx R, Ungar K (2006) Inter- and intra-continental transport of radioactive cesium released by boreal forest fires. Geophys Res Lett 33:L12806. https://doi.org/10.1029/2006GL026206
Yoschenko VI, Kashparov VA, Protsak VP, Lundin SM, Levchuk SE, Kadygrib AM, Zvarich SI, Khomutinin YV, Maloshtan IM, Lanshin VP, Kovtun MV, Tschiersch J (2006a) Resuspension and redistribution of radionuclides during grassland and forest fires in the Chernobyl exclusion zone: part I. Fire experiments. J Environ Radioact 86:143–163

Yoschenko VI, Kashparov VA, Levchuk SE, Glukhovskiy AS, Khomutinin YV, Protsak VP, Lundin SM, Tschiersch J (2006b) Resuspension and redistribution of radionuclides during grassland and forest fires in the Chernobyl exclusion zone: part II. Modeling the transport processes. J Environ Radioact 86:260–278

Zilinitkevich SS (1970) Dynamics of the atmospheric boundary layer. Gidrometeoizdat, Leningrad. (in Russian)

Part II
Behavior of Chernobyl-Derived Radionuclides in Soil–Water Environment

Chapter 3
Mobility and Bioavailability of the Chernobyl-Derived Radionuclides in Soil–Water Environment: Review

Alexei Konoplev

Abstract The chapter reviews the studies of the behavior of Chernobyl-derived radionuclides in the soil–water environment that have been carried out over more than 30 years after the accident at the Chernobyl nuclear power plant on 26 April 1986, the worst nuclear accident ever. As a result of post-Chernobyl investigations, the role of released fuel particles in the environmental behavior of radionuclides has been revealed. A conceptual model accounting for transformation of radionuclide chemical forms in soils and sediments is outlined and data on key kinetic and equilibrium parameters for this model are presented. Newly introduced after Chernobyl accident, parameters of radiocesium selective sorption by soils and sediments (capacity of frayed-edge sites [FES] and radiocesium interception potential [RIP]) and methodologics for their determination and application are discussed. Another important addressed issue is establishing the exchangeable distribution coefficient K_d^{ex}, which can be estimated on the basis of ion-exchange equilibrium from commonly accepted characteristics of soils and sediments. Development of methods to parameterize radionuclide bioavailability through soil and soil solution characteristics is also examined. The reviewed advances in post-Chernobyl studies help to reduce significantly the uncertainty associated with predicting radionuclide fate and transport in the environment.

Keywords Chernobyl · Radiocesium · Radiostrontium · Speciation · Fuel particles · Transformation · Sorption · Fixation

3.1 Introduction

In terrestrial and aquatic environment, behavior of accidentally released radionuclide is governed by the ratio of its chemical forms in fallout and site-specific environmental characteristics determining the rates of leaching and fixation–remobilization,

---

A. Konoplev (✉)
Institute of Environmental Radioactivity, Fukushima University, Fukushima, Japan

A. Konoplev et al. (eds.), *Behavior of Radionuclides in the Environment II*,
https://doi.org/10.1007/978-981-15-3568-0_3

as well as sorption–desorption of radionuclide (its solid–liquid distribution) (Konoplev et al. 1992; Beresford et al. 2016).

In terms of environmental conditions, the major part of the close-in areas contaminated due to the Chernobyl accident occurs in Pripyat Polesye. By terrain, this is outwash lowland plain. The vegetation is primarily broad leaved and pine woods located on drained terraces of alluvial origin. The climate is moderate continental, with the mean annual precipitation of 500–600 mm. The water table is 8 m at maximum, the groundwater outcropping in Pripyat floodplain depressions, forming lakes. In the Chernobyl region 45% of the territory is occupied by forest, bogs, and water bodies, while 54% are arable lands. The soil cover is represented by soddy weakly podzolic and medium podzolic soils (53% of arable lands), whereas soddy and meadow soils account for 30%, boggy soil—10%, and weakly sodded sands—7% (Konoplev 1998).

3.2 Speciation of Chernobyl-Derived Radionuclides in the Environment and Their Transformation

3.2.1 Fuel Particles: Major Distinguishing Feature of the Chernobyl Release

The speciation of radionuclides released to air from the damaged reactor varied with time after the accident (Konoplev and Bobovnikova 1991; Bobovnikova et al. 1991a, b). In the process of atmospheric transport, particles depositing on the surface were differentiated by size and density. Therefore, the fallout was not uniform and was a function of direction and distance from the ChNPP. All hot particles deposited on the surface after the accident could be attributed to two main types: fuel particles and condensation particles (Loschilov et al. 1991; Kashparov et al. 1996). Fuel particles were mainly particles of released uranium dioxide nuclear fuel containing its decay products at ratio close to irradiated fuel makeup, with some depletion of volatile fissile products of ^{131}I, $^{134,137}Cs$, ^{106}Ru, etc. (Bogatov et al. 1990; Konoplev and Bobovnikova 1991; Victorova and Garger 1991; Sandalls et al. 1993; Salbu et al. 1994, 2018; Kashparov et al. 1996; Salbu 2001; Beresford et al. 2016).

Due to high temperature in the damaged reactor (about 2000 °C), a large amount of volatile fission products (iodine, cesium, ruthenium, etc.) were released to the atmosphere, part of which were precipitated on dust particles—inert carriers in air. The resulting condensation particles were similar by their properties to those formed in the last stage of a nuclear explosion, characterized by surface contamination and relatively low activity concentration as compared to fuel particles (Kashparov et al. 1996). The condensation particles formed as a result of the Chernobyl accident contained 1–2 radionuclides and resembled global fallout following the nuclear weapons tests. For this reason, predicting behavior of these particles could be based on abundant data about the global fallout (Pavlotskaya 1974; Wicker and Schultz 1982). As regards fuel particles, prior to the Chernobyl accident, such particles were released to the environment only during the accidental release in Windscale (UK) in October 1957 and destruction of the satellite Kosmos 954 in 1978 (Sandalls et al. 1993; Salbu et al. 1994). These releases, however, were small scale and the environmental behavior of particles was not well understood. Because of this, the behavior of fuel particles determining the features of the Chernobyl footprint and occurring primarily within 60 km around the ChNPP was a major scientific challenge.

The occurrence of fuel particles in the Chernobyl depositions led to considerable nonuniformity of area radioactive contamination. Nonuniformity of soil contamination was seen over distances from several meters to macroscale (Izrael et al. 1990; Salbu et al. 1994; Ivanov 1997; Kashparov et al. 2018). Particle size and makeup were a function of direction and distance to the reactor, and therefore the behavior of radionuclides in soil and their characteristics determining transfer to plants and aquatic systems varied (Beresford et al. 2016).

Formation of particles at elevated temperatures in airflow was discovered in 1960 (Megaw et al. 1960). When heated to 600–1000 °C, particles with a median diameter of 100–600 μm were formed. Higher temperatures, in the authors' opinion, were conducive to formation of larger particles. This, however, has not been supported by the studies (Kashparov et al. 1996) in which oxidation of uranium fuel at 400 and 900 °C led to formation of particles with lognormal size distribution. The median size of particles was not dependent on temperature, decreased from 10 to 3 μm with increase in the annealing time from 3 to 21 h. The researchers (Kashparov et al. 1996) concluded that the main process of fuel particle formation during the accident was high-temperature oxidation of nuclear fuel.

The diameter of fuel particles in the first years after the accident varied from hundreds to fractions of micrometers (Loschilov et al. 1991; Sandalls et al. 1993; Kashparov et al. 1996). The particle density was found to be as high as 10^5 particles/m^2 within the 30-km zone of ChNPP (Victorova and Garger 1991). With distance from the reactor, the fraction of fuel component declined, and particle size distribution changed (Sandalls et al. 1993).

The presence of fuel particles in the Chernobyl release led to a lower mobility of radionuclides, as compared to the global fallout (Konoplev et al. 1992; Beresford et al. 2016), since fuel particles are insoluble in water. From environmental standpoint, this played a positive role in the early phase after the accident by reducing secondary contamination of water bodies and root transfer of radionuclides from soil to plants (IAEA 2006). With time, however, radionuclides tend to leach from fuel particles. Leaching of radionuclides from fuel particles is a crucial process because it influences long-term changes in mobility and bioavailability of radionuclides in terrestrial and aquatic ecosystems (Bobovnikova et al. 1991a; Avdeev et al. 1990a, b; Smith et al. 2009).

3.2.2 Basic Chemical Forms of Radionuclides and Their Transformation in the Environment

Physicochemical state of radionuclides controls their mobility and bioavailability in the environment. Ratio of different chemical forms is a function of the initial state of radionuclide in depositions and its changes due to transformation processes occurring in the environmental compartments (Konoplev and Bobovnikova 1991; Konoplev et al. 1992; Beresford et al. 2016).

Radionuclides can occur in the environment in the dissolved state and within the solid phase. When in solution, radionuclide can occur as a cation, as a complex compound with organic and inorganic ligands and colloid particles. In the solid phase, radionuclide can occur in the exchangeably sorbed state, i.e., at adsorption–desorption equilibrium with solution; be embedded in insoluble compound or insoluble hot particles (such as fuel particles), or fixed by clay mineral particles (nonexchangeably sorbed) (Konoplev et al. 1988, 1992; Bobovnikova et al. 1991a).

The long-lived radionuclides of radiocesium and radiostrontium, which were of most environmental significance after the Chernobyl accident, occur in solution mainly as cations and virtually do not form stable complex compounds. Conducted laboratory and field experiments (Konoplev et al. 1996; Matsunaga et al. 2004) have shown that the colloid form of these radionuclides, as a rule, does not play a major role in most of environmental conditions.

Therefore, with all the diversity of radiocesium and radiostrontium speciation in soil and water bodies, in terms of radionuclide transport, exchange between environmental compartments, and biological availability, it seems appropriate to distinguish the water-soluble, exchangeable, and nonexchangeable forms.

The ratio of key chemical forms of individual radionuclides is governing their mobility in the environmental compartments. This ratio, in turn, depends on the initial state of radionuclides in depositions and its change due to physicochemical processes occurring in soil or bottom sediments. For separation of different chemical forms, the method of sequential extraction by different chemical agents is usually used. In many works (Hilton et al. 1992; Oughton et al. 1992; Salbu et al. 1994; Askbrant et al. 1996; Tsukada et al. 2008), the technique proposed by Tessier et al. (1979), or its modification, was used to separate chemical forms of heavy metals, including the following extraction stages:

1. Deionized water for separation of water-soluble compounds (solid phase: liquid phase ratio 1:10).
2. 1 N ammonium acetate solution to separate exchangeable form of radionuclides (sorbed by ion-exchange mechanism), phase ratio 1:10.
3. Acetate buffer NaOAc + HOAc at pH = 5 for extracting carbonates.
4. 0.04 M solution of hydroxylamine $NH_2OH{\cdot}HCl$ in 25% acetic acid to extract easily reducible compounds (solid to liquid phase ratio 1:10).

5. Mixture of 30% hydrogen peroxide solution reduced to pH = 2 by nitric acid with 3.2 M ammonium acetate solution in 20% HNO_3 at ratio 3:1 (phase ratio 1:10) to extract radionuclides bound to organics.
6. 7 M nitric acid solution to extract acid soluble fraction of radionuclide (phase ratio 1:10).

Other authors (Pavlotskaya 1974; Konoplev et al. 1988, 2002; Konoplev and Bobovnikova 1991; Bobovnikova et al. 1991a) used the methodology for separating chemical forms of radionuclides in the solid phase of the environmental compartments (atmospheric aerosols, soil, bottom sediments suspension etc.) consisting of four stages:

– Extraction by distilled water at solid phase:water ratio 1:5 for 24 h.
– Extraction of 1 N NH4Ac solution at phase ratio 1:8 for 24 h.
– Extraction by solution 6 N HCl with boiling, phase ratio 1:5.
– Extraction by mixture of nitric and hydrofluoric acid until total decomposition.

In the process of sequential extractions, cations desorbed by ion-exchange mechanism and soluble complex compounds of radionuclides (neutral and anionic forms) pass into the aqueous extract. Concentration of cations in the aqueous extract solution is matching the position of ion-exchange equilibrium governed by soil cation-exchange capacity and concentration of exchangeable ions in solution. The key exchangeable ions in soil are known to be Ca^{2+}, Mg^{2+}, Na^+, K^+, NH_4^+, and H^+. Also extracted by water are colloid forms of radionuclides, i.e., particles of the size smaller than filter pores (ranging from 1 to 100 nm). These particles are normally transported by water flow, yet their biological availability is much lower than that of truly dissolved radionuclides.

Radionuclides extracted by ammonium acetate solution are those adsorbed in soil by ion-exchange mechanism, while nonexchangeable radionuclides, i.e., the forms, which do not normally pass to water in the environment, are extracted by acid. The nonexchangeable form, in turn, includes radionuclides incorporated in poorly soluble compounds, fuel particles, and the so-called irreversibly sorbed states (inclusions in the lattice of minerals, radionuclide-organic compounds with insoluble organic matter of soil, etc.) (Konoplev et al. 1988).

The initial ratio of radionuclide chemical forms in the environmental compartments is determined by speciation in atmospheric deposition. In the global fallout after nuclear tests, the water-soluble and exchangeable forms of ^{137}Cs and ^{90}Sr were predominant, as shown by Pavlotskaya (1974).

During the Kyshtym accident caused by explosion in a tank with high-level radioactive waste, radioactive materials were released in the form of liquid pulp to the height of 1–2 km and formed a plume of liquid and solid aerosols (Alexakhin et al. 2001; Il'in and Gubanov 2001). Data about the speciation of ^{90}Sr and other radionuclides in Kyshtym fallout do not exist. It is known, however, that radionuclides in the released aerosols were present primarily in the form of highly soluble compounds such as nitrates and acetates (Martyushov et al. 1995).

Table 3.1 Comparison of initial ^{90}Sr and ^{137}Cs mobility in fallout of different origin

Type of fallout	Part of mobile forms, %		Reference
	^{90}Sr	^{137}Cs	
Kyshtym	~80–90	~80–90	Pavlotskaya (1974)
Global fallout	>90	>50	Pavlotskaya (1974)
Chernobyl: 30-km zone	11.2	26.8	Konoplev and Bobovnikova (1991)
Western Europe	–	85	Hilton et al. (1992)

The radionuclides of ^{137}Cs and ^{90}Sr deposited on soil surface after the Chernobyl accident occurred mainly in nonexchangeable form, as follows from the analysis of fallout samples collected in April–May 1986 at Chernobyl meteorological station (Konoplev and Bobovnikova 1991; Bobovnikova et al. 1991a, b). This is also suggested by data on radionuclides speciation in soil at the same location in the first year after the accident (Bobovnikova et al. 1991a). Since ^{137}Cs tends to be easily fixed in soil, i.e., transferring to nonexchangeable form, the ratio of exchangeable and nonexchangeable forms in soil does not change much with time as a result of ^{137}Cs leaching from fuel particles.

Since the compounds of Cs isotopes are more volatile than those of Sr, essential part of deposited Cs occurred in mobile form embedded in condensation particles (see Table 3.1), and hence radiocesium was transported over longer distances than strontium. Radiostrontium, on the other hand, largely deposited within the 30-km zone and only a small part of this radionuclide was transported to southern areas of Belarus and Bryansk region (Izrael et al. 1990).

Hilton et al. (1992) analyzed ^{134}Cs and ^{137}Cs chemical forms in fallout in UK at the time of the Chernobyl accident. More than 85% of these isotopes were found to be in water-soluble and exchangeable forms, suggesting condensation origin of depositions on remote parts of the Chernobyl footprint. Unfortunately, there are no data available about radionuclide speciation in Chernobyl fallout at medium range, such as Bryansk region of the Russian Federation or central areas of Belarus. It can be assumed that in these regions the situation was “half-way” between the near zone (Chernobyl) and remote footprint (UK). The proportion of the mobile ^{137}Cs in depositions in Bryansk region can be expected to be 40–60% (Konoplev et al. 1996), which was actually supported by the analysis of chemical forms in soils in the first years after the accident (Konoplev et al. 1988; Konoplev and Bobovnikova 1991; Bobovnikova et al. 1991a). The initial forms of ^{90}Sr and ^{137}Cs in depositions of different types are compared in Table 3.1.

Therefore, when analyzing radionuclide speciation in soils of the Chernobyl-contaminated areas, two factors should be considered:

1. Nonuniformity of radionuclide speciation in the fallout as a function of distance from ChNPP.
2. Variety of soil sorption and fixation ability on different locations.

In the first years after the Chernobyl accident, ^{90}Sr speciation in the 30-km zone soils was notable for high proportion of nonexchangeable forms as compared to global fallout and East Ural Radioactive Trail (Konoplev et al. 1988; Konoplev and Bobovnikova 1991; Bobovnikova et al. 1991a).

A relatively high content of water-soluble ^{90}Sr forms was typical of sandy and boggy soils, which seems to be associated with a weak sorption capacity of sandy soils and high mineralization of aqueous extract in boggy soil (in particular, high concentration of Ca and possible formation of colloids: strontium-organic compounds) (Konoplev et al. 1988, 1992).

For ^{137}Cs the fraction of nonexchangeable form is even higher, being as high as 99% in some cases (Konoplev et al. 1988; Bobovnikova et al. 1991a). A higher proportion of ^{137}Cs nonexchangeable form, as compared to ^{90}Sr, was seen in the global fallout too (Pavlotskaya 1974), which is explained by fixation processes typical of radiocesium, i.e., transfer of exchangeable form to nonexchangeable due to incorporation into the interlayer space of clay mineral crystal lattice (Tamura and Jacobs 1960; Jacobs 1962; Sawhney 1972; Von Reichenbach 1968). Even though the part of ^{137}Cs mobile forms in the fallout was much higher than that of ^{90}Sr, as a result of ^{137}Cs fixation with time the proportion of ^{137}Cs mobile forms becomes much lower than that of ^{90}Sr. What is more, while leaching of ^{90}Sr from fuel particles leads to increase in its mobility in soil, leached ^{137}Cs immediately passes to nonexchangeable form, i.e., time scale of its fixation (several weeks or months) is much shorter than time scale of leaching from fuel particles (Konoplev et al. 1992; Beresford et al. 2016).

Subsequent to fallout on the surface of soil or water bodies, radionuclides become involved in physicochemical and biological processes, which lead to changes in their speciation. The dissolved fraction of radionuclide is subjected to sorption by solid particles, the main mechanism of sorption being ion exchange (Sawhney 1972; Konoplev et al. 1992). Exchangeable form undergoes fixation and transfer to colloid forms or complex compounds with organic and inorganic ligands. Fuel particles disintegrate, the effective surface of particle contact with water increases, and additional amount of radionuclide passes to solution. The main processes of radionuclide chemical forms transformation are shown in Fig. 3.1. In terms of speciation, fixation of radionuclides is transfer of their exchangeable form to nonexchangeable one.

Apart from the differences in radionuclide speciation between Chernobyl fallout and the global fallout, the radionuclides also differ by the pattern of their transformation in soil. The mobility of radionuclides from nuclear weapons testing decreased with time because of fixation by soil components, while in the Chernobyl zone in the first years after the accident the predominant process was leaching of radionuclides from fuel particles, resulting in increased mobility, especially for ^{90}Sr (Konoplev et al. 1992; Kashparov et al. 1999, 2004; Konoplev and Bulgakov 1999).

The main chemical forms of radionuclides shown in Fig. 3.1 differ by mechanisms and rates of migration in environmental compartments. Therefore, when predicting radionuclide transport in the environment and food chains, it is essential to measure, or predict, the ratio of different chemical forms for each radionuclide of

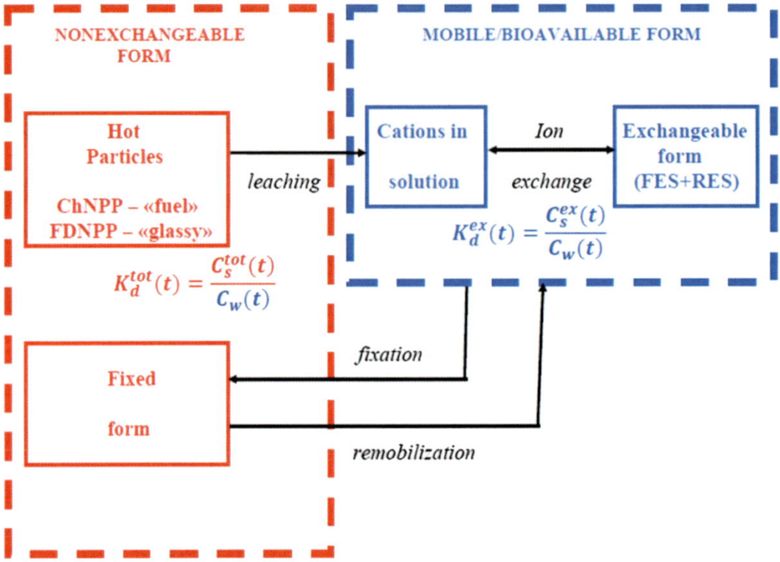

Fig. 3.1 Conceptual model of transformation processes for radiocesium speciation in soils/sediments (modified and updated after Konoplev et al. 1992, 1993; Konoplev and Bulgakov 2000b)

interest. In turn, the key parameters of radionuclide behavior used for migration prediction such as the distribution coefficient (K_d), wash-off coefficients, soil–plant transfer factor, and bioconcentration factors are governed by the speciation ratio to a great degree. Therefore, determination and parameterization of rates of transformation processes shown in Fig. 3.1, specifically leaching from fuel particles, sorption–desorption, fixation and remobilization, was an important task after the Chernobyl accident (Konoplev et al. 1992; Salbu et al. 2004).

3.2.2.1 Kinetics of Radionuclide Leaching from Fuel Particles

In the first years after the accident, radionuclide leaching from fuel particles was the key process with respect to radionuclide chemical form transformation in the environment. Radionuclides are released from fuel particles by three mechanisms: (1) due to diffusion from particle to the surface and passing into solution; (2) by mechanical disintegration of particle and correspondent increase in surface of interphase interaction, and (3) with dissolution of fuel matrix (Konoplev and Bulgakov 1999). Pure uranium dioxide is known to dissolve slowly even in concentrated acids (Vdovenko 1960). In natural conditions, however, four-valent uranium U(IV) is easily oxidized by air oxygen to six-valent U(VI) state, three-oxide of which is better dissolving in aqueous media. Crystals of UO_2, when in water, are quickly covered by oxide film of U_{2+x}, where $0 < x < 1$ and the rate of further oxidation is determined by diffusion of oxygen and surface area of crystal. In addition to oxygen, UO_2 can be oxidized by other oxidizers such as three-valent iron Fe (III) or hydrogen peroxide, molecules of which can be formed in the vicinity of hot particle as a result of water radiolysis. Dissolution of oxidized uranium is facilitated by the presence of inorganic carbonates and sulfates, as well as organic ligands (Vdovenko 1960).

Dissolution of fuel matrix has been confirmed experimentally for the first time by the presence of uranium in the aqueous extract from a soil (chernozem from the vicinity of the ChNPP) that was in contact with hot particles for 290 days (Avdeev et al. 1990a). In that case, (1–2) $\times$ 10^{-2}% of cesium and ruthenium isotopes, (3–4) $\times$ 10^{-3}% of ^{95}Zr and ^{144}Ce, (4–7) $\times$ 10^{-3}% of total α-activity, and (1–4) $\times$ 10^{-3}% of uranium were leached from hot particles. Better passage to solution was seen for radionuclides from graphite particles: 1.7–4.0% for cesium isotopes; 0.6–1.5% for ^{106}Ru, 0.9–1.7% for ^{95}Zr, and 1.2–1.8% for ^{144}Ce. It has also been shown that ^{144}Ce, ^{106}Ru, and to a lesser degree cesium isotopes are leached from high-density fuel particles more slowly as compared to particles with loose structure.

Bogatov et al. (1990) stated that the main mechanism of radionuclide release from fuel particles is dissolution of uranium matrix. In their study, the dissolution rate was estimated to be (0.57–16) $\times$ 10^{-5} g/cm^2 day at contact of particles of different size with solutions modeling aqueous media.

The diffusion release rate for radionuclides was determined by these researchers using the method of layer dissolution of particles (Bogatov et al. 1990). Based on depletion of surface layers the diffusion coefficients were estimated to be close and equal to (2–4) $\times$ 10^{-17} cm^2/s for ^{137}Cs, ^{90}Sr, ^{238,239,240}Pu, ^{241}Am. This value seems to be a slight underestimate, because besides diffusion the most depleted layers are subjected to dissolution in soil and bottom sediments and accordingly the radionuclide concentration gradient in particles is increasing.

The ability of fuel particles, being aggregates of fine crystals of uranium oxides, to break down easily at physical impact was indicated by many authors. Breakdown of particles was due to exposure of soil sample to ultrasound and cooling-heating cycles (Petryaev et al. 1993). The ability of fungi, widespread in the 10-km zone around the Ch NPP, to grow in immediate vicinity of hot particles and cause their breakdown was reported by Borisyuk et al. (1990). Breakdown of fuel particles is facilitated by a large number of radiation-thermal defects in their structure. As a result of disintegration, the mean size of hot particles decreases and the part of particles with high activity declines. Consequently, the rate of radionuclide release from fuel matrix by both mechanisms (diffusion and dissolution) should be growing.

Chernobyl-derived fuel particles did not dissolve in water or neutral solution of salts and therefore the proportion of mobile forms of radionuclides (exchangeable and water-soluble) in the soils of Chernobyl close-in areas in the first years after the accident was much lower than in similar soils after the global fallout or Kyshtym accident (Konoplev et al. 1988; Petryaev et al. 1993). This difference could be used for quantifying the part of radionuclides embedded in fuel particles (Konoplev and Bobovnikova 1991; Bobovnikova et al. 1991a; Konoplev et al. 1992; Kashparov et al. 1999, 2000, 2004; Konoplev and Bulgakov 1999). Data on ^{90}Sr speciation are preferable to be used because this radionuclide has sufficiently long half-life (28.8 years) enabling to follow changes in its exchangeability in soil over extended time period. Sorption equilibrium of radiostrontium in soils is achieved in several days (Frere and Champion 1967) and therefore its speciation, except for embedded in fuel particles, can be assumed to be at equilibrium. Furthermore, the steady-state fraction of exchangeable radiostrontium is 80–100% in most of soils and weakly dependent on soil characteristics (Pavlotskaya 1974).

The fraction of ^{90}Sr embedded in fuel particles in different points of the Chernobyl-contaminated zone can be calculated following the equation (Konoplev and Bulgakov 1999)

$$F_t = 1 - \frac{\alpha_{ex}^t}{\alpha_{ex}^\infty}, \qquad (3.1)$$

where F_t is the fraction of ^{90}Sr present in soil and embedded in fuel particles at the time moment t; α_{ex}^t and α_{ex}^∞ are the fractions of exchangeable ^{90}Sr at the time moment t and at equilibrium, e.g., after fuel particles decomposition in soil, respectively.

The exchangeable ^{90}Sr fraction at a given time moment (α_{ex}^t) was determined by extraction of 1 M ammonium acetate solution using the technique discussed above at phase ratio 1:8 and extraction time 24 h.

The equilibrium fraction of ^{90}Sr exchangeable form (α_{ex}^∞) can be determined by three different methods. The most accurate way is spiking of a tracer such as ^{90}Sr in the form of solution and determining the species ratio after equilibrium is achieved (Kashparov et al. 1999, 2004). Another method of α_{ex}^∞ determination consists in approximation with the stationary fraction of exchangeable ^{90}Sr in the lower soil layers not containing fuel particles having low mobility (Konoplev et al. 1992; Konoplev and Bulgakov 1999).

Modeling of the processes resulting in radionuclide release from hot particles is complicated by the diversity of their sizes, shapes, and chemical characteristics. For this reason, it would be appropriate to use integral parameters accounting for the rate of radionuclide leaching from fuel particles in different parts of the contaminated zone. Such parameters could be the first-order rate constants k_1 (year^{-1}) or fuel matrix dissolution rate v (g/cm^2 year) (Konoplev and Bulgakov 1999).

In the process of breakdown or dissolution of fuel particles, their size tends to decrease, and relative content of radionuclides embedded in most stable particles increases. The first process leads to an increase in the k_1 value, whereas the second one—to its decrease. The assumption of mutual compensation of these two processes is the basis for using the first-order kinetics for describing radionuclide-leaching rate from fuel particles. In this case, a decrease in the fraction of radionuclide embedded in particles as a function of time follows the equations (Konoplev and Bulgakov 1999)

$$\frac{\mathrm{d}F_t}{\mathrm{d}t} = -k_1 F_t, \tag{3.2}$$

and hence

$$F_t = F_0 \mathrm{e}^{-k_1 t}, \tag{3.3}$$

where F_t and F_0 are the fractions of radionuclides in fuel particles at time moment t and initial depositions, respectively; t is the time since the accident.

If particles have approximately the same size and dissolution resistance, the leaching rate will increase as a function of time due to increase in the ratio of surface area of particles and their weight. In this case, it is better to use dissolution rate normalized over particles surface area v (g/cm^2 year). By definition (Konoplev and Bulgakov 1999),

$$v = \frac{1}{S} \cdot \frac{\mathrm{d}P_t}{\mathrm{d}t}, \tag{3.4}$$

where P_t is the particle weight at time moment t; S is the particle surface area.

Assuming that the radionuclide is distributed uniformly in a particle volume and considering that

$$\frac{\mathrm{d}P_t}{\mathrm{d}t} = \rho S \frac{\mathrm{d}R_t}{\mathrm{d}t}, \tag{3.5}$$

and

$$\left(\frac{F_t}{F_0}\right)^{1/3} = \frac{R_t}{R_0}, \tag{3.6}$$

where R_t and R_0 are the current and initial radius of the particle, respectively, ρ is the fuel matrix density, F_0 is the initial fraction of the radionuclide embedded in fuel particles:

$$\left(\frac{F_t}{F_0}\right) = 1 - \left(\frac{v}{\rho R_0}\right) t. \tag{3.7}$$

Thus, with particles characterized by a single effective size and assuming the uranium matrix dissolution rate for unit surface to be constant under given conditions, the time dependence of $(F_t/F_0)^{1/3}$ should be linear. The analogue of the rate constant in this case is normalized dissolution rate $w = v/\rho R_0$.

On the other hand, when radionuclide leaching is described by the first-order kinetics, the time dependence of $\ln(F_t)$ should be linear in accordance with the following equation:

$$\ln(F_t) = \ln(F_0) - kt. \tag{3.8}$$

Applicability of these models for predicting radionuclide leaching from Chernobyl-derived fuel particles was validated by comparison of experimental and theoretical curves representing the changes in the concentration of ^{90}Sr embedded in fuel particles in soil samples collected in 1987–1991 near the settlement of Benevka and on the experimental plot near Korogod (Konoplev and Bulgakov 1999). Results of regressive analysis of the time changes in F_t/F_0 by Eqs. (3.7) and (3.8) showed a good agreement between experimental and theoretical dependencies (Konoplev and Bulgakov 1999). In this regard, Eq. (3.7) showed a slightly better agreement of calculation results and experimental data. The values of correlation coefficients, for the most part, are higher for Eq. (3.7).

What is also in favor of Eq. (3.7) is a good fit of the values of dissolution rate constants calculated by ^{90}Sr speciation changes in soil to those determined in laboratory experiment in which fuel particles separated from Chernobyl close-in area soils were exposed to solutions modeling natural aqueous media (Bogatov et al. 1990).

A great amount of data on radiostrontium speciation in soils of contaminated areas were used to obtain site- and soil-specific rate constants of radionuclide leaching from fuel particles k_l, which ranged from 10^{-3} to 10^{-4} day^{-1} (Konoplev and Bulgakov 1999) or 0.05–0.3 year^{-1} (Kashparov et al. 2004) depending on location and soil type.

Konoplev et al. (1992) used data on dynamics of ^{137}Cs speciation in soils of the 30-km zone to reveal the rate constant of its leaching with consideration of the rate of its subsequent fixation by clay minerals. The rate constants k_l for ^{137}Cs and ^{90}Sr were close, which indicated that the chemical nature of radionuclide has a minor effect on the rate at which it breaks loose from particles during their disintegration.

Rate of Radionuclide Leaching from Fuel Particles as a Function of Direction and Distance to ChNPP

Radionuclide-leaching rate from fuel particles is a function of the medium in which they are present, as well as characteristics of particles themselves. For example, in laboratory experiment on determining dissolution rate of fuel particles collected in the Chernobyl close-in area, when exposed to solutions modeling natural aqueous media, the fuel particles showed dissolution rates varying by two orders of magnitude from 5.7×10^{-6} to 1.6×10^{-4} g U/(cm^2 day) (Bogatov et al. 1990). It is common knowledge that in the initial phase of the Chernobyl accident radioactive materials released from the reactor were first transported north and northwest and then west, south, east, and northeast (Izrael et al. 1990; IAEA 2006; Talerko et al. 2020). Since the reactor state was changing, characteristics of deposited particles were most likely to be different in different directions. It is also clear that the particle mean size and density are decreasing with distance from the source. This being the

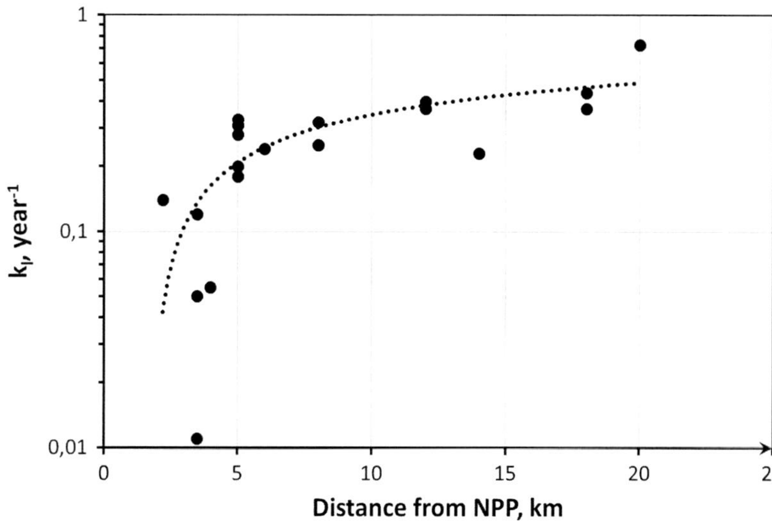

Fig. 3.2 Rate constants for ^{90}Sr leaching from fuel particles as a function of distance to the reactor for the south sector of the footprint (Konoplev 1998)

case, the impact of particles characteristics on leaching rate of radionuclides from these particles can be studied by comparison of rate constants determined at different locations in the contaminated zone. By the classification proposed by Izrael et al. (1990), the region contaminated after the Chernobyl accident can be divided into three sectors: west (northwest, west, and southwest), south (south, southeast, east), and north (north and northeast).

The leaching rate constants were ranging from 1×10^{-2} year^{-1} to 4×10^{-2} year^{-1}, the mean being 2.4×10^{-2} year^{-1} in the west and southwest directions, and from 4×10^{-2} year^{-1} to 2.9×10^{-1} year^{-1}, the mean being 1.1×10^{-1} day^{-1} in the northwest direction. In these directions, no dependence of leaching constants on distance from the reactor has been revealed (Konoplev and Bulgakov 1999; Kashparov et al. 2004) contrary to the south and southeast directions in which a clear trend for an increase in leaching rate constant with distance from the NPP was reported (Konoplev and Bulgakov 1999) (Fig. 3.2). At distances less than 5 km the mean constant value is estimated to be 7.7×10^{-2} year^{-1}, while at 5 km and more—3.3×10^{-1} year^{-1}. The dissolution rate constants have demonstrated similar behavior.

Since the soil cover in the 30-km zone is quite homogeneous, it may be reasonable to conclude that the particles more stable in terms of decomposition were deposited in the north, northwest, west, and southwest of the reactor, as well as south and southeastward in the immediate vicinity of the NPP, where the leaching rate constants varied from 4×10^{-3} to 4×10^{-2} year^{-1}. With distance from the NPP in the southern direction, the rate constants increase and at the distance of about 20 km are as high as 4×10^{-1} year^{-1}.

From this, it follows that the characteristics of fuel particles in the south direction changed with distance to ChNPP more drastically than in the west direction. Outside the 30-km zone, the leaching rate constants should be even higher due to smaller size of particles. In soils of Bryansk region of Russia, for example, speciation was determined for the first time in 1987. At that time, ^{90}Sr species were found to be practically at equilibrium, which leads us to the conclusion that the leaching rate constant was at least 1–2 year^{-1}. Similar conclusion can be drawn for the regions of Belarus occurring at the distance 200–250 km, where ^{90}Sr species in soils in 1987 were also close to equilibrium (Petryaev et al. 1993).

Therefore, depending on characteristics of fuel particles, rate constants of radionuclide leaching can vary by several orders of magnitude.

Rate of Radionuclide Leaching from Fuel Particles as a Function of Soil Properties

Since rate of radionuclide leaching from fuel particles largely depends on characteristics of particles themselves, impact of soil properties on their breakdown rate can be studied only using experimental plots on which soils of different types are in close proximity. In this case, differences in the nature of hot particles responsible for contamination in each point will be minimized. Comparison of the leaching rate constants determined at different points showed that there is no clear dependence on soil type, humus content, and cation-exchange capacity (Konoplev and Bulgakov 1999). This can be associated with the fact that the role of soil properties is impossible to discern against the influence of nature of particles and nonuniformity of their area distribution on leaching rate, as well as a truly weak dependence of leaching rate on characteristics of the environment in which they occur. The latter is possible when radionuclide leaching from particles is determined by processes of their oxidation by air oxygen or processes within particles resulting, for example, from radioactive decay of isotopes in them.

On the basis of large amount of statistically reliable data, Kashparov et al. (1999) established the dependence of k_l on the soil pH as follows (Kashparov et al. 2000, 2004):

For the Western sector:

$$\begin{aligned} k_l &= 0.6 \times 10^{-0.15 \cdot \mathrm{pH}} \text{at pH} < 7.0, \\ k_l &= 0.05 \text{ at } 7.0 < \mathrm{pH} < 7.5. \end{aligned} \tag{3.9}$$

For the Southern and Northern sectors:

$$\begin{aligned} k_l &= 40 \times 10^{-0.45 \cdot \mathrm{pH}} \text{at pH} < 6.5, \\ k_l &= 0.05 \text{ at } 6.5 < \mathrm{pH} < 7.5. \end{aligned} \tag{3.10}$$

During the years immediately after the accident, fuel particles primarily occurred in the upper centimeters of soil, both close to the reactor and at distances to 250 km (Konoplev 2001). Most of the particles were concentrated in the 0–1 cm soil layer and their proportion decreased markedly with depth. The vertical profile of radioactive particles in soil was found to be practically independent of the distance from the ChNPP and is mainly governed by soil type. The lack of dependence of the vertical distribution of particles on their size and chemical nature results from the fact that the primary mechanism of radionuclide migration in the upper soil layer is mixing by soil flora and fauna (bioturbation) (Bulgakov et al. 1991; Konoplev and Golubenkov 1991; Konoplev et al. 1992).

At present, fuel particles have almost completely disintegrated (Beresford et al. 2016) in terrestrial soils. Opposite situation is observed in the cooling pond (CP) of the Chernobyl NPP, where the vast majority of long-lived radionuclides deposited embedded in fuel particles. It is remarkable that in the cooling pond sediments the majority of ^{90}Sr activity still occurs in the form of fuel particles (Bulgakov et al. 2009; Kanivets et al. 2020). Due to low dissolved oxygen concentration and high pH, dissolution of fuel particles in the CP sediments is significantly slower than in soils. However, at the present time, these conditions are changing. After the cessation of water pumping from the Pripyat River to the pond in 2014, a significant part of the sediments is being drained and exposed to the air, which significantly enhances the dissolution rate of fuel particles in exposed sediments, and hence mobility and bioavailability of radionuclides will increase with time (Bulgakov et al. 2009). The rate constant was expected to increase almost instantaneously to the typical value for the aerated layer of neutral and slightly alkaline soils in the Chernobyl exclusion zone. After that, the dissolution rate may increase slowly due to potential acidification of the newly formed soils as a result of vegetation and microbiological activity. The dynamics of this acidification was estimated from the data on the decrease in soil pH as a function of time after soil liming ceased. Along with empirical equations relating the dissolution rate to soil pH, this allowed the prediction of fuel particle dissolution and ^{90}Sr mobility dynamics for different remediation scenarios. Model calculations have shown that in newly exposed sediments, fuel particles will be almost completely dissolved in 15–25 years, while in parts of the pond that remain flooded, fuel particle dissolution will take about 100 years (Bulgakov et al. 2009). Recent data on dynamics of ^{90}Sr and ^{137}Cs in the cooling pond confirm the predictions (Kanivets et al. 2020): dissolved ^{90}Sr activity concentration in the cooling pond increased from about 1 Bq/L in 2014 before the cessation of water pumping to 3–5 Bq/L in spring 2017. Also, as was predicted, dissolved ^{137}Cs in cooling pond water does not show any noticeable increase because of its fast fixation on clay minerals after release from fuel particles.

3.2.2.2 Radionuclide Sorption–Desorption by Soils and Sediments

One of the major processes governing migration of radionuclides in environmental compartments is their sorption by the solid phase of soil and bottom sediments

(Fig. 3.1). In a general form, sorption–desorption of a radionuclide from solution by sorption sites of the solid phase can be written by the equation

$$R + S \leftrightarrow (RS), \tag{3.11}$$

where R is the radionuclide in solution; S is the sorption site on the solid-phase surface; (RS) is the radionuclide sorbed form.

The position of sorption–desorption equilibrium is characterized by the equilibrium constant K (Orlov et al. 2005):

$$K = \frac{[RS]}{[R][S]}, \tag{3.12}$$

where [R], [RS], and [S] are the concentration of a radionuclide in solution, the radionuclide sorbed form concentration, and the capacity of sorption sites, respectively.

Many radionuclides are present in aqueous solution in the form of ions and then the main mechanism for their sorption–desorption is ion exchange (Timofeev-Resovsky and Titlyanova 1966; Cremers and Maes 1986; Konoplev et al. 1992). As a rule, the surface of naturally occurring solid particles in aqueous solution is characterized by excessive negative charge and hence cations are better sorbed by them than anions. The radionuclides of radiocesium and radiostrontium under consideration occur in solution primarily as cations. The cation-exchange equation for radionuclide cation R^{x+} with charge x+, in the general case, takes the form (Kokotov and Wilken 1969):

$$\frac{1}{x}R^{x+} + S\left(\frac{1}{y}M^{y+}\right) \overset{K_c}{\leftrightarrow} \frac{1}{y}M^{y+} + S\left(\frac{1}{x}R^{x+}\right), \tag{3.13}$$

where M^{y+} is the competing ion present in the medium. In environmental compartments, the key exchange ions are Ca^{2+}, Mg^{2+}, K^+, NH_4^+, and Na^+ (in acid medium also H^+ and Al^{3+}).

The equilibrium constant of reversible process (3.13) is referred to as the selectivity coefficient. It accounts for relative affinity of the cation R^{x+} to sorption sites with respect to exchange ion M^{y+} (Kokotov and Wilken 1969; Orlov et al. 2005):

$$K_c(R^{x+}/M^{y+}) = \frac{\left[S\left(\frac{1}{x}R^{x+}\right)\right]^{\frac{1}{x}}[M^{y+}]^{\frac{1}{y}}}{\left[S\left(\frac{1}{y}R^{x+}\right)\right]^{\frac{1}{y}}[R^{x+}]^{\frac{1}{x}}}. \tag{3.14}$$

It is now commonly believed that high ability of soils and sediments to sorb selectively radiocesium is determined by the presence of clay minerals with crystal lattice of group 2:1 (Sawhney 1972; Cremers et al. 1988; Comans et al. 1991). In work by Brouwer (Brouwer et al. 1983), based on the study of Cs sorption on illite, the multitude of ion-exchange sorption sites have been divided, at least, into three

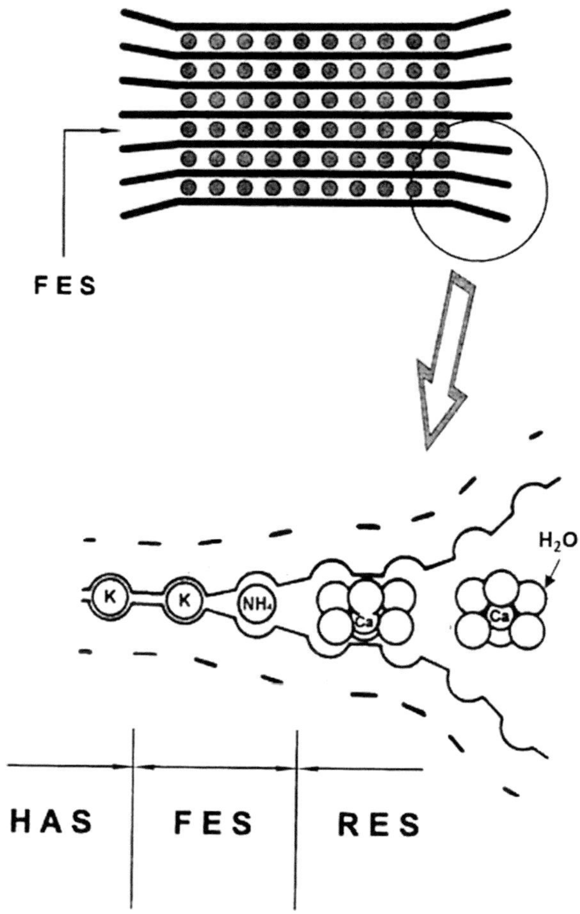

Fig. 3.3 Schematic structure of 2:1 micaceous clay mineral and frayed edges of interlayer (modified from Konoplev 1998)

main types with different selectivity with respect to cesium. The proportion of these sites was 0.25%, 2.5%, and 97.25% of total ion-exchange capacity in case of illite saturated with potassium; and 0.55%, 3.35, and 96.15% in case of illite saturated with calcium. The sorption sites of maximum capacity and minimum selectivity with respect to cesium were attributed to regular surface ion-exchange sites (RES). Two other types of sorption sites with smaller capacity, the first of which has extremely high (the highest) selectivity with respect to cesium, were attributed to sorption sites occurring in the area of frayed edges of neighbor layers of illite crystal lattice. Schematically, the structure of layer mineral of group 2:1 and its frayed edges are shown in Fig. 3.3. Specific location of frayed-edge sites (FES) is responsible for their high selectivity with respect to ions with low values of cation radius and low hydratation energy in solution.

Since most of cations in aqueous solution occur in hydrated form, i.e., surrounded by water molecules, they are too large to approach FES. Yet, ions with low hydration energy, such as K^+, Rb^+, NH_4^+, and especially Cs^+, can easily lose a hydratation water shell, penetrate to the frayed edges of mineral's interlayers, and be adsorbed on selective sorption sites FES. On the other hand, the ions Ca^{2+}, Mg^{2+}, and Sr^{2+}, surrounded by a large hydratation shell, are not capable of approaching FES and can be adsorbed only on regular exchange sites RES located on the surface of mineral particles. Quantitatively, the selectivity of sorption sites with respect to cesium is characterized by the selectivity coefficient. Radiocesium is adsorbed on surface sorption sites RES nonselectively, i.e., the selectivity coefficient of its sorption with respect to one-charge ions K^+, Na^+, NH_4^+, and others is close to 1. With that, the selectivity coefficient of Cs sorption on FES is equal to 1000 for K^+ and 200 for NH_4^+ (De Preter 1990; Wauters et al. 1996a). The selective sorption sites FES constitute a relatively small portion of ion-exchange sorption sites (1–5%) for most of soils and bottom sediments (De Preter 1990). Nevertheless, as follows from the published data (Brouwer et al. 1983; Comans et al. 1991; Konoplev et al. 2002), FES are nonuniform in terms of selectivity with respect to radiocesium and can be divided, at least, into two types: regular FES and sites of extremely high selectivity HAS (High-Affinity Sites).

Due to high selectivity of FES with respect to cesium and because radiocesium, and even stable cesium, occurs at trace concentrations in the environment, practically the entire radiocesium in most of soils and bottom sediments is adsorbed on FES.

The ability of soils and sediments to sorb radiocesium selectively is characterized by the capacity of selective sorption sites [FES] or by the so-called radiocesium interception potential (RIP), which is the product of [FES] and the selectivity coefficient of radiocesium in relation to the corresponding competitive ion (Sweeck et al. 1990).

Cremers and coworkers (Cremers et al. 1988; Sweeck et al. 1990; De Preter 1990; Valcke and Cremers 1994; Wauters et al. 1994) developed a method for quantitative determination of [FES] and RIP. The method is based on using silver thiourea Ag $(TU)^+$ as a masking agent for regular exchange sites, RES. RES selectively binds this complex, while $Ag(TU)^+$ does not interact with FES because of the molecule's large size. Thus, the masking blocks RES and allows the study of cesium sorption–desorption on FES. Later, the original method was simplified using blocking RES by sufficient concentration of hydrated Ca^{2+} cations (Wauters et al. 1996a, b).

It was found, however, that even during a relatively short laboratory procedure of [FES] and RIP determination (both original and modified) Cs was partly fixed by the solid phase and the method overestimated the amount of reversibly sorbed Cs (Konoplev and Konopleva, 1999a; Konoplev et al. 2002). Given highly organic soils and bottom sediments, collapse of the interlayers of micaceous clay prevents determination of [FES] when using this method. To avoid this drawback, modification of the procedure was proposed (Konoplev and Konopleva, 1999a; Konoplev et al. 2002), which included:

- An additional step of ammonium acetate extraction after equilibration time is used. This step allows avoiding the influence of fast fixation on [FES] or RIP determination.
- The [FES] is calculated using the linearized form of Langmuir isotherm and its value is given by the intercept on the ordinate of the plot of reversibly sorbed cesium inverse concentration versus inverse equilibrium concentration of cesium in solution.
- Taking the initial range of the isotherm at low cesium concentrations, the capacity of high-affinity sites [HAS] located between layers of micaceous clay particles can be determined. HAS have higher selectivity for cesium as compared to average FES and are therefore occupied by cesium in the first stage. Notice that HAS represent 1–10% of FES.

3.2.2.3 Radiocesium Fixation by Soils and Sediments

In terms of chemical forms, fixation of radionuclides is a transfer from their mobile to fixed form. Fixation represents long-term replacement of interlattice K^+ by Cs^+ ions due to collapse of expanded edges of mineral's crystal lattice interlayers. Models have been proposed in which radiocesium fixation is treated as an irreversible process (Comans and Hockley 1992). The available data suggest that this assumption is warranted at the initial stage of fixation and for processes of relatively short timescale. On the other hand, data about long-term transformation of radionuclides chemical forms in soil after nuclear weapon testing (Pavlotskaya 1974) and Kyshtym accident (Konoplev and Bobovnikova 1991) and 30-year studies after the Chernobyl accident are indicative of the existence of remobilization process that is the reverse of fixation (Konoplev et al. 1992; Smith and Comans 1996; Konoplev and Bulgakov 2000a). After deposition of radiocesium onto the soil, the fraction of its exchangeable form α_{ex} does not decrease to zero, as expected for irreversible fixation, but only decreases to a certain steady-state level, and then does not change significantly (Konoplev and Bobovnikova 1991; Konoplev et al. 1996).

Rate constant of radiocesium fixation k_f (see Fig. 3.1) for typical soils of the Chernobyl 30-km zone was found to be (1–4) $\times$ 10^{-2} day^{-1} (Konoplev et al. 1992, 1996). Remobilization was about an order of magnitude slower: k_r = (2–6)-$\times$ 10^{-3} day^{-1}, which is comparable with rate constant of radionuclide leaching from fuel particles (Fig. 3.1).

3.3 Radionuclide Distribution Between Solid and Liquid Phases in the Soil–Water Environment

Up to now, modeling and predicting of radionuclide behavior in the environment, and dose assessment is based on using, instead of specific rate and equilibrium constants, an integral parameter—distribution coefficient K_d (L/kg) (Zheleznyak et al. 1992, 2016; Borzilov et al. 1993; Konoplev et al. 1996a; Nair et al. 1996; Smith et al. 2005a, b; Monte et al. 2009), which is equal to the ratio of radionuclide equilibrium concentrations of particulate radionuclide $[R]_p$ (Bq/kg) and dissolved radionuclide $[R]_w$ (Bq/L):

$$K_d = \frac{[R]_p}{[R]_w}. \tag{3.15}$$

In case of radiocesium, for example, $[R]_p$ includes radiocesium embedded in fuel particles, exchangeably sorbed radiocesium and fixed or nonexchangeable radiocesium in solid phase.

3.3.1 Applicability of K_d in Nonequilibrium Conditions

The distribution coefficient K_d is by definition an equilibrium parameter and therefore should be measured at equilibrium or in close to equilibrium conditions. At the same time, using the distribution coefficient is appropriate only to predict processes for which time scale is much longer than the time of establishing equilibrium. In addition, nonequilibrium distribution of radionuclides can be caused by their deposition on the ground as water-insoluble particles, such as fuel particles following the Chernobyl accident. Part of radionuclides occurring in the solid phase will not take part in establishing equilibrium concentration in solution. The contribution of this part to total concentration of radionuclides in soil or depositions decreases with time because of disintegration of particles. As a result, the apparent distribution coefficient estimated as a ratio of the radionuclide concentration in the solid phase to its concentration in the liquid phase is a function of time even for radionuclides with a relatively fast sorption–desorption, such as ^{90}Sr (Konoplev et al. 1992; Beresford et al. 2016).

It has been proven that high retention of radiocesium in soil and bottom sediments is determined by two different processes: reversible selective sorption by micaceous clay minerals (Sawhney 1972; Cremers and Maes 1986; Cremers et al. 1988) and fixation/remobilization (Konoplev et al. 1992, 1996, 2002). Kinetics of the processes of leaching of radiocesium from fuel particles and its fixation/ remobilization determines dynamics of the integral parameter K_d.

3.3.2 Prediction of Solid–Liquid Distribution of Radionuclide Exchangeable Form: Exchangeable Distribution Coefficient

The existing methods for estimating the distribution coefficient have a common drawback, which is the absence of a clear division of radionuclide speciation based on their ability to exchange with the liquid phase. The problem can be resolved by using the concept of exchangeable distribution coefficient (K_d^{ex}). This concept provides that the total amount of the radionuclide in the solid phase of the soil–water system can be divided into two components: the exchangeable part, which occurs at instantaneous equilibrium with the liquid phase; and the nonexchangeable part, which does not contribute to the radionuclide concentration in solution (Fig. 3.1). In the immediate term, only the exchangeable form of radionuclide contributes to the solid–liquid interphase exchange. Therefore, the notion of the exchangeable distribution coefficient K_d^{ex} was proposed as a ratio of the concentration of the radionuclide in exchangeable form in the solid phase $[R]_{ex}$ to its concentration in solution at equilibrium (Konoplev et al. 1992, 2002; Madruga et al. 1996; Konoplev and Bulgakov 2000a, b):

$$K_d^{ex} = \frac{[R]_{ex}}{[R]_w} = \alpha_{ex} K_d^{tot}, \tag{3.16}$$

where α_{ex} is the fraction of exchangeable radionuclide, and K_d^{tot} is the apparent or total distribution coefficient.

The advantage of K_d^{ex} is that it is a function of environmental characteristics (sorption properties of the solid phase and cation composition in solution) and can be assessed/predicted on their basis (Konoplev and Bulgakov 2000a, b). On the other hand, the value of total K_d^{tot} is strongly dependent on the fraction of nonexchangeable radionuclide. Using the equation of ion-exchange equilibrium for Cs sorption on FES in the presence of M^+ ion, an equation can be written to relate the exchangeable distribution coefficient of radiocesium and the solid-phase FES capacity and M^+ concentration in solution:

$$K_d^{ex}(\mathrm{Cs}) = \frac{K_c(\mathrm{Cs}/\mathrm{M})[\mathrm{FES}]}{[\mathrm{M}^+]}, \tag{3.17}$$

where K_c(Cs/M) is the selectivity coefficient for ion exchange of Cs^+ on FES with respect to cation M^+. Or

$$K_d^{ex}[\mathrm{M}^+] = K_c(\mathrm{Cs}/\mathrm{M})[\mathrm{FES}] = \mathrm{RIP}^{ex}(\mathrm{M}). \tag{3.18}$$

Equation (3.18) accounts for the exchangeable radiocesium interception potential RIP^{ex}(M) (Konoplev and Konopleva, 1999a; Konoplev et al. 2002). The exchangeable radiocesium interception potential is a constant value for a given sorbent and characterizes its ability to sorb cesium selectively and exchangeably. The significance of the radiocesium interception potential is explained by the fact that it can be used to determine the exchangeable distribution coefficient of radiocesium.

In a general case, the value of ^{137}Cs distribution coefficient can be predicted based on separate consideration of sorption on FES and RES (Wauters et al. 1996c). Then

$$\begin{aligned} K_d^{ex} &= K_d^{ex}(\text{FES}) + K_d^{ex}(\text{RES}) \\ &= \frac{K_c^{\text{FES}}(\text{Cs}/\text{K})[\text{FES}]Z_{\text{K}}}{[\text{K}]_w} + \frac{K_c^{\text{RES}}(\text{Cs}/\text{K})[\text{K}]_{\text{RES}}}{[\text{K}]_w}, \end{aligned} \quad (3.19)$$

where K_c^{FES} is the selectivity coefficient for exchange of cesium and potassium cations on FES; [FES] is the capacity of selective sorption sites in the solid phase; Z_K is the FES fraction occupied by potassium cations; $[K]_{RES}$ is the concentration of potassium bound with nonselective adsorption sites.

For nonselective sorption sites, K_c^{RES}(Cs/K) is close to 1 and besides that the second term in the right part of Eq. (3.19) is usually negligible for most of soils and bottom sediments as compared to the first term. With this in mind, when potassium is a predominant ion competing with ^{137}Cs (Wauters et al. 1994, 1996a, b, c; Konoplev et al. 2002),

$$K_d^{ex}(\text{Cs}) = \frac{\text{RIP}^{ex}}{[\text{K}]_w}. \quad (3.20)$$

In those cases when ammonium cation is present in the system in noticeable amounts (Wauters et al. 1994, 1996c; Konoplev et al. 2002),

$$K_d^{ex}(\text{Cs}) = \frac{\text{RIP}^{ex}(\text{NH}_4)}{[\text{NH}_4]_w}. \quad (3.21)$$

In mixed cases, when the contributions of potassium and ammonium to competition with ^{137}Cs are comparable, the following ratios are applicable (Wauters et al. 1994, 1996a, b, c, d):

$$K_d^{ex}(\text{Cs}) = \frac{\text{RIP}^{ex}(\text{K})}{[\text{K}]_w + K_c^{\text{FES}}(\text{NH}_4/\text{K})[\text{NH}_4]}, \quad (3.22)$$

where $K_c^{FES}(NH_4/K)$ is the selectivity coefficient of ammonium in relation to potassium for FES sites. Post-Chernobyl studies showed that for most soils and sediments $K_c^{FES}(NH_4/K) \approx 5$, and simplified Eq. (3.22) can be written as follows (Wauters et al. 1994, 1996a; Konoplev et al. 2002):

$$K_d^{ex}(Cs) = \frac{RIP^{ex}(K)}{[K]_w + 5[NH_4]} \tag{3.23}$$

Knowing K_d^{ex} one can calculate the apparent ${K_d}^{tot}$ on the basis of Eq. (3.16); α_{ex} can be easily measured using 1 M ammonium acetate extraction or estimated on the basis of rate constants of radionuclide leaching from hot particles, fixation, and remobilization (Fig. 3.1) (Konoplev et al. 1992; Konoplev and Bulgakov 2000a, b).

Using this method enables reducing uncertainty involved in assessing the exchangeable distribution coefficient and the apparent ${K_d}^{tot}$. The downside of this method is its complexity and unavailability of data for many territories. For all that, measurement of RIP^{ex} can be recommended for critical regions, in order to improve reliability of predictions in emergency and remediation strategy development.

The methodology has been used for a number of water bodies contaminated as a result of the Chernobyl accident: Kiev reservoir, Glubokoe Lake (Ukraine), Pripyat and Dnieper rivers (Ukraine and Belarus), Svyatoe and Kozhanovskoe lakes (Russia), Devoke Water (UK), Lakes Constance and Vorsee (Germany), Lake Lugano (Switzerland), etc. (Konoplev et al. 1996, 2002; Konoplev and Konopleva, 1999a), and demonstrated its applicability to estimate and predict K_d^{ex} and ${K_d}^{tot}$ for radiocesium in various water bodies.

3.4 Radionuclide Bioavailability in Soil and Its Parameterization Through Soil Characteristics

Uptake of radionuclides from soil to plant is a significant pathway for transfer to humans and animals (IAEA 2006). For characterization of soil–plant transfer, the concentration ratio (CR) is normally used, which is the ratio of radionuclide concentration in plant and its concentration in the upper soil layer. Values of CR determined experimentally in different conditions for a particular plant can differ by hundreds or even thousands of times (IAEA 2010). That is why, using a plant-average CR leads to a considerable uncertainty in prediction. Numerous attempts were therefore made to develop methods for estimating site-specific values of CR. Empirical dependencies of radionuclides accumulation in plant on soil properties were derived (IAEA 2010), but the problem is that purely empirical characteristics are not all-purpose. In order to be applied for soils different from those for which they were obtained, they should be substantiated based on information about mechanisms of sorption desorption of radionuclides in soil and their soil–plant transfer.

3.4.1 Conceptual Model of Radionuclide Soil–Plant Transfer

It is generally agreed that a key parameter governing soil–plant transfer of radionuclide is its exchangeability in soil (Oughton et al. 1992; Konoplev et al. 1993). Yet it is not a single factor; account should also be taken of radionuclide and major cations concentrations in soil solution (Shaw 1993; Konoplev et al. 1996; Smolders et al. 1997; Absalom et al. 1999). In this respect, two hypotheses were proposed to account for the influence of soil solution composition on radionuclide transfer to plant. By the first hypothesis, radiocesium concentration in plant is proportional to the ratio of its concentration in soil solution and potassium concentration. Comparison with experimental data, however, has shown that there is no meaningful correlation between the concentration ratio and ^{137}Cs/K ratio in soil solution (ECP-2 1996). The other hypothesis, which seems more reasonable, is that the concentration ratio is proportional to the radionuclide fraction in the root-exchange complex as a function of composition of soil solution (Smolders et al. 1997). Using this hypothesis, a model of radiocesium bioavailability in soil was proposed (Konoplev et al. 1999; Konoplev and Konopleva 1999a).

Figure 3.4 shows the conceptual scheme of the equilibrium model of radiocesium soil–plant transfer. This model is based on the following assumptions:

1. Radiocesium is taken up by plant from soil solution and its concentration in plant is a linear function of radionuclide loading in the root-exchange complex, which is determined by cation composition of soil solution. The root-exchange complex is not selective, i.e., selectivity coefficient of Cs^+ in relation to K^+ and NH_4^+ is equal to unity (Konoplev et al. 1996b; Smolders et al. 1997);
2. Only the exchangeable fraction of ^{137}Cs inventory in soil is involved in immediate exchange with soil solution (Konoplev et al. 1992, 1993);
3. Exchangeable radiocesium interception potential (RIP^{ex}) and cation composition of soil solution control ^{137}Cs concentration in soil solution in line with ion-exchange equilibrium (Konoplev and Konopleva, 1999a).

The model includes ion-exchange sorption–desorption process in soil, ion exchange in the root-exchange complex, and uptake from root surface to plant. The fixed part of radionuclide is not involved in the immediate transfer to plant.

Prior to plant uptake, the radionuclide must pass through the cell wall free space characterized by specific cation-exchange capacity (Haynes 1980). This speculation differs from the approaches usually applied to radionuclide soil–plant transfer based on linearity of the radionuclide concentration in plant with its concentration in soil solution. The distinguishing feature of the proposed model is that Cs loading in the root-exchange complex is influenced by cationic composition of soil solution. The root-exchange complex is associated with carboxylic groups, and hence not selective, as opposed to soil exchange complex. This approach was used for modeling of radiocesium transfer in the soil solution-plant system (Smolders et al. 1997; Absalom et al. 1999). As the root-exchange complex is nonselective, the selectivity coefficients of both K/NH_4 and the Cs/K are equal to unity. The same can be

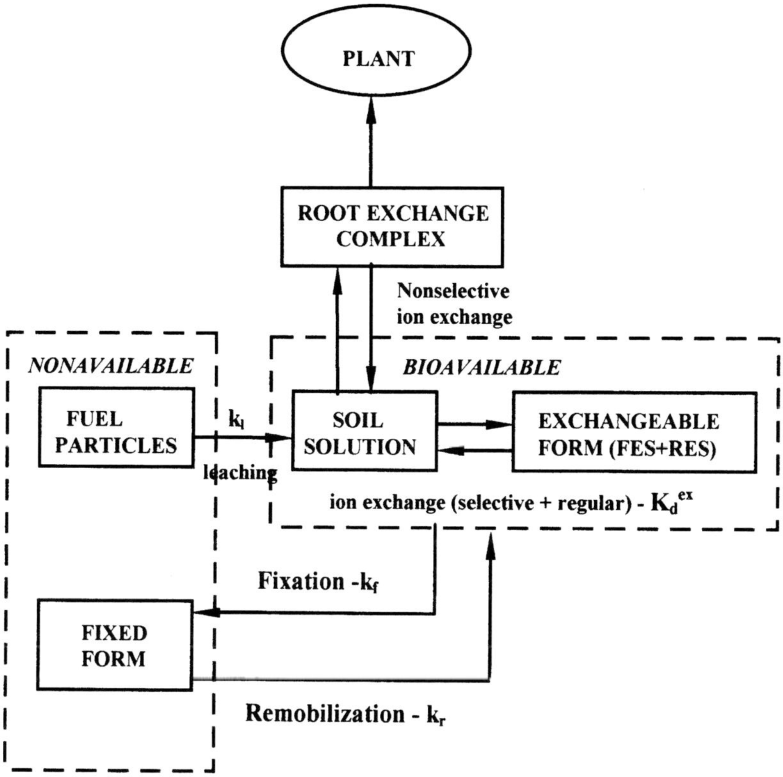

Fig. 3.4 Conceptual model of radiocesium soil–plant transfer (Konoplev et al. 1999)

assumed for the Ca/Mg pair. Accordingly, K^+ and NH_4^+ can be treated as one ion M^+, whereas Ca^{2+} and Mg^{2+} as one ion M^{2+}. On the basis of these assumptions (Smolders et al. 1997),

$$\left[{}^{137}\mathrm{Cs}\right]_{\mathrm{plant}} = k\frac{Z_{\mathrm{M}^+}}{\left[\mathrm{M}^+\right]_{\mathrm{w}}}\left[{}^{137}\mathrm{Cs}\right]_{\mathrm{w}}, \tag{3.24}$$

where Z_{M^+} is the fraction of M^+ in the root cation-exchange complex (potassium window); $[M^+]_w$ is the concentration of M^+ in soil solution; $[^{137}Cs]_w$ is the ^{137}Cs concentration in soil solution; $[^{137}Cs]_{plant}$ is the ^{137}Cs concentration in plant; k is the proportionality constant, which reflects efficacy of transport across plasmalemma. At the same time, the concentration of ^{137}Cs in soil solution is determined by the sorption–desorption equilibrium of the radionuclide in the "soil–soil solution" system. This means that radiocesium occurs in soil solution at simultaneous equilibrium with two cation exchangers: the selective exchanger of soil and the nonselective one of root surface. The ^{137}Cs concentration in soil solution at ion-exchange equilibrium of radiocesium on the FES can be presented as follows (Konoplev et al. 1996b, 1999, 2000):

$$\left[^{137}\mathrm{Cs}\right]_{\mathrm{w}} = \frac{\alpha_{\mathrm{ex}}\left[^{137}\mathrm{Cs}\right]_{\mathrm{tot}}}{\mathrm{RIP}^{\mathrm{ex}}(\mathrm{K})}\left([\mathrm{K}]_{\mathrm{w}} + K_{\mathrm{c}}^{\mathrm{FES}}(\mathrm{NH_4/K})[\mathrm{NH_4}]\right), \tag{3.25}$$

where α_{ex} is the portion of exchangeable ^{137}Cs in soil; $[^{137}Cs]_{tot}$ is the total concentration of ^{137}Cs in soil; $RIP^{ex}(K)$ is the radiocesium interception potential of soil in relation to potassium; $[K]_w$ and $[NH_4]_w$ are the concentrations of correspondent cations in soil solution; $K_c^{FES}(NH_4/K)$ is the selectivity coefficient of ammonium in relation to potassium at FES.

3.4.2 *Parameterization of Radiocesium Bioavailability Through Soil Characteristics*

The "potassium window" Z_{M^+} in the nonselective root-exchange complex can be expressed in terms of cation composition of soil solution. For most soils, $Z_{M^{2+}} \gg Z_{M^+}$, and Z_{M^+} is proportional to the following parameter $\frac{[K^+]_w+[NH_4^+]_w}{\sqrt{[Ca^{2+}]+[Mg^{2+}]}}$ of soil solution. Substituting this and (3.25) in (3.24) and taking into account that $CR = [^{137}Cs]_{plant}/[^{137}Cs]_{tot}$, at equilibrium conditions, CR should be a linear function of A (availability factor) (Konoplev et al. 1996b, 1999, 2000):

$$CR = B \times A, \tag{3.26}$$

where B is the parameter dependent on plant characteristics (first of all, capacity of the root-exchange complex). The availability factor A is taken to be proportional to the fraction of radiocesium in the root-exchange complex and is parameterized through the composition of soil solution and sorption properties of soils as follows (Konoplev et al. 1999, 2009; Konoplev and Bulgakov 2000b):

$$A = \frac{\alpha_{\mathrm{ex}}\mathrm{SPAR}}{\mathrm{RIP}^{\mathrm{ex}}}, \tag{3.27}$$

where SPAR is the sorption potassium–ammonium ratio (Konoplev and Bulgakov 2000b):

$$\mathrm{SPAR} = \frac{[\mathrm{K}]_{\mathrm{w}} + K_{\mathrm{c}}^{\mathrm{FES}}(\mathrm{NH_4/K})[\mathrm{NH_4}]}{\sqrt{[\mathrm{Ca}]_{\mathrm{w}}+[\mathrm{Mg}]_{\mathrm{w}}}}, \tag{3.28}$$

where $[K]_w$, $[NH_4]_w$, $[Ca]_w$, $[Mg]_w$ are the concentrations of corresponding cations in soil solution, $K_c^{FES}(NH_4/K)$ is the selectivity coefficient of potassium exchange for ammonium at selective adsorption sites of soil FES.

Equation (3.27) can actually serve as a basis for prediction of CR using soil and soil solution characteristics. In fact, availability factor A is a combination of three key soil parameters: fraction of exchangeable form of ^{137}Cs (α_{ex}) accounting for fixation ability of soil; exchangeable radiocesium interception potential (RIP^{ex}) representing ability of soil to adsorb radiocesium selectively and reversibly, and parameter of soil solution cation composition SPAR. A similar approach to model soil–plant transfer through soil and soil solution parameters was applied by Absalom et al. (1999, 2001).

The above method can be used for improving the accuracy of soil-specific estimates of CR. However, this requires knowing soil characteristics such as the content of exchangeable ammonium and exchangeable radiocesium interception potential (RIP^{ex}). Values of these parameters are known for a limited number of soils, methods for their evaluation have not been developed yet, and experimental determination of RIP^{ex} is rather complicated. In view of this, a simpler method was proposed for parameterization of the radiocesium availability factor using only those soil characteristics, which can either be taken from reference literature or estimated using known correlation ratios (Konoplev et al. 1999, 2000).

A simplified parameterization of the availability factor for "potassium scenario" can be obtained with the equation calculating radiocesium concentration in soil solution, as proposed in (Konoplev and Bulgakov 2000b):

$$\left[{}^{137}\mathrm{Cs}\right]_w = \frac{\left[{}^{137}\mathrm{Cs}\right]_{ex}[K]_w}{K_c^{eff}[K]_{ex}}, \tag{3.29}$$

where $[^{137}Cs]_{ex}$, $[K]_{ex}$ is the concentration of exchangeable ^{137}Cs (Bq/kg) and potassium (meq/kg) in soil, respectively; K_c^{eff} is the effective selectivity coefficient of the potassium cation exchange for caesium cation in the soil exchange complex.

In this case, the availability factor A^* can be written as

$$A^* = \frac{\alpha_{ex}\mathrm{PAR}}{K_c^{eff}(\mathrm{Cs/K})[K]_{ex}}, \tag{3.30}$$

where PAR is the potassium adsorption ratio ($mM^{1/2}$),

$$\mathrm{PAR} = \frac{[K]_w}{\sqrt{[Ca]_w + [Mg]_w}}, \tag{3.31}$$

Basically, Eq. (3.30) is similar to Eq. (3.27), the only difference is that Eq. (3.30) includes the effective selectivity coefficient, which allows estimation of the availability factor using the classification of soils by values of $K_c^{eff}(Cs/K)$ proposed in

(Konoplev and Bulgakov 2000b). A better accuracy can be achieved if the value of K_c^{eff}(Cs/K) is measured experimentally, which is much easier than to measure RIP^{ex}.

Using expression of PAR through EPR (exchangeable potassium ratio) equal to the ratio of the exchangeable potassium and the sum of exchangeable calcium and magnesium in soil (Richards 1954), A^* can be parameterized through the sum of exchangeable Ca/Mg and exchangeable K:

$$A^* = \frac{9.5\alpha_{ex}}{K_c^{eff}(Cs/K)} \left\{ \frac{1}{[Ca]_{ex} + [Mg]_{ex}} - \frac{0.036}{[K]_{ex}} \right\}. \quad (3.32)$$

Equation (3.32) enables parameterization of bioavailability via commonly used soil characteristics such as exchangeable cations Cd, L, and Mg that are available in conventional soil maps.

3.4.3 Radionuclide Bioavailability in Agricultural and Forest Soils

It is worth dwelling on limitations of the proposed method for parameterization of radiocesium soil–plant transfer. First, the presented model is based on the assumption that only exchangeable form of radionuclide is potentially involved in transfer to plant. At the same time, the exchangeable radionuclide fraction determined by the sequential extractions characterizes the state of dynamic equilibrium between radiocesium fixation and remobilization (see Fig. 3.1). In the long term, the total radionuclide inventory in soil can be potentially bioavailable as a result of remobilization. Therefore, the steady state of exchangeable radiocesium does not always correspond to the biologically available fraction of radionuclide. A more appropriate technique would be extraction from soil during time close to time scale of soil–plant transfer. Secondly, when estimating radionuclide uptake from soil, CR should be compared for the same vegetation species. Thirdly, an appropriate parameter of radionuclide soil–plant transfer needs to be selected. For example, for agricultural soils uniform distribution of radionuclides in the upper 20-cm horizon is normally taken, while in forest soils the upper soil layer is stratified, and the properties of layers L, Of, Oh, and Ah differ from each other. Therefore, for analysis of radionuclide bioavailability in forest soils, it is important to know in what layer the roots of plant under study occur. It is reasonable to compare CR_r, calculated specifically for root zone.

To test applicability of the proposed model, the data on radiocesium transfer from soil to understory in forests of Chernobyl exclusion zone (Ukraine), Baden Wuerttemberg (Germany), Bryansk region (Russia), Uppsala district (Sweden), and Ticino (Switzerland) were used (Konoplev et al. 1999, 2000; Victorova et al. 2000). As a reference plant for analysis, fern (*Pteridium aquilinum* (L.) Kuhn and *Dryopteris filix-mas* (L.) Schot) was chosen. For some sites, data on bilberry

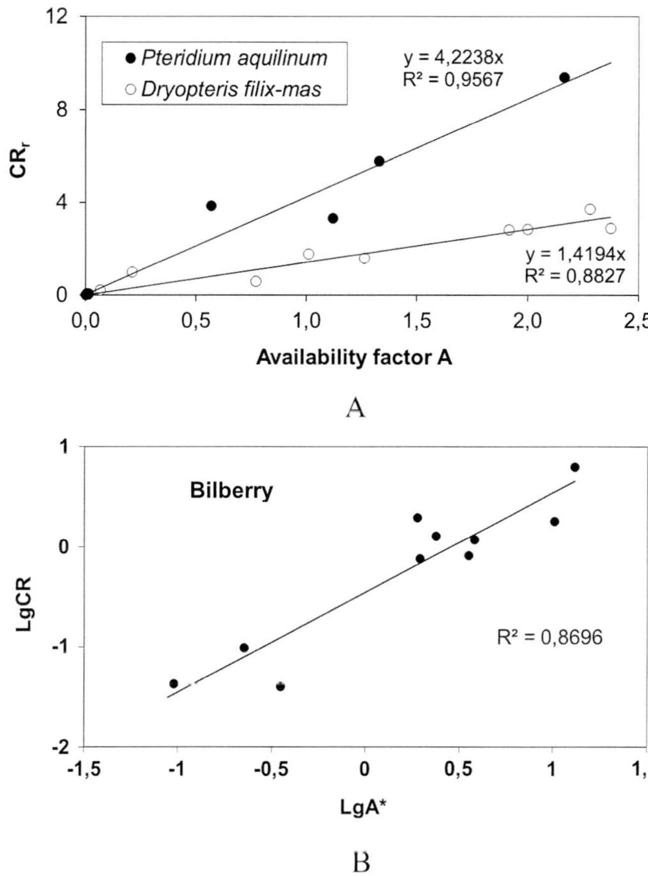

Fig. 3.5 Dependencies of CR_r for two species of fern on the availability factor A in different forest soils (**a**). Dependence of $LgCR_r$ for bilberry on LgA^1 (**b**) in case of simplified parameterization of bioavailability factor (Konoplev et al. 1999)

(*Vaccinium myrtillus*) and blackberry (*Rubus fruticosus*) were obtained and applied. Roots of these plants occur in the humus layer of soil (O_h or A_h). Since most of radiocesium was found in the same layers, soil characteristics involved in equations (Eqs. 3.27 and 3.30) were measured for these layers.

Figure 3.5 presents dependencies of CR_r on the availability factor A for two species of fern in forest soils (Konoplev 1998). As can be seen, they are very close to a linear function and the agreement of the theoretical dependence with experimental data can be considered as satisfactory.

Extended data sets on ^{137}Cs soil–plant transfer for maize and oat on arable soils have been used to validate the simplified model (Bulgakov and Shkuta 2004). For these plant species, satisfactory agreement between experimental values of CR and model predictions has been found. The model by Absalom et al. (1999, 2001) was validated against data sets for agricultural crops and grass (Smolders et al. 1997a;

Sanchez et al. 1999; Absalom et al. 2001) and also demonstrated ability to predict radiocesium dynamics across a wide range of arable and meadow soils.

The agreement in the theoretical and experimental dependencies (Konoplev et al. 1999, 2000, 2009; Konoplev and Konopleva, 1999b; Absalom et al. 2001; Bulgakov and Shkuta 2004) suggests that the described method can be used for estimation of site-specific concentration ratios of radiocesium in plants. Part of the parameters required for calculating the availability factor can be obtained by expert judgment or measured using a rather simple procedure (α_{ex}, K_c^{eff}), while others are important agrochemical indicators and can be taken from reference literature for many regions ($[Ca]_{ex}$, $[Mg]_{ex}$, $[K]_{ex}$). All this is made possible using the proposed method for mapping radiocesium phytoavailability. Incorporation of such maps into GIS can lead to essential reduction in uncertainty associated with dose and risk assessment.

Similar line of approach can be used for radiostrontium soil–plant transfer, but the situation is simpler since radiostrontium is not sorbed by soil selectively. Assuming that Ca is the main competitive cation for Sr in solution, the following equation was proposed by Sysoeva et al. (2005):

$$CR\left({}^{90}\mathrm{Sr}\right) \sim A\left({}^{90}\mathrm{Sr}\right) = \frac{\alpha_{ex}\left({}^{90}\mathrm{Sr}\right)}{K_c(\mathrm{Sr/Ca})[\mathrm{Ca}]_{ex}}, \tag{3.33}$$

where $\alpha_{ex}({}^{90}Sr)$ is the exchangeable fraction of ${}^{90}Sr$ in soil, $[Ca]_{ex}$ is the concentration of exchangeable Ca in soil. The authors tested the applicability of Eq. (3.33) for barley and lupine on a wide range of soils and showed reliability of such predictions (Sysoeva et al. 2005).

3.5 Conclusions

The studies of the behavior of Chernobyl-derived radionuclides became a basis for establishing a scientific framework for methods to estimate and predict mobility and bioavailability of accidentally released radiocesium and radiostrontium, which can be used for developing decision-support systems in case of nuclear emergency, environmental impact assessment for facilities of radiation hazard, and evaluating efficacy of countermeasures to reduce risk of living in contaminated areas.

Post-Chernobyl studies have allowed to reveal the role of hot particles in environmental behavior of radionuclides. In the case of Chernobyl, most of long-lived radionuclides were incorporated in nuclear fuel particles. Because of occurrence of fuel particles, initial mobility and availability of radionuclides in the near zone were lower than in similar conditions in the case of global fallout and Kyshtym accident. Moreover, deposition of fuel particles mainly in the near zone resulted in localized contamination by refractory radionuclides and strong dependence of initial mobility of more volatile radiocesium on distance to the accidental reactor. Disintegration of fuel particle resulted in leaching of radionuclides and hence changes in mobility and bioavailability. Studies of changes in radiostrontium exchangeability have made it possible to determine rates of decomposition and leaching of radionuclides from fuel particles in the soils of the Chernobyl 30-km zone. The time scale of radionuclide leaching from fuel particles in soil was determined to vary from 1 to 10 years and more.

Radionuclides leached from fuel particles become involved in the processes of transformation of radionuclide chemical forms in the soil–water environment. To account for major processes of radiocesium and radiostrontium transformation, a conceptual model was developed, including radionuclide leaching from fuel particles, sorption–desorption by ion-exchange mechanism, fixation, and remobilization. It was found that high retention of radiocesium in soils and sediments is associated with two distinct processes: high selective exchangeable sorption by micaceous clay minerals and fixation. In addition to radionuclide fixation, a reverse process of remobilization was shown to play a certain role and should be taken into account.

A major advancement in studies of radiocesium behavior following the Chernobyl accident consisted in introducing and using the concepts of frayed-edge sites (FES) capacity and radiocesium interception potential (RIP) for soils and sediments, which allowed parameterization of radiocesium behavior in the soil–water environment and prediction of its mobility and bioavailability.

Data have been generated regarding quantitative characteristics of radiocesium fixation kinetics on different soils for both laboratory and field conditions. The evidence suggests that the total, or apparent, distribution coefficient of radionuclide K_d can vary in a wide range, depending on deposition characteristics and environmental conditions. Its value is a function of contact time and depends on transformation rates of radionuclide chemical forms. Another important step forward was establishing the exchangeable distribution coefficient instead of the total K_d^{tot}. K_d^{ex} can be estimated on the basis of ion-exchange equilibrium from characteristics of soil and sediments such as cation-exchange capacity, RIP^{ex}, and cation composition of the solution. This allows reducing significantly the uncertainty involved in prediction of radiocesium behavior in the soil–water environment.

In terms of bioavailability, the achievement was the parameterization of radionuclide soil-to-plant concentration ratio CR through soil and soil solution characteristics. As a result, the uncertainty associated with estimates of soil–plant transfer in agricultural and seminatural environment can be reduced to a significant degree. This has been verified against numerous experimental data on radiocesium and radiostrontium soil-to-plant transfer.

Acknowledgments This research was supported by the Japan Society for the Promotion of Science (JSPS), Grant-in-aid for Scientific Research (B) (18H03389); by Science and Technology Research Partnership for Sustainable Development (SATREPS), Japan Science and Technology Agency (JST)/Japan International Cooperation Agency (JICA) (JPMJSA1603); and by bilateral project No. 18-55-50002 of Russian Foundation for Basic Research (RFBR) and Japan Society for the Promotion of Science (JSPS).

References

Absalom JP, Young SD, Crout NMJ, Nisbet AF, Woodman RFM, Smolders E, Gillett AG (1999) Predicting soil to plant transfer of radiocesium using soil characteristics. Environ Sci Technol 33:1218–1223

Absalom JP, Young SD, Crout NMJ, Sanchez A, Wright SM, Smolders E, Nisbet AF, Gillett AG (2001) Predicting the transfer of radiocaesium from or ganic soils to plants using soil characteristics. J Environ Radioact 52:31–43

Alexakhin RM, Buldakov AA, Gubanov VA, Drozhko EG, Il'in LA, Kryshev II, Linge II, Romanov GN, Savkin MN, Saurov MM, Tikhomirov FA, Holina YB (2001) Severe nuclear accidents: consequences and protective measures. IzdAT, Moscow, p 752. (In Russian)

Askbrant S, Melin J, Sandalls J, Rauret G, Vallejo R, Hinton T, Cremers A, Vandecastelle C, Lewyckyi N, Ivanov Y, Firsakova S, Arkhipov N, Alexakhin R (1996) Mobility of radionuclides in undisturbed and cultivated soils in Ukraine, Belarus and Russia six years after the Chernobyl fallout. J Environ Radioact 31(3):287–312

Avdeev VA, Krivohatsky AS, Savonenkov VG, Smirnova EA (1990a) Leaching of radionuclides from nuclear fuel particles and reactor graphite extracted from samples of Chernobyl NPP 30-km zone. Radiochemistry 32(2):55–59. (In Russian)

Avdeev VA, Biryukov EI, Krivohatsky AS, Selifonov VI, Smirnova EA (1990b) Leaching of radionuclides by various solutions from soil samples collected in the vicinity of Chernobyl NPP. Radiochemistry 32(2):59–63. (In Russian)

Beresford N, Fesenko S, Konoplev A, Skuterud L, Smith JT, Voigt G (2016) Thirty years after the Chernobyl accident: what lessons have we learnt? J Environ Radioact 157:77–89

Bobovnikova TI, Virchenko EP, Konoplev AV, Siverina AA, Shkuratova IG (1991a) Chemical forms of occurrence of long-lived radionuclides and their alteration in soils near the Chernobyl Nuclear Power Station. Sov Soil Sci 23(5):52–57

Bobovnikova TI, Makhon'ko KP, Siverina AA, Rabotnova FA, Gutareva VP, Volokitin AA (1991b) Physical-chemical forms of radionuclides in atmospheric fallout, and their transformations in soil, after the accident at the Chernobyl atomic energy plant. Sov At Energy 71:932–936

Bogatov SA, Borovoi AA, Dubasov YV, Lomonosov VV (1990) Form and characteristics of fuel release at the Chernobyl accident. At Energy 69(1):36–40. (In Russian)

Borisyuk LG, Gavrilyuk VI, Dotsenko IS (1990) "Hot" particles in biosphere. Vesti Belarus Acad Sci Ser Phys Energ Sci 4:38–41. (In Russian)

Borzilov VA, Novitsky MA, Konoplev AV, Voszhennikov OI, Gerasimenko AC (1993) A model for prediction and assessment of water contamination in emergency situations and methodology of determining its parameters. Radiat Prot Dosim 50:349–351

Brouwer E, Baeyens B, Maes A, Cremers A (1983) Cesium and rubidium ion equilibria in illite clay. J Phys Chem 87:1213–1219

Bulgakov AA, Konoplev AV, Popov VE, Bobovnikova TI, Siverina AA, Shkuratova IG (1991) Mechanisms of the vertical migration of long-lived radionuclides in soils within 30-kilometers of the Chernobyl Nuclear Power Station. Soviet Soil Science, 1991 23(5):46–51

Bulgakov AA, Shkuta OV (2004) Modelling of radiocesium soil-plant transfer. Radiat Biol Radioecol 44(3):351–360. (In Russian)

Bulgakov A, Konoplev A, Smith J, Laptev G, Voitsekhovich O (2009) Fuel particles in the Chernobyl cooling pond: current state and prediction for remediation options. J Environ Radioact 100:329–332

Comans RNJ, Haller M, De Preter P (1991) Sorption of caesium on illite: nonequilibrium behaviour and reversibility. Geochim Cosmochim Acta 55:433–440

Comans RNJ, Hockley DE (1992) Kinetics of caesium sorption on illite. Geochim Cosmochim Acta 56:1157–1164

Cremers A, Maes A (1986) Radionuclide partitioning in environmental system: a critical analysis. In: Application of distribution coefficients to radiological assessment models. Proceedings of an international seminar, Leuven, 7–11 Oct 1985, pp 4–14

Cremers A, Elsen A, De Preter P, Maes A (1988) Quantitative analysis of radiocesium retention in soils. Nature 335(6187):247–249

De Preter P (1990) Radiocaesium retention in aquatic, terrestrial and Urban environment: a quantitative and unifying analysis. Ph.D. thesis. K.V.Leuven, Belgium, p 93

ECP-2 (1996) The transfer of radionuclides through the terrestrial environment to agricultural products, including the evaluation of agrochemical practices. In: Rauret G, Firsakova S (eds) Final report. European Commission EUR 16528 en, Brussels, p 182

Frere MH, Champion DF (1967) Characterization of fixed strontium in sesquioxide gel-kaolimite system. Soil Sci Soc Am Proc 31:181–191

Haynes RJ (1980) Ion exchange properties of roots and ionic interactions within the root apoplasm: their role in ion accumulation by plants. Bot Rev 46(1):75–99

Hilton J, Cambray RS, Green N (1992) Fractionation of radioactive caesium in airborne particles containing bomb fallout, Chernobyl fallout and atmospheric material from the Sellafield site. J Environ Radioact 15:103–108

IAEA (2006) Environmental consequences of the Chernobyl accident and their remediation: twenty years of experience. IAEA, Vienna, p 166

IAEA (2010) Handbook of parameter values for the prediction of radionuclide transfer in terrestrial and freshwater environments. TRS-472. IAEA, Vienna, p 194

Il'in LA, Gubanov VA (2001) Severe radiation accidents: consequences and countermeasures. IzdAT, Moscow, p 751. (In Russian)

Ivanov (1997) Radioecological basis for long-term prediction of radioactive contamination in agriculture in case of severe nuclear accidents (Chernobyl NPP case study). Thesis for degree of Doctor of Science, RIARAE, Obninsk: 250 p.

Izrael YA, Vakulovsky SM, Vetrov VA, Petrov VN, Rovinsky AY, Stukin ED (1990) Chernobyl: radioactive contamination of the environment. Leningrad, Hydrometeoizdat, p 296. (In Russian)

Jacobs DG (1962) Cesium Exchange Properties of vermicullite. Nucl Sci Eng 12:285–292

Kanivets V, Laptev G, Konoplev A, Lisovyi H, Derkach G, Voitsekhovich O (2020) Distribution and dynamics of radionuclides in the Chernobyl Cooling Pond. In: Konoplev A, Kato K, Kalmykov S (Editors) Behavior of radionuclides in the environment II: Chernobyl. SPRINGER NATURE

Kashparov VA, Ivanov YA, Zvarich SI, Protsak VP, Khomutinin YV, Kurepin AD, Pazukhin EM (1996) Formation of hot particles during the Chernobyl nuclear power plant accident. Nucl Technol 114:246–253

Kashparov VA, Oughton DH, Zvarich SI, Protsak VP, Levchul SE (1999) Kinetics of fuel particles weathering and ^{90}Sr mobility in 30-km exclusion zone. Health Phys 76(3):251–259

Kashparov VA, Protsak VP, Ahamdach N, Stammose D, Peres JM, Yoschenko VI, Zvarich SI (2000) Dissolution kinetics of particles of irradiated Chernobyl nuclear fuel: influence of pH and oxidation state on the release of radionuclides in contaminated soil of Chernobyl. J Nucl Mater 279:225–233

Kashparov VA, Ahamdach N, Zvarich SI, Yoschenko VI, Maloshtan IM, Dewiere L (2004) Kinetics of dissolution of Chernobyl fuel particles in soil in natural conditions. J Environ Radioact 72:335–353

Kashparov V, Levchuk S, Zhurba M, Protsak V, Khomutinin Y, Beresford NA, Chaplow JS (2018) Spatial datasets of radionuclide contamination in the Ukrainian Chernobyl Exclusion Zone. Earth Syst Sci Data 10:339–353

Kokotov YA, Wilken SR (1969) Some issues of ion fixation by clays, micas and soils. In: Radioactive isotopes in soils and plants. Kolos, Leningrad, pp 35–43. (In Russian)

Konoplev AV, Borzilov VA, Bobovnikova TI, Virchenko EP, Popov VE, Kutnyakov IV, Chumichev VB (1988) Distribution of radionuclides in the soil-water system fallen out in consequence of the Chernobyl failure. Russ Meteorol Hydrol 12:42–49

Konoplev AV, Bobovnikova TSI (1991) Comparative analysis of chemical forms of long-lived radionuclides and their migration and transformation in the environment following the Kyshtym and Chernobyl accidents. Proceedings of seminar on comparative assessment of the environmental impact of radionuclides released during three major nuclear accidents: Kyshtym, Windscale, Chernobyl. Luxembourg, 1–5 Oct 1990, pp 371–396

Konoplev AV, Golubenkov AV (1991) Modeling of the vertical radionuclide migration in soil (as a result of a nuclear accident). Meteorol Hydrol 10:62–68. (In Russian)

Konoplev AV, Bulgakov AA, Popov VE, Bobovnikova TI (1992) Behaviour of long-lived radionuclides in a soil-water system. Analyst 117:1041–1047

Konoplev AV, Viktorova NV, Virchenko EP, Popov VE, Bulgakov AA, Desmet GM (1993) Influence of agricultural countermeasures on the ratio of different chemical forms of radionuclides in soil and soil solution. Sci Total Environ 137:147–162

Konoplev AV, Comans RNJ, Hilton J, Madruga MJ, Bulgakov AA, Voitsekhovich, Sansone U, Smith, Kudelsky AK (1996) Phisico-chemical and hydraulic mechanisms of radionuclide mobilization in aquatic systems. In: Karaouglou A, Desmet G, Kelly GN, Menzel HG (eds) Proceedings of the first international conference "The radiological consequences of the Chernobyl accident", Minsk, Belarus, 18–22 March 1996. ECSC-EC-EAEC, Brussels, pp 121–136

Konoplev AV, Bulgakov AA, Hoffman FO, Kanyar B, Lyashenko G, Nair SK, Popov A, Rascob W, Thiessen KM, Watkins B, Zheleznyak M (1996a) Validation of models of radionuclide wash-off from contaminated watersheds using Chernobyl data. J. Environ. Radioactivity 42(2–3):131–141

Konoplev AV, Drissner J, Klemt E, Konopleva IV, Zibold G (1996b) Parameterisation of radiocaesium soil-plant transfer using soil characteristics. Proceedings of XXVIth annual meeting of ESNA, working group 3: soil-plant relationships. Busteni (Romania), 12–16 Sep 1996, pp 147–153

Konoplev AV (1998) Mobility and bioavailability of accidentally released radiocesium and radiostrontium in soil-water environment. Dissertation of Doctoral Science. RIARAE, Obninsk, p 250

Konoplev AV, Bulgakov AA (1999) Kinetics of ^{90}Sr leaching from fuel particles in soils of the Chernobyl exclusion zone. At Energy 86(2):136–141

Konoplev AV, Konopleva IV (1999a) Characteristics of steady-state selective sorption of radiocesium on soils and bottom sediments. Geochem Int 37(2):177–184

Konoplev AV, Konopleva IV (1999b) Parameterisation of ^{137}Cs soil-plant transfer through key soil characteristics. Radiat Biol Radioecol 39(4):457–463. (In Russian)

Konoplev AV, Bulgakov AA, Popov VE, Avila R, Drissner J, Klemt E, Miller R, Zibold G, Johanson K-J, Konopleva IV, Nikolova I (1999) Modelling radiocaesium bioavailability in forest soils. In: Linkov I, Schell WR (eds) Contaminated forests: recent developments in risk identification and future perspectives. IAEA, Vienna, pp 217–229

Konoplev AV, Bulgakov AA (2000a) Transformation of ^{90}Sr and ^{137}Cs speciation in soil and bottom sediments. At Energy 88:56–60

Konoplev AV, Bulgakov AA (2000b) ^{90}Sr and ^{137}Cs exchangeable distribution coefficient of in soil-water system. At Energy 88:158–163

Konoplev AV, Avila R, Bulgakov AA, Johanson K-J, Konopleva IV, Popov VE (2000) Quantitative assessment of radiocaesium bioavailability in forest soils. Radiochim Acta 88 (9–11):789–792

Konoplev A (2001) Chemical forms of radiocaesium and radiostrontium in Chernobyl deposition. In: Van der Stricht E, Kirchman R (eds) Radioecology radioactivity & ecosystems. IUR, Belgium, pp 92–98

Konoplev AV, Kaminski S, Klemt E, Konopleva I, Miller R, Zibold G (2002) Comparative study of ^{137}Cs partitioning between solid and liquid phases in Lakes Constance, Lugano and Vorsee. J Environ Radioact 58:1–11

Konopleva I, Klemt E, Konoplev A, Zibold G (2009) Migration and bioavailability of ^{137}Cs in forest soils of Southern Germany. J Environ Radioact 100(4):315–321

Loschilov NA, Kashparov VA, Yudin EB, Protsak VP, Zhurba MA, Parshakov AE (1991) Physico-chemical characteristics of radioactive fallout after Chernobyl accident. In: Loschilov NA (ed) Problems of agricultural radiology. IAEA, Vienna, pp 8–12

Madruga MJ, Cremers A, Bulgakov A, Zhirnov V, Bobovnikova TI, Laptev G, Smith J (1996) Radiocesium sorption-desorption behavior in soils and sediments. In: Karaouglou A, Desmet G, Kelly GN, Menzel HG (eds) Proceedings of the first international conference "The radiological consequences of the Chernobyl accident", Minsk, Belarus, 18–22 Mar 1996. ECSC-EC-EAEC, Brussels, pp 209–211

Martyushov VV, Spirin DA, Bazylev VV, Fedorova TA, Martyushov VZ, Panova LA (1995) Radionuclide speciation in soils of East Ural radioactive trail. Ecology 2:110–113. (In Russian)

Matsunaga T, Nagao S, Ueno T, Takeda S, Amano H, Tkachenko Y (2004) Association of dissolved radionuclides released by the Chernobyl accident with colloidal materials in surface water. Appl Geochem 19:1581–1599

Megaw WJ, Chadwick RC, Wells AC, Bridges JE (1960) The ignition and oxidation of overheated model Calder fuel elements and the release of iodine-131 from them. UK AEA report AERE R3435. HMSO, London

Monte L, Perianes R, Boyer P, Smith JT, Brittain JE (2009) The role of physical processes controlling the behaviour of radionuclide contaminants in the aquatic environment: a review of state-of-the-art modelling approaches. J Environ Radioact 100:779–784

Nair SK, Hoffman FO, Thiessen KM, Konoplev AV (1996) Modeling the wash-off of Sr-90 and Cs-137 from an experimental plot established in the vicinity of the Chernobyl reactor. Health Phys 71(6):896–909

Orlov DS, Sadovnikova LK, Sukhanova NI (2005) Soil chemistry. Vysshaya Shkola, Moscow, p 558. (In Russian)

Oughton DH, Salbu B, Riise G, Lien H, Oestby G, Noeren A (1992) Radionuclide mobility and bioavailability in Norwegian and Soviet soils. Analyst 17:481–486

Pavlotskaya FI (1974) Migration of radioactive products of global fallout in soils. Atomizdat, Moscow, p 270. (In Russian)

Petryaev EP, Ovsyannikova SV, Lyubkina IY, Rubinchuk SY, Sokolik GA (1993) Speciation of Chernobyl-derived radionuclides in soils outside 30-km zone. Geochemistry 7:940–947. (In Russian)

Richards L (1954) Diagnosis and improvement of saline and alkali soils. USDA. USDA, New York, NY

Salbu B, Krekling T, Oughton DH, Ostby G, Kashparov VA, Brand TL, Day JP (1994) Hot particles in accidental releases from Chernobyl and Windscale nuclear installations. Analyst 119:125–130

Salbu B (2001) Hot particles – a challenge within radioecology. J Environ Radioact 53:267–268

Salbu B, Lind OC, Skipperud L (2004) Radionuclide speciation and its relevance in environmental impact assessments. J Environ Radioact 74:233–242

Salbu B, Kashparov V, Lind O, Garcia-Tenorio R, Johansen MP, Child DP, Roos P, Sancho C (2018) Challenges associated with the behaviour of radioacvtive particles in the environment. J. Environ Radioact 186: 101–115

Sandalls FJ, Segal MJ, Victorova N (1993) Hot particles from Chernobyl: a review. J Environ Radioact 18:5–22

Sanchez AL, Wright SM, Smolders E, Naylor C, Stevens PA, Kennedy VH, Dodd BA, Singleton DL, Barnett CL (1999) High plant uptake of radiocesium from organic soils due to Cs mobility and low soil K content. Environ Sci Technol 33:2752–2757

Sawhney BL (1972) Selective sorption and fixation of cations by clay minerals: a review. Clay Clay Miner 20:93–100

Shaw G (1993) Blockade by fertilisers of caesium and strontium uptake into crops: effects on the root uptake process. Sci Total Environ 137(1–3):119–133

Smith JT, Comans RNJ (1996) Modelling the diffusive transport and remobilisation of ^{137}Cs in sediments: the effect of sorption kinetics and reversibility. Geochim Cosmochim Acta 60 (6):995–1004

Smith JT, Belova NV, Bulgakov AA, Comans RNJ, Konoplev AV, Kudelsky AV, Madruga MJ, Voitsekhovich OV, Zibold G (2005a) The "AQUASCOPE" simplified model for predicting

$^{89,90}Sr$, and $^{134,137}Cs$ in surface waters after a large-scale radioactive fallout. Health Phys 89 (6):628–644
Smith JT, Voitsekhovich OV, Konoplev AV, Kudelsky AV (2005b) Radioactivity in aquatic systems. In: Smith JT, Beresford NA (eds) Chernobyl catastrophe and consequences. Springer-Praxis, Zürich, pp 139–190
Smith JT, Konoplev AV, Voitsekhovich OV, Laptev GV (2009) The influence of hot particle contamination on models for radiation exposures via the aquatic pathway. In: Radioactive particles in the environment. Springer, New York, NY, pp 249–258
Smolders E, Sweeck L, Merckx R, Cremers A (1997) Cationic interactions in radiocaesium uptake from solution by spinach. J Environ Radioact 34(2):161–170
Smolders E, Van den Brande K, Merckx R (1997a) The concentration of ^{137}Cs and K in soil solution predict the plant availability of ^{137}Cs in soil. Environ Sci Technol 31:2752–2757
Sweeck L, Wauters J, Valcke E, Cremers A (1990) The specific interception potential of soils for radiocaesium. In: Desmet G, Nassimbeni P, Belli M (eds) Transfer of radionuclides in natural and semi-natural environments. Elsevier Applied Science, Amsterdam, pp 249–258
Sysoeva AA, Konopleva IV, Sanzharova NI (2005) Bioavailability of radiostrontium in soil: experimental study and modeling. J Environ Radioact 81:269–282
Talerko M, Garger E, Lev T, Nosovsky A (2020) Atmospheric transport of radionuclides initially released as a result of the Chernobyl accident. In: Konoplev A, Kato K, Kalmykov S (Editors) Behavior of radionuclides in the environment II: Chernobyl. SPRINGER NATURE
Tamura T, Jacobs DG (1960) Structural implications in cesium sorption. Health Phys 2:391–398
Tessier A, Campbell PGC, Bisson M (1979) Sequential extraction procedure for the speciation of particulate trace metals. Anal Chem 51(7):844–851
Timofeev-Resovsky NV, Titlyanova AA (1966) Behaviour of radioactive isotopes in the system "soil-water". In: Radioactivity of soils and methods of its determination. Nauka, Moscow, pp 46–80. (In Russian)
Tsukada H, Takeda A, Hisamatsu S, Inaba J (2008) Concentration and specific activity of fallout ^{137}Cs in extracted and particle-size fractions of cultivated soils. J Environ Radioact 99:875–881
Valcke E, Cremers A (1994) Sorption-desorption dynamics of radiocaesium in organic matter soils. Sci Total Environ 157:275–283
Vdovenko VM (1960) Chemistry of Uranium and trans-uranium elements. Russian Academy of Science, Moscow, p 700. (In Russian)
Victorova NV, Garger EK (1991). Investigation of the deposition and spread of radioactive aerosol particles in the Chernobyl zone based on biological monitoring. Proceedings of the CEC seminar on comparative assessment of the environmental impact of radionuclides released during three major nuclear accidents: Kyshtym, Windscale and Chernobyl. Luxembourg, 1–5 Oct proceedings report EUR 13574, CEC, Luxembourg, pp 223–236
Victorova N, Voitsekhovich O, Sorochinsky B, Vandenhove H, Konoplev A, Konopleva I (2000) Phytoremediation of Chernobyl contaminated land. Radiat Prot Dosim 92(1–3):59–64
Von Reichenbach HG (1968) Cation exchange in the interlayers of expansible layer silicates. Clay Miner 7:331–341
Wauters J, Sweeck L, Valcke E, Elsen A, Cremers A (1994) Availability of radiocaesium in soils: a new methodology. Sci Total Environ 157:239–248
Wauters J, Madruga MJ, Vidal M, Cremers A (1996a) Solid phase speciation of radiocesium in bottom sediments. Sci Total Environ 187:121–130
Wauters J, Elsen A, Cremers A, Konoplev A, Bulgakov A, Comans RNJ (1996b) Prediction of solid/liqiud distribution coefficients of radiocesium in soils and sediments. Part one: a simplified procedure for the solid phase characterisation. Appl Geochem 11:589–594
Wauters J, Vidal B, Elsen A, Cremers A (1996c) Prediction of solid-liquid distribution coefficients of radiocaesium in soils and sediments. Part II: a new procedure for solid phase speciation of radiocaesium. Appl Geochem 11:595–599

Wauters J, Elsen A, Cremers A (1996d) Prediction of solid-liquid distribution coefficients of radiocaesium in soils and sediments. Part III: a quantitative test of a K_d predictive equation. Appl Geochem 11:601–603

Wicker FW, Schultz V (1982) Radioecology: nuclear energy and the environment, vol 1. CRC Press, Boca Raton, FL, p 195

Zheleznyak M, Demchenko R, Khursin VN, Kuzmenko S, Tkalich P (1992) Mathematical modeling of radionuclide dispersion in Pripyat-Dneper aquatic system after the Chernobyl accident. Sci Total Environ 112:89–114

Zheleznyak M, Kivva S, Ievdin I, Boyko O, Kolomiets P, Sorokin M, Mikhalsky O, Gheorghiu D (2016) Hydrological dispersion module of JRODOS: renewed chain of the emergencyresponse models of radionuclide dispersion through watersheds and rivers. Radioprotection 512:129–131

Chapter 4
Quantitative Assessment of Lateral Migration of the Chernobyl-Derived ^{137}Cs in Contaminated Territories of the East European Plain

Valentin Golosov and Maxim Ivanov

Abstract Results of detailed studies of Chernobyl-derived ^{137}Cs redistribution by fluvial processes in different landscape conditions of the East European Plain are presented. Lateral migration of ^{137}Cs is mostly associated with sediment transport along the pathways from the cultivated slopes through dry valleys to the river valley bottom. Most of the Chernobyl-derived ^{137}Cs redeposited in the low parts of the cultivated field and on the dry valley bottoms, where the total ^{137}Cs inventories exceeded initial fallout deposits after three decades from April to May 1986. Ponds and reservoirs, as well as low-level river floodplains, are also areas with high ^{137}Cs inventories due to high sedimentation rates. However, less than 0.3–0.5% of the total inventory of Chernobyl-derived ^{137}Cs was washed out from the areas with a high (>50%) proportion of arable lands over the past three decades. Climate and land-use changes observed during the last 30 years led to a considerable reduction of erosion rates on the arable lands and accordingly lateral migration of particulate ^{137}Cs.

Keywords Chernobyl-derived ^{137}Cs · Lateral migration · Water erosion · Sediment sinks · Climate change · Land-use changes · East European Plain

4.1 Introduction

The radioactive contamination resulting from the Chernobyl-derived ^{137}Cs fallout is inevitably transformed for two main reasons. The first cause is the process of radioactive decay contributing to a steady reduction of radionuclide content in the environment. Secondly, due to the denudation and accumulation of contaminated sediments, lateral migration of ^{137}Cs occurs. Those factors accelerated the reduction

V. Golosov (✉) · M. Ivanov
Faculty of Geography, Lomonosov Moscow State University, Moscow, Russia

Institute of Geography, Russian Academy of Sciences, Moscow, Russia

A. Konoplev et al. (eds.), *Behavior of Radionuclides in the Environment II*,
https://doi.org/10.1007/978-981-15-3568-0_4

of ^{137}Cs inventory in geomorphic units with the prevalence of erosion processes and led to an increase of ^{137}Cs inventories in the areas of accumulation. Simultaneously, there are areas in the landscape that are not affected by erosion, where reduction of ^{137}Cs inventory occurred only through its decay. The proportion of such territories depends on topography: an increase in ^{137}Cs inventory was observed within the lowlands, while reduction was found in the mountainous regions. Vegetation is another factor that limits the intensity of erosion and accumulation processes; its disturbance due to anthropogenic impact is believed to be critical. This chapter deals with the transformation of Chernobyl contamination of territories of Eastern Europe due to ^{137}Cs lateral migration for the past over 30 years after the accident. We do not discuss the transfer of ^{137}Cs associated with wind erosion since this is a separate issue requiring independent consideration. Moreover, wind erosion is argued to be negligible in the examined areas.

4.2 Spatial Variability of the Chernobyl-Derived ^{137}Cs Deposition

The fallout of Chernobyl-derived ^{137}Cs was associated with single-rainfall events during the period between 27 April and 15 May 1986 (Izrael 1998; De Cort et al. 1998). Broad areas in Europe and especially in Eastern Europe were subjected to radioactive contamination of different intensities (Table 4.1). (Izrael 1998; De Cort et al. 1998). A large-scale study of spatial variability of ^{137}Cs deposits in 1986 was undertaken across the most part of Europe. Soil sampling was carried out at numerous test sites for validation of the airborne survey results. Based on the survey

Table 4.1 Areas in each country with ^{137}Cs deposition in excess of specified levels in 1986 (De Cort et al. 1998)

Country	Area (10^3 km^2) with contamination level		Country	Area (10^3 km^2) with contamination level	
	>40 kBq/m^2	>1480 kBq/m^{2a}		>40 kBq/m^2	>1480 kBq/m^2
Austria	11	–	Poland	0.52	–
Belarus	46	2.6	Romania	1.2	–
Czech Republic	0.21	–	Russia (European)	60	0.46
Estonia	0.01	–	Slovak Republic	0.02	–
Finland	19	–	Slovenia	0.61	–
Germany	0.32	–	Sweden	24	–
Greece	1.2	–	Ukraine	38	0.56
Italy[a]	1.3	–	UK	0.16	–
Norway	7.1	–			

[a]Excluding sicily

results, an Atlas of the Chernobyl-derived ^{137}Cs contamination of Europe was compiled and published in 1998 (De Cort et al. 1998).

A comprehensive study of ^{137}Cs migration requires taking into account local variations and spatial trends of Chernobyl fallout. It is only possible to identify redeposition areas of sediments containing ^{137}Cs via depth incremental sampling in case when high coefficient of variation ($C_v > 25\%$) was found as, for example, within the near 40-km zone, where a large number of hot particles contributed to a very high soil contamination variability. Also, in terms of tracing ^{137}Cs lateral migration, it is essential to recognize that the Chernobyl-derived ^{137}Cs fallout was deposited on the soil containing bomb-derived ^{137}Cs since 1954, though it was already partly redistributed. For example, Higgitt et al. (1992) found that the spatial variability of Chernobyl-derived ^{137}Cs was very high in the Polish Carpathian Mountains. However, on a flat meadow in Salzburg, Austria, the initial spatial variability of the Chernobyl-derived ^{137}Cs fallout turned out to be close to 17–21% (Lettner et al. 2000). Those values are within the range of values determined for the bomb-derived ^{137}Cs fallout (Sutherland 1994, 1996; Owens and Walling 1996). Flat meadows with a relatively uniform grassy vegetation cover constitute an ideal reference site with the lowest spatial variability of ^{137}Cs inventories (C_v = 9–10%) (Golosov et al. 1999a, 2008a, 2013). It has been suggested that soil cultivation reduces random spatial variability (Owens and Walling 1996). Only the variable depth of plowing is most probably responsible for the nonuniform mixing of cultivated soils. Therefore, two or three bulk samples should be collected at each sampling point on the cultivated surface and be mixed into a single sample to decrease the possible effect of variability of the ^{137}Cs inventory (both at the reference sites and on eroded hillslopes) (Golosov et al. 2013).

Clay minerals of soil and sediment particles rapidly and strongly adsorb ^{137}Cs after its fallout. Particulate ^{137}Cs migrates in the landscape together with the transport of soil and sediment particles. Correct assessment of ^{137}Cs inventories for the reference sites is a key for investigation of radionuclide redistribution or lateral migration along the slopes or the entire catchment (Owens and Walling 1996; Quine 1989; Sutherland 1991; Walling and He 1999). A reference site is a location without any disturbance by erosion or deposition processes, given no trees or bushes that might influence the fallout of ^{137}Cs from the atmosphere on the soil surface. Each reference site is supposed to contain enough sampling points for appropriate evaluation of random (local) variations. Those variations in ^{137}Cs concentrations seem to depend on the intensity of contamination (Linnik et al. 2007). The highest variability was detected in the 100-km zone around the Chernobyl power plant with a deposition level of more than 1500 kBq/m^2. Soil samples, which were taken in this zone from the area of 100 m^2, showed variations of as much as three- to tenfold (Izrael et al. 1990, 1994).

Table 4.2 presents a summary of statistics for the local spatial variability of ^{137}Cs inventories within individual reference locations for the four key catchments located in areas with different levels of the initial Chernobyl-derived ^{137}Cs fallout. It can be seen that the coefficient of variation of ^{137}Cs inventories remains within 9–22% for the majority of reference sites. This range of values is close to that observed in the

Table 4.2 Statistics of ^{137}Cs inventory for the reference sites (modified after Golosov et al. 2008a, b)

Region, river basin, years of study	Reference site	Number of samples	Mean value, kBq/m^2	Range, kBq/m^2	C_v, %[a]
Plavsk, the Lokna River, 1996–1998	1	6	315 ± 20	274–363	9
	2	28	563 ± 28	380–828	16
	3	9	378 ± 41	296–505	20
	4	12	350 ± 16	295–402	10
Novosil, the Zusha River, 2001–2002	1	12	31.8 ± 1.3	29.2–39.2	9
	2	11	27.7 ± 2.6	18.2–37.1	19
	3	9	27.6 ± 1.7	23.2–32.1	11
	4	12	42.3 ± 3.4	31.7–52.5	16
Kursk, the Seim River, 2006–2007	1	12	9.3 ± 2.1	5.2–11.5	22
	2	12	7.5 ± 1.2	6.3–10.2	16
	3	12	9.1 ± 1.2	7.2–11.0	13
	4	12	8.5 ± 1.1	6.3–10.2	13
Stavropol, the Kalaus River, 2004–2005	1	12	4.9 ± 0.4	3.3–6.5	28
	2	11	3.6 ± 0.4	2.2–4.6	23
	3	13	2.1 ± 0.6	0.5–4.5	60

[a]C_v means the coefficient of variation

northern hemisphere for global ^{137}Cs fallout (Foster et al. 1994; Sutherland 1996; Owens and Walling 1996). However, a substantially higher spatial variability of ^{137}Cs inventories ($C_v = 23\%$, 28%, and 60%) was observed for reference sites in the Stavropol Region, which were characterized by relatively similar inputs of bomb-derived and Chernobyl-derived ^{137}Cs fallout (Table 4.2).

This increased variability of ^{137}Cs deposition can be partly explained by higher spatial variations of rainfall intensity in the southern part of the East European Plain and by deep cracking of topsoil during the dry summer period. Following this, accelerated migration of radionuclides through irregular infiltration during heavy rainstorms may occur (Belyaev et al. 2004). The Stavropol Region, with its climate conditions affected by heavy rainstorms in the total amount of precipitation, can be likened to territories prevailing in the Great Plains of the USA (Turnage et al. 1997). So a high initial variability of bomb-derived ^{137}Cs fallout was identified in the Great Plains also (Turnage et al. 1997). A very high variability ($C_v = 60\%$) of ^{137}Cs deposits at one of the sites in the Stavropol Region may be caused by bioturbation (Table 4.2).

Despite the aforementioned confirmation of relatively low local spatial variability of the Chernobyl-derived ^{137}Cs within individual reference sites, it is well known that a more significant spatial variability frequently occurred within small catchments. This irregularity is inherent to the nature of the Chernobyl-derived ^{137}Cs fallout, which was, in most cases, associated with a single rainstorm (Izrael 1998). It can be illustrated by the examples of different reference sites located only a few kilometers from each other within the Lokna River basin (Table 4.2) and the Zusha River basin (Table 4.2, Fig. 4.1A,B). Similar spatial variability is also documented

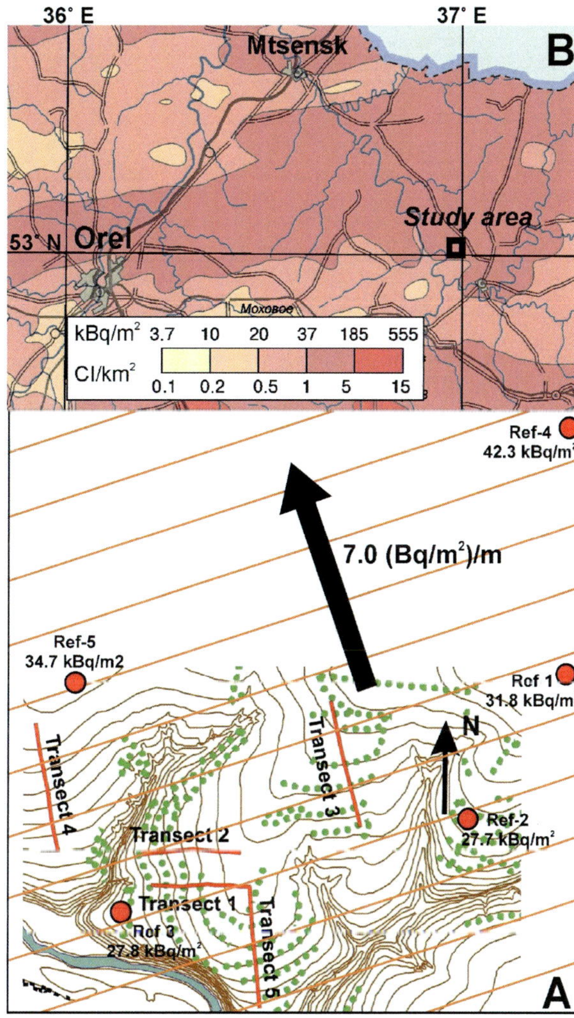

Fig. 4.1 The ^{137}Cs fallout inventory spatial trend (large arrow) established from field data (**a**) (Belyaev et al. 2009) and published information (**b**) (Izrael 1998)

for the bomb-derived ^{137}Cs fallout in the areas with Mediterranean climate: the Eastern part of Spain and the Southern part of Italy (Navas et al. 2007; Porto et al. 2009). Therefore, in some cases, several reference sites should be selected around the study area to evaluate the existence of spatial trend of the initial ^{137}Cs fallout. The Atlas of radionuclide contamination in European Russia, Belorussia, and Ukraine (Izrael 1998) may be used to select the appropriate reference site locations. For example, the Atlas map indicates the spatial trend of the initial Chernobyl-derived ^{137}Cs fallout in the central part of the Zusha River basin increased from southeast to northwest direction. The subsequent determination of local ^{137}Cs inventories at five reference locations confirmed this trend (Belyaev et al. 2009) (Fig. 4.1).

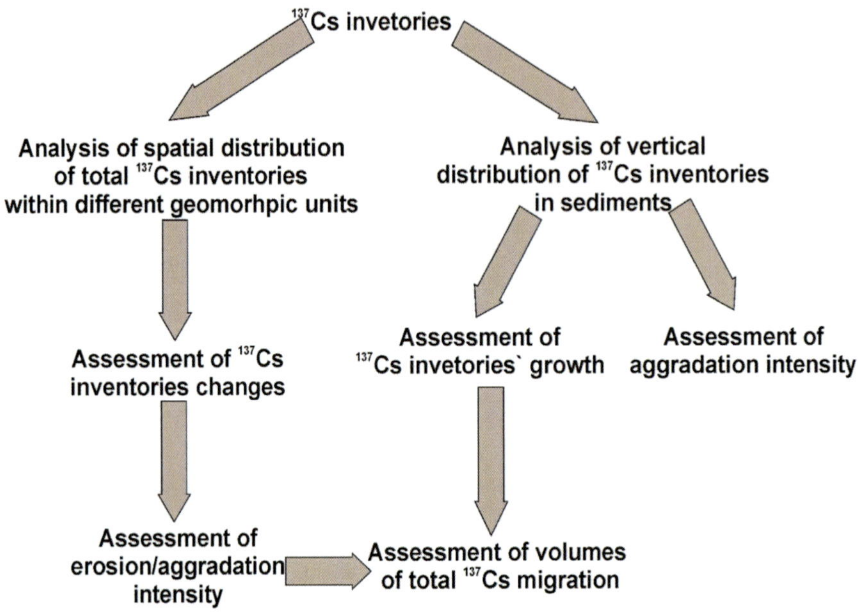

Fig. 4.2 Geomorphic approach for evaluation of ^{137}Cs migration caused by erosion in Chernobyl-affected areas

The precision of information related to the initial Chernobyl fallout decreased significantly for the areas with relatively low Chernobyl contamination levels due to the reduction of airborne gamma-spectrometric survey spatial resolution and the higher spatial variability of rainfall. In those cases, when the initial fallout trend cannot be detected, the intensity of ^{137}Cs migration can be quantified by using methods for estimating the rates of erosion and sedimentation processes. The processes of lateral migration of Chernobyl-derived ^{137}Cs is associated with erosion and deposition processes in various parts of the fluvial network and can be schematically represented as follows (Fig. 4.2).

The selected depth distribution curves of ^{137}Cs for uncultivated reference locations (Fig. 4.3a, b) demonstrate the typically observed exponential shape, with a maximum activity within the top 2–3 cm and most of the ^{137}Cs inventory contained within the upper 15–20 cm (Golosov et al. 1999a). It can be suggested that the depth of the Chernobyl-derived ^{137}Cs penetration into the soil profile is in the same range as that of the bomb-derived ^{137}Cs. For the cultivated sites, the major part of the ^{137}Cs deposits is contained within the plow layer. In contrast, deeper penetration of the ^{137}Cs is negligible (Fig. 4.3c, d).

Within the territories with very high deposits of Chernobyl-derived ^{137}Cs, field detectors can be used to determine the variability of the ^{137}Cs fallout as well as subsequent redistribution. It makes possible to determine the ^{137}Cs activity in an area of several square meters directly in the field with high accuracy (Fedin et al. 1997; Golosov et al. 2000a; Martynenko et al. 2003; Evrard et al. 2012) (Fig. 4.4).

It can be concluded that regions with high and moderate levels of Chernobyl-derived ^{137}Cs contamination are characterized by relatively low local spatial

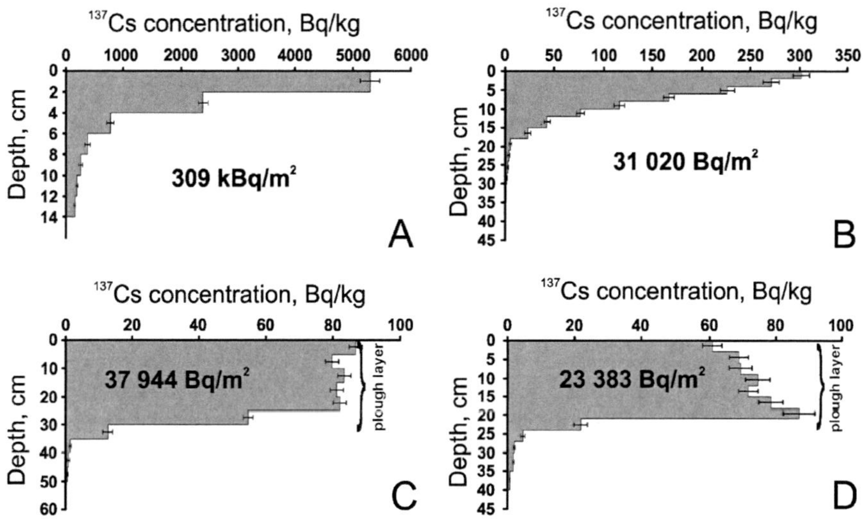

Fig. 4.3 Depth distribution curves of ^{137}Cs at selected reference locations (Table 4.2): (**a**) Lokna River basin, site 1; (**b**) Zusha River basin, site 1; (**c**) Zusha River basin, individual cultivated interfluve section; (**d**) Zusha River basin, site 2 (Golosov et al. 2013)

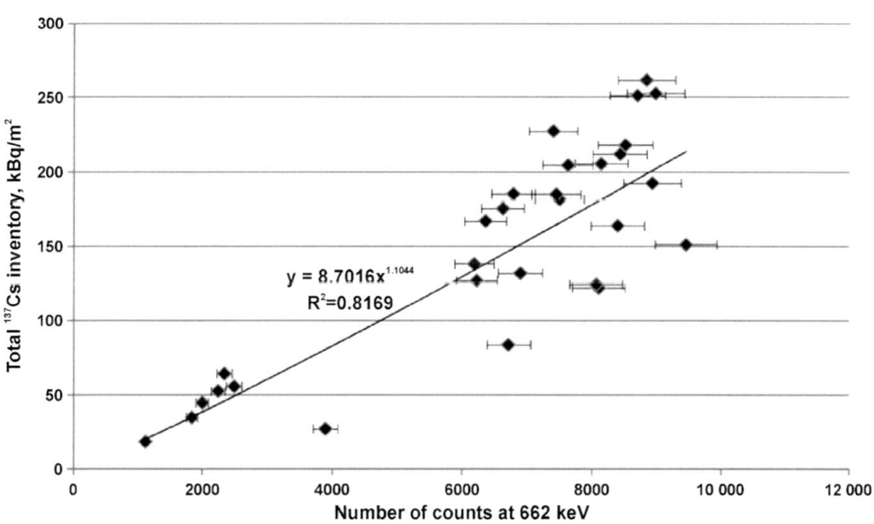

Fig. 4.4 The power function model developed to relate the field measurements of ^{137}Cs inventory to the laboratory measurements of ^{137}Cs inventories in soil, undertaken in the Plava River basin (August 2011). (Evrard et al. 2012)

variability (<20%). However, the detection of spatial fallout trend remains a crucial issue. In the subsequent sections of the chapter, we sequentially consider the features of the ^{137}Cs redistribution by water erosion in various parts of the fluvial network, starting from cultivated slopes and ending with the river basin as a whole.

4.3 Migration of the Chernobyl-Derived Particulate ^{137}Cs at The Cultivated Field Scale

Most of the Chernobyl fallout occurred in the central part of the East European Plain (Izrael et al. 1994; De Cort et al. 1998) within a temperate climatic zone with sufficient moisture, which determines the high projective coverage of the soil by vegetation in natural conditions unchanged by anthropogenic activity. As a result, lateral migration of ^{137}Cs occurred only in cases when vegetation was disturbed, and the projective coverage of the soil surface is reduced due to the overgrazing of livestock. ^{137}Cs lateral migration is observed in minimal areas and mainly contributes to the local transfer (the maximum distance is tens meters) of soil particles containing ^{137}Cs. The most significant lateral migration of particulate ^{137}Cs is detected on arable lands, whose share in the forest-steppe and steppe zones within the East European Plain (the territory of the Russian Federation and Ukraine) is more than 50%. In the forest zone (the territory of Republic of Belarus, the north of Ukraine, and the west of the Russian Federation on the border with the Republic of Belarus and Ukraine), the share of arable land, in general, does not exceed 40%, and the arable land is located significantly far from permanent watercourses. The share of arable land in the forest zone of the rest of the European part of Russia is even lower, decreasing gradually to 1–2% in the north of the taiga zone.

A high dissection characterizes the East European Plain. Sides and bottoms of dry valleys are used as pastures or meadows. At the same time, a significant part of the interfluve slopes is plowed up. At the time of the Chernobyl accident, most of the interfluves were plowed up, and the proportion of arable land reached its historical maximum. The lack of natural vegetation contributed to the development of sheet and rill erosion, which became the primary mechanism of Chernobyl-derived ^{137}Cs lateral migration together with soil particles. Topographical features of the terrain control the pathways that the sediments make after flushing from the arable land and entering the bottoms of the dry valleys and further into permanent watercourses (Panin et al. 2001).

The actual redistribution of ^{137}Cs can be examined based on in situ measurement by a field detector without soil sampling at small catchments with predominantly cultivated slopes of interfluves in areas with high levels of radioactive contamination (Table 4.3). It was established that a slight decrease in the ^{137}Cs inventories on arable slopes and its increase in the accumulation positions (the undisturbed sides of the valley and valley bottom) was observed already 10–12 years after the Chernobyl fallout in 1986 (Table 4.3). This is due to the processes of flushing the soil from arable land and redepositing a significant part of the washed-out soil on the covered by grass vegetation sides and bottom of the dry valley.

A more detailed pattern of ^{137}Cs inventory change within the arable slopes can be established by sampling along the runoff lines, from the upper part of the slope to bottom parts (Golosov et al. 1998a). The sections with a reduced and relatively high content of the ^{137}Cs are distinctive on the convex slopes, which are typical for the East European Plain (Fig. 4.5). A similar wave-like pattern of the ^{137}Cs

Table 4.3 Summary statistics and random spatial variability of ^{137}Cs inventories in geomorphic units of an area with Chernobyl deposits >40 kBq/m^2 (the Lokna River basin, Tula region) (Golosov 2003)

Geomorphic unit	Number of observations	Mean (kBq/m^2)	Standard deviation (kBq/m^2)	Coefficient of variation (%)
Cultivated slopes near watershed line with very low gradient	35[a]/42[b]	515/407	129/54	25/13
Cultivated slopes	96/103	372/403	60/48	16/12
Uncultivated valley sides and bottom	13/23	481/461	85/51	18/11

[a]Results of laboratory analysis
[b]In situ measurements

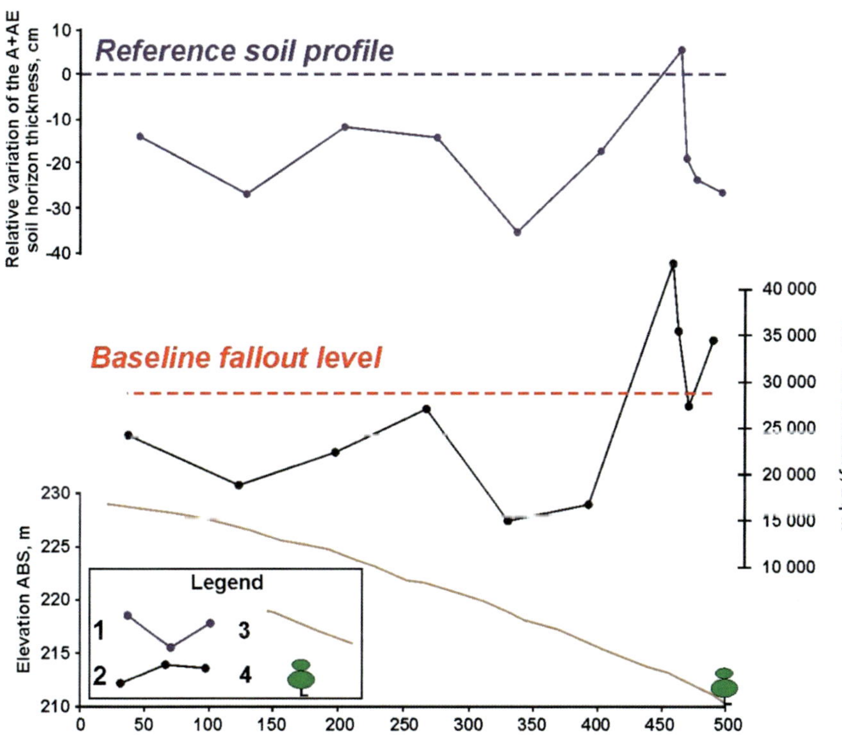

Fig. 4.5 Distribution of soil horizons A + AE relative thickness and ^{137}Cs inventories along the transect. Legend: (1) Relative variation of A + AE soil horizons thickness; (2) ^{137}Cs inventory; (3) Slope long profile along transects; and (4) Forest shelter belts (Belyaev et al. 2009)

redistribution was revealed on the arable slopes located in the Chernobyl-affected territories with different levels of initial contamination (Belyaev et al. 2003, 2009; Golosov et al. 2000b, 2011; Tsybul'ka et al. 2004; Markina et al. 2014; Zhidkin et al. 2015). The sediment accumulation is predominated at the bottom of a cultivated field nearby the tillage border, which contributes to a sharp increase in the ^{137}Cs inventory, usually exceeding initial fallout in 1986 (Fig. 4.5) (Markina et al. 2014; Zhidkin et al. 2015). Therefore, the bottom border of cultivated fields is the first significant sediment sink. A significant (20–90%) portion of the sediment-associated ^{137}Cs is redeposited here due to the presence of an embankment resulting from soil cultivation (Fridman et al. 1997; Golosov et al. 2000b). The rest of the cesium-137 washed away from the arable slopes is transported by temporary flows to the bottoms of dry valleys.

A strong correlation between the intensity of erosion on arable or pasture lands and the content of ^{137}Cs in soil allows its use (only for territories with variability of ^{137}Cs initial fallout $C_v < 20–22\%$) to estimate the rate of sediment redistribution on slopes for the period after precipitation (in the case of Chernobyl-derived ^{137}Cs since May 1986). The loss/gain of ^{137}Cs in individual sampling points can be used for recalculation of ^{137}Cs concentration in soil into erosion/deposition rates using conversion models (Walling and He 1999; Walling et al. 2002). Some patterns of initial ^{137}Cs fallout from the atmosphere should be considered, when Chernobyl-derived ^{137}Cs is used as a tracer for determining the rates of sediment redistribution (Golosov et al. 1998b; Golosov 2002). The assessments of erosion and accumulation rates within arable land based on the use of ^{137}Cs as a tracer need to include the combined effect of water and tillage erosion and soil losses during harvesting. Assessment of the intensity of sediment redistribution in the arable area in the Lokna River basin (Tula region) based on the use of the ^{137}Cs technique also confirmed the alternation of erosion and accumulation zones along the length of the slope section (Fig. 4.6).

The sediments and sediment-associated ^{137}Cs mobilised from cultivated slopes are deposited at the flat dry valley bottoms due to the high roughness of the grass cover, which leads to a gradual increase of total ^{137}Cs inventories in these morphological units (Golosov 2006; Golosov et al. 2018). Changes in sediment accumulation rates and ^{137}Cs inventories along the dry valley bottoms are defined by the morphological features of the entire catchment area as configuration: location of hollows and their distance from the valley bottom, the presence of artificial barriers along the transport pathways of sediments from the cultivated slopes to the dry valley bottoms, the shape of the longitudinal profile of the bottom, and secondary incisions of bottom gullies.

Usually, the layer with a maximum of ^{137}Cs content may be attributed to the time of Chernobyl fallout in 1986. The peak of the bomb-derived ^{137}Cs fallout was also usually detected in the ^{137}Cs depth profile in different undisturbed sediment sinks. The total inventories of Chernobyl-derived ^{137}Cs on the bottoms of dry valleys in the southern part of the East European Plain always exceeded the bomb-derived ^{137}Cs deposits everywhere (Golosov et al. 1999b, 2013, 2018). In the areas with a high level of Chernobyl contamination, in most cases, the bomb-derived ^{137}Cs peak was challenging to identify, as it was overlapped by the "tail" of Chernobyl-derived ^{137}Cs, due to its vertical migration via a range of pedogenic processes (Fig. 4.7).

The bottom gully head cut retreat is the primary mechanism for secondary mobilization of contaminated sediments within dry valleys (Fig. 4.8a). Due to global warming, the rate of linear increase of gullies` head cut retreat during the last two

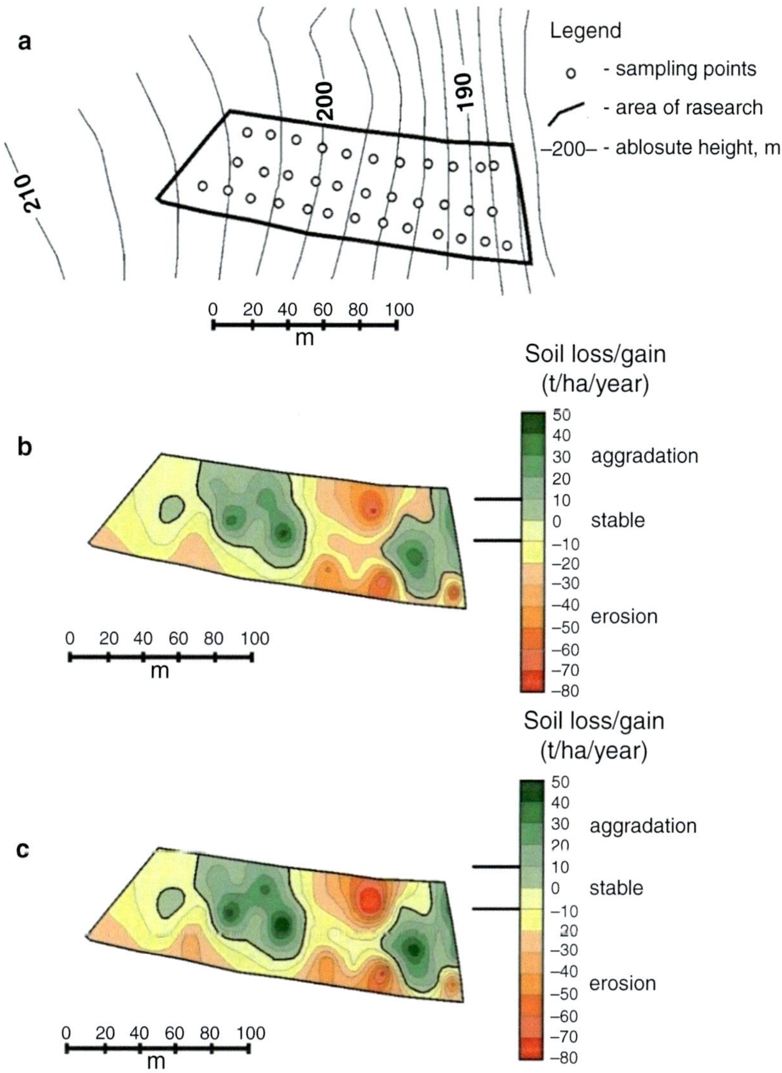

Fig. 4.6 Maps of soil redistribution for the period 1986–1997 within the cultivated slope of Chasovenkov Verh valley using ^{137}Cs as a tracer, based on conversion models: (**a**) proportional model (**b**) standard mass-balance model (**c**) (Golosov and Markelov 2002)

decades has decreased significantly and averaged 0.1–0.3 m/year (Rysin et al. 2017). The main reason for this trend is a decrease in surface runoff from the slopes during the snowmelt period due to the reduction of freezing of the soil during winter (Golosov et al. 2017). Moreover, most of the deposits formed during the bottom gully head retreat are redeposited at the points of several meters downstream (Panin et al. 2001). Therefore, the bottoms of dry valleys will remain significantly

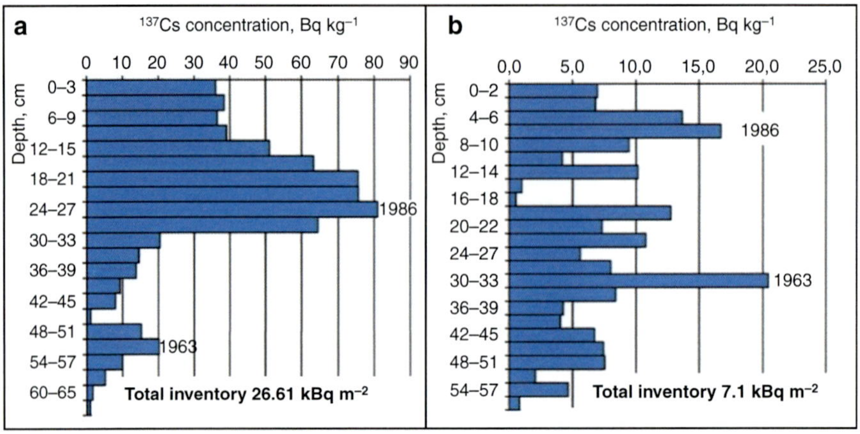

Fig. 4.7 ^{137}Cs profiles associated with the sediment cores collected from the bottoms of the first-order dry valleys, located in areas with different levels of the Chernobyl fallout: Center parts of European Russia. (**a**) Gracheva Loschina, Kursk region; (**b**) Temeva Rechka, Tatarstan (Golosov et al. 2018 with modifications). Legend: 1986—peak of Chernobyl-derived ^{137}Cs; 1963—peak of bomb-derived ^{137}Cs

contaminated by accumulated Chernobyl-derived ^{137}Cs as long as no sharp changes of climate and therefore no significant increase in the level of surface runoff from the slopes of interfluves will occur.

Quite often, artificial dams are built at the bottoms of dry valleys (Fig. 4.8b) (Baxter 1977; Golosov and Ivanova 1993; Smith et al. 2001; Belyaev et al. 2005, 2013b; Radoane and Radoane 2005). Such ponds, used as a source of water for irrigation, fish breeding, and recreation, are very typical in the southern part of the East European Plain (Golosov and Ivanova 1993; Golosov and Panin 2006; Belyaev et al. 2012). The increase of ^{137}Cs inventories due to sediment accumulation, even several decades after the accident, may exceed the losses associated with radioactive decay in the catchments of such reservoirs. The presence of given artificial water bodies contributes to the reduction in the flow of sediments and sediment-associated ^{137}Cs into permanent watercourses since reservoirs trapped up to 90% of sediment runoff. At the same time, the catchments of the reservoirs located in the upper links of the fluvial network are the most appropriate sites for quantitative assessment of the migration of sediments and pollutants. In this case, it is possible to more accurately estimate the sediment and Chernobyl-derived ^{137}Cs budget and to carry out an independent assessment of the rates of erosion from arable land and accumulation of sediments in various sediment sinks (Eakin and Brown 1939; Brune 1953; Renwick 1996; Verstraeten and Poesen 2001; Radoane and Radoane 2005; Nichols 2006).

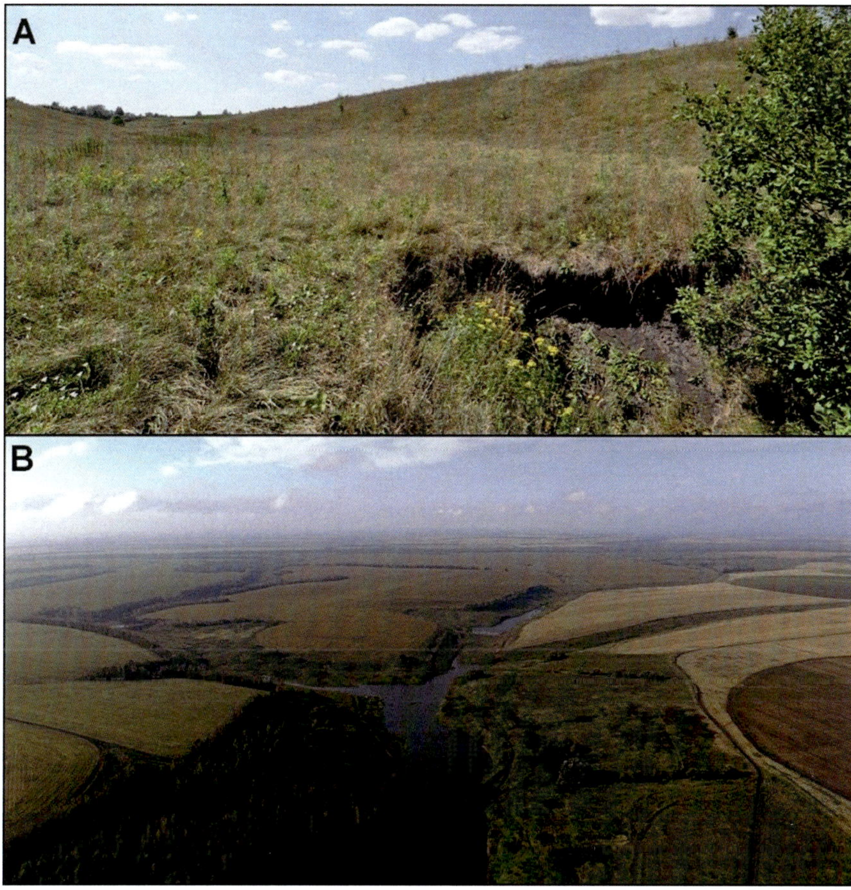

Fig. 4.8 (**a**) gully in the bottom of Upper Lokna catchment; (**b**) artificial reservoirs within Upper Lokna catchment

4.4 Lateral Migration of ^{137}Cs in River Valley Bottom

A significant part of ^{137}Cs involved in migration by the process of water erosion of soils is redeposited in the upper parts of the fluvial network, mainly directly inside the arable land and in the bottoms of dry valleys. Nevertheless, part of ^{137}Cs inventories is transported together with the sediments to the bottoms of river valleys, including directly into the channels of permanent streams or into reservoirs (Golosov et al. 1999a, b). The structure of the river valley bottoms is usually represented by several levels. Sediments and sediment-associated ^{137}Cs delivered at the valley bottom from the catchment area are redeposited at the foot of the valley sides or at the dry valley fans. Moreover, in the case of lateral migration of ^{137}Cs by sheet flows from the valley sides, all sediments reaching the river valley bottoms are redeposited on river terraces or floodplains. Sediments and sediment-associated ^{137}Cs,

Fig. 4.9 Different levels of the Lokna River's floodplain

transported by temporary streams flowing in the dry valley bottoms, only partially enter river channels.

In addition to the sediments coming from the catchment area, the sediment yield of rivers is formed due to the bank and channel bed erosion (Leopold and Wolman 1957; Blake and Ollier 1971; Nanson and Croke 1992; Chernov 2013). The main factor in the formation of floodplains of lowland rivers is the accumulation of suspended sediments during periods of flooding (Nanson and Croke 1992). The contribution of the basin sediment yield to the river sediment yield is more important for small rivers with a high proportion of cultivated lands in the watershed. The regular decrease of sediment delivery ratio coefficient (SDR) with increasing river basin area is observed for rivers flowing in the agricultural zone (Golosov 2006). The floodplain serves both as a sediment sink (mainly) and a sediment source for the river in the cases of bank erosion or destruction of vegetation cover on floodplain surface (Schumm 1977; Lewin 1978; Chalov and Chernov 1985). Redeposition of sediment-associated ^{137}Cs and other pollutants negatively influences alluvial soil quality (Chesnokov et al. 2000; Linnik et al. 2005; Markelov et al. 2012).

Accumulation of sediment-associated ^{137}Cs in floodplains was investigated on the rivers, which belong to the Upa River basin (right-bank tributary of the Oka River), the southern half of which was heavily contaminated by the Chernobyl fallout. In particular, it is the Lokna River basin, located in the zone with the maximum level (185–555 kBq/m^2) of radioactive contamination. The Lokna River is a left-bank tributary of the Plava River and flows into the Plava River in Plavsk. The area of the Lokna River basin is about 177 km^2. There are three floodplain levels in Lokna River valley bottom: low—directly adjacent to the river channel with height of 1 m above the low water level in the river, medium—up to 1.5 m high, occupying the central part of the floodplain, and high—with a height of more than 1.7 m (Fig. 4.9). The high floodplain has not been flooded since 1986. Most part of sediment and sediment-associated ^{137}Cs delivered from the valley sides is redeposited within the given level of floodplain. According to the results of the satellite image interpretation and field surveys, the total width of the low and middle floodplains along the whole Lokna River basin was within a range from 40 to 120 m. The low floodplain is a narrow strip (up to 5–7 m at maximum) along the channel. More often, it appears along one of the river banks (Fig. 4.9). Three sites were selected for investigation of the ^{137}Cs vertical distribution in the sections of alluvial soils in the floodplain of the Lokna River located at a different distance from the mouth of the river (Fig. 4.10).

The most significant accumulation of sediment-associated ^{137}Cs occurred on profiles LOK-1 and LOK-3 located at a low floodplain (Figs. 4.10 and 4.11). The total sediment deposition volumes and the sediment-associated ^{137}Cs inventories accumulated on the low and middle floodplains in the valley bottom were calculated for the period from 1986 to 2014, taking into account the areas that were occupied by each level of the floodplain and different rate of sediment accumulation (Table 4.4).

It was established that 47% of sediments and 46% of sediment-associated ^{137}Cs entering the bottom of the river valley were transported to the Plava River. This evaluation is based on the comparison of the total volume of sediment and ^{137}Cs delivered to the Lokna River bottom, and the assessment of the total sediment accumulation at different levels of the river floodplain. As a result, no more than 12–13% of sediment-associated ^{137}Cs, mobilized by soil erosion on the arable slopes of the basin of the Lokna River, are transported to the Plava River for the 30 years since the Chernobyl accident.

The Plava River (basin area is 1880 km^2) flows from the south to the north and crosses the most contaminated area within the "Plavsk hot spot" in the middle reaches. As a result, the concentration of sediment-associated ^{137}Cs delivered to the river channel from the arable slopes steadily increased from the upstream reaches of the Plava River to the confluence with the Lokna River (Fig. 4.12). Then concentration decreased toward downstream.

Seven sites were chosen along the length of the Plava River located from the upper reach to the river mouth for evaluation of the patterns of sediment and sediment-associated ^{137}Cs at different levels of the floodplain (Fig. 4.12). The main criteria for the selection of the floodplain section were the typical morphology of the site for a specific section of the bottom of the valleys of the Plava River and the level of radioactive contamination of the river basin (Ivanova et al. 2014).

The Plava River valley has two or three floodplain levels in the studied sites (Ivanova et al. 2014). In this case, a gradual increase in the relative altitudes of each floodplain level under the low water level in the river appeared from the upper reach to the river mouth and is observed to be a complementary feature to the increase in water discharge. In the upper reaches of the Plava River, where the floodplain is still relatively narrow, sediments washed off the sides of the valley often redeposited within a major part of the floodplain. Those features contributed to an increase in the total ^{137}Cs inventories with the magnitude comparable to sediment-associated ^{137}Cs in the low floodplain area. The depth profile of ^{137}Cs at S-6 (Fig. 4.13) shows that about 6 cm of the sediments eroded from the valley slopes was redeposited above the layer with the highest ^{137}Cs.

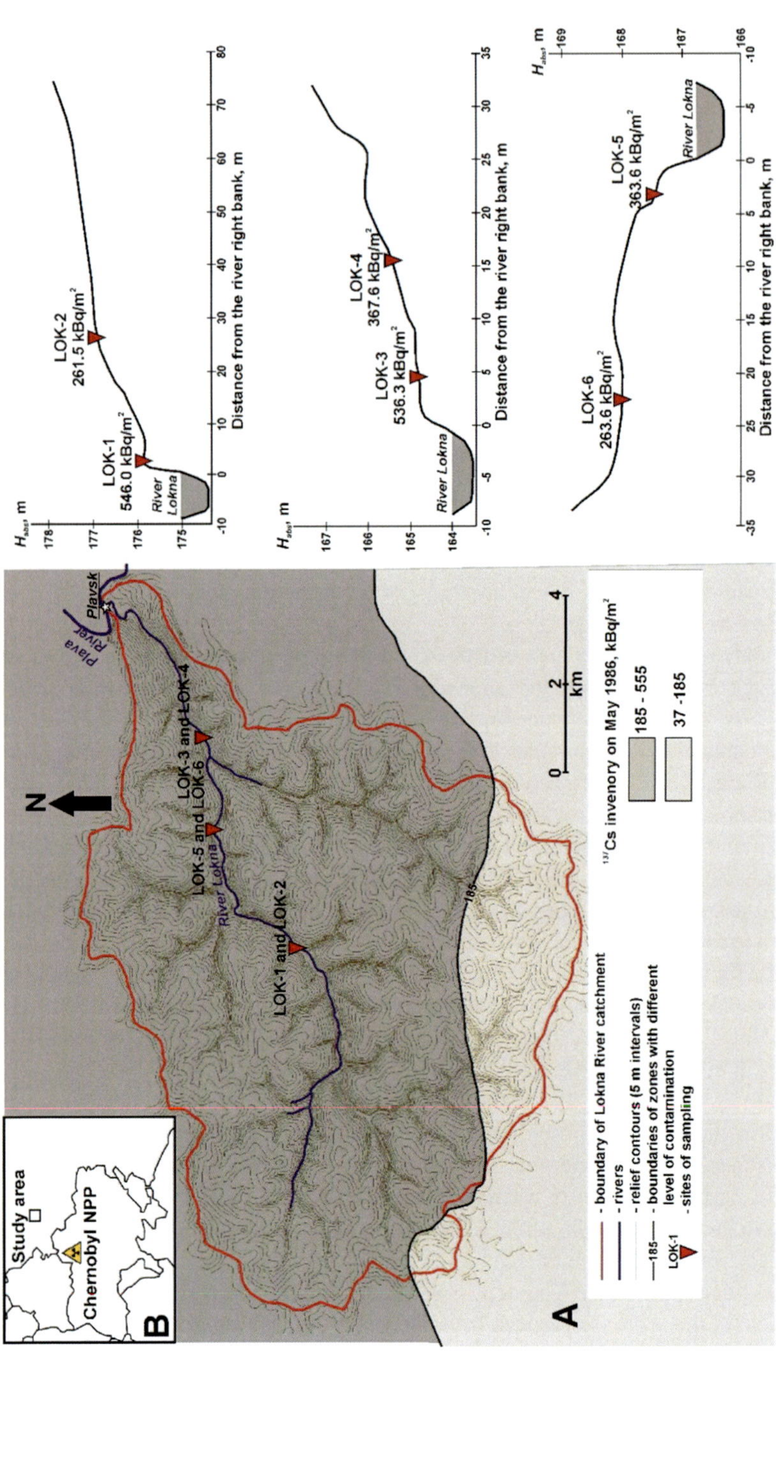

Fig. 4.10 Location and ^{137}Cs inventories in soil profiles in the Lokna River floodplain (Mamikhin et al. 2016)

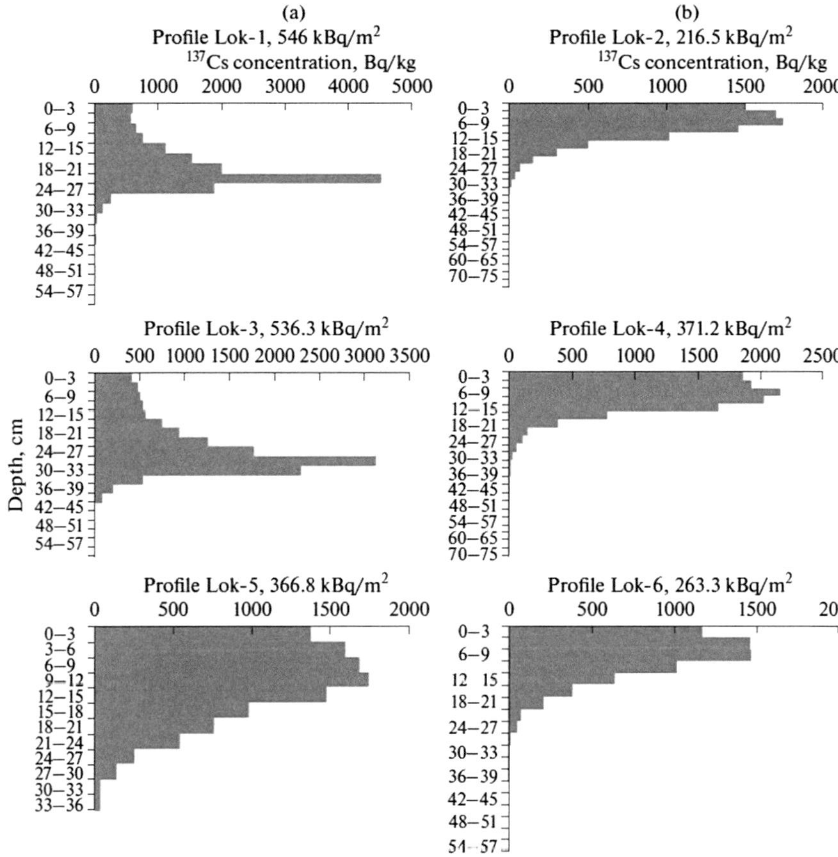

Fig. 4.11 Vertical distribution curves of ^{137}Cs in alluvial soils of the Lokna River floodplain: (**a**) low level; (**b**) medium level (Mamikhin et al. 2016)

Table 4.4 Accumulation of sediments and sediment-associated Chernobyl-derived ^{137}Cs on the Lokna River floodplain for the period 1986–2014

		Sediments[a], 10^3 t	^{137}Cs[b], 10^6 kBq
Floodplain level	Low	141.8–158	147.2
	Medium	45.1–62.6	67.9
Total		186.9–220.6	215.1

[a]Mean density 1100 kg/m^3
[b]Recalculated to 2012

The surface of the floodplain can be complicated by depressions inherited from the former river channel (oxbows). Due to accumulation in oxbows of silty particles (Chalov 2011), which sorbed ^{137}Cs more than coarser particle fractions (Korobova et al. 2007), the total inventories of sediment-associated ^{137}Cs in oxbows increased significantly (Fig. 4.14, section P-10). At all sites, the maximum rate of

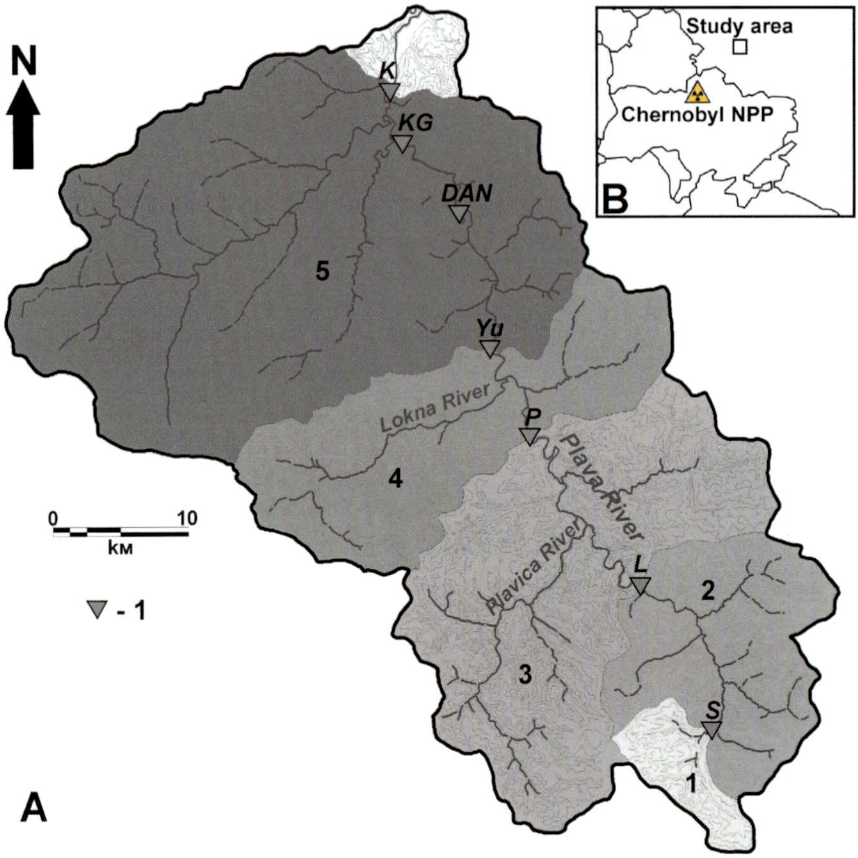

Fig. 4.12 Location of sampling sites on the Plava River floodplain: 1—sampling sites: S—"Streshnevo," L—"Lyapunovka," P—"Plavsk," Yu—"Yurievo," DAN—"Danilovo," KG—"Kosaya guba," K—"Krapivna Plava" basin segments: 1—Streshnevo, 2—Streshnevo-Lyapunovka, 3—Lyapunovka-Plavsk; 4—Plavsk-Yurievo; 5—Yurievo-Krapivna

accumulation of sediments and the increase in ^{137}Cs inventories were observed in the low floodplain, which is annually flooded many times during the passage of floods and spring high water.

To estimate the total accumulation of sediments and sediment-associated ^{137}Cs in the Plava River basin for the post-Chernobyl period, the area of each floodplain level was calculated based on the interpretation of satellite images and field verification of the interpretation results. It was also assumed that the mean values of accumulation of sediment and sediment-associated ^{137}Cs at each floodplain level are representative of the corresponding floodplain.

The total accumulation of sediment-associated ^{137}Cs on the floodplain of the Plava River is approximately twice lower than on the floodplain of its tributary, the Lokna River (2.15×10^8 kBq) (Table 4.5). Such a ratio between the inventories of

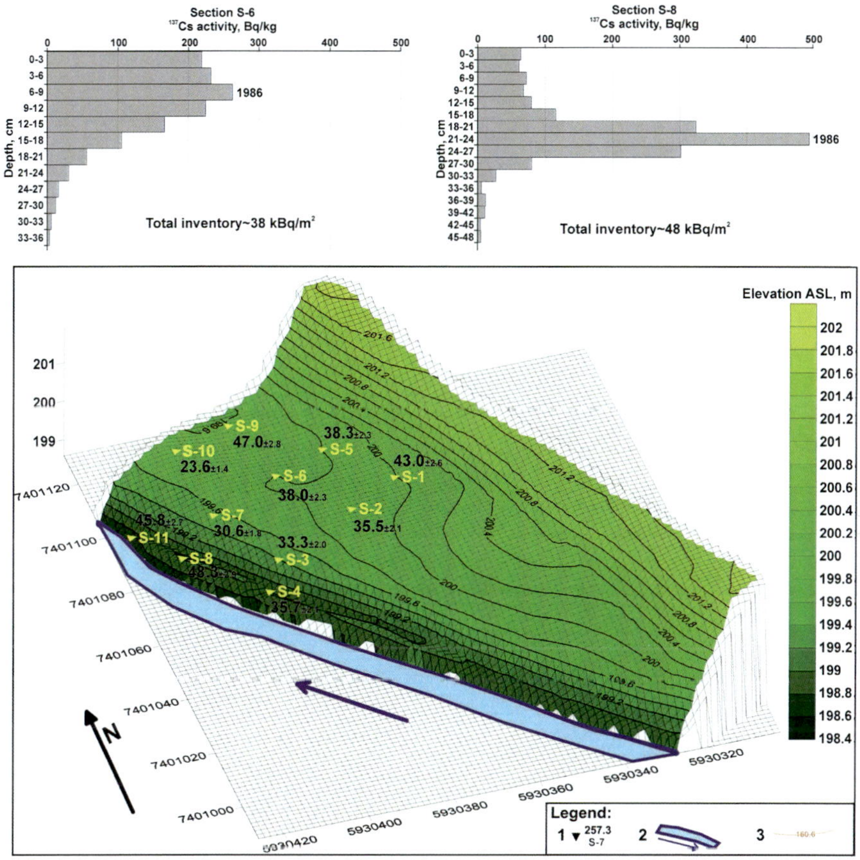

Fig. 4.13 The "Streshnevo" floodplain sampling site and detailed ^{137}Cs depth–distribution curves for sampling points S 6 and S 8. Legend: (1) sampling points with ID and ^{137}Cs inventory (kBq/m^2); (2) river channel and flow direction (not to scale); (3) contour lines with elevation asl, m (0.2-m interval). Coordinates *X,Y,Z* on the 3d diagram are in meters (Gauss–Kruger cartographic projection) (Belyaev et al. 2013a)

Chernobyl-derived ^{137}Cs on the floodplains of the Lokna and Plava Rivers is because practically the entire catchment area of the Lokna River is located within the territory with the maximum density of radioactive contamination for the whole of the Plava River basin. The Plava reaches the axial part of the "Plava hot spot" only in its middle section. The hydrological connectivity of the cultivated slopes and river valley in this area is rather low. Thus, it can be argued that lateral migration and following accumulation of Chernobyl-derived ^{137}Cs in the floodplain complex of the Plava River on the site below the confluence between the River Plava and the River Lokna occurred mainly due to sediment yield of the Plava River and deformations of its channel. The lateral migration of ^{137}Cs from the sides of the valley by the erosion had a secondary effect (Ivanova et al. 2014).

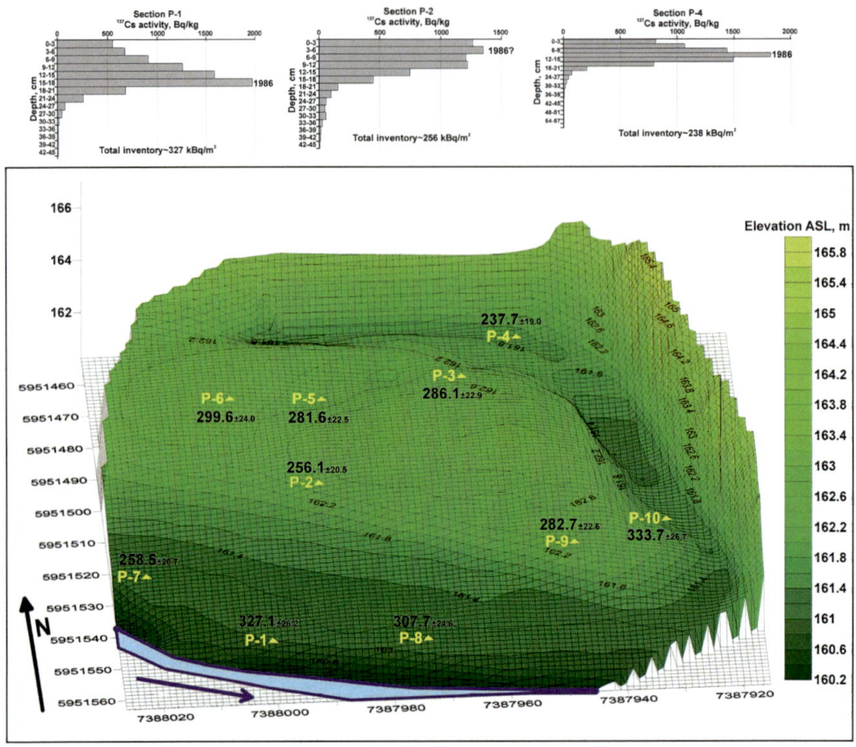

Fig. 4.14 The "Plavsk" floodplain sampling site and detailed ^{137}Cs depth–distribution curves for sampling points P-1, P-2, and P-4. For legend, see Fig. 4.13 (Belyaev et al. 2013a)

Table 4.5 Sediments and ^{137}Cs accumulation on the Plava River floodplain for the post-Chernobyl period

The Plava River basin reach	Low level		Medium level		High level	
	Sediments, 10^3 t	^{137}Cs, 10^6 kBq	Sediments, 10^3 t	^{137}Cs, 10^6 kBq	Sediments, 10^3 t	^{137}Cs, 10^6 kBq
Upstream from Streshnevo	9–10.3	1	–	–	7–10.5	2
Streshnevo—Lyapunovka	11	0.6	11.3–15	5,6	48.7–73.2	18.1
Lyapunovka—Plavsk	9.9–11.8	11.8	3.9–7.7	4,7	3.9–5.2	3.8
Plavsk—Yurievo	20–22.4	14.6	2.6–3.5	1,5	3.8–7.7	2.8
Yurievo—Krapivna	35.4–70.7	14.7	12.3–14.8	3,2	44.9–112.3	9.8
Total	Sediments, 10^3 t		^{137}Cs, 10^6 kBq			
	223.7–376.1		94.2			

Precise investigations on the accumulation of Chernobyl-derived ^{137}Cs in the typical floodplain sections with natural levees of the Iput River, which flows in the southwest of the Bryansk region, allow assessing the transformation of radioactive contamination of floodplains on rivers flowing in the forest zone (Linnik 2001, 2008). The general patterns of redeposition of the sediments and Chernobyl-derived ^{137}Cs obtained by the detailed study of the floodplain of the Iput River are similar to the Plava River. The maximum deposits of sediments and sediment-associated ^{137}Cs were observed in the low floodplain and depressions, regularly inundated. At the highest parts of the floodplain, accumulation of sediments was not identified at the time of the research in 2000, which indicates that this area of the floodplain was not inundated after 1986. However, there is also a specific feature associated with spring flood on the Iput River. The water level in the Iput River was relatively high at the time of precipitation from the atmosphere of Chernobyl-derived ^{137}Cs on April 28–29, 1986, because spring had not yet ended in this period. It is ascribable to the extensive coverage of the basin with forests, which contributes to slow down in snow melting and an increase in spring flood. As a result, low floodplains on the Iput River during the fallout of ^{137}Cs from the atmosphere were inundated, and this prevented the fixation of ^{137}Cs, which fell from the atmosphere, on alluvial soil (Linnik 2001). The maximum content of Chernobyl-derived ^{137}Cs at the level of the low floodplain was formed due to the fixation of ^{137}Cs dissolved in water, and therefore it differs in magnitude from the level of initial contamination on the rest of the valley bottom of the river. A similar situation was observed in other rivers of the western part of Russia, Belorussia, and the north of Ukraine with predominantly forested catchments.

In general, the peculiarity of the hydrological regime of small rivers of the European part of the former USSR in the post-Chernobyl period is a rare occurrence of high spring floods with inundation of the middle and high floodplains. Hence, at the level of the low floodplain, a significant increase in the total inventories of Chernobyl-derived ^{137}Cs occurred due to the accumulation of more than 10–15 cm of sediment (Linnik 2001; Troflmetz 2011; Markelov et al. 2012; Ivanova et al. 2014). A significant proportion of Chernobyl-derived ^{137}Cs entering river channels with sediments washed from the catchment slopes is transported along the river, given no artificial reservoirs intercepting the sediments. This contributes to the propagation of radioactive contamination on the bottoms of the river valleys. It was observed in the first years after the accident when a considerable amount of radionuclides from residential areas were transported to the rivers.

On the territory of the East European Plain, the most intensive fallout of radionuclides occurred mainly in the middle part of the Dnieper River basin (Zhukova et al. 1997), the upper reaches of the Don River basin, and the western part of the basin of the Volga River (mainly the basin of the Oka River) (Vetrov et al. 1990; Litvin et al. 1994). Studies conducted in the spring of 1987, one year after the fallout of ^{137}Cs from the atmosphere, showed that the highest concentration of ^{134}Cs and ^{137}Cs in the suspended sediment runoff was typical of the basins with the largest share of arable lands and, consequently, most intensive supply of material from cultivated slopes caused by water erosion (Vetrov et al. 1990). At that time, the models of transport of dissolved and sediment-associated ^{137}Cs in small rivers in the near zone were elaborated (Borzilov et al. 1993). It should be noted that in the early years after the accident, in some basins, the fraction of dissolved ^{137}Cs was higher in comparison with its content in the sediments.

Monitoring of the content of Chernobyl-derived ^{137}Cs and other radionuclides carried in both dissolved and particulate form was organized at gauging stations located in the cities of Belyov (the Oka River), Plavsk (the Plava River), Tula (the Upa River), and Kozelsk (the Zhizdra River). At those stations, samples of water and sediment were collected four times per year. A sharp drop in radionuclide content occurred already in 1988 when their total content was reduced threefold compared to their content in water and sediment in 1987 and subsequently continued to slowly decline until the mid-1990s (Vaculovskii et al. 1996). In the early years (1986–1988) after the accident, the potential removal of radionuclides outside the river basin was estimated using the coefficient M_c calculated as follows (Vetrov et al. 1990):

$$M_c = Cb^{-1},$$

where C is the average concentration of ^{137}Cs in water (Bq/m^3) and b is the average contamination level of the river basin, Bq/m^2.

This approach was acceptable in the early years after the accident when a large amount of dissolved ^{137}Cs came from the catchments to the rivers (Bulgakov and Konoplev, 2004). But later on, when the vast majority of ^{137}Cs began to be transported together with the sediments, and ^{137}Cs content in the water and sediment decreased sharply, more accurate estimates could be obtained based on using other approaches (Bulgakov et al. 1992). These approaches account for the specificity of redistribution of sediments in the river basin, the intra-annual distribution of water and sediment runoff, and several geomorphological and climatic characteristics that determine the intensity of erosion-accumulation processes in various chains of the fluvial network. Of great importance is an accurate estimate of the heterogeneity of the ^{137}Cs initial deposition over the basin.

The budget studies performed for individual typical river basins and based on quantitative estimates of the lateral migration of Chernobyl-derived ^{137}Cs in various parts of the fluvial network comprise one of the approaches to assessing the redistribution of ^{137}Cs in the river basin (Golosov et al. 1998a, b). Such estimates were made for the part of the Plava River basin heavily contaminated after the Chernobyl accident (Fig. 4.15) (Ivanov 2017).

The Plava River basin is characterized by a very homogeneous composition of the soil cover represented by gray forest soils and northern chernozems. There are two main sources of river sediments: erosion of riverbank and channel bottoms and soil erosion on interfluve slopes. The interfluve component of the sediment yield includes sheet and rill erosion on cultivated lands and degraded pasture, erosion of linear relief forms (hollows; bottom, bank, and slope gullies) and different types of erosion from residential areas (Ivanova et al. 2014). The Plava River has a typical small basin in the center of the agricultural zone of European Russia. A number of

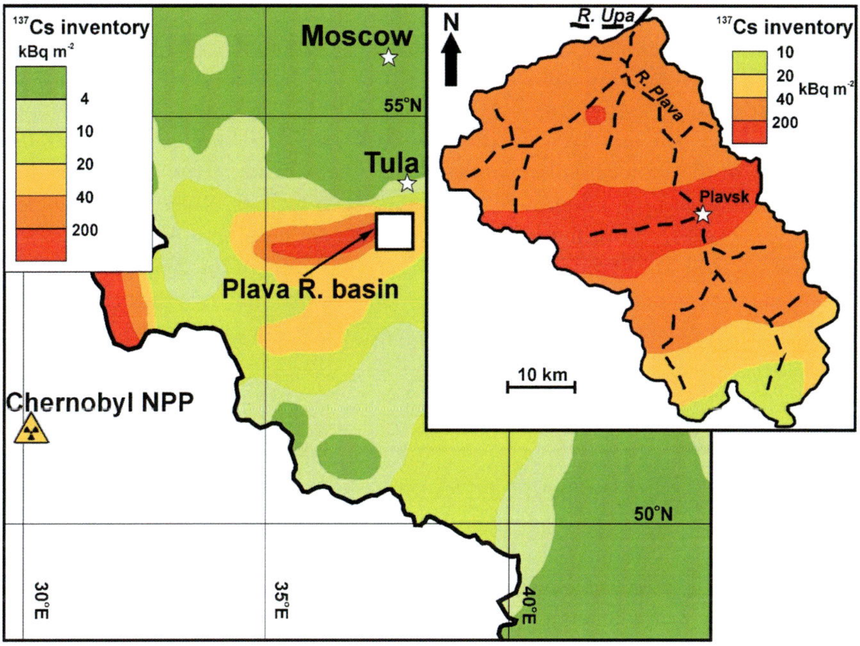

Fig. 4.15 The location of Plavsk hot spot (based partly on Izrael 1993 and Walling et al. 2000)

Table 4.6 ^{137}Cs budget of Plava River basin for the period 1986–2012 (2014)

	Mean erosion rate 6.4 t/ha/year
Total ^{137}Cs mobilization by water erosion[a]	6.7 × 10^9 kBq/100%
Total ^{137}Cs delivery from the basin area to river bottoms	2.1 × 10^9 kBq/31.6%
Storage in river valleys	0.7 × 10^9 kBq/10.7%
Transfer to the Upa River	1.4 × 10^9 kBq/20.9%

[a]Excluding parts of the basin separated by reservoirs

investigations of Chernobyl-derived ^{137}Cs lateral and vertical migration within the various components of the landscape have been conducted for the given region (Kryshev et al. 1991, Fridman et al. 1997; Izrael et al. 1999; Panin et al. 2001; Linnik et al. 2004; Gennadiev et al. 2008; Romantsova 2012; Romantsova et al. 2012; Paramonova et al. 2017, Kvasnikova et al. 1998; Ivanova et al. 2014; Ivanov et al. 2016; Shamshurina et al. 2016; Mamikhin et al. 2016).

For interpolation of the estimation results for sediments and Chernobyl-derived ^{137}Cs redistribution within small watersheds of various orders, all small basins in the Plava River were allocated and classified using a topographic map of scale 1:100,000 (Ivanov et al., 2017). This allowed us to rank all small catchments depending on their SDRs and sediment-associated ^{137}Cs transported together with them, taking into account the initial radioactive contamination of the basin territory, the total volume of sediments, and sediment-associated ^{137}Cs transported to the river valleys (Table 4.6). It was assumed that on the other tributaries of the Plava River, accumulation of sediments at different levels of the floodplain and their variations along the river's length are similar in volume to the same reaches of valley bottoms of the thoroughly studied rivers. This made it possible to estimate the total accumulations of ^{137}Cs on the river floodplains of the basin, taking into account the differences in the radioactive contamination of the tributary river basins. Finally, the contribution of each tributary to ^{137}Cs transfer to the Upa River was calculated (Table 4.6). As a result of the performed calculations, it became possible to determine the removal of Chernobyl-derived ^{137}Cs from the basin for the period that has elapsed since the Chernobyl accident. These calculations may overestimate erosion rates since they account for the situation in the Plava River basin in the late 1980s. Since then, there has been a significant reduction in the erosion rates due to a sharp decrease in runoff from slopes during the spring snowmelt because of the general climate warming and, as a consequence, a reduction in the depth of soil freezing (Litvin et al. 2017; Golosov et al. 2017). What is more, the values for arable lands used in the calculations are those valid for the period 1986–1991. After the collapse of the USSR, arable land areas in the basin were reduced by about 15% by the beginning of the 2000s and only after 2010 they began to increase again.

At the same time, the calculated share of sediment yield outside the basin of the Plava River, which amounted to about 21% (Table 4.6), can roughly be described by the empirical relationship between the SDR and the basin size of the rivers of the northern forest-steppe zone (Sidorchuk 1996):

$$\mathrm{SDR} = aA^{-0.45},$$

where *SDR* is the Sediment Delivery Ratio for the river basin; *a* is the empirical coefficient (for the Volga and Dnieper River basins—0.25), *A* is the basin area (km^2).

According to the relationship between SDR and river basin area (Fig. 4.16), the SDR of the Plava River basin is expected to be about 18%. However, in this case, it should also be taken into consideration that the area of arable land after 1991 in the Plava River basin has decreased, which has inevitably contributed to a decrease in the SDR. Nevertheless, the determined and quantitatively characterized relationships between the volume of Chernobyl-derived ^{137}Cs mobilized by erosion processes and its redeposition in various parts of the fluvial network can be used to identify features of lateral migration in small river basins of the forest-steppe zone with a high proportion of arable land. SDR for sediments and sediment-associated ^{137}Cs should be significantly lower for river basins in the forest zone, where the share of arable land is decreasing and the area of forest is quite large. The SDR also decreases with the growth of the basin area (Fig. 4.16).

In many cases, rivers that had received sediment discharge and sediment-associated ^{137}Cs from river basins with high levels of radioactive contamination, upstream of the confluence of such rivers, are characterized by low concentrations of ^{137}Cs in the sediment yield. The contamination levels of sediment discharge of rivers

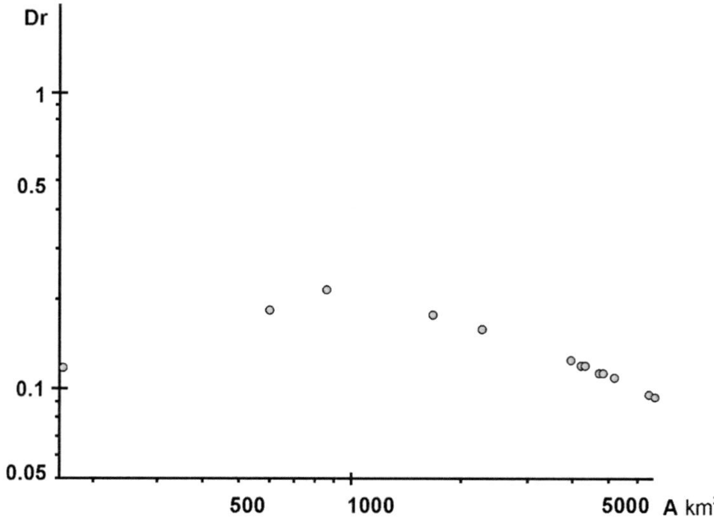

Fig. 4.16 Delivery ratio changes along the main channel of the Zusha River for conditions of high human impact (modified from Sidorchuk 1996). *Dr* delivery ratio coefficient, *A*—river basin area

can be easily determined by depth increment of sediments at the low floodplain level, which is inundated annually during the floods. Changes in ^{137}Cs concentration from the surface to the horizon with a maximum content of ^{137}Cs, which corresponds to the floodplain surface at the time of ^{137}Cs deposition in April–May 1986, reflect the time dynamics of the ^{137}Cs integral concentration from 1986 to the sampling time. As an illustrative example, we can present curves of the vertical distribution of ^{137}Cs in the floodplain sediments of the Upa River upstream and downstream from the confluence of the Plava River and in the floodplain sediments of the Plava River in its low reach (Fig. 4.17). Opposite, in the southern half of the basin, the Plavsk hot spot is extending from west to the east. As a result, only the runoff from the left-bank tributaries of the Upa River is characterized by high levels of ^{137}Cs content in sediments. At the same time, most of the contaminated sediments entering the Upa River upstream from the confluence of the Plava River is redeposited in the Shatskoye and Shchekinskoye reservoirs. Therefore, the content of ^{137}Cs in the sediment of the Upa River upstream from the mouth of the Plava River does not show sharp differences from the level of contamination of adjacent relatively weakly contaminated areas (Belyaev et al. 2013a; Ivanova et al. 2014).

The trend of a slight decrease in ^{137}Cs concentration from a layer of 60–65 cm (which was the floodplain surface in 1986) to that of 30–35 cm and subsequent low content of ^{137}Cs, which almost does not change up to the surface is clearly in evidence. Right downstream from the confluence of the Plava River and the Upa River, the content of ^{137}Cs in the sediments increases considerably (Fig. 4.17b). The maximum values (in 1986) practically coincide with the maximum ^{137}Cs content at low floodplain of the Plava River in its lower reach (Fig. 4.17d). The peak values of the ^{137}Cs content in the sediments of the Upa River floodplain (Fig. 4.17b) change in

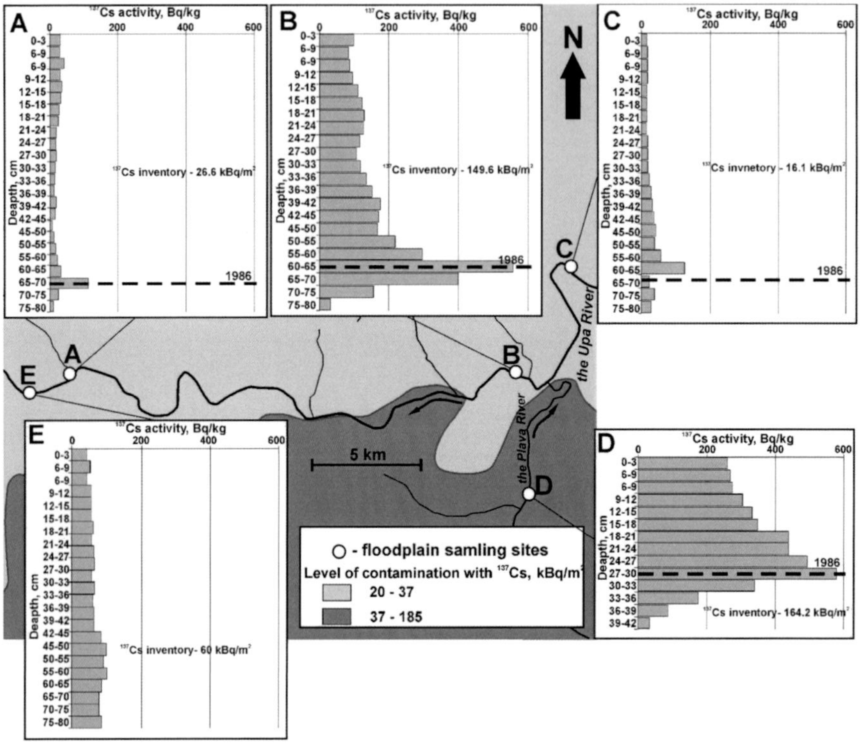

Fig. 4.17 Depth distributions of ^{137}Cs associated with soil cores collected from floodplains of the Plava and the Upa Rivers near and at a distance from the Plava River mouth

accordance with the ratio between the ^{137}Cs content in the sediment of the Upa River and Plava River and turbidity of runoff. Since the Plava River is the last large tributary of the Upa River up to its confluence with the Oka River, and the low floodplain is stretched by a narrow strip along both banks of the river, up to the mouth, the content of ^{137}Cs in the sediments of the Upa River changes very slightly (Fig. 4.17e). Although downstream from the confluence of small right-bank tributaries, the content of ^{137}Cs may decrease to the values typical in the floodplain deposits upstream from the confluence of the Plava River (Fig. 4.17a, c). Likely, at the first dozen of meters from the confluence, there is still no mixing of the tributary and receiving watercourse flows. As a result, the distribution of radionuclide inventories on the floodplains of rivers, the tributaries of which have different contamination density, is extremely heterogeneous.

4.5 Conclusion

The most significant lateral migration of Chernobyl-derived ^{137}Cs is observed on the arable slopes in the temperate climatic zone. Most of the sediment-associated ^{137}Cs transported outside the arable lands are redeposited along the bottom parts of cultivated slopes and on the bottoms of dry valleys, which are typical elements of the relief of the nonglacial zone of the East European Plain. The bottoms of dry valleys and especially artificial reservoirs located in the upper reaches of the fluvial network are the main sink of Chernobyl-derived ^{137}Cs deposition. Not more than one-third of the sediments eroded from the arable lands and sediment-associated ^{137}Cs enters the bottoms of river valleys. At the same time, about one-third of sediment-associated ^{137}Cs transported by the rivers is redeposited on floodplains of small rivers with a catchment area of less than 2000–3000 km^2. The proportion of sediments accumulated at different levels of floodplain was limited in the period from 1986 to the present due to a significant reduction in floodwaters. The most significant increase in the inventories of Chernobyl-derived ^{137}Cs is observed on a low floodplain, usually stretching a relatively narrow, sometimes interrupted strip along both banks from the river bed.

In general, even in areas with a high (>50%) proportion of arable land, no more than 0.3–0.5% of the total inventory of Chernobyl-derived ^{137}Cs was transported with river sediment yield over the past three decades. Such small values demonstrate that the lateral migration of ^{137}Cs, associated with sediment transport, does not necessarily result in a significant change of ^{137}Cs contamination of the basin. The primary transformation occurs in the upper reaches of the fluvial network (hollows on cultivated slopes), where surface runoff is conducive to maximum soil losses.

References

Baxter RM (1977) Environmental effects of dams and impoundments. Ann Rev Ecol Syst 8:255–283

Belyaev VR, Markelov MV, Golosov VN, Bonte P, Ivanova NN (2003) The use of cesium-137 to assess modern agrogenic transformation of soil cover in the areas subjected to Chernobyl fallout. Eurasian Soil Sci 36(7):788–802

Belyaev VR, Wallbrink PJ, Golosov VN, Murray AS, Sidorchuk AY (2004) Reconstructing the development of a gully in the Upper Kalaus basin, Stavropol Region (Southern Russia). Earth Surf Process Landforms 29:323–341

Belyaev VR, Golosov VN, Ivanova NN, Markelov MV, Tishkina EV (2005) Human-accelerated soil redistribution within an intensively cultivated dry valley catchment in southern European Russia. Sediment Budgets I (Proceedings of symposium S1 held during the seventh IAHS scientific assembly at Foz do Iguacu, Brasil, April 2005), IAHS Publ. 291. IAHS Press, Wallingford, pp 11–20

Belyaev VR, Golosov VN, Kuznetsova Yu S, Markelov MV (2009) Quantitative assessment of effectiveness of soil conservation measures using a combination of ^{137}Cs radioactive tracer and conventional techniques. Catena 79:214–227

Belyaev VR, Shamshurina EN, Markelov MV, Golosov VN, Ivanova NN, Bondarev VP, Paramonova TA, Evrard O, Lio Soon Shun N, Ottle C, Lefevre I, Bonte P (2012) Quantification of river basin sediment budget based on reconstruction of the post-Chernobyl particle-bound ^{137}Cs redistribution. Erosion and sediment yields in the changing environment (proceedings of a symposium held in Chengdu, China, Oct, 2012), IAHS Publ. 356. IAHS Press, Wallingford, pp 394–403

Belyaev VR, Golosov VN, Markelov MV, Evrard O, Ivanova NN, Paramonova TA, Shamshurina EN (2013a) Using Chernobyl-derived ^{137}Cs to document recent sediment deposition rates on the River Plava floodplain. Hydrol Process 27(6):807–821

Belyaev VR, Golosov VN, Markelov MV, Ivanova NN, Shamshurina EN, Evrard O (2013b) Effects of landuse and climate changes on small reservoir siltation in the agricultural belt of European Russia. Considering hydrological change in reservoir planning and management, proceeding of H09, IAHS-IAPSO-IASPEI assembly, Gothenburg. Sweden, Jul 2013, pp 134–145

Blake DH, Ollier CD (1971) Alluvial plains of the Fly River, Papua. Z Geomorphol 12:1–17

Borzilov VA, Novitsky MA, Konoplev AV, Voszhennikov OI, Gerasimenko AC (1993) A model for prediction and assessment of surface water contamination in emergency situations and methodology of determining its parameters. Radiat Prot Dosim 50(2–4):349–351

Brune GM (1953) Trap efficiency of reservoirs. Trans Am Geophys U 34:407–418

Bulgakov AA, Konoplev AV (2004) Modeling the long-term dynamics of 137Cs in rivers with different structures and types of watershed contamination. Russ Meteorol Hydrol 1:23–32. (in Russian)

Bulgakov AA, Konoplev AV, Popov VE (1992) Prediction of ^{90}Sr and ^{137}Cs behavior in soil-water system after the Chernobyl accident. In: Ecological and geochemical aspects of nuclear accidents. Hydrometizdat, Moscow, pp 21–42. (in Russian)

Chalov RS (2011) Ruslovedenie: teoriya, geographiya, praktica. Morphodinamika rechnykh rusel. Izd-vo Krasad. Moscow 2:955. (in Russian)

Chalov RS, Chernov AV (1985) Geomorphological classification of plain river floodplains. Geomorphologiya 3:3–11. (in Russian)

Chernov AV (2013) Typification of river valleys and river channels (an attempt of compatible typification). Geomorphol RAS 2:15–22. https://doi.org/10.15356/0435-4281-2013-2-15-22. (in Russian)

Chesnokov AV, Govorun AP, Linnik VG, Shcherbak SB (2000) ^{137}Cs contamination of the Techa river flood plain near the village of Muslumovo. J Environ Radioact 50(3):181–193

De Cort M, Dubois G, Fridman SD, Germenchuk MG, Izrael YA, Jones AR et al (1998) Atlas of caesium deposition on Europe after the Chernobyl accident, European Commission report EUR 16733. European Commission, Luxembourg

Eakin HM, Brown CB (1939) Silting of reservoirs. USDA Tech Bull 524:174

Evrard O, Belyaev V, Chartin C, Ottle C, Ivanova N, Markelov M, Lefevre I, Golosov V, Bonte Ph (2012) Tracing the dispersion of sediment contaminated with radionuclides in catchments exposed hyto Chernobyl and Fukushima fallout. Erosion and sediment yields in the changing environment. IAHS Publ. vol. 356. Wallingford, pp 412—417

Fedin AV, Govorun VI, Ivanov AP, Liksonov VI, Potapov VN, Smirnov SV, Scherbak SB, Urutskoev LI (1997) Collimated detector technique for measuring a ^{137}Cs in soil under a clean protected layer. Appl Radiat Isot 48:1265–1272

Foster IDL, Dalgeish H, Dearing JA, Jones ED (1994) Quantifying soil erosion and sediment transport in drainage basins: Some observations on the use of Cs. Int Assoc Hydrol Sci 224:55–64

Fridman SD, Kvasnikova EV, Golosov VN, Ivanova NN (1997) Caesium-137 migration in the geographical complexes of the central Russian hills. Russ Meteorol Hydrol 5:45–55

Gennadiev AN, Golosov VN, Markelov MV, Chernyansky SS, Kovach RG, Belyaev VR (2008) The development of a method of different age tracers to evaluate soil erosion stages. Vestnik Moskovskogo universiteta, serija 5. Geogr 3:24–31. (in Russian)

Golosov VN, Ivanova NN (1993) Some reasons of river net aggradation in the conditions of intensive agricultural development of lands. Water Resour 6:684–688
Golosov VN, Kvasnikova EV, Panin AV, Ivanova NN (1998a) Radionuclide migration in the Chernobyl contamination zone. In: Enviromental radioactivity and its application in environmental studies, proceeding of the international symposium, 16–20 Feb 1998. Christchurch, New Zealand, pp 38–49
Golosov NV, Markelov MV, Panin AV, Walling DE (1998b) Cs-137 contamination of river systems in Central Russia as a result of the Chernobyl incident. In: Hydrology in a changing environment (Proceedings of the hydrological society international conference, Exeter, UK, Jul 1998). Wiley, Chichester, pp 535–546
Golosov VN, Walling DE, Panin AV, Stukin ED, Kvasnikova EV, Ivanova NN (1999a) The spatial variability of Chernobyl-derived ^{137}Cs inventories in a small agricultural drainage basin in central Russia. Appl Radiat Isot 51:341–352
Golosov VN, Panin AV, Markelov MV (1999b) Chernobyl Cs-137 redistribution in the small basin of the Lokna river, Central Russia. Phys Chem Earth A Solid Earth Geodesy 24(10):881–885
Golosov VN, Walling DE, Kvasnikova EV, Stukin ED, Nikolaev AN, Panin AV (2000a) Application of a field-portable scintillation detector for studying the distribution of Cs-137 inventories in a small basin in Central Russia. J Environ Radioact 48(1):79–94
Golosov VN, Walling DE, Panin AV (2000b) Post-fallout redistribution of Chernobyl-derived caesium-137 in small catchments within the Lokna River basin. The Role of Erosion and Sediment Transport in Nutrient and Contaminant Transfer (Proc. Symp. Waterloo, 2000). IAHS Publication 263:49–57
Golosov VN, Markelov MV (2002) Application of Chernobyl-derived Cs for assessment of soil redistribution in agricultural catchments of central Russia. Environ Changes Radioact Tracers 2002:367–383
Golosov VN (2002) Special considerations for areas affected by Chernobyl fallout. In: Zapata F (ed) Handbook for the assessment of soil erosion and sedimentation using environmental radionuclides. Kluwer, Dordrecht, pp 165–183
Golosov VN (2003) Application of Chernobyl-derived ^{137}Cs for the assessment of soil redistribution within a cultivated field. Soil Tillage Res 69(1–2):85–98
Golosov VN (2006) Erosional-accumulative processes in river basins of developed plains. GEOS, Moscow, p 296. (in Russian)
Golosov V, Panin A (2006) Century-scale stream network dynamics in the russian plain in response to climate and land use change. Catena 66:74–92
Golosov V, Belyaev V, Kuznetsova Y, Markelov M, Shamshurina E (2008a) Response of a small arable catchment sediment budget to introduction of soil conservation measures. In: Schmidt J, Cochrane T, Phillips C, Elliott S, Davies T, Basher L (eds) Sediment dynamics in changing environments, international association of hydrological sciences, IAHS Publication no, vol 325. IAHS Press, Wallingford, pp 106–113
Golosov VN, Markelov MV, Belyaev VR, Zhukova OM (2008b) Problems in determining spatial inhomogeneity of Cs fallout for estimating rates of erosion-accumulative processes. Russ Meteorol Hydrol 4:217–227. (in Russian)
Golosov VN, Gennadiev AN, Olson KR, Markelov MV, Zhidkin AP, Chendev Yu G, Kovach RG (2011) Spatial and temporal features of soil erosion in the forest-steppe zone of the East-European plain. Eurasian Soil Sci 44(7):794–801
Golosov VN, Belyaev VR, Markelov MV (2013) Application of Chernobyl-derived ^{137}Cs fallout for sediment redistribution studies: lessons from European Russia. Hydrol Process 27:807–821
Golosov V, Gusarov A, Litvin L, Yermolaev O, Chizhikova N, Safina G, Kiryukhina Z (2017) Evaluation of soil erosion rates in the southern half of the Russian plain: methodology and initial results. In: Collins A, Stone M, Horowitz A, Foster I (eds) ICCE symposium 2016 – integrating monitoring and modelling for sediment dynamics, Okehampton, UK, 11–15 Jul 2016. Proceedings of the IAHS, vol 375. Copernicus Publications, Göttingen, pp 23–27

Golosov VN, Walling DE, Konoplev AV, Ivanov MM, Sharifullin AG (2018) Application of bomb-and Chernobyl-derived radiocaesium for reconstructing changes in erosion rates and sediment fluxes from croplands in areas of European Russia with different levels of Chernobyl fallout. J Environ Radioact 186:78
Higgitt DL, Froehlich W, Walling DE (1992) Applications and limitations of Chernobyl radiocaesium measurements in a Carpathian erosion investigation, Poland. Land Degrad Rehabil 3:15–26
Ivanov MM, Ivanova NN, Golosov VN, Shamshurina EN (2016) Assessing the accumulation of sorbed isotope ^{137}Cs within the upper components of the fluvial network in the zone of Chernobyl contamination. Geogr Nat Resour 37(4):355–361
Ivanov MM, Golosov VN, Belyaev VR (2017) Analysis of topography structure for the evaluation of sediment delivery ratio within the Plava River basin (Tula oblast). Moscow University bulletin, series 5. Geography 3:14–23. (in Russian)
Ivanov MM (2017) Erosion-accumulation processes as a factor of spatial transformation of radioactive contamination of the Plava River basin. Extended abstract of Cand. Sci. (Geogr.) Dissertation. Moscow State University, Moscow (in Russian)
Ivanova NN, Shamshurina EN, Golosov VN, Belyaev VR, Markelov MV, Paramonova TA, Evrard O (2014) Assessment of ^{137}Cs redistribution by exogenic processes in the Plava River valley bottom (Tula oblast) after the Chernobyl accident. Moscow University bulletin, series 5. Geography 1:24–34. (in Russian)
Izrael YA, Vakulovsky SM, Vetrov VA, Petrov VN, Rovinsky FY, Stukin ED (1990) Chernobyl: radioactive contamination of the environment. Hydrometeoizdat, Leningrad. (in Russian)
Izrael YuA (ed) (1993) Map of the caesium-137 radioactive contamination of the European part and the ural region of Russia (in Jan 1993) 1:500000. Russian State Cartographic Service (in Russian)
Izrael YA, Kvasnikova EV, Nazarov IM, Fridman SD (1994) Global and regional radioactive contamination ^{137}Cs European territory of former USSR. Meteorol Hydrol 5:5–10. (in Russian)
Izrael YA (1998) Atlas of radioactive contamination of European Russia, Belarus and Ukraine. Federal Service of Geodesy and Cartography of Russia, Moscow, p 143. (in Russian)
Izrael YuA, Kvasnikova EV, Nazarov IM,Stukin ED, Tsaturov YuS (1999) Radioactive contamination of territory of CIS and Europe. In: Ecologicheskaya bezopastnost' na poroge XXI veka. Proceeding of international conference, VSEGEI, Sankt-Peterburg, pp 88–89 (in Russian)
Korobova EM, Chizhikova NP, Linnik VG (2007) Distribution of ^{137}Cs in the particle-size fractions and in the profiles of alluvial soils on floodplains of the Iput and its tributary Buldynka Rivers (Bryansk oblast). Eurasian Soil Sci 4:404–417
Kryshev II, Alexakhin RM, Ryabov IN, Smirnov VV, Prister BS, Sanzharova NI, Perepelyatnikova LV, Astasheva NP (1991) Radioecologicheskie posledstviya Chernobylskoi avarii. Nauka, Moskva, p 190. (in Russian)
Kvasnikova EV, Stukin ED, Golosov VN, Ivanova NN, Panin AV (1998) Caesium-137 behaviour in small agricultural catchments on the area of the chernobyl contamination. Czechoslov J Phys 48:109–115
Leopold LB, Wolman MG (1957) River channel patterns: braided, meandering, and straight. US Government Printing Office, Washington, DC
Lettner H, Bossew P, Hubmer AK (2000) Spatial variability of fallout caesium-137 in Austrian alpine regions. J Environ Radioact 47(1):71–82
Lewin J (1978) Floodplain geomorphology. Prog Phys Geogr 2(3):408–437
Linnik VG (2001) A landscape- hydrological framework of Cs-137 distribution in the Iput River floodplain (the Bryansk Region). Soil Erosion Channel Proc 13:120–132. (in Russian)
Linnik VG, Govorun AP, Volosov AG (2004) Radionuclide contamination of floodplain soils of the Plava River. Contemporary problems of soil pollution, Proceeding of the international conference, In, pp 63–65. (in Russian)
Linnik VG, Brown JE, Dowdall M, Potapov VN, Surkov VV, Korobova EM, Volosov AG, Vakulovsky SM, Tertyshnik EG (2005) Radioactive contamination of the Balchug (upper

Enisey) floodplain, Russia in relation to sedimentation processes and geomorphology. Sci Total Environ 339(1–3):233–251

Linnik VG, Saveliev AA, Govorun AP, Ivanitsky OM, Sokolov AV (2007) Spatial variability and topographic factors of ^{137}Cs soil contamination at a field scale. Int J Ecol Dev 8(7):8–25

Linnik VG (2008) Spatial-temporal scales of landscape differentiation of technogenic radionuclides. Doctor of science thesis, Moscow, p 45. (in Russian)

Litvin LF, Kiryukhina ZP, Krasnov SF, Dobrovol'skaya NG (2017) Dynamics of agricultural soil erosion in European Russia. Eurasian Soil Sci 50(11):1343–1352

Litvin LF, Golosov VN, Dobrovolskaya NG, Ivanova NN, Kiryukhina ZP, Krasnov SF (1994) Redistribution of ^{137}Cs by the processes of water erosion of soil. Water Resour 23(3):286–291

Mamikhin SV, Golosov VN, Paramonova TA, Shamshurina EN, Ivanov MM (2016) Vertical distribution of ^{137}Cs in alluvial soils of the Lokna river floodplain (Tula oblast) long after the Chernobyl accident and its simulation. Eurasian Soil Sci 49(12):1432–1442

Markelov MV, Golosov VN, Belyaev VR (2012) Changes in the sedimentation rates on the floodplains of small rivers in the central Russian Plain. Moscow University bulletin, series 5. Geography 5:70–76. (in Russian)

Markina ZN, Shoshin VV, Vecherov VV (2014) Condition of 137Cs in soils of radioactive contaminated protective forest of the Bryansk region. Forest Eng J 2:42–51. (in Russian)

Martynenko VP, Linnik VG, Govorun AP, Potapov VN (2003) Comparison of the results of field radiometry and sampling in the investigation of ^{137}Cs soil content in the Bryansk Region. At Energy 95:727–733

Nanson GC, Croke JCA (1992) Genetic classification of floodplains. Geomorphology 4 (6):459–486

Navas A, Walling DE, Quine T, Machín J, Soto J, Domenech S, López Vicente M (2007) Variability in Cs inventories and potential climatic and lithological controls in the central Ebro valley, Spain. J Radioanal Nucl Chem 274:331–339

Nichols MH (2006) Measured sediment yield rates from semiarid rangeland watersheds. Rangeland Ecol Manag 59(1):55–62

Owens PN, Walling DE (1996) Spatial variability of caesium-137 inventories at reference sites: an example from two contrasting sites in England and Zimbabwe. Appl Radiat Isot 47(7):699–707

Panin AV, Walling DE, Golosov VN (2001) The role of soil erosion and fluvial processes in the post-fallout redistribution of Chernobyl-derived caesium-137: a case study of the Lapki catchment, Central Russia. Geomorphology 40:185–204

Paramonova TA, Shamshurina EN, Belyaev VR, Komissarova OL (2017) Homo/heterogeneity of Cs-137 distribution within ploughed horizon of arable chernozems, 30 years after Chernobyl accident. Radiat Appl 2(3):192–199

Porto P, Walling DE, Callegari G, Capra A (2009) Using caesium-137 and unsupported lead-210 measurements to explore the relationship between sediment mobilisation, sediment delivery and sediment yield for a Calabrian catchment. Mar Freshw Res 60:680–689

Quine TA (1989) Use of a simple model to estimate rates of soil erosion from caesium-137 data. J Water Res 8:54–81

Radoane M, Radoane N (2005) Dams, sediment sources and reservoir silting in Romania. Geomorphology 71:112–125

Renwick WH (1996) Continent-scale reservoir sedimentation patterns in the United States. Erosion and sediment yield: global and regional perspectives (proceedings of the exeter symposium, Jul 1996). IAHS Publ. 236. IAHS Press, Wallingford, pp 513–522

Romantsova NA (2012) Estestvennaya i technogennye radionuklidy v pochvakh Plavskogo radioactivnogo pyatna Tul'skoi oblasti. Agrokhem Vestn 6:34–36. (in Russian)

Rysin II, Grigoriev II, Zaitseva MY, Golosov VN (2017) Dynamic of the linear retreat of gully heads within the Vyatka-Kama interfluve at the turn of 20th century. Moscow University bulletin, series 5. Geography 1:63–72. (in Russian)

Romantsova NA, Paramonova TA, Matveev YV, Semenikhin AI (2012) Modern features of the accumulation of cesium-137 in various phytocenoses of the Plavsky radioactive spot in the Tula region. In: Actual problems of ecology and nature management: Proceedings of the annual all-russian scientific conference. publishing house of rudn university. Moscow 14:206–214. (In Russian)

Schumm S (1977) A. The fluvial system. MLA

Shamshurina EN, Golosov VN, Ivanov MM (2016) Spatial-temporal reconstruction of Chernobyl-derived Cs fallout on the soil cover in the upper reach of the Lokna River. Radiat Biol Radioecol 4:414–425. (in Russian)

Sidorchuk AY (1996) Sediment budget change in the fluvial system at the central part of the Russian Plain due to human impact. In: Walling D, Webb B (eds) Erosion and sediment yield: global and regional perspectives, vol 236. IAHS Press, Wallingford, pp 445–452

Smith SW, Renwick WH, Buddemeier RW, Crossland CJ (2001) Budgets for soil erosion and deposition for sediments and sedimentary organic carbon across the conterminous United States. Glob Biogeochem Cycles 15(3):697–707

Sutherland RA (1991) Examination of caesium-137 areal activities in control (uneroded) locations. Soil Technol 4:33–50

Sutherland RA (1994) Spatial variability of Cs and the influence of sampling on estimates of sediment redistribution. Catena 21:57–71

Sutherland RA (1996) Caesium-137 soil sampling and inventory variability in reference samples; literature survey. Hydrol Process 10:34–54. (in Russian)

Trofimetz LN (2011) Application of radiocaesium method to the investigations of wash-out and accumulation processes in the River Vytebet valley. Uchenye zapiski Orlovskogo Universiteta, seriya: estestvennye, tekhnicheskie I meditsinskie nauki 5:259–268. (In Russian)

Tsybul'ka NN, Chernysh AF, Tishuk LA, Zhukova II (2004) Horizontal migration of ^{137}Cs with water erosion of soils. Radioekologiya 44(4):473–477. (in Russian)

Turnage KM, Lee SY, Foss JE, Kim KH, Larsen IL (1997) Comparison of soil erosion and deposition rates using radiocesium, RUSLE, and buried soils in dolines in East Tennessee. Environ Geol 29(1–2):1–10

Vaculovskii SM, Nikitin AI, Bovkun LA, Chumichev VB, Tasiev YAL, Nazarov LE, Kryshev II (1996) ^{137}Cs and ^{90}Sr contamination of Russian Federation water objects in the influence zone of Chernobyl NPS. Meteorol Hydrol 4:18–24. (in Russian)

Verstraeten G, Poesen J (2001) Factors controlling sediment yield from small intensively cultivated catchments in a temperate humid climate. Geomorphology 40:123–144

Vetrov VA, Alexeenko VA, Poslovin AL, Chereminisov AA, Nikitin AA, Bovkun LA (1990) Radionuclide washout from natural catchments in the Dnieper river basin. J Hydrol Meteorol 2:120–123. (in Russian)

Walling DE, He Q (1999) Improved models for estimating soil erosion rates from cesium-137 measurements. Journal of Environmental Quality 28(2):611–622

Walling DE, Golosov VN, Panin AV, He Q (2000) Use of radiocaesium to investigate erosion and sedimentation in areas with high levels of Chernobyl fallout. In: IDL F (ed) Tracers in Geomorphology. Wiley, Chichester, pp 183–200

Walling DE, He Q, Appleby PC (2002) Conversion models for use in soil-erosion, soilredistribution, and sedimentation investigations. In: Zapata F (ed) Handbook for the assessment of soil erosion and sedimentation using environmental radioactivity. Kluwer, Dordrecht, pp 111–164

Zhidkin AP, Golosov VN, Svetlichny AA, Pyatkova AV (2015) An assessment of load on the arable slopes on the basis of field methods and mathematic models. Geomorphol RAS 2:41–53. https://doi.org/10.15356/0435-4281-2015-2-41-53. (in Russian)

Zhukova OM, Matvienko II, Myshkina NK, Shrovarov GA, Shiryaeva NM (1997) Formation and dynamic of distribution of radionuclide contamination in the Belarus Rivers after Chernobyl Atomic Energy Plant. Inzhenerno-fizichesky Zhurnal 70(1):73–80. (in Russian)

Part III
Behavior of Radionuclides in Agricultural and Forest Ecosystems

Chapter 5
Behavior of the Chernobyl-Derived Radionuclides in Agricultural Ecosystems

Boris Prister

Abstract In this chapter, the key aspects of agricultural radioecology on the territory of Ukraine, Belarus, and Russia, following the severe accident at the Chernobyl NPP, specifically, the iodine attack and population protection, the contamination of arable lands with long-lived radionuclides ^{90}Sr and ^{137}Cs, and their entry into the food chain, are considered. The factors governing the behavior of ^{90}Sr and ^{137}Cs in soil and their uptake by plants are examined in detail. A model has been proposed to study the behavior of ^{90}Sr and ^{137}Cs in the "soil–plant" system to identify the contamination of crops. Countermeasures for reducing the accumulation of radionuclides in crops and animal-derived food products are discussed, and an assessment of their effectiveness is provided.

Keywords Chernobyl · Monitoring · Soil–plant transfer · Modelling · Countermeasures effectiveness

---

Foreword from editors: The author worked at Experimental Scientific-Research Station (ESRS) on the territory of East Urals Radioactive Trail (EURT) for 19 years from 1962 to 1979 after graduation from Timiryazev Moscow Agricultural Academy in 1962. In 1967, Boris Prister defended Ph.D. thesis "The behavior of uranium in soil and biological chains" and in 1978 received a degree of Doctor of Science majoring in "Problems of agricultural radiobiology and radioecology in the areas contaminated with fresh mixture of nuclear fission products." An awardee of USSR State Prize (1974) for the development and implementation of "Recommendations on agricultural practices on radioactively contaminated areas." In 1986, immediately after the Chernobyl accident, on his initiative together with Academician Bogdanov G.A., State Program "Agricultural Radiology" was established by the USSR State Committee for Science and Technology and he was appointed as its scientific supervisor. He was awarded, together with a group of scientists, the State Prize of Ukraine in the field of science and technology in 2004 for a series of works on an integrated study of the consequences of the Chernobyl disaster and development of mitigation measures. Since 2004 he has been a principal research scientist of Institute of Safety of Nuclear Power of National Academy of Agricultural Science of Ukraine. Boris Prister has summarized his experience focusing on agricultural aspects of the Chernobyl accident.

B. Prister (✉)
Institute of Nuclear Safety Problems of National Academy of Science, Kiev, Ukraine

A. Konoplev et al. (eds.), *Behavior of Radionuclides in the Environment II*,
https://doi.org/10.1007/978-981-15-3568-0_5

5.1 Introduction

The accident at the Chernobyl nuclear power plant (ChNPP) was unprecedented in scale and has led to the contamination of a significant part of Europe (Borzilov and Klepikova 1993; De Cort 1998). During the Chernobyl accident, among numerous radionuclides released were two isotopes of major environmental significance: ^{134}Cs and ^{137}Cs with the half-lives of 2.06 and 30.17 years, respectively. Due to the core fire, 0.15×10^{18} Bq of ^{134}Cs and 0.05×10^{18} Bq of ^{137}Cs were released to the environment within fuel particles and in the gaseous form, which was 20–60% of the entire amount in the core (Guntay et al. 1996; IAEA 2001). The release of radioactivity following the Chernobyl accident has led to ^{137}Cs contamination of extensive arable lands in Belarus, Russia, and Ukraine (Table 5.1).

Long-lived radionuclides of Sr and Pu deposited as part of heavy fuel particles and therefore did not disperse over significant distances. Their deposition on the soil surface was above the established limits mostly within the 30-km zone. Pu is known to have a very low ability to penetrate biological membranes, and therefore the concentration factors for its plant uptake from soil and transfer from the gastrointestinal tract (GIT) to blood were several orders of magnitude lower than those for Sr and Cs. At the same time, the inhalation pathway of α-emitting radionuclide intake was and continues to be a major hazard in the near zone, at the ChNPP site, and in the Sarcophagus structure (Garger 2008).

After the Chernobyl accident, ^{90}Sr played a much smaller role as compared to ^{137}Cs in terms of incorporation into food chains. Only in several areas of Ukraine and Belarus, the contribution of ^{90}Sr to the overall radiation dose was high, whereas on the territory of Russia, this radionuclide did not present any hazard. In contrast to the EURT in Chelyabinsk region, Chernobyl-derived ^{90}Sr was deposited in the near zone in a water-insoluble form within fuel particles, from which it was released due to their oxidation and decomposition (Salbu et al. 2001; Konoplev 2020). Due to ^{90}Sr leaching, its soil–plant transfer factor in the near zone increased several fold during 8–10 years, the process of particle destruction increased faster on acidic soils (Kashparov et al. 2004). Eventually, 30 years after the accident, almost entire ^{90}Sr in soils was leached from fuel particles to the soil solution.

In Belarus, 262 thousand hectares of arable land was contaminated by ^{137}Cs at levels from 185 to 1480 kBq m^{-2}; of them, 163 thousand hectares of land was also contaminated by ^{90}Sr at levels from 11 to 111 kBq m^{-2} (0.3–3.0 Ci km^{-2}) (Shevchuk and Gurachevsky 2001, 2003).

Table 5.1 Territories of Ukraine, Russia, and Belarus where ^{137}Cs deposition was more than 37 kBq m^{-2}

Country	Area, thousand km^2	% of total territory of the country
Russia	57.9	0.5
Belarus	46.5	23
Ukraine	41.9	4.8

In Ukraine, the areas affected by strontium contamination, besides the 30-km zone of ChNPP, include Kiev and Chernigov regions. The ^{90}Sr soil contamination density from 3.7 to 37 kBq m^{-2} (0.1–1.0 Ci km^{-2}) in these areas occurred on 35 thousand km^2 (about 35% of the total territory), and 3.7 thousand km^2 was contaminated at levels above 37 kBq m^{-2} (1 Ci km^{-2}) (Baloga et al. 2006).

In several dozens of settlements in Ivankov district of Kiev region and Kozeletsky district of Kiev region, countermeasures had to be taken in producing grain and milk to comply with relevant standards of Ukraine. In Gomel oblast of Belarus, in which a significant part of ^{90}Sr-contaminated lands occurred, a major part of grain yield (from 23 to 47 thousand tons per year) could only be used as fodder or processed into alcohol (Bogdevich 1996).

In case of a radiation-related accident characterization of physical and chemical properties of radionuclides in depositions, their solubility, bioavailability from soil solution for plants, and cattle GIT-blood absorbability should be part of the first-stage radioecological monitoring on agricultural lands.

The implementation of countermeasures has demonstrated that a radical reduction of ^{90}Sr uptake by plants can be achieved only through long-term soil cultivation, which actually requires significant investments. Monitoring-based optimization of crop placement on the fields in "critical" farms of Mogilev oblast has made the prevention of producing grains exceeding the ^{90}Sr permitted regulatory levels possible.

The region of the accident is an intensive agro-industry zone, with agriculture being the central economic sector. The population of Belarus, Russia, and Ukraine living in the accident-affected region, has a "rural" diet, consuming local products without any processing. The main pathway responsible for the intake of radionuclide for humans is the consumption of agricultural products and "forest products" (berries, mushrooms, etc.), and as result, radiation doses of the rural population are higher than those of urban dwellers. Over extensive territories, the contamination density was so high that the production and use of agricultural products was not possible without applying countermeasures, at least in the first few years. Estimates suggest that by 2005–2007, ^{137}Cs- and ^{90}Sr-contaminated lands as compared to 1986 were reduced: by 1.2 times in Belarus; two times in Russia, and 1.7 times in Ukraine, yet remaining quite large.

In Ukraine, more than eight million hectares of land in 74 districts of 12 regions has been contaminated primarily with radioiodine and radiocesium. The worst affected was Polesye[1]—the northern areas of Volyn, Rovno, Zhytomyr, Kiev, and Chernigov oblasts (Baloga et al. 2006). The distribution of the arable lands contaminated by ^{137}Cs, as of 1997, in five worst affected oblasts of the Ukrainian Polesye is given in Table 5.2.

A total of 3.2 million people, including 600 thousand children, were living in 2.2 thousand villages in the contaminated territories (Baloga et al. 2006). Social

[1]Polesye is a natural and historical region starting from the farthest edges of Central Europe into Eastern Europe.

Table 5.2 Areas in five regions of Ukrainian Polesye and total contaminated by ^{137}Cs, thousand km^2 (as of 1986)

		^{137}Cs soil contamination density, kBq m^{-2}			
Oblast	Total area, km^2	<37	37–185	185–555	>555
Volyn	20.2	20.0	0.2		
Zhytomyr	29.9	18.9	8.7	1.7	0.64
Kyiv	28.9	17.5	8.8	1.6	1
Rivne	20.1	12.2	7.8	0.1	
Chernihiv	31.9	29.7	2.1	0.1	
Total	603.7	560.9	37.5	3.7	1.64

consequences were particularly severe for the population of Polesye, where a significant part of agricultural products was produced on natural landscapes (Prister 1998).

The consequences of the accident were aggravated by the demographic features of Ukrainian Polesye. Densely populated territories where agriculture was highly developed, especially milk and meat cattle breeding and sheep raising, were contaminated at significant levels. The accident occurred at the beginning of the grazing period when fodder reserves were practically used up, both on public and private farms. In the overwhelming majority of villages, the cattle grazed on pastures or received fresh fodder, and as a result, they were exposed to radioactivity deposited on pasture vegetation and haystacks. The international conference "Fifteen Years after the Chernobyl Accident. Lessons Learned" recognized the Chernobyl accident to be "a communal rural accident" that has disrupted customary way and mode of life of the affected rural population (Proceedings 2001).

5.2 Behavior of Radioiodine in Agricultural Environment: Exposure and Protection of the Population from Iodine Attack

5.2.1 *Nature of Hazard*

Half-lives of the short-lived iodine radionuclides ^{133}I and ^{131}I are 20.3 h and 8.1 days, respectively, and therefore their behavior is difficult to study when these radionuclides are actually released. The accuracy of data obtained immediately after the release is usually not very high. Studies after the nuclear weapon tests showed that countermeasures had to be developed to protect humans from iodine radionuclides. In 1968–1973, experiments were carried out at "Mayak" experimental and scientific research station to study crop contamination with fission products, doses and biological effects of radiation exposure of plants and animals. Information on biological impacts and behavior of fission products in cattle was so important for population radiation safety that, in spite of the defense relevance of these studies, outcomes of experiments on agricultural radiobiology and radioecology were decided not to classify and make public. First materials were published in the monograph (Fedorov et al. 1973a). The results of the experiments were then used for developing recommendations for farming and forestry in case of environmental radioactive contamination (Fedorov et al. 1973b). Some of the relevant aspects were also discussed in subsequent publications (Prister et al. 1978; Sirotkin et al. 1978), and findings were presented in a holistic way in Prister (2008).

^{131}I is characterized by a very high rate of absorption from GIT into blood: the maximum concentration of the nuclide in the blood (0.23% of injected amount per L) was reported as soon as 3 h after injection (Russell 1962; Prister 2008). The absorption of this radionuclide in GIT is 100%, i.e., all ingested iodine is transferred to the blood via the gastrointestinal wall. Likewise, it is fully absorbed in the lungs in case of inhalation pathway.

Iodine radionuclides are known to enter the human body with food, first of all, milk and leaf vegetables (up to 90%). The behavior of iodine radionuclides in the agricultural food chain deserves more detailed consideration, particularly, from the standpoint of human protection. When Chernobyl accident occurred, the reason for the accident and the occurrence of the radioactive iodine was not disclosed during the critical first days, and as a consequence, the use of contaminated foods could not be totally prevented.

In adults, about 20% of ingested iodine accumulates in the thyroid, whereas in children younger than 1 year 40% accumulates in the thyroid. In general, an average adult of 70 kg has the thyroid of 18 g, but it is only 4 g in children; for this reason, iodine intake is particularly harmful to children under 1 year. If the same amount of the radionuclide enters an adult and a child's body, its concentration and absorbed dose will be ten times higher in the child's thyroid. Interestingly, in cows weighing 5–6 times more than humans (400–500 kg), the thyroid structure and mass practically do not differ from an average human adult: 18 g with two lobes, 9 g each.

Following the Chernobyl accident, the major part of the absorbed thyroid dose (80–90%) was formed in the first three days after the intake of radionuclides by the human body. With one-time intake of fission products, the contribution of ^{133}I to the overall dose was significant in the first days; however, it became below 10% on the fifth day after the accident and only a few tenths of a percent by the tenth day (Prister 2008).

The critical importance of the issue of environmental iodine contamination became apparent in the process of mitigating the consequences of ChNPP accident in 1986. In the first days and weeks after the accident, the primary hazard for humans, particularly children, was the exposure of the thyroid to iodine radionuclides. The contribution of ^{133}I to the thyroid dose in the population was much lower as compared to that of fresh fission products in case of nuclear explosion or immediate release because of radiation accident. The Chernobyl radioactivity release continued for 10 days after the termination of nuclear fission, and the fallout on the second and subsequent days was depleted in ^{133}I as a result of radioactive decay. Misunderstanding of this fact has led to many accusations of dosimetry experts who allegedly did not take into account ^{133}I presence in the fallout.

5.2.2 Countermeasures to Protect the Population from Iodine Attack

Humans and animals can be protected from iodine attack by food chain intervention, thereby preventing or minimizing the entry of iodine radionuclides into the body (agricultural countermeasure). The second method is iodine prophylaxis, aimed at reducing the uptake of ingested iodine by the thyroid, which makes it possible to avert a significant part of the thyroid dose (medical countermeasure).

An important feature of the behavior of iodine in cattle body is the ability of the radionuclide to be excreted in significant amounts with milk, which aggravates the hazard of iodine fallout (Korneev and Sirotkin 1970). Parameters of radionuclide excretion with milk are shown in Table 5.3.

Iodine can be detected in milk 30 min after its administration, and 3 h after administration, the concentration of ^{131}I in milk is 0.7–0.8% per L of the amount ingested for 24 h. After 12 h, its concentration is maximum—1.3% per L. Our data are in agreement with the results of many experiments (Squre et al. 1961; Russell and Possingham 1966; Korneev and Sirotkin 1970). As soon as 1 h after administration, the radionuclide concentration in milk is 0.36% of the administered amount per liter, and in 12 h it becomes maximum—0.8–1.47%. The larger part of ingested ^{131}I is excreted from the cow body in the first few days. Therefore, banning milk consumption during the first few days after iodine fallout is an effective countermeasure to be introduced immediately upon receiving a signal about possible iodine fallout. In recommendations regarding agriculture and forestry management in case of radioactive contamination of the environment (Fedorov et al. 1973b), it is advised that upon receiving a warning about a possible accident or notification of the occurrence of accident, before details of the situation are known, cattle should be fed with uncontaminated fodder from the stock or should fast for 3–4 days. The ban for grazing and using contaminated milk can be lifted only when radiation monitoring data are received. Unfortunately, following the Chernobyl accident, none of the above measures were introduced or implemented in a centralized way by order of authorities, with few exceptions.

During 30 days from the time of pasture contamination and uptake of fission products by cows, milk contamination is primarily due to iodine radionuclides, their

Table 5.3 Parameters of ^{131}I excretion with milk for cows after the one-time administration of radionuclide mixture with feed, $\delta \leq \pm 30\%$ (Prister 2008)

Time of maximum excretion, τ (h)	C_0, % administered amount per L	Biological half-life, days		Fraction excreted with	Excreted with milk per day, % of administered amount	
		T_b	T_{ef}	T_{ef}	4	12
12	1.3	1.1	1.0	0.97	12.1	12.8
		5.9	3.4	0.03		

Table 5.4 Parameters of ^{131}I behavior in plants with aerial contamination and in cow's body in case of intake with feed (Prister 2008)

Parameter	Notation, units			
Herbage interception factor	10–20% at $M = 1$ kg m^{-2}			
	$T_b^{q\,a}$, days	T_{ef}^{q}, days	$T_b^{s\,b}$, days	$T_{ef}^{s\,s}$, days
Half-lives from plants: biological T_b and effective T_{ef}	4–6	3.0	12–15	5.0
Biological half-lives for milk	1.1	0.96	5.9	3.4
Half-lives to be excreted from muscles	1.7	1.4	9.0	4.2
Half-life to remove from the thyroid	63	7.1	–	–

[a]Quick
[b]Slow

concentrations decreasing quickly in timc. ^{133}I is a major contributor to milk contamination only in the first three days after radioactive cloud occurrence.

On the year of the accident, the contamination of crops and pasture vegetation was mainly via aerial pathway as a result of direct deposition and resuspension. For predicting foliage contamination of plants, we used the model developed in the course of model experiments in South Urals (Prister 2008).

It is extremely important to avoid consuming contaminated food crops and fodder in the first hours and days after the accident, and during this time period, apart from radioactive decay, plants tend to self-purify by a factor of 2 during several hours or days (Table 5.4).

The approach to iodine prophylaxis was developed long before the Chernobyl accident (Il'in et al. 1972). Essentially, iodine prophylaxis consists of the administration of stable iodine (about 150 mg) several hours prior to a potential iodine attack so that the thyroid will be saturated with stable iodine isotope. This ensures that the intensity of iodine exchange between blood and the thyroid is reduced, and accordingly the thyroid intake of iodine isotopes decreased sharply by 50 100 times. Given iodine deficiency in human and cattle diet, the efficacy of this measure is 2–3 times higher, which was confirmed by the Chernobyl experience. In the case of stable iodine administration simultaneously with the intake of iodine radionuclides, the effectiveness of this measure is 2–3 times lower, but still quite high (Table 5.5). For this reason, the term "iodine prophylaxis" is more appropriate than "iodine blockade," and this measure should be applied immediately in the event of a hazard alert about radioactivity release to the environment.

5.2.3 *Effectiveness of Protecting the Population from Iodine Attack on the Territory of Former USSR*

Concealment by the government of the accident occurrence and its scale did not allow the implementation of iodine prophylaxis in due time. Even in the town of

Table 5.5 Expected and actual effectiveness of iodine blocking following the Chernobyl accident in Ukraine (Prister et al. 2016)

Expected effectiveness of iodine blocking as a function of time after the intake of the radionuclide		Actual efficacy of iodine blocking in terms of the thyroid dose reduction in Ukraine, times		
Time after intake of ^{131}I, h	Dose reduction, times			
2 h prior to intake	50–100	Pripyat, 45,000 people	Twofold administration	2.3
1	10–12		Single administration	1.6
2	4.0	Evacuated villages		<1.4 (children)
3	2.0			
15	1.2			

Pripyat, in which NPP workers lived, people were not warned of the iodine attack hazard immediately after the accident, and both adults and children stayed outdoors. Nevertheless, the evacuation of residents from Pripyat and surrounding settlements, even though delayed, prevented the intake of significant amounts of radioactive iodine by humans.

In Kiev, organizational measures (Prister 1999) were taken to protect the population from thyroid exposure. In all milk and meat processing factories of Kiev, competent spectrometry monitoring of products was set up by the veterinary authority jointly with Nuclear Research Institute of National Academy of Sciences (NAS) of Ukrainian Soviet Socialist Republic. In many batches of milk, the concentrations of radionuclides were above the temporary maximum permitted levels and were as high as tens of kBq kg^{-1}, the regulatory limit being 0.37 kBq kg^{-1}.

The incoming inspection at milk receiving stations and outcoming inspection at dairy factories were organized by the institute of National Academy of Agrarian Sciences of Ukraine (NAASU) jointly with Veterinary and Dairy Industry Authority of Ukrainian Ministry of Agricultural Produce (Prister 1999). Each milk tank vessel was subjected to inspection. The analysis of measurements, despite nonuniformity of land contamination levels, enabled within several days, dividing farms into groups by radiological quality of their products and developing a plan for milk delivery and processing.

The city of Kiev with a population of about four million people was supplied with milk and dairy products from three dairy factories. Milk from the farms not meeting relevant requirements was processed into condensed or dry milk and butter and kept in storage or refrigerators until iodine radionuclides decayed. Children care, infant feeding centers, and hospitals received dairy products from the factory that received milk from the least contaminated farms. The implemented measures allowed the reduction of milk contamination in Kiev and suburban areas by 7–8 times (Prister 1999). By the decision of the USSR government, about 35 large-capacity tank vessels were provided for milk delivery from 13 least contaminated regions.

The second hazard in the initial time period after the fallout of fission products, other than milk contamination, was ^{131}I contamination of vegetables. Consuming fresh vegetables, provided no radiological monitoring, presents a hazard on the territory on which milk consumption is not permitted because of contamination levels.

In the first days after the fallout, the concentration of radionuclides in vegetative parts of plants was 2–3 orders of magnitude higher than that in productive parts. Concentrations decreased with time quickly due to the radioactive decay and field losses of radionuclides (Table 5.4). Even in the first week, the concentration of fission products in vegetable biomass decreased by 1–2 orders of magnitude. Harvesting should not be allowed in this time period. If deposited fission products occur in the soluble form, washing and cleaning of vegetables will allow diminishing contamination level by 1.5–3 times.

Removing from the diet vegetables grown in the open and termination of cattle grazing on pasture during the first 10–14 days after contamination decreases the intake of radionuclides significantly (Prister 2008).

Unfortunately, right after the Chernobyl accident, the population, especially villagers, consumed mainly unprocessed milk and greenery, ignoring the ban to give fresh milk to children, even though delayed. After several weeks, the mistrust translated into panic and the consumption of dairy products declined sharply, at that point unnecessarily.

Table 5.6 and Fig. 5.1 show averaged thyroid doses for residents in six districts of three Ukraine regions, as well as areas of Russia and Belarus. The average doses vary from tens to thousands of mGy. Such doses can induce cell mutagenesis, not causing, however, lethal mutations. Actually, the survival of mutating cells is responsible for the impairment of the thyroid and depression of its function—hypothyroidism.

The consequence of the iodine attack was a sharp increase in the number of thyroid cancers among children and adolescents. Prior to the accident, only 2–4 thyroid cancer cases per year were diagnosed in children in Ukraine. By the end of 1998, a total of 1120 persons from 0 to 18 years were operated for thyroid cancer (clinical evidence available). By the end of 2008, the number of operated cancers was 5009 and 11,000 cases as of the beginning of January 2015.

It should be noted that when the accident occurred the expertise regarding radiation hazard in the event of the radioactive iodine release to the environment was already available in the country and appropriate recommendations were in place,

Table 5.6 Thyroid doses of ^{131}I for children and adolescents in selected areas of Ukraine (Prister et al. 2016)

Region	District	Number of settlements	Population 0 18 years as of 1986, persons	Dose, mGy		
				Mean	Min	Max
Zhytomyr	Narodychi	76	7000	1559	119	6879
	Olevsk	60	19,000	213	44	1259
Kyiv	Ivankov	67	7100	199	55	632
	Polesye	61	8100	778	16	7269
Chernihiv	Repki	112	9900	236	34	1471
	Chernigov	125	18,200	427	43	6528

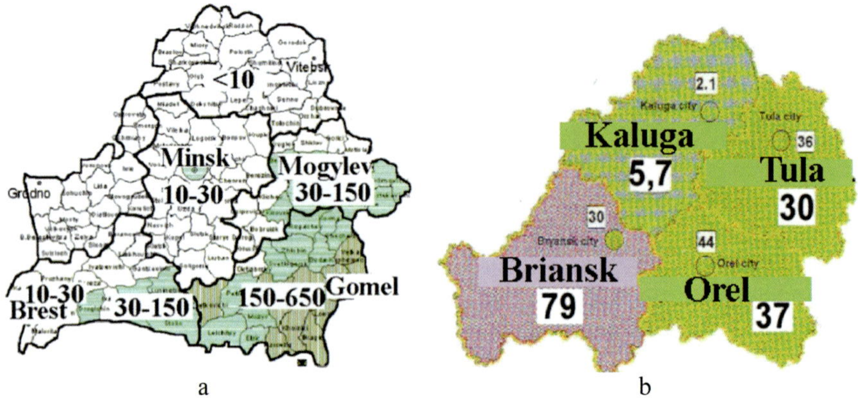

Fig. 5.1 District and region averaged thyroid doses in children (mGy) on the territory of (**a**) Belarus and (**b**) Russia contaminated as a result of the Chernobyl fallout (Prister et al. 2016)

banning uncontrolled consumption of milk and fresh vegetables following a radioactivity release. These recommendations were in the possession of the heads of territorial entities of the USSR, but they have not been enforced by the authorities who tried to conceal the reason for the accident. A lesson should be learnt from what happened for the benefit of emergency preparedness in all countries.

Credit should be given to the teams of scientists in the three republics who have done much for determining the scale of iodine attack and spatial distribution of this radionuclide, namely the groups directed by IA Likhtarev and ND Tronko (Ukraine), YaE Kenigsberg and YuE Kruk (Belarus), and MI Balonov and VK Ivanov (Russia). Hundreds of thousands of direct measurements of ^{131}I content in the thyroid were made in members of the public and clean-up workers, resulting in the reconstruction of absorbed doses and substantiation of policies for delivering health care services.

An analysis of the above materials shows high effectiveness of radiation protection of the population from the exposure of iodine radionuclides by way of implementation of protective measures in the agricultural sector. These measures should be realized irrespective of medical actions such as thyroid blocking or be complementary to them.

5.3 ^{137}Cs and ^{90}Sr: Concepts Used for Predicting Contamination of Agricultural Products and Substantiating Countermeasures

5.3.1 Polesye: A Geochemical Province with Low Content of Clay Minerals

The phase of iodine hazard being quite short, 45 days after the Chernobyl accident, ^{131}I was practically not found in milk and other foodstuffs. As ^{131}I contamination levels in the environment were declining, cesium radionuclides became a greater contributor to radiation exposure, first of all, internal doses from consumption of locally produced foodstuffs. The second phase after a nuclear accident or nuclear explosion is often referred to as the "cesium phase." Given the biomass of about 1 kg m^{-2}, the vegetation is capable of retaining 10–40% of depositions depending on the characteristics of radionuclides and the composition of herbage species (Prister 2008).

The hey made in the first year after the Chernobyl accident showed particularly high concentrations of ^{134,137}Cs, and concentrations in milk were hundreds of times in excess of the prescribed limits in more than 1000 settlements. In the second year after the accident, the contamination levels in agricultural products were still high in many areas and often exceeded the permitted regulatory levels (Table 5.7).

By the beginning of the 1987 growing season, the radionuclides intercepted by vegetation cover were transferred to soil or meadow sod and began interacting with the soil adsorption complex (SAC). Binding of radionuclides to the soil adsorption complex resulted in a decrease of their soil–plant transfer. In the second year after the accident, plant contamination was primarily due to ^{137}Cs transfer via the root pathway.

In Ukraine, the areas worst contaminated with ^{90}Sr were Kiev and Chernigov regions. A total area of 35 thousand km^2 was contaminated at ^{90}Sr deposition from 0.1 to 1.0 Ci km^2 and an area of 3.7 thousand km^2 at depositions higher than 1 Ci km^2 (about 35% of the total territory).

The territory of Polesye is generally made up of forest and meadows, as well as frequently occurring boggy landscapes formed on hydromorphic organogenic meadow-boggy and peat-boggy soils. The data of model experiments (Fedorova 1968) and results of radiation monitoring in Belarussian and Ukrainian Polesye during the intense nuclear weapons testing in 1954–1963 (Marey et al. 1974, 1984) are indicative of higher accumulation of cesium nuclides by peaty soils. The ^{137}Cs radiation doses in humans in Polesye, on the average, turned out to be ten times higher than those in humans living in other areas of the country with the same deposition, which is associated with endemic features such as mineral composition of soils, influencing not only absorption capacity but also strength and selectivity of cesium ions absorption. This trend is also obvious from a comparison of the concentration ratios of ^{137}Cs uptake by plants in Ukrainian and Belarussian Polesye on the soils with practically identical content of organic matter (2.4%), but different mineral composition (Table 5.8).

Generally, higher humus content in the soil leads to reduced uptake of radionuclides by plants. This is explained not only by increased overall absorption capacity and quality changes in the soil absorption complex, resulting in increased energy of cation absorption and reduction in active and hydrolytic acidity of the soil, but also by the formation of complexed or organoelement compounds, within which radionuclides are unavailable for plant root uptake. On peaty soils, the organic matter is quite specific, composed predominantly of fulvic acids capable of forming soluble compounds with monovalent cations, which significantly increases cesium mobility in the soil–plant–milk system as compared to soils with a predominance of humic acids (Pronevich 2013, 2014).

Table 5.7 ^{137}Cs concentration in the crop yield in Ukraine in 1987 (root uptake) and the temporary permitted levels (TPL)

Oblast, rayon	Soil type	Soil contamination density, kBq m^{-2}	^{137}Cs concentration, Bq kg^{-1} of natural product			
			Natural herbs TPL = 370[a]	Vegetable TPL = 740	Potato roots TPL = 740	Cereal crops TPL = 370[b]
Rivne Dubrovitsk	Peat-bog	120–230	44,100	–	2000–2200	350–400
	Soddy-podzolic	110–210	4200	520–660	200–300	160–220
Zhytomyr Ovruchsk	Peat-boggy	120–220	18,130	–	600–900	400–500
	Soddy-podzolic	200–300	4400	500–580	180–250	130–140
Kyiv, Borodyansk	Soddy-podzolic	390–550	11,020	600–1000	450–500	220–300
Kyiv, Boguslav	Chernozem podzolized	110–330	–	150–190	15–50	20–40

[a]Milk, dairy products
[b]Bread, bread products

Table 5.8 ^{137}Cs uptake by pasture vegetation as a function of mineral composition in soil

Soil type	Soil mineral composition	^{137}Cs concentration ratio
Belarusian–Ukrainian Polesye		
Peat-bog sandy	No clay minerals	16
Meshcherskaya lowland		
Muck peat gley	Illites	2.2
Alluvial flood land	Illites and montmorillonites	0.2

5.3.2 Milk and Meat as the Main Pathways of ^{137}Cs Intake for People on the Territory of Ukrainian Polesye

One of the valuable natural resources of Ukrainian Polesye is livestock feed, primarily grasses produced on natural and cultured pastures. They form a basis for livestock breeding, particularly cattle and sheep raising. About 80% of animal-derived food products in Polesye had been produced by farms using natural hey fields and pastures. For the rural population living on Chernobyl-contaminated areas of Ukrainian Polesye, the radiation hazard was associated with the consumption of cow milk and meat—foodstuffs that have always been a staple of their diet. The daily intake of radionuclides with milk and meat can be as high as 90% (Los 1993) (Table 5.9). Even 30 years after the accident, the situation in Polesye still continues to exist: 75–90% of the internal dose for people is due to the consumption of locally produced milk and dairy products.

Radionuclides are transferred to the animal body with feed and absorbed into the blood stream through the GIT walls. The absorption factor, i.e., the part of nuclide transferred to the blood stream, decreases with age. In young animals ^{137}Cs absorption is as high as 70–90%, whereas in adult animals of 1 year and older, it is 70%, on the average (Sirotkin et al. 1978; Sirotkin 1992). Radionuclides are transported by the blood stream and dispersed all over the body to be excreted with milk and to accumulate in muscles and internals.

Table 5.10 includes estimates of transfer of long-lived radionuclides from animal daily diet to animal-derived food products. It should be pointed out that ^{137}Cs concentrations in muscle tissue are strongly dependent on the characteristics of animal species.

Digestibility of feed, with which radionuclides enter the animal body, has only a minor influence on their absorption, even though the inverse relationship has been revealed between fiber content in cow diet during indoor maintenance and cesium transfer to milk. According to Prister (1998), an increase in fiber content from 1.3–1.8 to 3 kg per day results in a decrease in the cesium transfer factor to milk from 0.9% to 0.6%.

In comparison with cattle, the uptake of ^{137}Cs in muscles of pigs, goats, and sheep is higher due to more intense mineral metabolism. The feed of pigs and chickens, however, mainly consists of concentrated and succulent stuff (cereal grain, tubers, and roots), and daily radionuclide intake is lower than that with grass or hay for cows. After the Chernobyl accident, practically no cases of contamination of poultry such as geese, duck, and hens above the permitted levels were reported.

Table 5.9 ^{137}Cs intake with foodstuffs at the soil contamination density 37 kBq m^{-2} (1 Ci km^{-2})

Foodstuff	Average consumption, kg day^{-1}	^{137}Cs average intake in humans	
		Bq day^{-1}	%
Milk	0.800	80.0	81.0
Meat	0.220	8.0	8.1
Potato	0.550	5.5	5.6
Vegetables	0.250	1.5	1.5
Fish	0.030	0.9	0.9
Mushrooms	0.008	1.8	1.9
Bread	0.500	1.0	1.0
Total	2.338	98.7	100

Table 5.10 Average radionuclide transfer from daily diet to animal-derived food products, % of the content in the diet, 1 kg (L) of the product (Prister 1998; Prister et al. 2016)

Products		^{137}Cs	^{90}Sr
Cow milk	Stall feeding period	0.7–1.0	0.14
	Grazing period	0.9–1.3	0.14
Beef		4.0	0.04
Pork		15	0.10
Lamb		450	0.10
Chicken meat		3.5	0.20
Hen's eggs		–	3.20

Thus, restricting the dietary intake of radionuclides by animals is the main measure of radiation protection of humans. This can be achieved by an allowance for the key features of soil-to-plant transfer of radionuclides, making possible a proper organization of production on forage lands and efficient placement of crops on soils with different agrochemical properties and contamination levels.

5.3.3 *Radiation Dose or Deposition Density*

When the Chernobyl accident occurred, the behavior of ^{137}Cs in natural food chains was much less understood as compared to ^{90}Sr, and therefore approaches had to be developed to establish standards for permitted levels of this radionuclide in the environment. One of the fundamental mistakes after the accident was the adoption by the Superior Soviet of Ukrainian SSR, under the pressure of some members of the parliament, of the deposition, rather than radiation dose for humans, as the main criterion of radiation hazard. The value of 555 kBq m^{-2} (15 Ci km^{-2}) was adopted as a limiting ^{137}Cs deposition for the territories on which living was allowed and 185 kBq m^{-2} (5 Ci km^{-2}) for agricultural countermeasures. It was not taken into account that spatial distribution of internal radiation dose in the region is determined to a greater extent by environmental factors, rather than ^{137}Cs deposition density. Given the same deposition, ^{137}Cs contamination concentration in grass and cow milk can differ by two orders of magnitude.

Table 5.11 ^{137}Cs concentration in the sod and grass at soil contamination density of natural meadow 1 kBq m^{-2} (1988–1989)

Soil	Meadow type	^{137}Cs concentration, Bq kg^{-1} air dry mass	
		sod	grass
Meadow chernozem loamy	Wet flood plain	3.0	0.6
Meadow sandy	Normal dry valley	10–14	2.0–3.0
Meadow sandy	Wet flood plain	12–15	8.0–11
Soddy-podzolic loamy	Normal dry valley	4.0–14	1.0–4.0
Soddy-podzolic sandy	Normal dry valley	40–63	5.0–9.0
Soddy-podzolic sandy	Wet flood plain	53–75	25–39
Peat gley	Dried peat	77–90	30–45
Peat gley	Flooded peat	123–172	58–82
Peat gley	Lowland peat	170–198	135–189

As to animal feeding management on the contaminated territory, restricting radionuclide intake by animals is the main way of radiation protection of humans. On natural meadows, the majority of radiocesium (70–98% depending on soil type) occurred in the sod layer, i.e., the upper soil horizon enriched by plant residues, where a greater part of herbaceous plant roots can be found. The presence of sod, in which ^{137}Cs concentration was up to 14 times higher than that in plants (Table 5.11), was often responsible for the higher uptake of radionuclides by grass (Perepelyatnikov 2012a).

The analysis of data on Table 5.11 shows that ^{137}Cs activity concentration varies by two orders of magnitude, whereas agrochemical soil properties are responsible for changes within only one order of magnitude. This leads us to the conclusion about a crucial role of landscape conditions: position with respect to aquatic landscape, groundwater level, duration of excessive moistening, etc. Most meadows are flooded regularly, which not only is conducive to the intense movement of radionuclides down the soil profile but also promotes increased ^{137}Cs uptake by meadow plants. For this reason, the soil–plant transfer factors ^{137}Cs on overmoistened soils are 3–8 times higher than those on dry meadows. This phenomenon was visible for the global fallout in the Baltic countries, where ^{137}Cs and ^{90}Sr concentrations in the grass were 7–8 times higher on meadow soils with excessive moistening than those on moderately moist soils. Monitoring of global fallout ^{137}Cs in Polesye and Meshchera lowland has demonstrated that on same-type soils a groundwater decline from 0.5 to 1.0 m resulted in the reduction of ^{137}Cs concentration in plants by a factor of 3–10 (Marey et al. 1974).

Radionuclide concentration in plants is governed by multiple factors and can vary in a very wide range, even with the same deposition. For example, ^{137}Cs accumulation can vary by a factor of 20 on peaty meadows due to specific characteristics of grasses. These differences, as a rule, are even more significant because of nonuniform distribution of radioactive contamination on soil surface characterized by gradients or spotty character depending on conditions and environment.

The actual concentrations of radionuclides in yield are determined by the combined influence of soil contamination level, landscape, and environmental conditions, and therefore the picture of ^{137}Cs concentration distribution in biomass and yield of agricultural crops and natural plants is very mottled. In view of this, a methodology for predicting the concentrations of radionuclides in plants in specific conditions is required,and countermeasures should be implemented in a prioritized way for most critical components of the territory, in terms of radiation. Identifying critical areas in a timely manner enables the optimization of expensive countermeasures, based on a targeted, rather than area-wide, approach.

The data of radiation monitoring of radionuclide concentrations in milk and crops from Polesye showed that even in the first weeks after the accident in many cases they were much in excess of those predicted on the basis of deposition density data. There were no reliable methods in place for predicting the radiation levels with due allowance for the environmental conditions. The results of research by scientists and experts confirmed a complicated radiation situation, and high doses were expected for the population specifically due to ^{137}Cs intake with milk and dairy products. The legislations adopted by the Supreme Soviet of Ukraine placed a priority on deposition density in the assessment of radiation hazards for the public, which was mainly valid for the conditions of Kiev and some parts of Zhytomyr and Chernigov oblasts, where mineral soils were predominant. Considering the leading role of milk in dose formation, in 1991, Ministry of Chernobyl Affairs (set up in 1990) took a decision regarding national dosimetry passportization of the settlements in the Chernobyl-affected zone, using a unified methodology. For assessing internal radiation dose, ^{137}Cs average concentration in milk and potato collected at least from ten households were used (Kovgan et al. 2009). External radiation dose was estimated based on ^{137}Cs average deposition density in the settlement (Likhtarev 1996).

Figure 5.2 shows the distribution of internal dose due to the consumption of ^{137}Cs with milk and potatoes in settlements in the western part of the Chernobyl trail (Kovgan et al. 2009). It can be seen that as far as 300 km from the ChNPP in the settlements with the deposition level only 37 kBq m^{-2} but located on peaty soils, the dose appeared to be 6–10 times higher than that in the vicinity of the accident epicenter on mineral soil with the deposition level 370 kBq m^{-2}. The total absorbed radiation dose in people is deemed to be an objective criterion showing the hazard of living on the contaminated territory and the need for countermeasures.

As a consequence of neglecting the environmental factor in Rovno and Volyn regions, where the deposition densities were 111–185 kBq m^{-2} for ^{137}Cs (3–-5 Ci km^{-2}), the regulatory permitted levels for milk and meat were exceeded even on the territories considered to be safe officially. It has taken a lot of efforts from NAASU scientists to persuade the country's administration and Supreme Soviet of Ukraine to designate Rovno and Volyn regions as "affected areas," as the soil–plant–milk transfer factors in these oblasts were high. Unfortunately, the government's decision was not taken until 1988, and only since then agricultural measures started to be implemented toward the reduction of product contamination. In the first two most critical years, countermeasures in these regions were not taken under the state-run program at all. If a model for radionuclide behavior in soil–plant–milk chain in

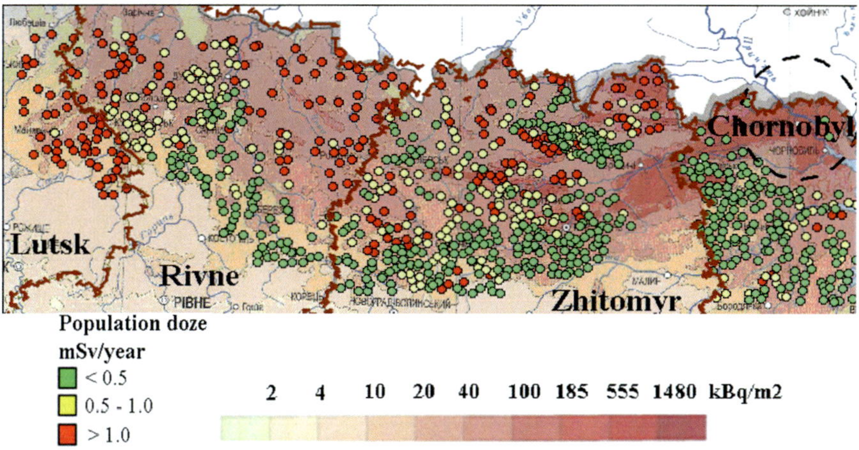

Fig. 5.2 Spatial distribution of ^{137}Cs soil contamination density (kBq km^{-2}) and settlement-averaged internal radiation dose (mSv year^{-1}) on the western part of the Chernobyl footprint in 1997 on the map of the soil type. Dotted circle corresponds to the 30-km zone around Chernobyl NPP

the initial period after the accident had been available, the mistakes would have been avoided and radiation doses for the population would have been lower in this radio ecologically critical region (Prister et al. 2018).

5.4 Science-Based Radioecological Monitoring of Agricultural Lands

5.4.1 Monitoring Framework and Methods

For assessment and management of the radiological situation in the agricultural sector, a scientific radioecological monitoring network was deployed on the Chernobyl-contaminated territories (Prister et al. 1996). The experimental data of long-term scientific monitoring from 1987 to 2007 in the Ukrainian territory contaminated after the Chernobyl accident were used as a basis for modeling the behavior of radionuclides in the soil–plant system (Prister et al. 2003a). The network of agricultural radioecological monitoring over extensive contaminated territory (Fig. 5.3) was organized in such a way to embrace a wide spectrum of soils, biological and climate conditions influencing the soil–plant transfer of radionuclides.

The monitoring was established to include the western and southern parts of the radioactive footprint. The western footprint went through Chernigov, Kiev, Zhitomir, Rovno, and Volyn oblasts, whereas the southern footprint crossed the southern part of Kiev and northern part of Odessa oblasts, touching Cherkassy oblast. Studies were conducted until 1990 on the experimental area of Ukrainian Polesye—

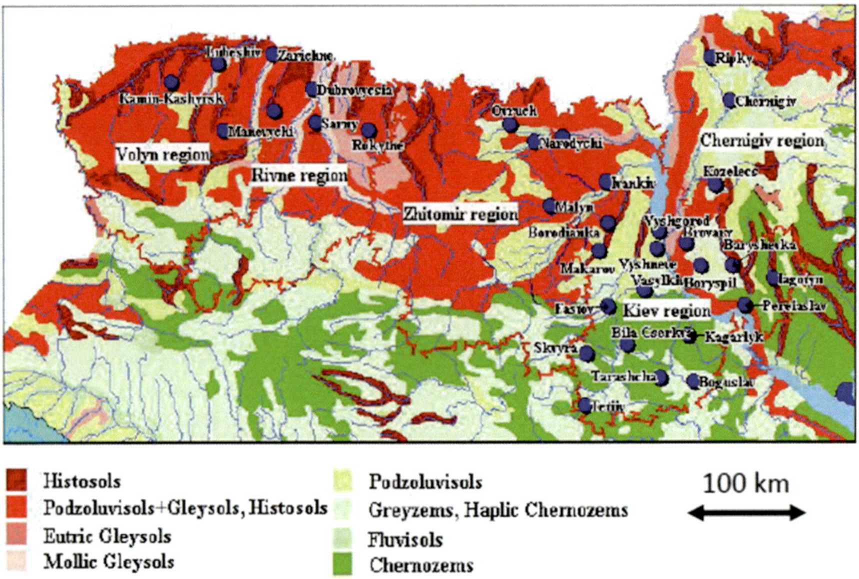

Fig. 5.3 Network of many-years radioecological monitoring on the map of soil types based on FAO-UNESCO classification. The blue circles are farms locations

marshland covered with shrubs and wooded steppe (Fig. 5.3). These were 31 collective farms of five most contaminated oblasts of Ukraine, representing all main soil types that are characteristic of Ukraine and Europe. The soil contamination density varied in a wide range from 100 to 3000 kBq m^{-2} for ^{137}Cs and from 3.7 to 40 kBq m^{-2} for ^{90}Sr on the year of fallout. The soils represent four soil types specific for Polesye and wooded steppe: peat-boggy, soddy-podzolic, gray forest, and chernozem, within which 3–15 soil catenas are identified. Thereby a large spectrum of soil conditions are covered, which enables extending the spatial range of model application and investigating the relationship of soil–plant transfer coefficient and soil properties. A total of 16 crops of different families and varieties were studied representing almost all components of animals and human diet and primarily responsible for the intake of radionuclides by humans.

When sampling, soil and plant samples were sought to be complementary: plant samples were collected from the same parts of the field where soil samples were taken. Representativeness of soil and plant samples was achieved by preparing an average sample from 25 individual areal samples with allowance for terrain features.

Databases with ^{137}Cs and ^{90}Sr soil–plant transfer factors TF_{ag} were compiled and used for building the model accounting for long-term behavior of radionuclides in the soil–plant system. For developing the model, more than 3000 values of ^{137}Cs soil–plant transfer factor TF_{ag} (Bq kg^{-1})/(kBq m^{-2}) and more than 500 values for ^{90}Sr were used. The data were retrieved and samples were formed with the values TF_{ag} "soil–plant–time after radionuclide entry into soil," with processing performed as described above.

The analysis of a large data array has shown that the dependence of radionuclide activity concentration in a plant on its deposition was linear for a given crop for the same soil type at a given time (Prister et al. 2003a). The linear dependence of activity concentration on σ was considered as an identity test of soil conditions in monitoring points with different deposition and validity of attributing value of TF_{ag} to the sample "soil–plant–time after the accident." Using the developed algorithm for processing the experimental data allows the application of the model to a wider range of soil conditions and biological factors. The ^{137}Cs concentration in soil and biota samples was measured using the γ-spectrometer ORTEC with semiconductor detector and CANBERRA pulse analyzer. ^{90}Sr concentration in samples was determined by the radiochemical method after dry incineration and measuring ^{90}Y activity on a low-background β-counter.

5.4.2 *Relationship of ^{137}Cs TF_{ag} with Biological Characteristics of Plants and Soil Properties*

The analysis of the radiation situation in the regions with abundant organic peaty and peaty-boggy soils, occupying as much as 30% of arable lands in some areas, has demonstrated an extremely high plant uptake of ^{137}Cs on these lands, particularly pastures. For example, in 1997, the proportionality factor between the ^{137}Cs deposition and dose rate (mSv year^{-1}/kBq m^{-2}) was 0.02–0.15 on peaty soils in Volyn and Rivne regions and 0.002–0.007 on soddy-podzolic soil of Kiev and Chernigov regions (Kovgan et al. 2009).

The scale of contamination and deep incorporation of ^{137}Cs in biological chains, alongside the concealment of the accident fact and data about radiation levels, caused anxiety and mistrust to the information from Soviet scientists. The European Commission and Belarus, Russia, and Ukraine initiated cooperation projects to map contaminated areas, study environmental migration of radionuclides, dosimetry, and prediction of biological effects. The guidance was provided by Scientific Coordination Board, and dozens of institutions of Europe and CIS took part in the projects. Out of 16 projects, three research and experimental projects (ECP) were aimed to investigate radioactivity migration in seminatural and agrarian systems (ECP-2 1996, ECP-5 1996, ECP-9 1996). In 1992, comparison and intercalibration of the methods for field work and analytical studies was undertaken which has shown adequacy and consistency of all techniques and measurement methods used by the project participants. Outcomes of these studies were concurred, published in EC proceedings, and presented at First International Conference in Minsk in 1996 (Karaoglu et al. 1996). The obtained results have been widely used by International Atomic Energy Agency (IAEA) to develop "Guidance on the use of countermeasures in agriculture in the event of the release of radionuclides into the environment" (IAEA 1994) and to compile a unified database for the Chernobyl accident under the French–German Initiative (Deville-Cavelin et al. 2002).

It has become obvious that methods ensuring reliable predictions of radionuclide uptake by crops were required. Above all, parameters influencing the relevant processes most of all were to be selected. The physical and chemical properties of the soil and the biological characteristics of a plant are among the soil–plant system characteristics that control the behavior of radionuclides.

For estimating the soil–plant transfer of radionuclides, we use TF_{ag}—the aggregated transfer factor:

$$TF_{ag} = \frac{\text{Radionuclide activity concentration in plant, Bq kg}^{-1}}{\text{Radionuclide depositon, Bq m}^{-2}} \tag{5.1}$$

Using this parameter makes it possible to compare data for agrarian (cultured) and natural landscapes with similar deposition but a different vertical distribution of radionuclide in the soil.

5.4.2.1 Role of Biological Characteristics of Plants

For the studied crops, the variation of ^{137}Cs TF_{ag} exceeds by a factor of 30 on peaty-boggy soils and by a factor of 16 on gray forest soils (Table 5.12), declining from natural herbage to cereal crops. The order of sequence of crops in this series is practically the same for all soil types. The values of TF_{ag} for the same crop on different soil types decrease in the following order: peat-boggy, gray forest, soddy-podzolic, and chernozem, which correlates with the strength of Cs^{+} binding by soil. The differences in TF_{ag} for different plant varieties are primarily governed by two factors of biological nature: different content of potassium (stable analogue of ^{137}Cs)

Table 5.12 ^{137}Cs TF_{ag} values, as of 1987, Bq kg^{-1}/kBq m^{2}

	Soil type			
Crop	Peat-bog	Soddy-podzolic	Gray forest	Chernozem
Natural grasses	98 ± 9	21 ± 3.3	6.3 ± 0.5	–
Sown cereal grasses	50 ± 5	3.4 ± 0.6	2.8 ± 0.4	1.9 ± 0.2
Clover	–	2.9 ± 1.0	1.7 ± 0.20	1.1 ± 0.2
Lucerne	–	3.0 ± 1.0	1.4 ± 0.2	1.0 ± 0.1
Maize	21 ± 4	1.9 ± 0.3	1.1 ± 0.3	0.75 ± 0.06
Cabbage	–	2.2 ± 0.2	1.2 ± 0.14	0.75 ± 0.07
Tomato	–	1.0 ± 0.1	1.0 ± 0.01	1.0 ± 0.1
Cucumbers	–	2.0 ± 0.5	1.2 ± 0.1	0.8 ± 0.2
Onion	–	1.1 ± 0.2	0.31 ± 0.01	–
Beetroot	5.9 ± 0.5	1.1 ± 0.2	0.33 ± 0.06	0.49 ± 0.05
Potato	3.7 ± 0.3	0.7 ± 0.1	0.4 ± 0.2	0.14 ± 0.04
Winter wheat	–	0.70 ± 0.14	0.43 ± 0.11	0.14 ± 0.06
Barley	3.1 ± 0.3	0.63 ± 0.14	0.54 ± 0.14	0.25 ± 0.09
Rye	2.6 ± 0.2	0.49 ± 0.10	0.30 ± 0.03	–

and mutual disposition of root vertical profile and radionuclide vertical distribution in soil.

As a result of the synergetic influence of the above factors, given the same soil deposition density, radionuclide concentration in produce can vary by more than 380 times. By changing food crops in human and animal diet, allowing for soil type on which these are growing, and placing crops on soils with different contamination levels, radionuclide fluxes in food chains can be controlled to a great extent.

Metabolic processes of radionuclide transfer to different plant organs have not been formalized until now because they are not well understood and have not been described quantitatively yet. The $TF_{ag}(t)$ should be determined experimentally for a pair "soil–crop." The formation of reproductive organs is a plant-specific function and highly genetically determined, and therefore for developing the model, $TF_{ag}(t)$ values were used, which were determined at the time of productive ripeness, characterizing all metabolism processes in an integrated way.

5.4.2.2 The Link Between TF_{ag} and Soil Properties

When the accident happened, no models describing analytically the relationship between TF_{ag} and soil properties were available. The study of all available data has shown that the soil–plant ^{137}Cs transfer factor for a specific crop varies significantly if data are grouped by the soil type regardless of soil properties (Prister 2006). For instance, the scatter in the TF_{ag} values for winter rye and potato, even within one soil subtype, is as high as one to two orders of magnitude.

The main agrochemical properties of soils in Ukrainian Polesye are shown in Table 5.13. In the northern parts of Polesye (Volyn, Rivne, Zhitomir, Kiev, and Chernigov regions), the typical soils are soddy-podzolic sandy soils with a humus content of about 1% and pH below 5.5. A major part of the contaminated territory in Kiev region has gray forest soils with leached chernozems and signs of podsolization. In the Volyn and Rivne regions, different kinds of peaty soil occurring on floodplains and depressions in Zhytomyr, Kiev, and Chernigov regions can be found.

As can be seen from the above, the soils in the study area are characterized by a very broad range of agrochemical properties and grain size characteristics: from sand to clay. This enables investigating the role of key soil properties in radionuclide uptake by plants but makes it difficult to select an area for applying uptake models. Samples of soils attributed to the same type have shown large variations in parameters such as pH, cation exchange capacity (CEC), and humus content. That is why methods for integrated evaluation of soil properties responsible for radionuclide mobility need to be developed.

In the initial phase of monitoring, on the basis of 1988 data, the relationship between ^{137}Cs TF_{ag} and key soil characteristics was studied in detail. For ^{137}Cs, the relationship between TF_{ag} and all studied parameters has been found to be linear in a narrow range. It is only TF_{ag} and pH that show a functional relationship, the value of TF_{ag} being equally sensitive to pH in a wide range from 4.7 to 7.6 (Fig. 5.4). The

Table 5.13 Average values of agrochemical characteristics for the main soil types of Ukrainian Polesye

Soil type	pH	Humus, %	CEC, meq per 100 g of soil	K_2O, mg per 100 g of soil	Silt, %
Peaty-boggy	3.8–6.6	–	2.3–19.0	2.0–18.5	–
Soddy-podzolic sandy and loamy	4.4–6.0	0.40–1.2	6.0–18	2.0–12	5.0–10
Soddy-podzolic sandy and loamy	4.5–6.2	0.53–1.4	8.4–18	5.5–8.7	6.8–12
Soddy-podzolic clay-sandy	5.1–6.5	1.0–1.8	7.2–20	7.8–13	10–14
Soddy-podzolic gleyed	4.9–7.2	1.5–2.2	10–20	9.2–20	14–20
Gray forest	5.3–6.6	0.92–1.9	8.4–18	5.8–16	12–22
Podzolized chernozem	4.9–6.6	1.2–2.1	7.5–20	14–22	26–32
Typical chernozem	6.4–7.5	1.5–3.9	18–36	22–38	25–38

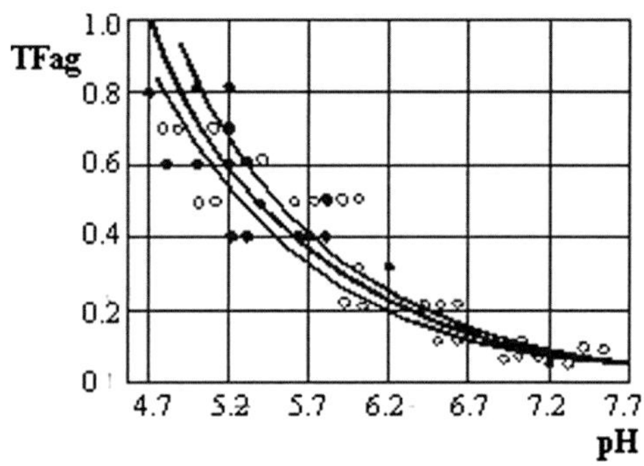

Fig. 5.4 ^{137}Cs TF_{ag} (m^2 kg^{-1}) for the grain of winter wheat as a function of soil pH (Prister 2006)

inverse relationship can also be clearly seen for the pairs TF_{ag}–pH, hydrolytic acidity, humus, absorption capacity, and clay content. At the same time, the scatter of experimental points is too wide for the considered relationships to be written in analytical form (Prister 2006). Considerable variability in TF_{ag} associated with differences in pH, potassium concentration, clay content, and organic matter has been noted by many authors (Absalom et al. 1995; Arapis and Perepelyatnikova 1995; Sanzharova et al. 1996; Bogdevitch 2012). They showed that the spread in ^{137}Cs TF_{ag} values as a function of the applied amount of potassium can be up to 3.5 times, whereas the correlation factor is rather low. The empirical equations describing the relationship between ^{137}Cs TF_{ag} and agrochemical parameters do not reflect the nature of processes and have a very narrow area of application.

A rigorous relationship between TF_{ag} and specific soil characteristic can be obtained only if soils are grouped properly by subtypes. That is why it is difficult to use the UNESCO soil classification based on soil grouping by larger categories, which results in smoothing the soil cover motley and diversity observed in the accident region. Due to combining different subtypes, one group appears to include soils differing by agrochemical and water-physical characteristics, which strongly influence radionuclide uptake by plants. A thorough analysis of the models accounting for soil–plant transfer of radionuclides was undertaken in Müller and Pröhl (1993); however, the analytical relationship between soil properties and TF_{ag} was not considered by them either.

With this in mind, we deem it appropriate to use an integrated indicator of soil properties S_{ef}. What soil properties are to be used to define S_{ef}? We consider soil as a three-phase system: solid phase-mineral matrix, liquid phase-soil solution, and quasicrystal or quasiliquid phase-organic matter in the form of colloidal micelles, films, etc.

The most important characteristic of the solid phase, in our view, is adsorption capacity E representing the total number of sorption sites on the surface of the mineral matrix and organic colloids. It is characterized by quality parameters such as the degree of base saturation, the sum of absorbed bases (SAB), and partial content in the exchangeable state of individual cations. In most experiments, SAB is of primary interest for researchers. This parameter in terms of informativeness is not inferior to adsorption capacity, as the behavior of macroelements in the soil–plant system is, to a great extent, governing the behavior of trace quantities of radionuclides of alkaline and alkaline earth elements. Another key characteristic of the soil is organic matter (OM), which determines, among other things, the ability of soils to absorb cations.

The soil solution composition is determined by the exchange of macro- and microelement cations between the liquid and solid phases of soil. An integral criterion for the liquid phase in the equilibrium system "soil–solution" is pH, which can be regarded as the third important indicator controlling the behavior of ion in the "soil–plant" system.

The fraction of ion adsorbed from soil solution by different soils should be proportional to the soil adsorbing capacity of the soil adsorbing complex (SAC), a measure that can be used for the consideration of the effective cross-section of the three-dimensional space S_{ef}. For any soil, S_{ef} can be presented as a square area of the triangle, the vertexes of which lie at the points corresponding to vector values on the axes pH, OM, and E (or SAB). It is very difficult to assess S_{ef} experimentally, and therefore we suggested a simple empirical method for calculating the square area of the cross section of a 3-D space, whose dimensions are mutually perpendicular axes (Fig. 5.5)—soil properties determining ion absorption (Prister 2002, 2006; Prister et al. 2003b).

The parameters used for S_{ef} calculation have different units; therefore, their values should be normalized to parameters of standard soil. The pH value is worth normalizing to the value corresponding to neutral reaction of soil solution pH 7, and the content of OM and E (SAB) to the highest values for a series of soils under study, in our case 6% of humus and 40 meq/100 g of soil, respectively.

The dependence of radiocesium soil–plant transfer factor (TF_{ag}), extrapolated to deposition time, on S_{ef} (Fig. 5.6) can be approximated by the exponent

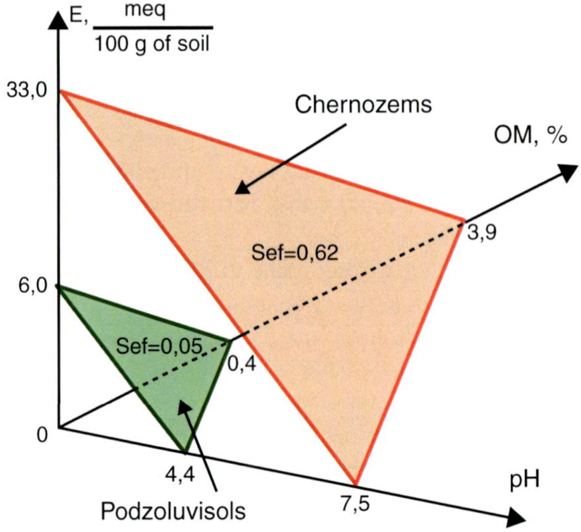

Fig. 5.5 Graphical depiction of S_{ef} as the cross-section area of the three-phase system

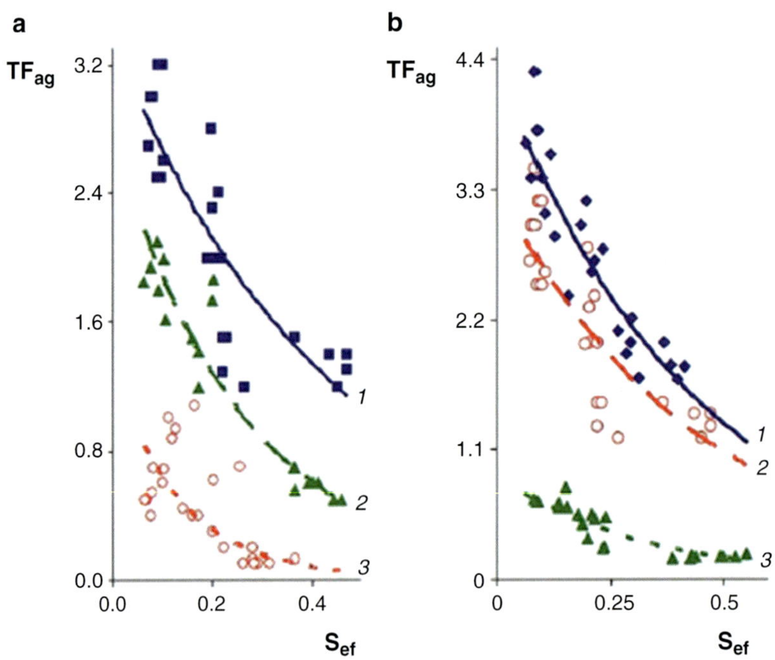

Fig. 5.6 Transfer factor TF_{ag} ($m^2\ kg^{-1}$) as a function of the integrated estimate of soil properties S_{ef} (relative units) for the triad OM–pH–SAB: (**a**) different crops in 1989: 1—sown grass hay, 2—tomato, 3—potato tuber; (**b**) sown grass hay in different years: 1—1987, 2—1989, 3—1994

Table 5.14 Parameters of Eq. (5.2) for natural grass and agricultural crops, as of 1987

Crop	Number of samples	TF^0_{ag}	λ	R^2
Natural grass	13	51	8.9	0.63
Sown grass	22	4.3	2.4	0.86
Maize	25	2.4	3.6	0.77
Cabbage	17	2.9	3.3	0.79
Potato	21	1.2	5.4	0.79
Winter wheat	17	1.1	6.4	0.78

$$TF_{ag} = TF^0_{ag} \cdot e^{-\lambda \cdot S_{ef}}, \qquad (5.2)$$

where TF^0_{ag} is the extrapolated ^{137}Cs soil–crop transfer factor with $S_{ef} = 0$ and λ is the exponent parameter for the crop accounting for plant response to the change in soil properties.

The values of the parameters of relationship (5.2) are shown in Table 5.14. The values of the parameter showing plant response to the changes in soil properties λ differ for all crops by a factor of 4.3.

The regression coefficient for ^{137}Cs TF_{ag} dependence on S_{ef} for the majority of studied crops in different time periods after deposition is above 0.6, suggesting high reliability of the approximation by Eq. (5.2).

5.4.3 *Time Dependence of TF_{ag}: From Empirical to Dynamic Model*

On the third year after the accident, it became obvious that for all crops TF_{ag} was a function of time for all studied soils ranging from peat to chernozems (Fig. 5.7) (Prister et al. 1991, 2003a; Prister 1996; IAEA 2006). In the first few years, the radionuclide soil–plant transfer factor decreased by 3–5 times. The TF_{ag} decline with time was also noted in Sanzharova et al. (1996).

Mathematical analysis of the TF_{ag} time series for more than 20 years has enabled us to describe the transfer factor dynamics as the two-exponential Eq. (5.3)

$$TF_{ag}(t) = TF_{ag}(0) \cdot \left\{ a^q(0) \cdot \left(-\frac{0.693}{T^q_e} \cdot t \right) + a^s(0) \cdot \exp\left(-\frac{0.693}{T^s_e} \cdot t \right) \right\}, \qquad (5.3)$$

where $TF_{ag}(0)$ is the transfer factor extrapolated to 1986; a^q (0) and a^s (0) $= 1 - a^q$ (0) are the fractions of the initial radionuclide concentration in soil with TF_{ag} effective half-lives T^q_e and T^s_e, respectively ("quick" and "slow" forms of ^{137}Cs, also referred to as exchangeable and fixed forms) (Prister 2006).

Tables 5.15, 5.16, and 5.17 show the values of the parameters in Eq. (5.3) for different crops on different soil types. Quantitative analysis of the parameters in Eq. (5.3) has shown that $TF_{ag}(0)$ values extrapolated to the deposition time (1986)

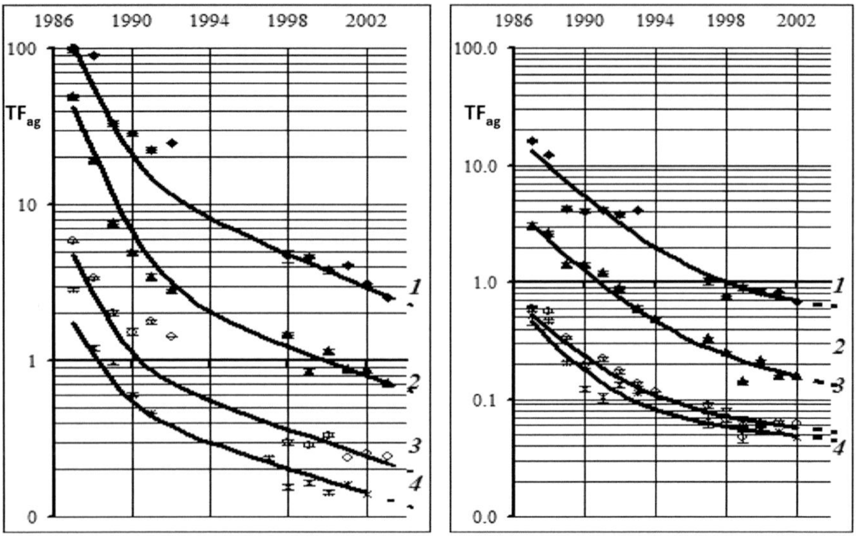

A B

Fig. 5.7 Dynamics of ^{137}Cs TF_{ag} ($m^2\ kg^{-1}$) for transfer from peat-boggy (**a**) and soddy-podzolic (**b**) soils to agricultural crops: *1*—natural grass hay; *2*—cereal grass hay; *3*—maize green mass; *4*—potato tubers

Table 5.15 TF_{ag0} values extrapolated to the deposition time (1986), Bq kg^{-1}/kBq m^{-2} ($\delta \leq \pm 25\%$)

	Peat bog		Soddy-podzolic		Gray forest		Chernozem	
Crop groups	$TF^q_{ag}(0)$	$TF^s_{ag}(0)$	$TF^q_{ag}(0)$	$TF^s_{ag}(0)$	$TF^q_{ag}(0)$	$TF^s_{ag}(0)$	$TF^q_{ag}(0)$	$TF^s_{ag}(0)$
Hay of natural grasses	220	22	29	0.78	10	0.49	–	–
Hay of sown grasses	95	4.7	6.0	0.38	4.9	0.11	3.7	0.037
Green forage: *Mays, Lucerna, Trifollium*	35	1.4	3.4	0.37	1.5	0.18	1.9	0.039
Vegetables: *Brassica, Licopersicum, Cucumis*	–	–	3.0	0.10	2.0	0.03	1.4	0.014
Root, tubers: *Allium, Beta, Potatoes*	10	0.84	1.5	0.10	0.60	0.06	0.56	0.017
Grains: *Triticum, Hordeum, Secale*	7.0	0.81	0.89	3.5	0.60	0.05	0.35	0.019
Range, times	33	27	31	7.8	17	17	11	2.8

account for biological features of plants with respect to their ability to uptake ^{137}Cs (IAEA 2006).

The analysis of TF_{ag} effective half-lives for radionuclides T^q_e and T^s_e (Table 5.16) has shown that they are equal for different plant species grown on the soil of the same type. This implies that the TF dynamics for radionuclides is governed by the

Table 5.16 Average values of effective half-lives of TF_{ag} for the exchangeable form T_e^q and fixed form T_e^s of ^{137}Cs, years ($\delta \leq \pm 20\%$)

Group of crops	Peat bog		Soddy-podzolic		Gray forest		Chernozem	
	T_e^q	T_e^s	T_e^q	T_e^s	T_e^q	T_e^s	T_e^q	T_e^s
Hay of natural grasses	0.87	5.7	2.1	28	1.6	26	–	–
Hay of sown grasses	0.92	6.6	2.0	11	1.9	35	1.3	53
Green forage: *maize, lucerne, clover*	0.97	7.1	1.8	24	1.7	>60	1.2	>60
Vegetables: *cabbage, tomatoes, cucumbers*	–	–	1.5	14	1.7	26	1.3	>60
Tubers, roots *onion, beet, potatoes*	0.88	6.4	2.2	21	1.7	46	1.3	>60
Cereal grain: *wheat, barley, rye*	0.88	6.9	1.8	39	1.8	>60	1.2	>60
Average	0.89	6.6	1.8	28	1.8	50	1.3	>60
Standard deviation ±	0.04	0.48	0.23	9.4	0.09	11	0.05	–

Table 5.17 Mean values for the initial fraction of the quick component of ^{137}Cs in soil a^q (0) at $t = 0$ ($\delta \leq \pm 20\%$)

Group of crops	Peat bog	Soddy-podzolic	Gray forest	Chernozem
Hay of natural grasses	0.91	0.97	0.95	–
Hay of sown grasses	0.95	0.94	0.98	0.99
Green forage: *maize, lucerne, clover*	0.96	0.89	0.89	0.98
Vegetables: *cabbage, tomatoes, cucumbers*	–	0.95	0.98	0.99
Tubers, roots: *onion, beet, potatoes*	0.92	0.93	0.89	0.97
Cereal grain: *wheat, barley, rye*	0.89	0.87	0.91	0.95
Average	0.92 ± 003	0.92 ± 0.04	0.93 ± 0.04	0.98 ± 0.02

transformation of their forms in soil, rather than biological features of plants. As a result of selective sorption and fixation of ^{137}Cs (Konoplev 2020), the absolute values of TF_{ag} have significantly decreased since 1986: by 17–23 times on peaty-bogs and soddy-podzolic soils, and up to 40 times on chernozem soils (Prister 1999). According to our calculations, the major part of ^{137}Cs in the fallout for remote areas outside 30-km zone, where condensation particles dominated in the fallout, occurred in the bioavailable form (Table 5.16), which explains a quick decrease in the availability of radionuclide to plants from the soil.

The dynamics of ^{90}Sr availability for crops differs from that of ^{137}Cs: in the early years after the accident, TF_{ag} did not change quickly and was masked by natural variability associated with climatic and other factors (Table 5.18).

Credible conclusions about ^{90}Sr TF_{ag} were possible to draw only 15 years after the accident when a trend for decline appeared. Table 5.18 includes data on ^{90}Sr TF_{ag} and the half-reduction time for the studied crops, on the average, in 5–7 farms in which soil and plant samples were collected, one sample in each location. The half-

Table 5.18 Dynamics of mean TF_{ag} for ^{90}Sr to productive organs of crops from chernozem, Bq kg^{-1}/kBq m^{-2}($\delta \leq \pm 25\%$) and T^s_e, years

Year	TF_{ag}			
	Potato	Cabbage	Tomato	Winter wheat
1987	0.090	1.1	0.39	0.11
1988	0.13	2.2	0.37	0.083
1989	0.98	0.79	0.37	0.073
1994	0.073	1.7	0.43	0.073
2000	0.023	0.44	0.24	0.053
2001	0.027	1.4	0.28	0.027
T^s_e, years	6.2	24	7.8	9.8

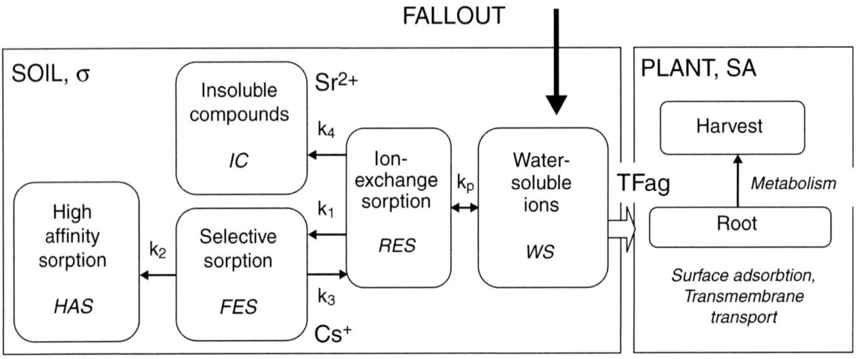

Fig. 5.8 Conceptual dynamic model of radionuclide transfer from soil to plants for agricultural plants, σ—radionuclide deposition

reduction time of T^s_e for ^{90}Sr for the studied crops on chernozem varied from 6.2 (potato) to 24 (cabbage) years and were 10 on the average.

The comparison of TF_{ag} half-lives shown in Table 5.16 with the kinetic parameters of radionuclide sorption and fixation by soils and minerals (Konoplev 1998; Konoplev and Konopleva 1999) provided a basis for suggesting a conceptual dynamic model accounting for radionuclide behavior in soil–plant system (Fig. 5.8) (Prister and Vinogradskaya 2009, 2011).

In the immediate term, only radionuclides and nutrients occurring in the soil solution are available for root plant uptake. Sorption sites of soil can be divided into three main groups (Fig. 5.9) (Cremers et al. 1988; De Preter 1990): regular (nonselective) exchange sites (RES); selective sites for K^+, NH_4^+, Cs^+, and interlayers occurring in micaceous clay minerals or frayed edge sites (FES) and highly selective sites localized inside interlayers or high-affinity sites (HAS). Rate constants of transformation processes k_i can be determined from the time dependence of various chemical forms in soil and TF (Konoplev 1998, 2020: Konoplev and Bulgakov 1999).

Interphase partitioning of ions in soil starts from establishing exchangeable sorption equilibrium of radionuclide ions in the system “soil solution—ion exchange sites RES.” Time of equilibrium setting is taken to be zero (Prister and

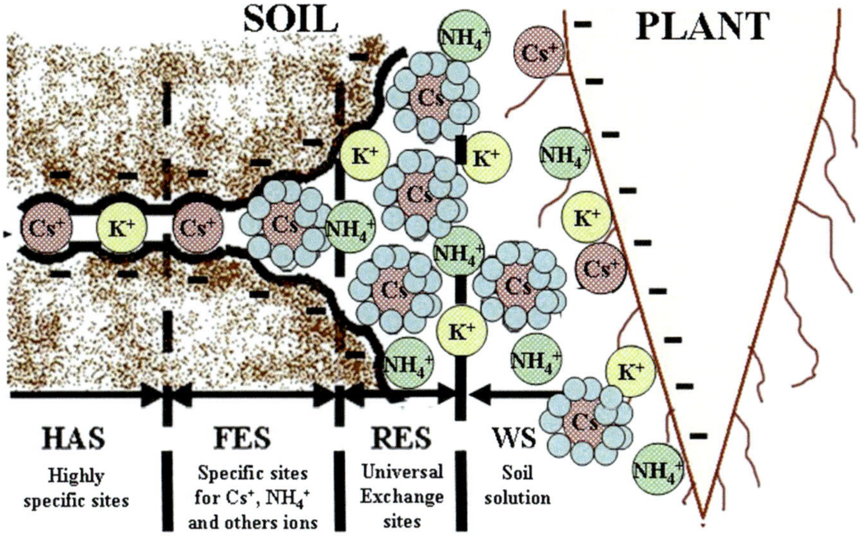

Fig. 5.9 Schematic representation of the key processes of ^{137}Cs ions distribution in "soil–plant root" system

Vinogradskaya 2011) because this process is very fast compared to the plant growing process. The proportion of exchangeable radionuclide at sites RES at time moment $t = 0$ is governed by the soil solution composition and cation exchange capacity. Then, at $t > 0$, Cs^+ transfers from RES to FES with the rate constant k_1. At the solid phase interface, the ion Cs^+ may lose a hydrate shell, which makes it possible for the Cs^+ to be selectively sorbed on FES and then to migrate to HAS with the rate constant k_2. Cs^+ ions are strongly adsorbed on FES, but with time they can transfer partly to RES with the rate constant k_3 (Prister and Vinogradskaya 2011). As the dehydrated cesium ion is moving deeper in the crystal lattice, the interlayers may collapse and Cs gets fixed by clay mineral.

Radionuclide uptake by plants occurs directly from the soil solution and its concentration in plant is proportional to the concentration in soil solution, the dynamics of which is determined by the kinetics of Cs^+ redistribution between the sorption sites.

The transformation kinetics of radionuclide chemical forms in soil, in accordance with the conceptual scheme (Fig. 5.8), is written as a system of differential equations, the analytical solution of which describes the transformation of ^{137}Cs forms in the soil as a function of time (Prister and Vinogradskaya 2009, 2011):

$$\mathrm{TF_{ag}}(t) = \mathrm{TF_{ag}}(0) \cdot \left\{ \frac{k_1 - \left(2 \cdot k_2 + \frac{k}{2}\right)}{k_1 - k_2} \cdot \mathrm{e}^{-\left(k_1 + \frac{k_3}{2}\right) \cdot t} + \frac{k_2 + \frac{k}{2}}{k_1 - k_2} \cdot \mathrm{e}^{-\left(k_2 + \frac{k_3}{2}\right) \cdot t} \right\}, \quad (5.4)$$

where k_1 is the rate constant of the Cs^+ transfer from RES to FES, k_2 is the rate constant of the Cs^+ transfer from FES to HAS, k_3 is the rate constant of the Cs^+ transfer from FES to RES.

Values of the parameters of Eq. (5.4) for agricultural crops on soddy-podzolic soil are shown in Table 5.19. The rate constants, calculated from the data on TF_{ag} dynamics, do not differ much for various crop types, which allows us to consider them as characteristics of soils, not depending on biological features of plants.

This implies that plants act as an indicator of radionuclide mobility and bioavailability in soil, which allows averaging parameters k_1, k_2, and k_3 for each of the soil groups and using them as a soil type characteristic (Table 5.20). Simultaneously, $TF_{ag}(0)$ extrapolated to deposition time, which accounts for biological plant features, varies by more than 30 times.

The rate constant k_3 of Cs transfer from FES to RES and the rate constant k_2 is the highest on peatbogs and tends to decrease in the range of mineral soils from soddy-podzolic to chernozems by 6–7 times. The rate constant k_3 is 10–15 times higher than k_2 for all soil types. The values of the ratio coefficients k_3/k_2 are close for all soil types, so that suggests the same mechanism for the transfer of cesium ions into the intermediate layers of the crystal lattice irrespective of the soil type in which this mineral occurs.

Absolute values of rate constants for ^{137}Cs transformation processes on sites of different nature differ significantly for soil types under study. The highest rate constant is reported for radiocesium transfer from RES to FES (k_1). The rate constant k_1 for acid peaty-boggy soil is 2.2 times higher than that in mineral soils.

Table 5.19 Average values of $TF_{ag}(0)$ for the main groups of crops and kinetic parameters of the model for ^{137}Cs behavior in the system "soddy-podzolic soil–plant," ($\delta \leq \pm 25\%$)

Group of crops	TF_{ag} (0), Bq kg^{-1}/kBq m^{-2}	Rate constants, year^{-1}		
		k_1	k_2	k_3
Hay of natural grasses	29	0.35	0.0031	0.032
Hay of sown grasses	5.8	0.34	0.0025	0.036
Green forage: *maize, lucerne, clover*	3.8	0.33	0.0022	0.041
Tubers, roots: *onion, beet, potatoes*	1.6	0.33	0.0029	0.037
Cereal grain: *wheat, barley, rye*	0.89	0.34	0.0028	0.039
Mean	–	0.34 ± 0.01	0.0026 ± 0.0004	0.038 ± 0.003

Table 5.20 Mean values of rate constants for ^{137}Cs ion transfer between various sorption sites for major soil types in Chernobyl-contaminated areas, year^{-1} ($\delta \leq \pm 25\%$)

Soil type	k_1	k_3	k_2	k_1/k_3	k_3/k_2
Peaty-boggy	0.76	0.082	0.0085	9.27	9.64
Soddy-podzolic	0.34	0.038	0.0026	8.95	14.6
Gray forest	0.38	0.026	0.0017	14.6	15.3
Chernozem	0.48	0.013	0.0011	36.9	11.8
Average	–	–	–	–	12.8 ± 1.2

The fact that the value k_1 for different types of mineral soils differs only slightly is indicative of the low-binding ability of RES for all of them. The determination of the kinetic parameters of ^{137}Cs forms transformation in soil provides to allows predict a change in ^{137}Cs TF_{ag} for a wide range of crops and soil properties.

The comparative estimation of rates of sorption processes with soil properties has shown that k_1, k_2, k_3 are linearly dependent on the integrated indicator of soil properties S_{ef}. With allowance for the dependence of kinetic parameters of radionuclide sorption on S_{ef}, the kinetic model for ^{137}Cs behavior in the "soil (S_{ef})–plant" system is written as follows:

$$TF_{ag}(t, S_{ef}) = TF_{ag}(0,0) \cdot e^{-\lambda \cdot S_{ef}} \cdot \left\{ \begin{array}{l} (1+f \cdot \ln(S_{ef})) \cdot e^{-q \cdot (1+Sef) \cdot t} + \ldots \\ \ldots + (1+f \cdot \ln(S_{ef})) \cdot e^{-s \cdot (1+Sef) \cdot t} \end{array} \right\}, \quad (5.5)$$

where f is the fraction of exchangeable radiocesium in soil with S_{ef} at $t = 0$; q and s are the integrated rates of decrease in the content of radionuclide ions available for plants (Table 5.21).

The ^{137}Cs soil–plant transfer factor $TF_{ag}(0,0)$ extrapolated to deposition time decreased by a factor of 61 in a series of crops from natural herbage to cereals. The difference between the values of f, q, and s for all crops is not more than 2.3 times, allowing averaging over all crops. The derived average values of f, q, and s are "constants" of the transformation processes in soil because they do not depend on either soil types or biological characteristics of plants. The plots for the dependence

Table 5.21 Parameters of the kinetic model for the transfer of radionuclides from mineral soils to crops with allowance for soil properties

		^{137}Cs				
Crop	Plant part	$TF_{ag}(0,0)$	λ	f	q	s
Natural grasses	Hey	55	6.1	0.030	0.32	0.072
Sown grasses		7.2	2.2	0.032	0.30	0.067
Clover	Green mass	9.2	5.6	0.033	0.29	0.043
Lucerne		6.2	3.3	0.032	0.31	0.051
Maize		4.2	4.2	0.029	0.34	0.054
Cabbage	Head	4.5	3.0	0.032	0.32	0.071
Tomato	Fruit	4.0	4.9	0.026	0.33	0.087
Cucumbers		2.7	1.9	0.030	0.36	0.065
Onion	Bulb	2.9	3.9	0.031	0.29	0.049
Beet	Root	2.2	2.6	0.033	0.28	0.049
Potato	Tuber	2.0	5.3	0.032	0.33	0.040
Winter wheat	Grain	1.7	6.8	0.029	0.31	0.043
Barley		1.0	1.6	0.030	0.32	0.046
Rye		0.90	2.2	0.032	0.26	0.038
Mean				0.031	0.31	0.055
Standard deviation ±				0.002	0.03	0.01

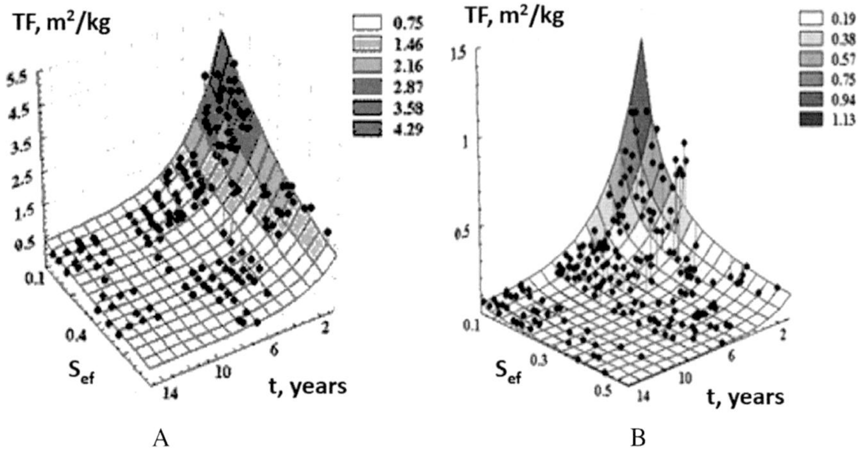

Fig. 5.10 ^{137}Cs TF_{ag} values as a function of integrated indicator S_{ef} of soil properties and time after deposition: (**a**) sown grass hay, (**b**) potato tuber. S_{ef} is presented in relative units

of ^{137}Cs transfer factor on soil properties and time since fallout for each crop can be presented in a 3-D space with axes TF_{ag}, S_{ef}, and t (Fig. 5.10).

The dependence of transfer factor on parameter S_{ef} with allowance for ^{137}Cs transformation kinetics in the soil is described by formula (5.5) (Prister and Vinogradskaya 2011), which can be viewed as a working equation for prediction purposes.

5.4.4 Verification of the Model and Assessing Its Applicability on Other Territories

For verifying the model applicability to predict plant contamination beyond the exclusion zone in Ukrainian Polesye, the database (DB) compiled within Franco-German Initiative (FGI) (Project 2, subproject 3a) was employed (Deville-Cavelin et al. 2001). The DB includes more than 3000 records of "soil–plant" pairs obtained by the agrarian monitoring of the contaminated territories of Belarus and Russia and 3000 records for Ukraine. Of all these data, 14 pairs of soil (soddy-podzolic loamy sand)–plant (potato tuber) for five areas of Belarus and Russia for 1989 were selected (Table 5.22). When the model was developed, these data were not used and therefore they were appropriate for model verification.

The average values of parameters pH and OM within each soil group were quite equal, whereas the concentration of exchangeable potassium in the second group was almost twice as high, and in the third group, three times higher as compared to the first. These differences may be caused by uncertainties associated with soil type (for example, degree of podsolization or particle size category was not taken into

Table 5.22 Expert evaluation of FGI—3a DB records, calculated and experimental values of ^{137}Cs TF_{ag} for the set "soddy-podzolic soil–potato–1989"

Settlement	pH	OM,%	K_2O	S_{ef}	EE[a,b]	TF_{ag}, Bq kg^{-1}/kBq m^{-2} Estimated $A^{O:2}$	Estimated $B^{O:3}$	Experiment $C^{O:4}$	$\delta(A-C)$, %	$\delta(B-C)$, %
Bragin (BY)	5.4	1.5	9.8	0.17	3	0.39	0.30	0.16	+143	+88
Novozybkov (RUS)	5.5	1.3	5.9	0.12			0.42	0.30	+30	+40
Klintsy (RUS)	5.3	1.6	6.2	0.13			0.38	0.40	−3	−5
Novozybkov (RUS)	5.3	2.0	7.9	0.17			0.30	0.30	+30	–
Mean	*5.4*	*1.6*	*7.5*	*0.15*		*0.39*	*0.35 ± 0.060*	*0.29 ± 0.099*	*+34*	*+20*
Novozybkov (RUS)	5.9	2.3	16	0.29	2	0.39	0.14	0.20	+95	−30
Novozybkov (RUS)	5.3	2.0	12	0.21			0.23	0.26	+50	−12
Mean	*5.6*	*2.2*	*14*	*0.25*		*0.39*	*0.19 ± 0.063*	*0.23 ± 0.042*	*+70*	*−18*
Klintsy (RUS)	5.4	2.4	19	0.32	1	0.39	0.12	0.10	+290	+20
Novozybkov (RUS)	5.9	2.3	23	0.39			0.072	0.20	+95	−64
Novozybkov (RUS)	5.9	2.3	20	0.34			0.098	0.10	+290	−2
Novozybkov (RUS)	5.9	2.3	25	0.41			0.065	0.10	+290	−35
Novozybkov**	6.1	2.6	23	0.42			0.063	0.10	+290	−37
Novozybkov**	6.1	2.6	29	0.51			0.036	0.098	+290	−62
Krasnogorsk*	5.8	2.4	19	0.34			0.10	0.40	+98	−75
Ulyanovsk**	4.9	1.5	25	0.32			0.11	0.15	+160	−25
Mean	*5.8*	*2.3*	*23*	*0.38*		*0.39*	*0.083 ± 0.028*	*0.16 ± 0.11*	*+143*	*−49*

[O:1]Expert estimate; [O:2]$A = TF_{ag}$, calculated on the basis of data of Table 5.18; [O:3]$B = TF_{ag}$, calculated on the basis of data of Table 5.20; [O:4]C is experimentally determined TF_{ag}; K_2O, mg/100 g of soil; $\delta(A - C) = (A - C)\cdot 100/C$, %; $\delta(B - C) = (B - C)\cdot 100/C$, %; *RUS*—Russia, *BY*—Belarus

* Expert evaluation, ** Russia, *** Belarus

account), or more likely no regard was taken of application of potassium fertilizers as a countermeasure after the accident. Information on implemented countermeasures was not included in the database. All these differences are allowed for in calculations using the integrated estimate of soil properties S_{ef}.

The experimental values of ^{137}Cs TF_{ag} taken from the DB were compared with the values of ^{137}Cs TF_{ag} estimated by the dynamic model for radionuclide behavior in the system "soil–plant." The model estimation was done by two methods: using the parameters accounting for soil type and plant species (Table 5.19) and using the parameters for plant species only with allowance for S_{ef} calculated based on specific values of agrochemical properties of soil (Table 5.21).

The results of estimating ^{137}Cs TF_{ag} using the tabulated parameters for soil type and plant species, given expert estimates 3, 2, 1 (Table 5.22), differed from the experimental values by 34%, 70%, and 143%, respectively. At the points with low values of "expert estimate," the calculated and experimental values TF_{ag} differed by a factor of 2–4. The discrepancies between soil agrochemical properties and soil classification name and between the predicted TF_{ag} and actual values were most likely explained by neglecting the application of large amounts of potassium fertilizer as a countermeasure. In this case, using only the classification name of soil type could lead to higher uncertainty in prediction.

The accuracy of prediction with allowance for soil agrochemical properties through S_{ef} is much higher. When using the actual value obtained by the integrated assessment of soil properties for each "soil–plant" pair, even with the mismatch of soil agrochemical properties and soil classification name, the relative deviation of the predicted values of ^{137}Cs TF_{ag} from the experimental ones was not more than ±50%. This accuracy is sufficient for taking a decision on countermeasures.

For estimating the accuracy of predicting ^{137}Cs behavior in the "soil–plant" system in the long term using the model, in 2006 and 2008, a total of 186 combined samples of soil and plants were collected in 93 sites of the radioecological monitoring network. The obtained experimental data were statistically processed.

The values of the soil–crop transfer factor for radiocesium obtained in the experiments and estimated by the model are shown in Table 5.23. Practically for all studied soil types within each crop group, the values of ^{137}Cs TF_{ag} decreased from 2006 to 2008, even though the decrease was not significant as compared to the first years after the accident, which corresponds to the second "slow" exponent in the empirical model (see above).

The crops on the same soil type in terms of ^{137}Cs TF_{ag} are in descending order: natural herbs, sown herbs, green forage, vegetable crops, potato, roots, and cereals. The differences between the first and last crops in the row are up to 30 times on boggy soils, ten times on soddy-podzolic soil, and five times on chernozem, which is in agreement with the ratios used for building the model for radiocesium behavior in the soil–plant system. The differences in ^{137}Cs TF_{ag} values for natural grasses on peaty-boggy soil and chernozem are 15 times and more, as in the previous years, and from 2 to 7 times for other crops.

The prediction for 20 years (1987–2008) has adequately reflected the decrease in ^{137}Cs availability for plants. It should be noted that the results were obtained under

Table 5.23 Verification of the model for ^{137}Cs TF_{ag} dynamics for crops on different soil types based on the experimental data of 2006 and 2008

Crop	Soil type	Year	n[a]	TF_{ag}, (Bq kg^{-1})/(kBq m^{-2}) TF_{age}[b] ± δ	TF_{agm}[c]	$\delta(TF_{ag})$, %
Natural grasses	Peaty-boggy	2006	7	2.8 ± 0.92	1.6	−42
		2008	3	2.1 ± 0.31	1.4	−33
	Soddy-podzolic	2006	2	0.49 ± 0.33	0.59	+20
		2008	7	0.52 ± 0.33	0.55	+6
	Chernozem	2006	3	0.16 ± 0.044	0.18	+13
Sown grasses	Soddy-podzolic	2008	2	0.16 ± 0.010	0.11	−31
Forage grass: *clover, lupine, maize*	Soddy-podzol	2006	3	0.31 ± 0.051	0.32	+3
		2008	2	0.29 ± 0.083	0.28	−3
	Chernozem	2006	3	0.021 ± 0.007	0.021	0
Vegetables: *cabbage, cucumber*	Soddy-podzolic	2006	3	0.11 ± 0.006	0.091	−17
		2008	8	0.079 ± 0.030	0.086	+9
	Chernozem	2006	2	0.019 ± 0.013	0.018	−5
Tubers, roots: *Beet, potato*	Peaty-boggy	2008	3	0.11 ± 0.028	0.13	+15
	Soddy-podzolic	2006	8	0.044 ± 0.016	0.042	−5
		2008	16	0.035 ± 0.0042	0.037	+6
	Chernozem	2006	7	<Measurement sensitivity		
Cereals: *Oats, barley, rye*	Soddy-podzolic	2006	3	0.075 ± 0.037	0.063	−16
		2008	5	0.053 ± 0.028	0.058	+9
	Chernozem	2006	6	<Measurement sensitivity		

[a]Number of pairs soil–plant
[b]Experimental value
[c]Model calculated value

agricultural production conditions and major uncertainty in prediction data was associated with the estimation of deposition density, which is intrinsically a random value. The decrease in ^{137}Cs TF_{ag} was not as significant as in the first years after the accident. As can be seen from the above, by 2006–2008, all general trends of ^{137}Cs behavior in the soil–plant system remained unchanged. Widely differing environmental conditions in the monitoring area have made it possible to derive representative estimates for ^{137}Cs TF_{ag} and other model parameters for a wide range of crops and soil types. Twenty years after the accident, the ^{137}Cs activity concentrations in potato, beet, and cereals were below the sensitivity of the measurement method.

The verification of the model for ^{137}Cs soil–plant transfer based on 2006–2008 experimental data has demonstrated that the prediction accuracy $\leq \pm$ 20% in 17 out of 26 variants of "soil–crop year." In five variants, the accuracy was ±20–35%, and only in one case for natural grass on peatbogs, the ^{137}Cs TF_{ag} relative deviation from the actual value was ±42%. Parameters of the dynamic model for this variant can be adjusted based on the experimental data of 2006–2008.

Generally, the analysis of Tables 5.21 and 5.22 leads us to conclude that the robustness of estimates of ^{137}Cs concentration in plants based on the developed kinetic model is not worse than 30%, which can be recognized as appropriate for predicting and management purposes in case of radiation-related emergency.

The verification of the model for ^{137}Cs soil–plant transfer has shown that in the long time after the Chernobyl accident the accuracy of predicting radionuclide concentration in plants, based on the model allowing for kinetics of nuclide binding by soil and soil-specific agrochemical properties through S_{ef}, is within ±30% in 75% of cases. This accuracy can be deemed to be adequate, in general, for predicting a radiological situation in areas influenced by NPP in routine and emergency conditions, as well as for decision-making with respect to countermeasures, so that permissible levels in agricultural products are not exceeded. If no data are available about agrochemical properties of soil, model parameters for soil types and plant species can be used. In doing so, it should be borne in mind that prediction results will be conservative and actual values of plant contamination can be much lower. The proposed model enables predicting TF_{ag} for all crops on all soil types under study.

5.5 Countermeasures

Prediction of the behavior of radionuclides in the soil–plant system using the model shows that unless comprehensive countermeasures in the agricultural sector are implemented, the situation can continue to exist over several decades. Actions to substantiate countermeasures, development of their techniques, and assessment of their effectiveness were done in the three countries in parallel with allowance for specific local conditions. In doing so, each country took the lead in specific research and application area.

In Belarus, the framework for liming acidic soils was established; a set of regulatory standards was developed in support for use of liming materials, ranging from raw ground dolomite to defecate (Ageyets 2001; Bogdevich 1996; Bogdevich et al. 2001, 2002). Technology for application of ferrocene boluses was tested and implemented for milk and meat production (Il'yazov et al. 2002).

In Russia, *the* rationale was provided for heavy fertilizing on soddy-podzolic and meadow soils (Korneev et al. 1977; Aleksakhin et al. 2001b). The importance of applying mineral fertilizers and liming was demonstrated to decrease ^{137}Cs uptake by plants in different time periods after radioactive fallout. It has been shown that in the initial period after the accident, biogeochemical fixation processes are the major contributors to decrease ^{137}Cs uptake by plants, whereas 5–10 years after fallout its contribution was declining to 55%, on the average, with radioactive decay playing much more significant role. Later on (third period after the ChNPP accident), reduction in ^{137}Cs uptake by plants was mainly determined by radioactive decay, its contribution being about 50%, even with heavy fertilizing (Aleksakhin et al. 2006).

In Ukraine, a radioecological classification of meadows was developed, a system for grassland management was established, methods for surface and radical amelioration of pastures were developed, and the application of zeolites and sapropels was substantiated and implemented (Prister et al. 1993a, b; Perepelyatnikov 2012a, b). In the three CIS countries, the application of NPK on ^{137}Cs contaminated soils at optimal ratio 1:1.5:2.0 was rationalized and implemented (Prister et al. 1998; Prister et al. 2001). The kinetic model for radionuclide behavior in the soil–plant system was developed and used as a theoretic basis for countermeasures and their assessment and management of their effectiveness (Prister and Vinogradskaya 2011). With the proportion of mobile ions in the soil solution decreasing, the effectiveness of countermeasures, especially their cost-effectiveness, tends to decline. Controlling the processes of radionuclide distribution between soil phases is the way to control radionuclide uptake by plants. The availability of radionuclide ions from the solution remains practically constant, whereas their concentrations vary as a function of time and soil properties. This premise is key for developing countermeasures (Konoplev et al. 1993; Nisbet et al. 1993; Bogdevich et al. 2001). Methods for animal feeding and management, as well as techniques for applying natural mineral sorbents in milk and meat production, have been developed and implemented (Bogdanov et al. 1996; Prister et al. 1996).

5.5.1 *Countermeasures for Crop Production*

Countermeasures regarding plant cultivation can be grouped as follows: soil treatment, agrochemical measures, and organizational measures.

Agrochemical measures are meant to facilitate a shift in radionuclide equilibrium in the soil solution–soil absorbing complex system (SAC) toward SAC, i.e., the highest possible transfer **radionuclide** to the absorbed state (as well as insoluble state). When calculating fertilizer application rate, account should be taken of the content of main nutrients in the soil and the crop demand for them, with allowance for planned crop yield and impact of fertilizers on soil properties such as increase in acidity and concentration of NH_4^+ ions capable of displacing Cs^+ ions from soil to the soil solution. Taking countermeasures after the Chernobyl accident, of major importance was the recommendation to change the ratio of main nutrients. For the soils contaminated with ^{137}Cs, the optimal NPK ratio 1:1.5:2.0 was recommended. Also, phosphorus and potassium were recommended for fertilization instead of nitrogen (Prister et al. 2001; Prister 2007). The lower the soil fertility, the higher the effectiveness of fertilizer application. Generally, radionuclide uptake by crops can be reduced by 2–3 times by applying effective agrochemical countermeasures in the absolute majority of cases (Table 5.24).

Soil treatment. High effectiveness of countermeasures has been achieved by deep ploughing (35–40 cm) and meadow amelioration, which is of particular importance as cow milk is a major contributor to internal exposure of humans. Soil treatment should aim at isolating the root system of plants from the radionuclide-containing

Table 5.24 Reduction in radioactive contamination of crop production in case of implementing the principal countermeasures, times (aggregated data) (Prister et al. 2016)

	^{137}Cs		^{90}Sr	
	Soils			
Countermeasure	Mineral	Organic	Mineral	Organic
Agrochemical				
Liming, 4–6 t/ha	1.5–3.0	1.5–2.0	1.5–2.6	–
Treatment with NPK at optimal ratio 1:1.5:2.0	1.5–2.0	1.5–3.0	0.8–1.2	–
Manure, 50 t/ha	1.5–3.0	–	1.2–1.5	–
Manure + NPK	1.5–3.0	–	–	–
Liming + NPK	1.8–2.7	2.5–4.0	–	–
Liming + NPK + manure	2.5–4.0	–	1.5–1.8	–
Zeolites	1.5–2.5	–	1.5–1.8	–
Sapropel	2.4–4.0	–	1.0–1.5	–
Soil treatment				
Ploughing, 20–25 cm	2.5–3.0	3.0–4.0	2.0–4.8	5.0–7.0
Ploughing, 35–40 cm	8–12	10–16	2.0–3.0	–
Organizational (agronomy)				
Selection of crops and varieties with low uptake of radionuclides	On the average, 3–30 times			

layer. This layer should be thrown by skimmer to furrow bottom and covered by clean soil layer. Common ploughing and rotary tillage allow the contaminated layer to be mixed with clean soil, resulting in the reduction of nuclide concentration in root habitable layer and their binding to SAC (Table 5.24). The cost of countermeasures in crop farming is estimated to be 95–509 Euros per hectare.

The most effective countermeasure is pasture amelioration, its effectiveness ranging from 4 to 16 times. When this countermeasure is repeated, however, effectiveness is not so drastic, reducing down to 3–5 times only. It should also be noted that radical amelioration is highly expensive, given ploughing with clod overturn, application of fertilizers and ameliorants, and planting feed grasses.

High effectiveness of agronomy countermeasures is associated with the different ability of plant species and varieties to accumulate stable mineral nutrients—nonisotopic and isotopic carriers (potassium and calcium for ^{137}Cs and ^{90}Sr), it equals 10–30 times, on the average. The selection of species and varieties of forage and production crops with low uptake of radionuclides makes it possible to reduce their concentrations in human food and animal feed.

Of course, the above factors contribute jointly to radiation dose for humans, as shown in Fig. 5.11. The role of ^{137}Cs flux from seminatural and natural ecosystems is much higher than that from arable lands. As can be seen, radiation dose would be several times higher for individuals consuming mushrooms as compared to those who do not eat mushrooms. Preservation of mushrooms and berries in large quantities is common practice in Polesye, and these are usually collected on boggy soils, resulting in high cesium concentrations (Table 5.25). Economic circumstances and

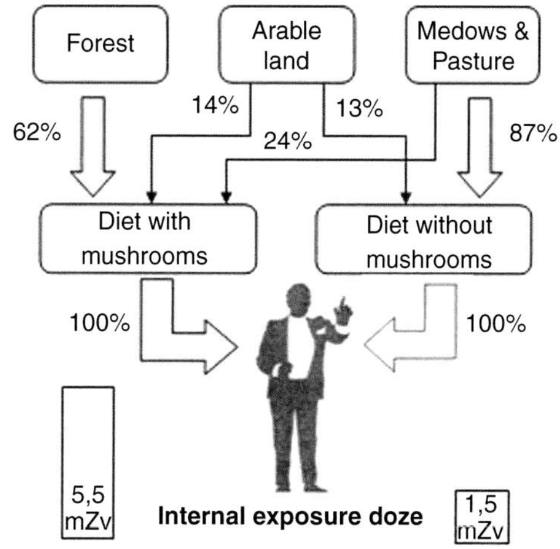

Fig. 5.11 Structure of mean annual internal radiation dose in the village of Milyachi of Rivne oblast (^{137}Cs contamination density is 37–185 kBq m^{-2}) (Prister et al. 2016)

Table 5.25 ^{137}Cs concentration in berries and mushrooms in 1995, Bq kg^{-1}

Oblast	Forest berries	Mushrooms
Zhytomyr	500–3500	1300–5000
Rivne	800–8000	5000–32,000

inertia of ethnic customs do not let the population stop consuming these products in spite of anxiety about health safety.

The experimental data available after the Chernobyl accident have demonstrated that planning of countermeasures should be based on comprehensive environmental and radiation monitoring. Environmental factors associated with agrochemical properties of soils and their water regime have even a greater impact on exposure than nonuniformity of contamination. Landscapes with organogenic hydromorphic soils turned out to be of higher hazard not only in Ukrainian and Belorussian Polesye, but also in some parts of contaminated regions of Russia, Sweden, and the UK. Sheep breeding common in the areas of Belarus, Ukraine, and the UK contaminated after the Chernobyl accident was jeopardized due to sheep grazing on natural pastures showing high soil-to-grass ^{137}Cs transfer factor (Howard 1993).

In some settlements of Ukrainian Volyn region, as many as 10–20 different pastures may be used for grazing, their contamination levels ranging from 5 to 20 kBq m^{-2}. Nevertheless, as a result of high moisture content and acidity of peaty soils (pH from 3.7 to 5.0), the values of transfer factor TF_{ag} can be as high as 328 Bq kg^{-1}/kBq m^{-2}. Therefore, countermeasures are required even on the territories on which gamma dose rates practically do not differ from the background level prior to the accident Prister et al. 2018. Aggregated data on the effectiveness of actions on meadow pasture lands are shown in Table 5.26.

Organizational countermeasures consist of changing land use. For developing and evaluating the effectiveness of organizational measures, diverse information needs to be aggregated such as soil contamination levels and soil properties, features

Table 5.26 Effectiveness of countermeasures on grasslands

Countermeasure	^{137}Cs concentration reduction in grass, times	
	Mineral soils (sandy, loamy)	Organic soils (peaty)
Drainage	–	2–4
Disking or rotary cultivation	1.2–1.5	1.8–3.5
Ploughing	1.8–2.5	2.0–3.2
Ploughing with clod overturn and deepening to 35–40 cm	8–12	10–16
Liming	1.3–1.8	1.5–2.0
Application of nitrogen fertilizers and increase of the dosage of phosphoric and potassium fertilizers	1.2–3.0	1.5–3.0
Surface improvement	1.6–2.9	1.8–14.0
Amelioration	3.0–12.0	4.0–16.0

of radionuclide uptake, and crop yield. Effectiveness of countermeasures can be assessed by experts with consideration of impact of agrochemical measures, changes in planting pattern, soil cultivation, and targeted use of produce (food, forage). Changing land use is a long-term countermeasure including selection of crop species and varieties with different uptake of radionuclides. The data analysis suggests that changing crop species is having a more noticeable effect as compared to changing varieties: the accumulation of radionuclides decreases by 2–11 times with changing plant species and 1.5–4.0 times with changing varieties.

5.5.2 Countermeasures for Livestock

Methods of feeding cattle. After the Chernobyl accident, countermeasures for livestock were widely applied in Belarus, Russia, and Ukraine. In Russia, organizational measures were taken: cattle of private farms were evacuated from the population centers with ^{137}Cs contamination density more than 555 kBq m^{-2} (15 Ci km^{-2}) and cleaner foodstuffs from collective farms or noncontaminated territories were supplied (Aleksakhin et al. 2001a). These measures were applied in 200 settlements, and a total of 8813 heads of cattle and over 15 thousand of sheep, goats, and pigs were evacuated as soon as 1986. In 1987–1988, further 6881 heads of cattle and more than 10 thousand of other agricultural animals were withdrawn from farming (Prister 2007; Prister and Aleksakhin 2008; Prister 2011).

The absorption coefficient from feed to animal body is influenced by factors such as type of feed and degree of its digestion, age, and physiological condition of the animal, which are, however, contributing just a few percent. There are three radically different ways to reduce the radionuclide content in milk and meat (muscle).

The first way is reducing the uptake of the radionuclide by diminishing its concentration in the diet through selecting fodder. For this purpose, the above countermeasures in crop farming can be applied or fodder with low ^{137}Cs concentration can be selected. The dynamics of ^{137}Cs contamination of fodder is governing radionuclide concentration in milk which, similarly to crops, can be approximated by exponential dependence with two components. During the grazing period, the radionuclide concentration in milk declines more slowly than in stall feeding period, which is associated with higher ^{137}Cs fixation rate by arable lands on which succulent fodder is grown such as silage, roots, and sown grass, making a major part of fibrous feed in nongrazing season. This explains higher concentration in milk during grazing time as compared to no grazing. For meat cattle, ^{137}Cs concentration in silage diet is 18% of that in hay diet and for milk cattle, 57% (Prister 1999). By changing the diet, the radionuclide concentration can be reduced by 3–5 times in milk and meat. Unfortunately, the recommendations of scientists regarding the separation of stored feeds with different contamination levels and their separate use practically have not been translated to practice.

The second way to reduce radionuclide concentration in milk and meat is to transfer radionuclide in the animal rumen from the ionic state to the bound state prior to intestinal absorption. Long before the Chernobyl accident, methods for reducing ^{137}Cs transfer to agricultural products had been available, including the application of ferrocene, which is forming an insoluble compound with ^{137}Cs or mineral sorbents to GIT (intestinal sorption). These methods have been widely used after the Chernobyl accident (Sirotkin 1992). Ferrocene makes it possible to reduce ^{137}Cs transfer to milk and meat by 3–10 times. Sorbents are less effective, reducing contamination of animal-derived food products by 1.5–5 times, nevertheless, playing an important part due to binding ammonia and some heavy metals in GIT and serving as a source of microelements (Bogdanov et al. 2002). Intestinal sorbents should be given within compound feed. In Russia and Belarus, various formulations of ferrocene have been devised: within feed supplements, bound to inert carriers, as only several grams per head is required daily. In all three countries, boluses containing ferrocene, barium sulfate, and wax developed in Norway are widely used. They have been implemented with involvement of the author Hove et al. (1995). In Belarus, the countermeasure is obligatory for the regions in which ^{137}Cs permissible level in milk and meat is thought to be exceeded. Production, supply, and administration of boluses is the responsibility of the state. In Ukraine, the salt lick blocks of 4–5 kg, containing intestinal sorbents dispensed in feeders or on pastures, were widely used.

In the course of activities to minimize the Chernobyl accident consequences, the rationale was provided for the role of intestinal sorption in immunologic and metabolic homeostasis of animals and in reducing contamination of animal-derived food products in case of using ^{137}Cs contaminated feed. Also, the justification was provided and the methodology and criteria were developed for integrated evaluation of applying natural sorbents as a feed supplement for the production of milk, beef, poultry, and rabbit meat in the farms on contaminated areas (Bogdanov et al. 1996).

Sorption properties were investigated in a comprehensive way and evidence was obtained for the safety of Ukrainian natural mineral silicates to be used as a countermeasure for livestock, reducing ^{137}Cs uptake in the animal body and animal-derived food products (1.3–4.4 times in milk and 1.6–7.4 times in meat) and facilitating removal of incorporated radionuclides.

In Ukraine, zeolites containing clay minerals and having high affinity to cesium were tested as intestinal sorbents, namely vermiculite (Zaporozhskaya region), palygorskite (Cherkasskaya region), and clinoptilolite (Khust Zakarpatskaya region). The efficacy of these minerals in decreasing order is as follows: palygorskite, vermiculite, and clinoptilolite. They allow reducing radiocesium transfer to milk by 3–9.7 times. Effectiveness of zeolites is increasing with an increase in dosage and degree of fineness. They can be used as part of compound feed or salt lick blocks (Bogdanov et al. 2002; Pronevich 2014). The modification of zeolites and ferrocene by other compounds leads to a several-fold increase in ^{137}Cs selective sorption of (Il'yazov 2006).

Based on many thousand livestock, considerable effect has been confirmed from using sorbents as part of the feed and giving animals clean feed before slaughter. In the finishing period, 2–3 months before slaughter, contaminated feed can be replaced by clean one, and in this case feed with practically any contamination level can be used during 12–16 months. With sorbents added, this measure becomes even more effective. In Rivne oblast, salt lick blocks with mineral nutrients and ferrocene proved to be effective. The use of zeolites prior to slaughter allows reducing ^{137}Cs uptake in animal muscles by 2–2.4 times (Loschilov et al. 1994).

The Sokornitsky zeolites milling plant, established by Ministry of Emergencies of Ukraine, has produced 1362 t of zeolite powder to be supplied to the feed processing plants in three oblasts. Specialization of farms has been changed for meat cattle breeding and raising of reproduction pig in five regions. The breeding flock was rejuvenated for meat production stock, facilities and equipment were upgraded, and scientific method support was provided.

Enhancing the fodder base is key for the production of clean milk and meat. Dairy cattle operations should be supplied with feedstuffs added with ferrocene and zeolites. Under the program "Children of Ukraine," a project to ensure radiation safety of children on the territories affected by the Chernobyl accident has been implemented by UIAR. It has been demonstrated that comprehensive measures in feed production and cattle husbandry (grassing and regressing and use of sorbent added feedstuff) make it possible to produce milk and meat with radiocesium concentration below the permitted regulatory level in all contaminated farms and settlements of Ukrainian Polesye. Unfortunately, the program on beef cattle breeding stopped to be funded shortly after it was started.

The third way to reduce radionuclide content in animals is to give "clean" feedstuffs before slaughter. About 3–10 weeks before slaughter, the diet of animals can be changed to feeds with a low concentration of ^{137}Cs and then ^{137}Cs is excreted from the animal body rather quickly. Three stages of ^{137}Cs excretion from animal body have been identified. The first stage is excretion of contaminated feed mass from GIT (T_1), the second stage is removal of more mobile ^{137}Cs (intercellular) from the body (T_2), and the third stage is excretion of intracellular immobile ^{137}Cs (T_3). For cattle, T_1 is 6–7 days and T_2 is 8–9 days. T_1 and T_2 are 8–9 and 7–8 days for horses and 12–13 and 5–6 days for sheep. The length of T_3 is determined by the degree of muscle contamination (Prister 2007).

In 1986, slaughtering and meat processing plants were often supplied with cattle having ^{137}Cs muscle concentration above the permissible level. Contamination levels were monitored by sampling from the carcass and if the permissible level was exceeded, the carcass had to be disposed of by processing into tankage. It became necessary to determine the concentration of cesium radionuclides in cattle body prior to slaughter so that cattle could be returned for keeping on clean feed. A method has been developed by UIAR to determine cesium in live animal body using a field whole-body counter based on scintillation gamma radiometer SRP-68 (produced in USSR, later in Russia) with a lead collimator detector. This method enabled selecting animals for slaughter, both on farms and at slaughtering plants (Loschilov et al. 1994). The unjustified slaughter of cattle thereby was prevented and additionally about 5000 uncontaminated meats were produced annually. Starting from 1994, no contaminated meat was supplied to meat-processing plants almost on all contaminated territories (Astasheva et al. 1997). Batch production of the instrument for monitoring ^{137}Cs concentration in the live animal body was organized by Ministry of Chernobyl Affairs of Ukraine.

Parameters for ^{137}Cs metabolism in the body of different animals and nuclide transfer to milk and meat have been estimated in actual conditions of animal keeping and feeding on contaminated feed. The ^{137}Cs biological half-life was 7 and 30 days for cattle, and 30, 10, and 4 days for pigs, sheep, and chicken, respectively. It was proved that if the clean feed was given for 40–60 days, a significant part of ^{137}Cs was excreted from the body and radionuclide concentration decreased by six- to tenfold. This provided the basis for developing and implementing the method for three-phase feeding of animals (Prister et al. 1993a, b). The method for ^{137}Cs determination in muscles of live animals allowed controlling when animals were to be placed on cleaner feed. For 14–16 months, animals were given feed with contamination level unrestricted, and after that they were eating feed with the contamination level 4–5 times lower for 1–2 months. Then, after 1–1.5 months, the cattle was placed on cleanest feed available on a farm (Prister et al. 1993a, b). In 1996, a total of 1600 heads of cattle were placed on additional feed in Zhitomir and Kiev oblasts. In these animals, the ^{137}Cs concentration in muscles was 3000 Bq kg^{-1}, and in 2–3 months, it decreased to 130 Bq kg^{-1}, i.e., 20 times. The effectiveness of this measure increased when combined with the use of sorbents. The concept of phase feeding was widely used in the UK, where sheep in Cambria were moved to cleaner pastures prior to slaughter on scientists' recommendation (Howard 1993).

The use of zeolites when feeding prior to slaughter also allows reducing ^{137}Cs uptake in muscles by a factor of 2–2.4 (Bogdanov et al. 1996, 2002). Based on phase feeding with measurement of ^{137}Cs concentration in live animals, a program was developed in Ukraine for beef cattle breeding on contaminated territories. It provided for using sorbent supplements in the final feeding phase and using Polesye feed resources for meat production without restrictions as to contamination levels. Aggregated assessment of the effectiveness of the discussed countermeasures is shown in Table 5.27.

Table 5.27 Reduction of radioactive contamination of animal-derived food products as a result of countermeasures, times

	^{137}Cs		^{90}Sr
Countermeasures	Milk	Meat	Milk
Zootechnical measures			
Clean feed before slaughter	–	2.0–15	
Feed supplements	1.2–1.5	1.5–3.1	1.3–1.5
Prudent use of hay fields and pastures	1.5–15	3.0–4.0	
Selection of feedstuff	2.0	30	
Separate storage and use of feed with different contamination levels	Depending on crops, agrochemical properties, and soil contamination levels		
Veterinary measures			
Use of Cs-binding compounds	1.5–6.0	1.5–2.1	
Use of sorbents	5.0	4.5	1.5

High radiological effectiveness of beef cattle raising can be seen from the figures below. The ^{137}Cs flux due to the consumption of meat produced on the area, where feeding stuff is grown, on the average, is only 2% of that coming from the feed. In the case of giving clean feed prior to slaughter, the radionuclide flux with meat can be reduced to 0.05–0.5%. The ^{137}Cs flux with milk is, on average, 8.3% of that coming with feed and decreases to 2.8% after milk processing. Thus, beef cattle breeding on contaminated areas allowed the reduction of individual and collective radiation doses for the population.

5.5.3 Organization of Beef Cattle Breeding on ^{137}Cs Contaminated Territories of Ukrainian Polesye

The first link in the "soil–plant" food chain is crucial for contamination of all agricultural products. Specific features of ^{137}Cs metabolism for different animals have been studied in detail and the transfer factor for milk and meat has been assessed for actual feeding conditions with contaminated feed. It has been demonstrated that when feeding animals with clean feed during 40–60 days, a major part of accumulated ^{137}Cs is excreted from the body, with radionuclide concentration in muscles decreasing by 6–10 times. To use effectively feed resources, a methodology was developed including three phases: initial, intermediate, and final (Prister et al. 1993a). Justification, development, and implementation of the methodology of three-phase feeding has provided a basis for beef and dairy cattle breeding on the contaminated areas (Prister 2007).

Table 5.28 shows the permissible ^{137}Cs concentrations in cattle diet and meat in different feeding phases, which made it possible, in the finishing phase, to achieve

Table 5.28 ^{137}Cs concentration in the diet and muscles of cattle in different feeding phases

Feeding phase	Diet, kBq day^{-1}	Muscles, Bq kg^{-1}	
		Phase beginning	Phase end
Initial	74	No restrictions	2960
Intermediate	33	2960	1300
Finishing	15	1300	600

Table 5.29 Length of feeding phases for cattle, days

Feeding phase	Animal age at slaughter, years		
	1.5	2.5	5.5–9.0
Initial	Prior to placing on intermediate phase		
Intermediate	15	15	30
Finishing	50	60	60–120

the radionuclide concentration of 600 Bq kg^{-1} in meats, adopted as a permissible level in Ukraine in 1992.

Table 5.29 shows recommended feeding length with consideration for features of animal metabolism. In the finishing feeding phase, feedstuffs from fodder crop rotation were usually used.

In the initial feeding phase, animals can graze on pastures with the highest contamination level of herbage. In fact, in the initial feed phase, which is more than a year, cattle can be grazing without restrictions on territories on which farming is permitted, i.e., with the level up to 555 kBq m^{-2}.

The accumulation of radiocesium is four times higher in cow muscles than in milk. Therefore, beef is a major contributor of cesium to human body after milk. As to pig raising, the radionuclide content in pig daily diet should be <3 kBq (Prister et al. 1993a).

The rationale was provided for the role of intestinal sorption in maintaining the immunological and metabolic homeostasis of animals and in the reduction of product contamination with the use of ^{137}Cs contaminated feedstuff (Bogdanov et al. 1996, 2002; Zubets et al. 2011). The methodology and criteria were developed for integrated evaluation of applying natural sorbents as a feed supplement in the production of milk, beef, poultry, and rabbit meat in the farms on contaminated areas. Sorption properties of sorbents were investigated comprehensively, and the safety of Ukrainian natural mineral silicates has been proved to be used as a countermeasure for livestock, reducing ^{137}Cs uptake in the body and animal-derived food products (1.3–4.4 times in milk and 1.6–7.4 times in meat) and facilitating removal of incorporated radionuclides. Even though the effectiveness of natural sorbents is 1.5–2 times lower than that of ferrocene, they boost digestion processes in animals and promote health and productivity of animals (Bogdanov et al. 1996, 2002; Prister 1999).

Resources of mineral silicates on the territory of Ukraine and feasibility of their use as feed supplement have been assessed. A plant for producing fine-fraction clinoptilolite in the town of Khust, Zakarpatskaya region, has been built, the use of zeolites as part of mixed feed has been tested. Implemented measures have enabled the rehabilitation and economic revival of the traditional livestock sector in Polesye, including sheep farming (Prister 2007; Zubets et al. 2011).

The system of radiation monitoring of animal-derived food products has been developed and implemented. If such measures had been realized immediately after the accident, the population averted dose would have been much higher.

The applied countermeasures have resulted in the reduction of ^{137}Cs concentration in milk and meat by a factor of 4–12 and averted collective dose were of several thousand man·Sv in each of the three affected countries.

Comparison of the effectiveness of countermeasures for crops and livestock leads us to the conclusion about their close radioecological effectiveness. It should be noted that the costs of averted similar collective dose due to implemented countermeasures in livestock and crop farming are from 1:10 to 1:100 (Prister et al. 1996).

An effective method for decreasing ^{137}Cs uptake by the human body is milk processing (Table 5.30). To reduce the concentration of radionuclides in milk and meat to the permissible levels, the effectiveness of countermeasures of 4–10 times was sufficient practically in all population centers of Ukraine. Therefore, milk processing can be effectively used for reducing radionuclide intake with dairy products for humans. In this regard, state support is to be provided to the population and sales of products.

5.5.4 Assessment of Countermeasures Effectiveness

Combined countermeasures in crop and livestock farming and processing of products have made it possible to obtain agricultural products meeting the national permitted regulatory levels almost on all Chernobyl-contaminated territories of Belarus, Russia, and Ukraine on which the external radiation dose allows people to live.

Of late, the ^{137}Cs concentrations in milk in 10–50 population centers of Ukraine were in excess of 100 Bq L^{-1} on different years. The radiation situation will not be back to normal due to natural rehabilitation processes only, and it is imperative to conduct countermeasures on these areas. In this regard, priority should be given to countermeasures in cattle breeding showing much higher cost-effectiveness. Given countermeasures implemented in both agricultural production sectors, their

Table 5.30 Reduction of radioactive contamination of products with milk processing, times (Prister et al. 2007, 2016)

Product	^{137}Cs concentration reduction, times
Cottage cheese	1.1–1.4
Hard cheese	2.4
Dry milk	Increases by eight times
Sour cream	1.2–1.3
Butter	3.6–5.6

radioecological effectiveness is multiplied and the total effectiveness of 4–10 times is sufficient for reducing radionuclide concentration in milk and meat to the levels prescribed by DR-97 practically in all population centers of Ukraine, Russia, and Belarus.

When developing a strategy for countermeasures, account should be taken of not only a specific type of products but also fluxes of radionuclides from soil to plant and along the food chain. For example, the amount of radiocesium transferred from soil to cereal crops in Polesye did not exceed 1–2% of its total amount transferred to all plants. By estimates, the use of cereal crops as cattle fodder and for bread production would result in the collective dose of about 10–20 man Sv, whereas the use of contaminated hay for cattle feeding would lead to a dose 50–70 times higher. Therefore, irrespective of radioecological effectiveness of countermeasures for cereal crops, they will not lead to a noticeable reduction of total collective dose (Prister 1998, 2007; Prister et al. 2007).

For comparative evaluation of the effectiveness of countermeasures and decision-making regarding their implementation, three components have been identified (Prister 1998). The first is *radioecological effectiveness* which shows how many times the contamination level of products can be reduced by implementation of a countermeasure.

The main criterion for the effectiveness of agricultural countermeasures, however, should be the total dose averted by taking countermeasures. We refer to this parameter as *dose effectiveness*. This component is governed by many factors such as the amount of produced product and time and way of product use. For example, if meadow amelioration allows making hay with low ^{137}Cs concentration and this hay is used for feeding young cattle, rather than dairy cows, the dose effectiveness of this measure will be zero. The overall strategy of countermeasures should be dictated by dose effectiveness, but radioecological effectiveness can play a key role for decision-making on countermeasures, given radionuclide concentration in product is in excess of the maximum permitted level.

The same dose can be averted by different countermeasures, their cost differing significantly. The third component of the overall effectiveness *is cost-effectiveness*, a qualitative measure of which is the cost of unit dose averted due to countermeasures: additional expenses US$ per man·Sv of averted dose. The acceptable cost of man·Sv averted dose is 1–20 thousand US$.

The analysis has shown that whereas radiological effectiveness due to applied countermeasures may be similar, for example, in case of measures leading to a twofold reduction in product contamination levels, their dose effectiveness for livestock is much higher than that for crops and cost-effectiveness even more so, differing by orders of magnitude. Effectiveness of a countermeasure will be higher for higher trophic levels in the "soil–plant–animal (meat and milk)—processing" chain on which it is applied.

The international forum "Chernobyl's Legacy: Health, Environmental and Socio-Economic Impacts" in Vienna on 5–6 September 2005 has recognized that the measures taken by the governments of the affected CIS countries to mitigate the Chernobyl accident consequences were generally timely and appropriate.

5.6 Conclusion

The Chernobyl accident has resulted in the contamination of extensive agricultural lands where more than ten million people lived in the three CIS countries. For the majority of the population cohort, agricultural products were the main pathway for the intake of radionuclides. It has rightly been referred to as a communal rural accident demonstrating that the agrarian sector should be given priority in monitoring and conducting countermeasures. The Chernobyl accident has clearly demonstrated that concealment of the fact of the accident and the delayed information to the population led to harmful consequences.

For assessing and predicting the actual radiation hazard, conducting emergency measures, such as population and cattle evacuation and iodine prophylaxis, as well as providing a rationale for long-term agricultural countermeasures, knowledge of radionuclide migration in the environment, and biological chains and models for predicting the radiation situation, is not sufficient. Large amounts of radioecological information about the contaminated territory are required to calculate doses for the population, identify territories of higher hazard in terms of absorbed doses and planning frameworks for comprehensive countermeasures. Based on the experience gained following the Kyshtym and Chernobyl accidents, it should be emphasized that the accumulation of such information would take many years of monitoring, and gathering of relevant data needs to be carried out in a preventive mode within emergency preparedness programs.

The long-term radioecological monitoring has revealed the high role of processes of natural rehabilitation of the contaminated territories. For instance, as a result of ^{137}Cs sorption by soil clay minerals, radionuclide uptake by plants has decreased by tens of times. None of the developed countermeasures have shown such high effectiveness. At the same time, provinces with low content of clay minerals can be attributed to the areas of higher hazard, showing higher contamination level of products, especially feed, milk, and beef. Therefore, given the same deposition, human exposure appears to be tens or hundreds of times higher on peat soil than on chernozem. In spite of the applied countermeasures and natural rehabilitation processes, in Ukraine, even 30 years after the accident, the ^{137}Cs concentrations in cows' milk were exceeding the permitted regulatory level of 100 Bq L^{-1} in 11–30 population centers on different years.

Aggregation of large amounts of experimental data has provided a basis for developing the model to account for ^{90}Sr and ^{137}Cs behavior, which analytically describes the soil-to-plant transfer factors as a function of soil properties, soil contamination levels, and biological features of plants. The model enables predicting the spatial distribution of radiation doses for the population in an emergency, optimizing systems of radiation and product contamination monitoring, and proposing agricultural countermeasures with allowance for specific characteristics of territory such as terrain, soils, and land use.

When mitigating the consequences of the Chernobyl accident, high effectiveness of countermeasures for crops and livestock has been demonstrated, provided they were applied timely. This has been confirmed by independent data of Ukraine, Belarus, and Russia, as well as international projects of the European Commission, FAO, IAEA. It is essential that the gained experience of mitigating the negative consequences of the accident be widely available and translated into working documents to be put into action automatically in case of a severe radiation-related and nuclear accident in affected countries.

References

Absalom JP, Young SD, Crout NMJ (1995) Radiocaesium fixation dynamics: measurement in six Cumbrian soils. Eur J Soil Sci 46:461–469

Ageyets VY (2001) System of radioecological countermeasures in the agro-sphere of Belarus. RIIPE Institute of Radiology, Minsk, p 241. (In Russian)

Aleksakhin RM, Buldakov LA, Gubanov VA, Drozhko EG, Il'in LA, Kryshev II, Linge II, Romanov GN, Savkin MN, Saurov MM, Tikhomirov FA, Holina YB (2001a) Major radiation accidents: consequences and protective measures. Moscow, Izdat, p 752. (In Russian)

Aleksakhin RM, Sanzharova NI, Fesenko SV, Kurganov AA, Mosharov VN (2001b) The main results of the work to eliminate the consequences of the Chernobyl accident in the field of agro-industrial production. NV Gerasimov, Moscow, pp 105–141. In Russian

Aleksakhin RM, Sanzharova NI, Fesenko SV, Spirin EV, Spiridonov SI, Panov AV (2006) Chernobyl, agriculture, the environment. Materials for the 20-th anniversary of the Chernobyl nuclear power plant accident in 1986. Obninsk, UNO, p 24. (In Russian)

Arapis G, Perepelyatnikova L (1995) Influence of agrochemical countermeasures on the yield of crops grown on areas contaminated by Cs-137. In: Kotsaki-Kovatsi V (ed) Aspects on environmental toxicology. Springer, Thessaloniki, pp 228–232

Astasheva NP, Kashparov VO, Semenyutin AM, Lazarev NM, Loschilov NA, Romaov LN (1997) Method of life-time determination of specific activity of cesium in muscle tissue of farm animals: Directory for radiological services of the Ministry of Agriculture of Ukraine. Nora-Print, Tanjung Karang, p 176. (In Ukrainian)

Baloga VI, Holosha VI, Evdin AT (eds) (2006) 20 Years after the Chernobyl accident. National report of Ukraine. Attica, Kiev, p 223. (In Russian)

Bogdanov G, Prister B, Lazarev N (1996) Radioecological aspects of radiosorbents application and their place in the system of countermeasures on the contaminated territory of Ukraine. The radioecological consequences of the Chernobyl accident. EUR 16544 EN, Luxembourg, pp 229–235

Bogdanov GO, Prister BS, Strelko VV, Sjaskiy SV, Pronevich VA (2002) Methodology, endoecological justification and criteria for the integrated assessment of the use of sorbents in the production of milk in radionuclide-contaminated areas. Biol Anim 4(1–2):169–187. (In Russian)

Bogdevich IM (1996) Fundamentals of agriculture. In: Bogdevich IM (ed) Ecological, mediko-biological and social-economic consequences of the Chernobyl catastrophe in Belarus. Institute of Radiobiology of the Academy of Sciences of Belarus, Minsk, pp 52–102. (In Russian)

Bogdevich I, Sanzharova N, Prister B, Tarasiuk S (2001) Countermeasures on natural and agricultural areas after Chernobyl accident. Role of GIS in lifting the cloud of Chernobyl. Academic, Minsk, pp 147–158. (In Russian)

Bogdevich IM, Putyatin YV, Malyshko AV (2002) Influence of acidity of Sod-podzolic soil and doses of potassium fertilizers on ^{137}Cs and ^{90}Sr transition to clover meadow. Soil Sci Agrochem 32:219

Bogdevitch I (2012) Fertilization as a remediation measure on soils contaminated with radionuclides ^{137}Cs and ^{90}Sr. In: Bogdevitch I, Mikhailouskaya N, Mikulich V (eds) Fertilizing crops to improve human health: a scientific review. Volume 3: risk reduction IPNI, Norcross, GA, USA. IFA, Paris, pp 275–290

Borzilov VA, Klepikova NV (1993) Effect of meteorological conditions and release composition on radionuclide deposition after the Chernobyl accident. In: Merwin SE, Balonov MI (eds) The Chernobyl papers. Research Enterprises, Richland, WA, pp 47–68

Cremers A, Elsen A, De Preter P, Maes A (1988) Quantitative analysis of radiocesium retention in soils. Nature 335(6):247–249

De Cort M (ed) (1998) Atlas of cesium deposition on Europe after the Chernobyl accident. EUR16733, Brussels, p 65

De Preter P (1990) Radiocesium retention in aquatic, terrestrial and urban environment: a quantitative and unifying analysis. Ph.D. thesis, K. V. Leuven, p 93

Deville-Cavelin G, Biesold H, Bogdevitch I, Sanzharova N, Prister B (2002) FGI project № 2. "Radioecological consequences of the Chernobyl accident" Sub-project N 5 "Countermeasures on natural and agricultural areas". Final report. Minsk

Deville-Cavelin G, Alexakhin RM, Bogdevitch IM, Prister BS (2001) Countermeasures in agriculture: assessment of efficiency. Proceeding of the international conference "fifteen years after the Chernobyl accident. Lessons learned", 18–20 Apr, 2001, Kiev, pp 118–128

ECP-2 (1996) Experimental collaboration project № 2. The transfer of Radionuclides trough the terrestrial environment to agricultural products, including the evaluation of agrochemical practices. In: Rauret G, Firsakova S (eds) Final report. EUR 16528 en, Brussels, p 182

ECP-5 (1996) Experimental collaboration project № 5. Behavior of radionuclides in natural and semi-natural environments. In: Belli M, Tikchomirov F (eds) Final report. EUR 16531 en, Brussels, p 147

ECP-9 (1996) Experimental collaboration project № 9. In: Howard B (ed) Final report. EUR 16539 en, Brussels, p 249

Fedorov EA, Prister BS, Romanov GN, Burov NI, Fedorova MN (1973a) Biological effect of young fission products on dairy cattle and their transition to livestock production. In: Annenkov BN, Dibobes IK, Aleksakhin RM (eds) Radiobiology and radioecology of agricultural animals. Atomizdat, Moscow, pp 70–140. (In Russian)

Fedorov EA, Romanov GN, Prister BS, Arkchipov NP, Alexakhin RM, Tikchomirov FA, Dibobes IK, Povaljaev AP (1973b) Recommandation on the management of agriculture and forestry in the radioactive contamination of the environment. Springer, New York, NY, p 101. (In Russian)

Fedorova TA (1968) Assimilation of plants by strontium and calcium depending on soil properties. Agrochemistry 6:108–114. (In Russian)

Garger EK (2008) Secondary rise of radioactive aerosol in the ground layer of the atmosphere. Institute of Nuclear Safety Problems of the NAS of Ukraine, Chernobyl, p 192. (In Russian)

Guntay S, Powers DA, Devill L (1996) The Chernobyl reactor accident source term: development of a consensus view. One decade after Chernobyl summing up the consequences of the accident. Vol. 2. IEAE – TECDOC-964. p 183

Hove K, Strand P, Salbu B, Oughton D, Astasheva N, Vasilev A, Ratnikov A, Averin V, Firsakova S, Crick M, Richards JI (1995) Use of cesium binders to reduce radiocesium of milk and meat in Belarus, Russia and Ukraine. Environmental input of radioactive releases (Proceedngs of the international symposium, Vienna, 1995). IAEA, Vienna

Howard BI (1993) Management methods of reducing radionuclide contamination of animal food production in semi-natural ecosystems. Sci Total Environ 137:249–260

IAEA (1994) Guidance on the use of countermeasures in agriculture in the event of the release of radionuclides into the environment. IAEA-TECDOC-745. IAEA, Vienna, p 104

IAEA (2001) Present and future environmental impact of the Chernobyl accident TECDOC-1240. IAEA, Vienna, p 128
IAEA (2006) Environmental consequences of the Chernobyl accident and their remediation: twenty years of experience. Report of the UN Chernobyl Forum Expert Group "Environment" (EGE). UN Chernobyl Forum Expert Group "Environment" (EGE), Vienna, p 166
Il'in LA, Arkhangelskaya GV, Konstantinov YA, Likhtarev IA (1972) Radioactive iodine in the problem of radiation safety. Atomizdat, LA Ilyin, Moscow. (In Russian)
Il'yazov RG, Sirotkin AN, Kruglikov BP, Levakchin VI, Pjatnov UN, Michalusev VI, Yunusova RM, Cigvincev PN, Gvozdik AF, Gulakov AV (2002) Ecological and radiobiological consequences of the Chernobyl catastrophe for livestock and ways to overcome them Ed Il'yazov Kazan, Phaethon, p 330. (In Russian)
Il'yazov RG (2006) Veterinary and radiological aspects of the Chernobyl disaster and the consequences of radioactive contamination in animal husbandry (dedicated to the 20th anniversary of the Chernobyl accident). Agric Biol 2:3–8. (In Russian)
Karaoglu A, Desmet G, Kelly GN, Menzel HG (eds) (1996) The radiological consequences of the Chernobyl accidents. European Commission and the Belarus, Russian and Ukrainian Ministries on Chernobyl Affaire, Emergency Situations and Health. EUR 16544en ESSC – EC – EAEC, Brussels, p 1194
Kashparov VA, Ahamdach N, Zvarich SI, Yoschenko VI, Maloshtan IM, Dewiere L (2004) Kinetics of dissolution of Chernobyl fuel particles in soil in natural conditions. J Environ Radioact 72:335–353
Konoplev AV, Viktorova NV, Virchenko EP, Popov VE, Bulgakov AA, Desmet GM (1993) Influence of agricultural countermeasures on the ratio of different chemical forms of radionuclides in soil and soil solution. Sci Total Environ 137(1993):147–162
Konoplev AV (1998) Mobility and bioavailability of radiocesium and radiostrontium of accidental origin in the "soil-water" environment. Thesis Dr. biological sciences. Obninsk, p 250. (In Russian)
Konoplev AV, Bulgakov AA (1999) Kinetics of ^{90}Sr leaching from fuel particles in soils of the Chernobyl exclusion zone. At Energy 86(2):129–134. (In Russian)
Konoplev AV, Konopleva IV (1999) Parametrization of ^{137}Cs transition from soil to plants based on key soil characteristics. Radiat Biol Radioecol 39(4):455–461. (In Russian)
Konoplev A (2020) Mobility and bioavailability of the Chernobyl-derived radionuclides in soil-water environment: review. In: Konoplev A, Kato K, Kalmykov SN (eds) Behavior of radionuclides in the environment II: Chernobyl. Springer Nature, Singapore, pp 157–193
Korneev NA, Sirotkin AN (1970) The excretion of ^{131}I with milk in cows. Distribution, kinetics of metabolism and the radiobiological effect of iodine isotopes. Medicine 1970:16–19. (In Russian)
Korneev NA, Sirotkin AN, Korneeva NV (1977) Decrease radioactivity in plants and livestock products. Kolos, Moscow, p 208. (In Russian)
Kovgan LN, Likhtarev IA, Perevoznikov ON (2009) Experience of general dosimetric passportization of territories exposed to intensive Chernobyl fallout in Ukraine. Radiat Hyg 2 (3):32–37. (In Russian)
Likhtarev I.A. (1996) Radiation-dosimetric passportization of settlements of the territory of Ukraine, exposed to radioactive contamination as a result of the accident at the ChNPP, including thyroid-dosimetric passportization. Instructive and methodological instructions. Kiev. (In Russian)
Los IP (1993) Hygienic assessment of dose-forming sources of ionizing radiation of natural and man-made origin and doses of irradiation of the population of Ukraine: Author's abstract. dis... biological sciences. Kiev, p 41 (In Russian)
Loschilov NA, Prister BS, Perepelyatnikova LV (1994) Management agricultural production in areas contaminated with radioactive elements. Recommendations. IAEA, Kiev, p 182. (In Russian)

Marey AN, Barkhudarov PM, Novikova HY (1974) Global precipitation of cesium-137 and man. Atomizdat, Moscow, p 168. (In Russian)

Marey AN, Zykova AS, Saurov MM (1984) Radiation communal hygiene. Energoatomizdat, Moscow, p 177. (In Russian)

Müller H, Pröhl G (1993) ECOSYS-87: a dynamic model for assessing radiological consequences of nuclear accidents. Health Phys 64(3):232–252

Nisbet AF, Konoplev AV, Shaw G, Lembrechts JF, Merkx R, Smolders E, Vandecasteele CM, Lonsjo H, Carini F, Burton O (1993) Application of fertilisers and ameliorants to reduce soil to plant transfer of radiocesium and radiostrontium in the medium to long term – a summary. Sci Total Environ 137:173–182

Perepelyatnikov GP (2012a) Fundamentals of general radioecology: monograph, 2nd edn. Atika, Kiev, p 440. (In Russian)

Perepelyatnikov GP (2012b) Radioecological substantiation of rational crop management in case of pollution of territories by radioactive emissions after nuclear and radiation incidents. Thesis doctor of agricultural sciences, Kiev, p 32. (In Russian)

Prister BS, Burov NI, Buldakov LA, Panchenko IYA, Sirotkin AN (1978) Recommendations for early prediction of radiation damage, sorting and conservation of cattle when animals receive a mixture of young products of nuclear fission. Radioecology of Vertebrate Animals. Sat Art Moscow Sci:138–148. (In Russian)

Prister BS, Loschilov NA, Nemets OF, Poyarkov VA (1991) Fundamentals of agricultural radiology, 2nd edn. Harvest, Kiev, p 472. (In Russian)

Prister BS, Perepelyatnikov GP, Il'in MI (1993a) Actual problems of forage production in conditions of radioactive contamination of the territory. Report National Academy of Sciences of Ukraine 1: 153–163. (In Russian)

Prister BS, Perepelyatnikov GP, Perepelyatnikova LV (1993b) Countermeasures used in the Ukraine to produce forage and animal food products with radionuclide levels below intervention limits after the Chernobyl accident. Sci Total Environ 137(1–3):183–198

Prister BS (ed) (1996) Agricultural aspects of the Chernobyl disaster. Sb Sci Tr 4:3–9. (In Russian)

Prister BS, Belli M, Sanzharova NI, Perepelyatnikov GP (1996) Behavior of radionuclides in meadows including countermeasures application. Proceedings of the first international conference "The radiological consequences of the Chernobyl accident". EC, Brussels, pp 59–69

Prister BS (1998) Agricultural management in conditions of radioactive contamination of the territory of Ukraine as a result of the Chernobyl accident for the period 1999–2002. In: Prister BS, Kashparov VA, Perepelyatnikova LV, Bogdanov GA, Kashparov VA, Lazarev NM, Mozhar PP Methodological recommendations. Kiev: Atika-N 137. (In Russian)

Prister BS (1999) Consequences of the Chernobyl disaster for agriculture in Ukraine. Research CEPR, no. 20, Kiev, p 104. (In Russian)

Prister B., Howard B., Vinogradskaya V. (2001). Regularities of Chernobyl ^{137}Cs and ^{90}Sr behavior in a soil-plant system. International congress on the radioecology-ecotoxicology of continental and estuarine environments. Aix-en-Provence-France. 3–7 Sep. Ref. P2T11 (575)

Prister BS (2002) Radioecological Consequences of the accident. Project № 2. Sub-project 3a: "SOIL – Plant Transfer. Franco-German Initiative for Chernobyl, Kiev

Prister BS, Baryakhtar VG, Perepelyatnikova LV, Rudenko VA, Ivanova TA (2003a) Experimental substantiation and parameterization of the model describing ^{137}Cs and ^{90}Sr behavior in a soil-plant system. Environ Sci Poll Res 1:126–136

Prister BS, Biesold G, Deville-Cavelin G (2003b) A method for the complex assessment of soil properties for predicting the accumulation of radionuclides by plants. Radiat Biol Radioecol 43 (6):39–42. (In Russian)

Prister BS (2006) Problems of predicting the behavior of radionuclides in the soil – plant system. In: Il'ljazov RG (ed) Agrosphere adaptation to technogenesis conditions. Academy of Science of Tatrstan Republik, Fan, Kazan, pp 85–124

Prister BS (ed) (2007) Farm management agricultural production in the territories contaminated as a result of the Chernobyl disaster in the remote period. Methodological recommendations. Kiev, Atika-N, p 196
Prister BS, Aleksakhin RM, Bebeshko VG, Bogdevich IM, Likchtarev IA, Ivanov VK (2007) The Chernobyl disaster: the effectiveness of population protection measures, the experience of international cooperation. In: Prister BS (ed) Power and electrification. Atika, Kiev, p 103. (In Russian)
Prister BS (2008) Problems of agricultural radiobiology and radioecology in the environment contamination with a young mixture of nuclear fission products: monograph, vol 320. Institute of Problem for Nuclear Safety of the NAS of Ukraine Chernobyl, Pripyat. (In Russian)
Prister BS, Aleksakhin R.M. (2008) Radiation protection of the population - lessons from the Kyshtym and Chernobyl accidents. Ed. Acad RM Aleksakhin (ed) XXXVI radioecological readings devoted to the actual member of the Vaskhnil VM Klechkovsky (28 Nov 2007, VNIISKhRAE, Obninsk). pp 47–74. (In Russian)
Prister BS, Vinogradskaya VD (2009) A model for predicting the dose of internal exposure of the population in the soil pathway for the inclusion of long-lived radionuclides in food chains. Probl Saf Atom Power Chornobyl 11:128–135. (In Russian)
Prister BS (2011) Radioecological consequences. Dynamics of radioactive contamination of terrestrial ecosystems and the effectiveness of protective measures. The National Report of Ukraine "Twenty-five years of the Chernobyl disaster. Security of the future". KIM, Kiev, pp 37–57. (In Russian)
Prister BS, Vinogradskaya VD (2011) Kinetic model of ^{137}Cs behavior in the "soil–plant" system, taking into account the agrochemical properties of the soil. Probl Saf Nucl Power Chornobyl 16:151–161. (In Russian)
Prister BS, Kluchnikov AA, Barjakchtar VG, Shestopalov VM, Kuchar VP (2016) The safety problems of the nuclear power. In: Prister BS (ed) The lessons of chernobyl. Supplemented, 2nd edn. Monograf, Chernobyl, p 356
Prister BS, Vinogradskaya VD, Lev TD, Talerko MM, Garger EK, Onisha Y, Tischenko OG (2018) Preventive radioecological assessment of territory for optimization of monitoring and countermeasures after radiation accidents. J Environ Radioact 184–185:140–151
Proceedings of the International Conference "Fifteen years of the Chernobyl disaster. Lessons learned" (2001). Kiev: Chernobylinform. Kiev, Ukraine, 18–20 Apr, 2001. p 144. (In Russian)
Pronevich VA (2013) Transformation of organic matter of drained peat soils under the influence of structural melioration. Agroecol J 1:50–54. (In Ukrainian)
Pronevich VA (2014) Scientific bases of rehabilitation of drained peatlands and radioecological safety in agroecosystems of Polissya. Ref. Dissertation for obtaining sciences degree doc of agricultural sciences, Kiev. Insttitute of Agroecology and Nature Management NAANU, p 39 (In Ukrainian)
Russell R (1962) Radioactivity and human food. Atomizdat, Moscow, p 375. (In Russian)
Russell RS, Possingham IV (1966) Physical characteristics of fallout and its retention on herbage. The entry of fission products into food chains. Pergamon Press, Oxford, p 2
Salbu B, Krekling T, Lind DH (2001) High energy X-ray microscopy for characterization of fuel particles. Nucl Instr Meth Phys Res 467-468:1249–1252
Sanzharova NI, Fesenko SV, Kotik VA, Spiridonov SI (1996) Behavior of radionuclides in meadows and efficiency of countermeasures. Radiat Prot Dosim 64(1/2):43–48
Shevchuk VE, Gurachevsky VL (eds) (2001) 15 Years after the Chernobyl disaster: consequences in the Republic and their overcoming. National report. Committee on the Problems of the Consequences of the Chernobyl Disaster, Minsk, p 201. (In Russian)
Shevchuk VE, Gurachevsky VL (eds) (2003) Consequences of Chernobyl in Belarus: 17 years later. National report. Propylaea, Minsk, p 747. (In Russian)

Sirotkin AN, Burov NI, Fedorov EA, Koldaeva KA (1978) The receipt and exchange of radioisotopes in agricultural animals. Radioecology of vertebrates. Nauka, Moscow, pp 103–123. (In Russian)

Sirotkin AN (1992) Metabolism of radionuclides in the organism of agricultural animals. Agricultural radioecology. In: Aleksakhin RM, Korneev, NA (eds) Ecology. pp 92–106. (In Russian)

Squre HM, Middleton LI, Russel RS, Taylor R (1961) In: Loutit IF, Russel RS (eds) The entry of fission products into food chain. Pergamon Press, Oxford, p 3

Zubets MV, Prister BS, Aleksakhin RM, Bogdevich IM, Kashparov VA (2011) Actual problems and tasks of scientific support of production of agricultural products. Agroecol J 1:5–23. (In Russian)

Chapter 6
Behavior of the Chernobyl-Derived Radionuclides in Forest Ecosystems and Effects of Radiation

Vasyl Yoschenko, Valery Kashparov, and Tatsuhiro Ohkubo

Abstract This chapter describes the consequences of the Chernobyl accident in forest ecosystems. We focus mainly on forests in the near zone of the accident at the territories of Ukraine and Belarus. Within this area, especially in the Ukrainian part of the 30-km Chernobyl zone, forests were exposed to high doses during the acute phase of the accident resulting in the slight to severe (up to lethal) damages to the ecosystems. Three decades after the accident, Chernobyl forests in the near 10-km zone remain heavily contaminated with long-lived radionuclides such as ^{137}Cs, ^{90}Sr, and isotopes of transuranium elements, which excludes the possibility of their economical utilization in the long term. At the most contaminated places, dose rates to the tree species may reach the level of mGy h^{-1}, that is, exceeding the safe levels for terrestrial ecosystems. We review the data on the radioactive contamination levels in Chernobyl forests, dynamics of radionuclides in the typical tree species at the early and late stages after the deposition, and the effects of acute and chronic radiation to plant species and analyze the contribution of forests to the doses to humans. Many important issues, such as radioactive contamination and effects on animals and others, are out of the scope of this chapter. Although this volume of the book is focused on the consequences of the Chernobyl accident, in some parts of this chapter, we find it necessary to refer to the situation after the Fukushima accident in order to emphasize the generic processes and their mechanisms in the radioactively contaminated forests. Materials of this chapter were partly presented in our recent reviews (Yoschenko et al., J Forest Res 23:3–14, 2018b; Yoschenko et al., Radiocesium dynamics in a Japanese forest ecosystem. Initial stage of contamination

---

V. Yoschenko (✉)
Institute of Environmental Radioactivity, Fukushima University, Fukushima, Japan
e-mail: r705@ipc.fukushima-u.ac.jp

V. Kashparov
Ukrainian Institute of Agricultural Radiology, National University of Life and Environmental Sciences of Ukraine, Kyiv, Ukraine

T. Ohkubo
School of Agriculture, Utsunomia University, Utsunomia, Japan

A. Konoplev et al. (eds.), *Behavior of Radionuclides in the Environment II*,
https://doi.org/10.1007/978-981-15-3568-0_6

after the incident at Fukushima Daiichi Nuclear Power Plant, 2019); here, we present the extended version of the review.

Keywords Chernobyl · Forest ecosystems · Radionuclide dynamics · Radiation effects · Forest fires · Doses on population

6.1 Introduction

According to the International Nuclear Event Scale (INES), the accident at the Chernobyl Nuclear Power Plant (ChNPP) in Ukraine on April 26, 1986 was classified as the largest nuclear accident (INES Level 7) in the history of humankind, causing "widespread health and environmental effects" (IAEA 2001). The accident resulted in large-scale radioactive contamination of the terrestrial and aquatic environments. The presence of long-lived radionuclides in the Chernobyl release made/brought the effects not only widespread but also long term: the Pu isotope deposition exceeds 4 kBq m^{-2} over an area of about 500 km^2 (Kashparov et al. 2003). Residing and economic activities were prohibited in this area (Verkhovna Rada of Ukraine 1991) and will remain so far into the future due to the long half-lives of ^{239}Pu (24,000 years) and ^{240}Pu (6500 years).

After the accident, evacuation efficiently protected people from high doses of radiation or reduced the doses received. The contaminated environments remain exposed to radiation and may be sources of spreading radioactivity. This is of particular concern with regard to forests, which currently cover about 60% of the territory in the Chernobyl exclusion zone (Ministry of Emergencies of Ukraine 2006). Without being decontaminated, Chernobyl forests will remain the long-term environmental repositories of radionuclides. It is necessary to understand the effects of radiation on forests and the processes by which radionuclides are redistributed in forests and dispersed to inhabited areas in order to evaluate the possibility of utilizing these forests and to develop related strategies.

The Chernobyl accident stimulated the development in study of forest radioecology. This resulted in the accumulation of large amounts of data on radionuclide transfers into forest compartments, which were compiled in the Chernobyl Forum Report (IAEA 2006) and the IAEA Handbook (IAEA 2010), and in the creation of physically based predictive models describing the behavior of radionuclides in forests (IAEA 2002).

In the present chapter, we review key findings concerning the consequences of the Chernobyl accident in forest ecosystems. We focus on the effects of acute and chronic radiation on tree species, the impacts of the accident on forestry and forest products, and the long-term radionuclide dynamics in forests. Though many important problems related to radionuclides in Chernobyl forests (e.g., radioactive contamination of animals and related radiation effects, contamination of game, etc.) are not addressed in this chapter.

6.2 Characterization of Forests in the Chernobyl Zone, Scale of Their Contamination, and Regulations in Forestry

The forested area in the Ukrainian part of the Chernobyl exclusion zone significantly changed during the last centuries. Due to extensive anthropogenic impacts from 19th to the beginning of twentieth centuries, it decreased from almost 100% of the total area in the past to 11–12% in 1913 (Ministry of Emergencies of Ukraine, National Academy of Sciences of Ukraine 1996; Nepyivoda 2005; Kuchma et al. 1997). In the period of the USSR, systematic reforestation began in the 1920s because of the sharp decrease of the land fertility (Nepyivoda 2005); the most extensive recovery of forests was performed in the 1950–1960s, when the forest rate reached about 50%. At the time of the accident, forests covered about 40% of the territory of the present exclusion zone, agricultural and meliorated lands occupied 28% and 14%, respectively, and grassland and marsh area amounted to 14% (Davydchuk 1994). The dominant tree species, Scots pine (*Pinus sylvestris* L.), was distributed at 80% of the afforested territory (Davydchuk 1994). This species was mainly used for reforestation of the territory in the soviet period. Silver birch (*Betula pendula*) and common oak (*Quercus robur*) occupied 8–10% and 5–6% of the forest area, respectively (Davydchuk 1994).

Currently, in three decades after the accident, the forest area in the exclusion zone reaches 1500 km^2. It increased from the above-cited value of approx. 40% in 1986 to about 58% of the total territory of the zone (Nikonchuk 2015) due to both establishing the new plantations and ecological succession of the forest stands onto adjacent abandoned agricultural areas. Species distribution is presented in Table 6.1.

The Chernobyl accident resulted in the release of huge amounts of radionuclides into the environment. In addition to radioisotopes of noble gases (Xe and Kr) and volatile elements (H, Te, I, Cs, and, to a much lesser extent, Sr and Ru), the release included radioisotopes of refractory elements, such as Zr, Eu, Ce, Pu, Cm, Am, and others. The refractory radionuclides and radiostrontium were deposited mainly in the near zone of the accident in the form of fuel particles (Kashparov et al. 1996). Being

Table 6.1 Current composition of forests in the Chernobyl exclusion zone (modified from Nikonchuk (2015))

Species	Area, ha	Wood volume, thousand m^3
Conifers (mainly Scots pine)	89,247	22,216
Deciduous hardwood	8148	1510
Birch	38,494	4226
Alder	9984	1761
Other	4141	451
Total	150,013	30,164
Ripe and overripe stands	8625.7	2202

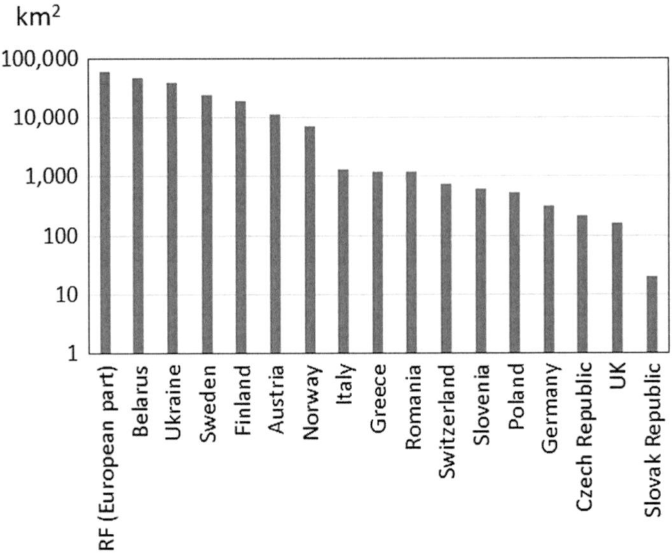

Fig. 6.1 Areas in Europa contaminated with ^{137}Cs above 40 kBq m^{-2} (Data by De Cort et al. 1998)

Table 6.2 The territory contamination with ^{137}Cs in Belarus, Russia, and Ukraine on 10.05.1986 (in thousands km^2)

Country	Contamination density, kBq m^{-2}					Total	
	40–100	100–185	185–555	555–1480	>1480		
Russia	44	7.2	5.9	2.2	0.46	59.8[a]	31.1[b]
Belarus	21	8.7	9.4	4.4	2.6	46.1	
Ukraine	29	4.3	3.6	0.73	0.56	38.2[c]	21.5[d]

Estimated in 1998 (data by Ministry of Ukraine of Emergencies 2011; De Cort et al. 1998; Izrael and Bogdevich 2009)
[a]65,100 km^2 according to the estimation provided in 2006
[b]On 2006
[c]42,800 km^2 according to Ministry of Emergencies of Ukraine (2011)
[d]On 2011

encapsulated in the fuel particle (FP) matrix, these radionuclides were not available for migration in soil and uptake into plants. With time, FPs gradually dissolved and the radionuclides leached into mobile and bioavailable forms (Kashparov et al. 1999, 2000a, 2004; Kashparov 2002).

The total magnitude of the Chernobyl release (excluding noble gases) was about 5300 PBq (Steinhauser et al. 2014). The released activities of the long-lived radionuclides were 85 PBq of ^{137}Cs (UNSCEAR 2008), 4 PBq of ^{90}Sr (Kashparov et al. 2003), 0.046 PBq of $^{238\text{-}240}$Pu (Kashparov et al. 2003; UNSCEAR 2008), 0.0024 PBq of ^{241}Am (Kashparov et al. 2003), and 2.6 PBq of ^{241}Pu (UNSCEAR 2008). As a result, the vast areas in Europa were heavily contaminated with the

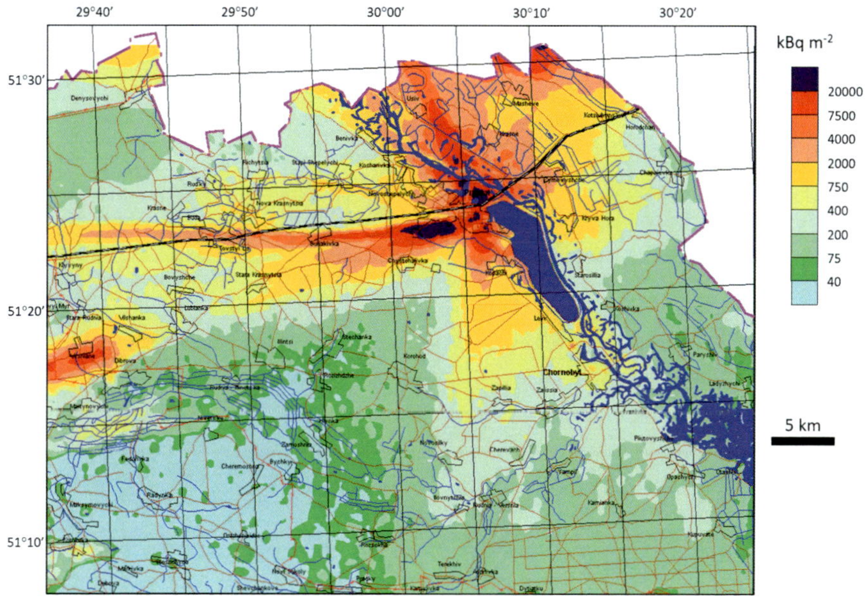

Fig. 6.2 Density of contamination with ^{137}Cs in the Chernobyl exclusion zone (This map was created using the dataset from Kashparov et al. (2018) (freely available at https://catalogue.ceh.ac.uk/documents/782ec845-2135-4698-8881-b38823e533bf))

Table 6.3 The forest territory contamination with ^{137}Cs in Belarus, Russia, and Ukraine (in thousands km^2)

	Contamination density, kBq m^{-2}				
Country	37–185	185–555	555–1480	>1480	Total
Belarus	12.9	3.2	1.7	0.14	17.9
Russia	9.3	1.1	0.36	0.026	10.8
Ukraine	10.9	1.06	0.31	0.095	12.3

Data by Izrael and Bogdevich (2009), Nadtochy et al. (2003)
[a]On 1.01.2006
[b]On 1.01.2006
[c]On 1.01.1993

long-lived ^{137}Cs (Fig. 6.1, Table 6.2) with the highest deposition levels found in the Ukrainian part of the Chernobyl exclusion zone (Fig. 6.2).

Territories of the radioactively contaminated forests in three most affected countries are presented in Table 6.3.

^{90}Sr and isotopes of transuranium elements (e.g., Pu) deposited mainly within the exclusion zone in Ukraine (Figs. 6.3 and 6.4).

The deposition maps (Figs. 6.2, 6.3, and 6.4) were created based on the large-scale sampling and measurement campaign carried out in the exclusion zone in 1997 (Kashparov et al. 2001, 2003, 2018). The measurement results can be used to derive

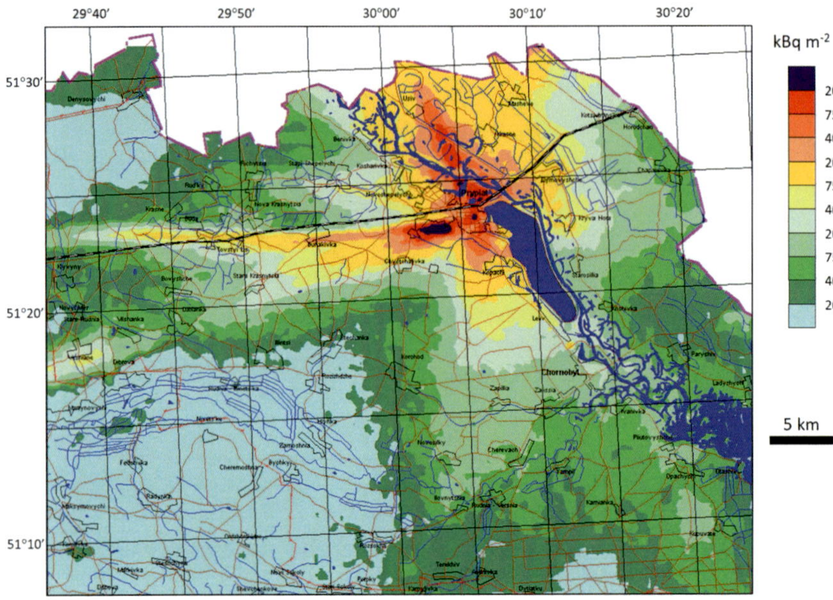

Fig. 6.3 Density of contamination with ^{90}Sr in the Chernobyl exclusion zone (This figure is modified from Kashparov et al. (2018))

the distributions of the forest area according to the levels of deposition of the measured radionuclides (Tables 6.4 and 6.5).

Zoning of the radioactively contaminated territories in Ukraine, Belarus, and Russia was established in 1991, before the collapse of the USSR, based on the calculated effective equivalent dose from the Chernobyl radionuclides. The State Laws of the three countries set the restrictions on the habitation and economic activity, including forestry, in the contamination zones. As an example, the Ukrainian regulations are presented in Table 6.6. The Chernobyl zones 1 and 2 are the exclusion zone and the zone of unconditional (mandatory) resettlement, respectively; according to the Ukrainian law (Verkhovna Rada of Ukraine 1991), commercial forestry at these territories is prohibited, any decontamination measures are not planned, and these forests will remain contaminated in the long term, carrying out a barrier function for the radioactivity spreading from the zones.

The State legislations also set the permissible levels (PL) of the radionuclide concentrations in wood for its utilization for various purposes (Tables 6.7, 6.8, and 6.9). The Belarusian legislation establishes PL only for ^{137}Cs concentration; besides ^{137}Cs, PL for the ^{90}Sr concentrations are established for firewood in Ukraine and for all categories of wood in Russia. However, there are no international standards of radionuclide concentrations in timber. A generic methodology that may be used as a basis for the establishment of such standards was developed by IAEA (2003). In the absence of the international or national standards, the limits for the Chernobyl-derived radionuclides in wood may be set by industry. IKEA (Sweden) performs

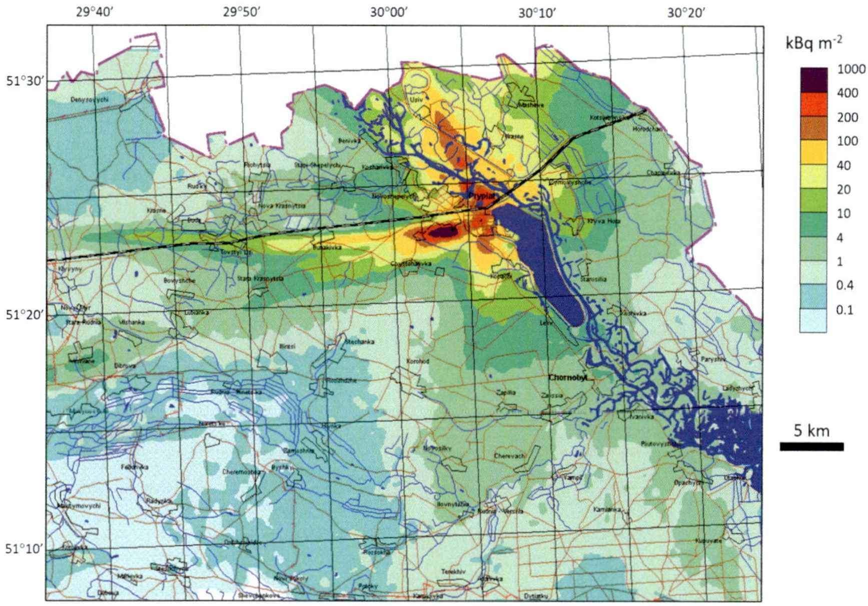

Fig. 6.4 Density of contamination with 239 + 240Pu in the Chernobyl exclusion zone (This map was created using the dataset from Kashparov et al. (2018) (freely available at https://catalogue.ceh.ac.uk/documents/782ec845-2135-4698-8881-b38823e533bf))

Table 6.4 Distribution of forest area in the Ukrainian part of the Chernobyl exclusion zone according to deposition of ^{90}Sr (as of 1997)

Deposition, kBq m^{-2}	Area, km^2	% of the total area
<20	147	14.5
20–40	251	24.8
40–75	234	23.1
75–200	168	16.6
200 400	70	6.9
400–750	54	5.3
750–2000	57	5.6
2000–4000	21	2.1
4000–7500	8	0.8
7500–20,000	2	0.2
> 20,000	0.005	0.0005
Total	1012	100

tests of wood supplied from the affected areas in Sweden, RF, Belarus, Ukraine, Eastern Europe, and Finland. Affected areas were defined as those that received deposition of ^{137}Cs above 37 kBq m^{-2}, and the limit of its concentration in wood was chosen as 300 Bq kg^{-1} (IKEA 2011).

Table 6.5 Distribution of forest area in the Ukrainian part of the Chernobyl exclusion zone according to deposition of $^{239+240}Pu$ (as of 1997)

Deposition, kBq m^{-2}	Area, km^2	% of the total area
0.1–0.4	162	16
0.4–1.0	365	36.1
1–4	282	27.9
4–10	88	8.7
10–20	46	4.5
20–40	40	3.95
40–100	23	2.3
100–200	5.1	0.5
200–400	0.68	0.067
Total	1012	100

Table 6.6 ^{137}Cs deposition levels, radioactive contamination zones, and restrictions on forestry in Ukraine (Verkhovna Rada 1991; Ministry of Emergencies of Ukraine, Intellectual systems GEO Ltd 2008)

^{137}Cs deposition, kBq m^{-2}	Radioactive contamination zone	Restrictions on forestry
>1480	1. Exclusion zone (zone from which population was evacuated in 1986)	Development of the special regime for forestry. Working time limited
555–1480	2. Zone of unconditional (mandatory) resettlement	
370–555	3. Zone of guaranteed voluntary resettlement	Use of wood for people's needs restricted
259–370		Use of wood as fuel and to manufacture domestic goods and food storage facilities prohibited
185–259		Use of wood as fuel and consumption of wild animal meat restricted. Hunting prohibited
74–185	4. Zone of enhanced radiological control	Consumption of wild berries and mushrooms prohibited. Use of medicinal plants and wild animals restricted
37–74		Use of mushrooms, wild berries, and some medicinal plants restricted
<37		Forest products may be used freely

[a]Or if ^{90}Sr deposition exceeds 111 kBq m^{-2}, or if sum of Pu isotopes deposition exceeds 3.7 kBq m^{-2}, and the calculated effective equivalent dose from the Chernobyl radionuclides exceeds 5 mSv $year^{-1}$

[b]Or if ^{90}Sr deposition ranges 5.55–111 kBq m^{-2}, or if sum of Pu isotopes deposition ranges 0.37–3.7 kBq m^{-2}, and the calculated effective equivalent dose from the Chernobyl radionuclides exceeds 1 mSv $year^{-1}$

[c]Or if ^{90}Sr deposition ranges 0.74–5.55 kBq m^{-2}, or if sum of Pu isotope deposition ranges 0.185–0.37 kBq m^{-2}, and if the calculated effective equivalent dose from the Chernobyl radionuclides exceeds 0.5 mSv $year^{-1}$

Table 6.7 Belarusian Hygienic Norms of the ^{137}Cs concentrations in wood and wood products (Ministry of Health 2000)

Wood products	^{137}Cs, kBq kg^{-1}
Round timber for construction of house walls	0.74
Other round timber	1.48
Raw materials	1.48
Firewood	0.74
Treated timber	
Timber for internal lining of walls of residential houses	0.74
Other treated timber	1.85
Other production of forestry (except food)	1.85

Table 6.8 Ukrainian Hygienic Norms of the radionuclide concentrations in wood and wood products (Ministry of Health 2005)

	kBq kg^{-1}	
Wood products	^{137}Cs	^{90}Sr
Raw wood products		
Wood with bark	1.5	
Barked wood	1.0	
Raw materials for plywood and veneer	1.0	
Wood for industrial constructions and for temporary buildings	1.5	
Materials for fixing in mining (stanchions)	3.0	
Wood for technological needs	1.5	
Treated timber		
Unedged timber, box board	1.0	
Edged timber, beam, parquet, materials for furniture	0.74	
Wood for social and living needs		
Firewood	0.60	0.06
Fence material	1.0	
Souvenirs, arms for tools, kitchen cutting boards, etc.	0.74	

6.3 Long-Term Dynamics of the Chernobyl-Derived Radionuclides in Forest Ecosystems

Based on the results of the long-term observations in the zones of the Kyshtym and Chernobyl accidents, Tikhomirov and Shcheglov (1994) summarized the general patterns of the radionuclide dynamics in forests and identified the dominant processes redistributing radionuclides at various stages following their deposition.

At the first stage, radionuclides from the accidental release are deposited to the tree canopies and to the soil surface. The intercepted by the tree canopies fraction depends on the forest type and age, plantation density, season of vegetation period, type, and physical–chemical forms of deposition. For the Chernobyl release, it ranged from 40% of the total deposition to the deciduous forests (Melin and

Table 6.9 Russian national permissible levels of ^{137}Cs and ^{90}Sr concentrations in wood products (Ministry of Health 1999)

Wood products	kBq kg^{-1}	
	^{137}Cs	^{90}Sr
Alive stand for industrial use		
Wood with bark	11.1	5.2
Barked wood	3.1	2.3
Alive stand for domestic use		
Instruments for outdoor use	3.1	2.3
Products for indoor and personal use (furniture, parquet, musical instruments, etc.)	2.2	0.5
Fuel	1.4	0.4
Civil construction	0.4	5.2
Secondary forest resources	2.2–3.1	0.5–2.3
Fresh needles for making green fodder	0.6	0.1

Wallberg 1991) to 70–90% in the conifer stands (Melin and Wallberg 1991; Bunzl et al. 1989; Tikhomirov and Shcheglov 1994). Interestingly, despite the difference between the tree species, a similar range of 70–100% for the initial interception fraction by the canopies of the conifer stands was reported after the Fukushima accident (Kato et al. 2012, 2017; Teramage et al. 2014).

The radionuclides intercepted by the aboveground forest biomass are transported from its compartments to the ground surface with precipitations (via throughfall and stemflow) and with litterfall (Shaw 2007). For various conditions, after the Chernobyl accident, the radionuclide half-loss period from the tree canopies varied from 3–4 weeks to 4–6 months (Tikhomirov and Shcheglov 1994; Sombre et al. 1990). Bunzl et al. (1989) reported the two-exponential dependence describing the removal of the Chernobyl-originated radiocesium from the aboveground biomass in the mature Norwegian spruce forest with the half-loss periods of 97 days (0.4 of the total deposition) and 195 days (0.26 of the total deposition). Obviously, the radionuclide removal rate depends on the species physiology: for the Fukushima forests, removal of the initially deposited radiocesium from the tree canopies is described by the two-exponential dependence with the fast half-loss period of 87 days (19%, 20%, and 29% of initial deposition to the ecosystems of mature Japanese cedar, young Japanese cedar, and broad-leaved species, respectively) and slow half-loss periods of 550 days and 780 days (40% and 50%) for mature and young cedar stands, respectively (Kato et al. 2017). The slower removal of radiocesium from the aboveground biomass in the artificial Fukushima forests as compared to Chernobyl is explained by the longer persistence of foliage in the main plantation species, Japanese cedar and Japanese cypress.

In Chernobyl, the next stage begins in 2–4 years after deposition, when its major fraction is removed from the tree canopies and is distributed between the forest litter and soil (Tikhomirov and Shcheglov 1994). The intensity of the root uptake of radionuclides into the aboveground plant biomass depends, among other factors, on the distribution of the roots in soil (Fesenko et al. 2001). At this stage, the radionuclides migrate to the root-inhabited soil layer, which results in the significant increase of the root uptake flux and, consequently, can lead to an increase of the radionuclide activities in the aboveground biomass after the period of their decrease. In 10–15 years, the quasi-equilibrium between uptake and removal should be reached (Tikhomirov and Shcheglov 1994). Perevolotsky (2006) found that the radiocesium transfer factors to the pine forest compartments in Belarus reached the maximum in 1992–1994; as compared to 1988, they increased several times. Similar regularities for ^{90}Sr might be explained by the fact that it was deposited in the FP form and therefore was not bioavailable in the first years; however, radiocesium at the "far" traces of release in Belarus was deposited mainly in the bioavailable forms. Mamikhin et al. (1997) and Shcheglov et al. (1996, 2001) observed increase of the radiocesium inventories in the aboveground biomass of the mixed forests at hydromorphic soils in 1990–1994; at automorphic soils, maximum of the radiocesium inventories in this period was observed not always. Shcheglov et al. (2001) summarized that the radiocesium transfer factors had the tendencies to increase and decrease in hydromorphic and automorphic soils, respectively.

Shcheglov et al. (2014) stated that in 25 years after the deposition of the Chernobyl-originated ^{137}Cs and ^{90}Sr, their "annual involvement in the biological cycle in forest ecosystems exceeds their total withdrawal (downward migration) from the root abundant soil profile."

The radioactive and stable isotopes of the same chemical elements are distributed in the ecosystem by the same processes. Therefore, in the long term, when the residues of the initial deposition are removed from the aboveground biomass and the quasi-equilibrium is reached between the radionuclide root uptake and removal fluxes, the distributions of radioactive and stable isotopes of the same chemical elements in the ecosystem should become similar and magnitudes of their soil to plant transfer factors should become equal. This was clearly shown for ^{137}Cs and ^{133}Cs (stable natural isotope) for the range of the forest species studied in 1998 at the radioactively contaminated territory of Belarus (Yoshida et al. 2004, 2011) and at other locations in Europe and in Japan (Yoshida et al. 2002). Moreover, at the quasi-equilibrium stage, the basic concepts used for modeling of the cycling of nutrients, namely, potassium and calcium, in forest ecosystems (Cole and Rapp 1981) may also be applied for radioisotopes of their chemical analogs, cesium and strontium, respectively. Based on that, the model of the radionuclide cycling was developed and successfully applied for quantitative assessments of the annual radionuclide fluxes and for prediction of the long-term dynamics in Chernobyl forests at the late stage after the accident (Myttenaere et al. 1993; Goor and Thiry 2004; Goor et al. 2007; Thiry et al. 2009; and others).

Observations of Ukrainian Institute of Agricultural Radiology (UIAR) show that currently in pine forests of the Chernobyl zone, up to 10% of total ^{137}Cs inventory may be found in the aboveground tree biomass, and up to 40% in litter (Figs. 6.5a and 6.6), while these amounts are much lower in the biomass and fast decomposing litter of deciduous forests (Figs. 6.5b and 6.6).

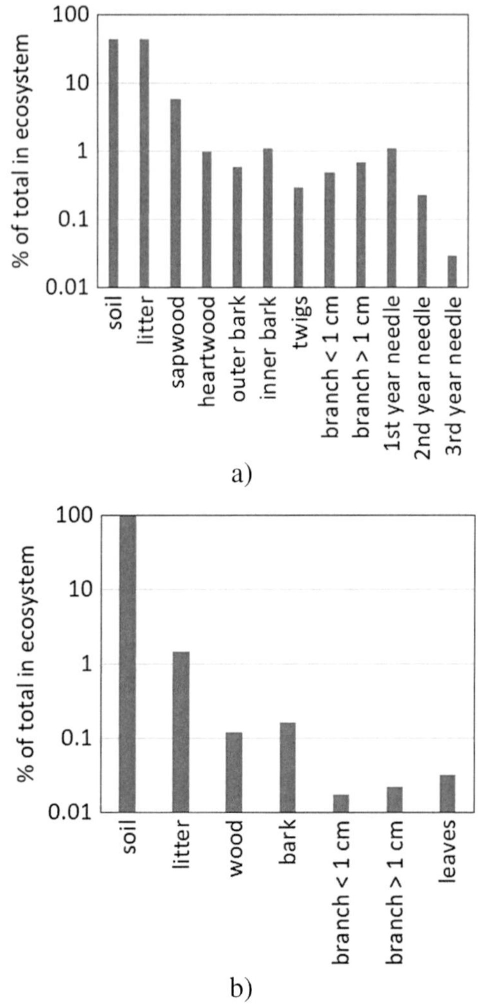

Fig. 6.5 Typical distributions of ^{137}Cs in the Chernobyl forest ecosystems in 30 years after the deposition. (**a**) Pine, (**b**) birch forest (Data by authors (UIAR))

The radiocesium soil-to-plant transfer factors (T_{ag}) in forests of the Chernobyl zone in the present time vary from 0.1 (Bq kg^{-1}) (kBq m^{-2})$^{-1}$ in dry fertile soils to 3 (Bq kg^{-1}) (kBq m^{-2})$^{-1}$ in wet low fertile soils (Krasnov et al. 2007).

Root uptake of ^{90}Sr into wood depended on its leaching from the fuel particles as well as on content of exchangeable calcium in soil (Fig. 6.7). In 30 years after the deposition, in forests growing at soils with low humus content, more than half of all ^{90}Sr inventory in the ecosystem can be localized in the aboveground tree biomass, while a large fraction, about 20%, has already migrated beneath the root-inhabited soil layer (data by UIAR). Thus, ^{90}Sr T_{ag} into wood increased by orders of magnitude during the postaccidental period, and the T_{ag} values obtained at a certain moment could not be used for long-term predictions. Based on the dependence of ^{90}Sr T_{ag} on the exchangeable Ca content in soil (Fig. 6.7), data on the radionuclide

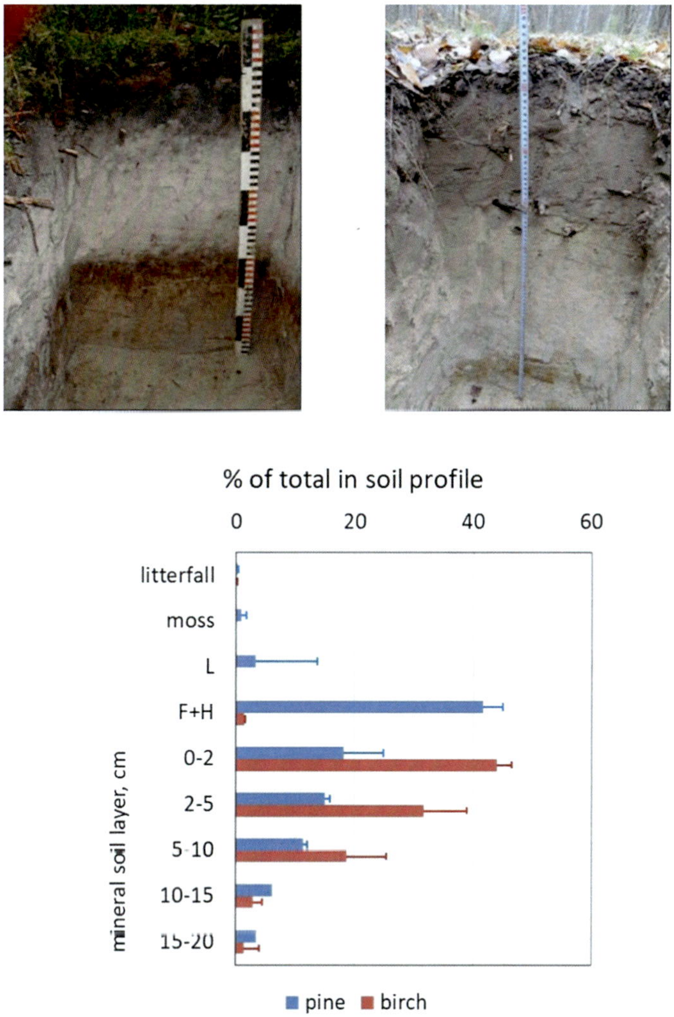

Fig. 6.6 Soil cross sections in the pine (top left) and birch (top right) forests, and typical vertical distributions of ^{137}Cs in soil profiles in the Chernobyl forest ecosystems in 30 years after the deposition (Data by authors (UIAR))

deposition (Fig. 6.3) and dynamics of the radionuclide leaching from FP into bioavailable forms depending on the FP oxidization level and soil acidity (Kashparov et al. 1999, 2000a, 2004), UIAR modeled the future dynamics of radiostrontium concentrations in wood of Scots pine in the Chernobyl zone (presented at http://uiar.org.ua/). The model predictions along with the experimental data on ^{137}Cs transfer into wood and information on the taxation characteristics of forests of the Chernobyl zone were used for analysis of compliance of wood to permissible levels (Table 6.8; Fig. 6.8).

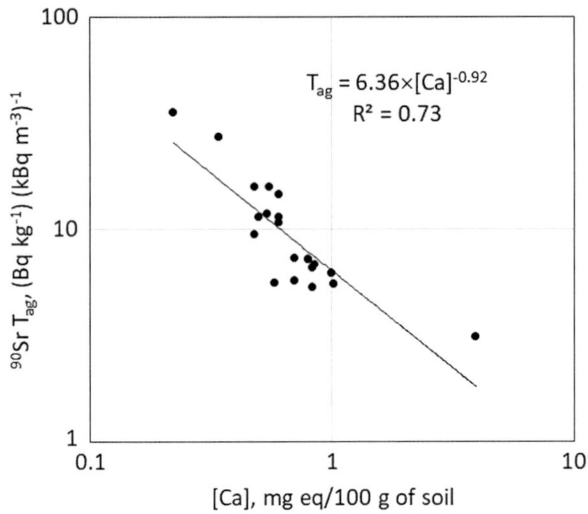

Fig. 6.7 Dependence of ^{90}Sr T_{ag} into pine wood on exchangeable calcium content in soil (Modified from Yoschenko et al. (2019))

According to the modeling results (Fig. 6.8), radiocesium concentrations in wood in the present time comply with the permissible levels at the large part of the Chernobyl zone, and in 2100, it can be harvested at almost whole territory of the zone. However, ^{90}Sr concentrations almost everywhere exceed the permissible level (60 Bq kg^{-1} in firewood), and such a situation will last for a long time. Taking into account the current concentrations of ^{90}Sr in wood and assuming its 100-fold concentration during incineration, all ash will be classified as radioactive waste (>10 kBq kg^{-1}).

The above-presented regularities of the radiocesium dynamics in the Chernobyl forests and modeling approaches are rather universal and might be applied for prediction of the further fate of radiocesium in the contaminated forests in Fukushima. However, adjustment to the local conditions and species composition is necessary. Our recent observations (Yoschenko et al. 2017, 2018a) revealed an increase of the radiocesium concentrations in some compartment of the Japanese cedar forest in 2016 following the period of their decrease that should be related to the increased root uptake due to migration of the radionuclide to the root-inhabited soil layer. Radioactive and stable cesium isotopes gradually approach the quasi-equilibrium state (Yoschenko et al. 2018a).

6.4 Effects of Radiation in the Chernobyl Forests

6.4.1 Effects of Acute Radiation and Early-Stage Impacts

The best-known example of acute radiation impacts on a forest ecosystem in the Chernobyl zone is the so-called Red Forest. This is an area measuring about 10 km^2

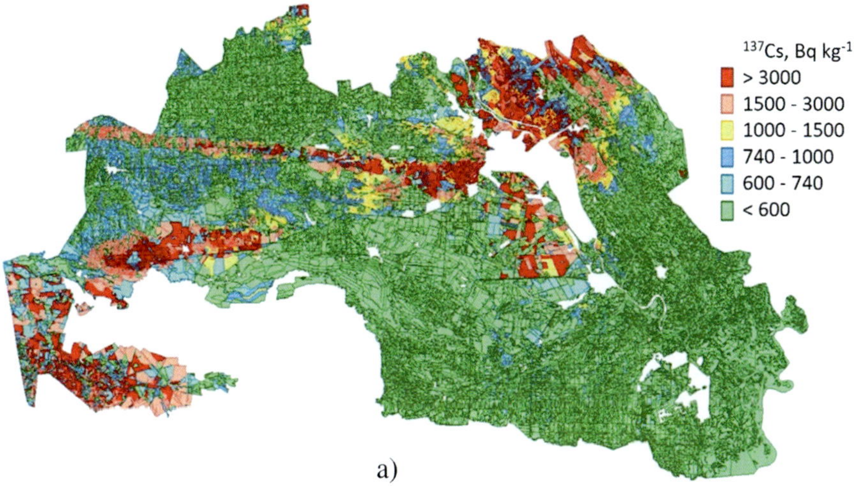

a)

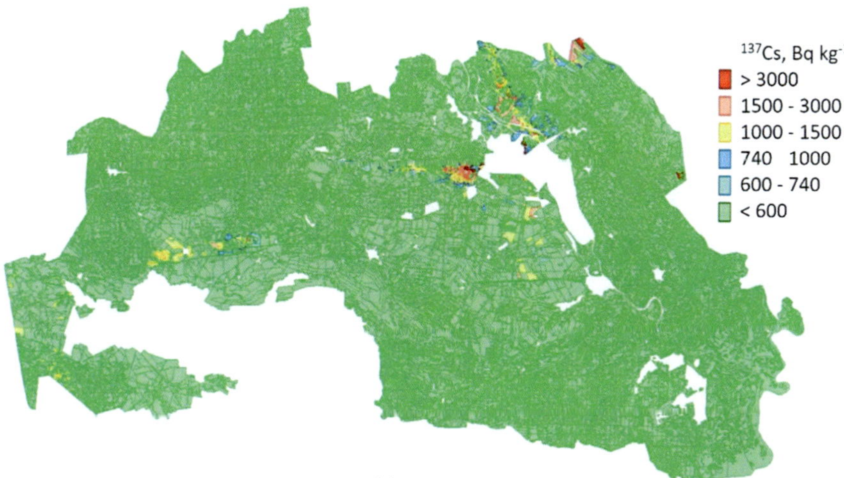

b)

Fig. 6.8 Predicted concentrations of ^{137}Cs (**a**, **b**) and ^{90}Sr (**c**, **d**) in pine wood in 2020 (**a**, **c**) and 2100 (**b**, **d**) (Data by authors (UIAR; presented at http://uiar.org.ua/))

located at a distance up to 5 km west of ChNPP unit #4 (Kashparov et al. 2012). After the first explosion at the reactor, a radioactive cloud containing very short-lived radionuclides passed over the mature Scots pine (*Pinus sylvesris* L.) forest growing in this area. Detailed information about dose rates in that period is absent. According to some estimates, the initial exposure dose rates in the radioactive cloud may have reached 8000–10,000 R h^{-1}, or 80–100 Sv h^{-1} (IAEA 1991). In addition to irradiation from the radioactive cloud, trees were irradiated by FPs intercepted by their crowns. These FPs contained high specific activities of β- and α-emitters, such as isotopes of Sr, Zr, Nb, Y, Pm, Pu, Am, and Cm. The α-particle ranges in

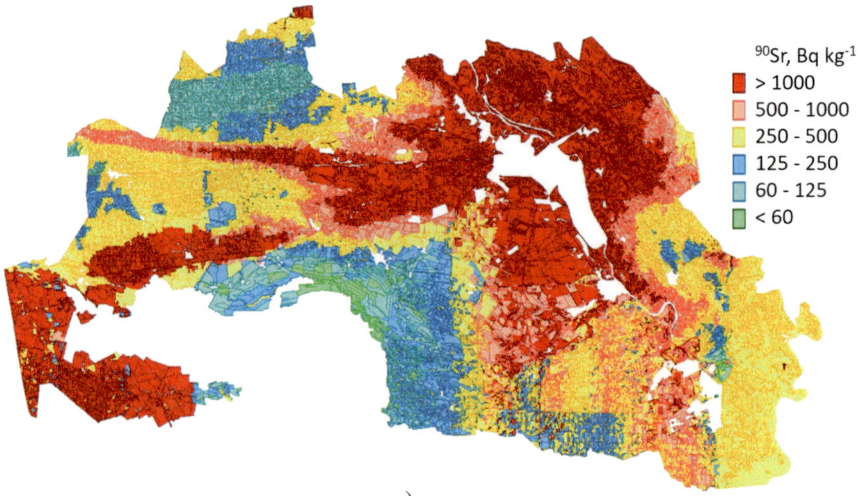

c)

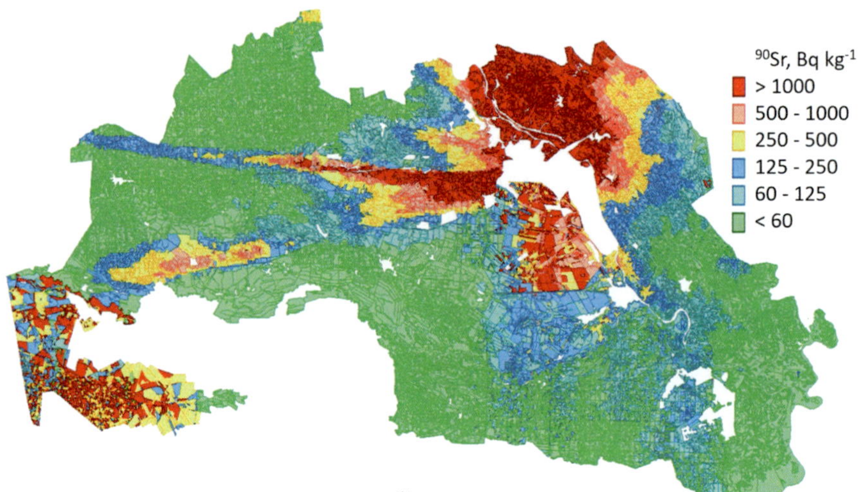

d)

Fig. 6.8 (continued)

biological tissues are limited to several tens of micrometers, but FP can create zones of necrotic cell damage because of the very high dose rates formed within the mentioned ranges; such damage was observed in pine leaves irradiated in 1986 (Kozubov and Taskaev 2002). The doses received by the pine forest from the radioactive cloud and the radionuclides deposited in the ecosystem in various forms (including FPs) were high enough to kill the forest. Soon after the accident, the color of the pine tree foliage changed to red, which indicated the death of the trees and gave the site its present name, the Red Forest. The pine trees died

Table 6.10 Summary of pine forest damage zones in Chernobyl (from Kozubov et al. 1991)

Zone	Area, ha, total/damaged forests	Absorbed dose rate[a], mGy h^{-1}	Estimate of the absorbed dose[a,b], Gy	Description of the early-stage radiation impacts
I—Lethal damage	4400/600	>5	80–100	Complete perishing of the pine populations till 1989
II—Sub-lethal damage	12,500/3800	2–5	30–40	Perishing of 25–75% of tress, necrosis of buds, and twigs in 1986, suppression of growth processes, numerous morphological abnormalities in 1987–1989
III—Moderate damage	43,400/11900	0.2–2	5–6	Suppression of the growth processes, loss of foliage, and numerous morphological abnormalities in 1986–1987
IV—Slight impact		<0.2	0.5–1	Some suppression of the growth processes, 2–2.5-fold increase of DNA aberration rate

[a]From γ-radiation only. On June 1, 1986

[b]According to Abaturov et al. (1991), in the lethal and sub-lethal damage zones, up to 90% of the dose could be received in a few hours after the accident

throughout an area of about 4.5 km^2. Later, the dead trees were cleared by bulldozers and placed, along with heavily contaminated forest litter and topsoil, into subsurface storage sites (trenches) for radioactive waste. The entire Red Forest was then covered with an approx. 30-cm-thick layer of clean sand. To prevent wind resuspension of sand, a new plantation was established, consisting of Scots pine, birch, oak, and some shrub species (Kashparov et al. 2012)

The Red Forest is the best known but not the only example of lethal damage to forest ecosystems in the Chernobyl zone. Similarly, damaged areas consisted of entirely pine trees were located on the left bank of the Prypiat river, 7–8 km north of the ChNPP, and in two small spots adjacent to the power plant on the north-west (Kozubov et al. 1991). The pine populations in other areas received serious or moderate damage including death of parts of the trees, suppression of development, and formation of the morphological abnormalities. Based on observations in 1986, Kozubov et al. (1987) classified the Scots pine forest ecosystems into four zones according to the level of damage (Table 6.10). Later, the borders of the zones were defined based on dose rates measured on June 1, 1986, and a map was created (Kozubov et al. 1991). This classification was also supported by other researchers (Abaturov 1990; Tikhomirov and Sidorov 1990; Kuchma et al. 1994) with some disagreements concerning the dose assessments.

Exposure dose rates and absorbed doses in Table 6.10 are presented only for γ-radiation. It is clear, however, that a part of the total doses to the trees could have been caused by β- and α-radiation from radionuclides deposited on the tree surfaces as FPs, or by β-radiation of the radionuclides deposited on the forest floor (Tikhomirov and Sidorov 1990; Kozubov and Taskaev 2002). In the Chernobyl

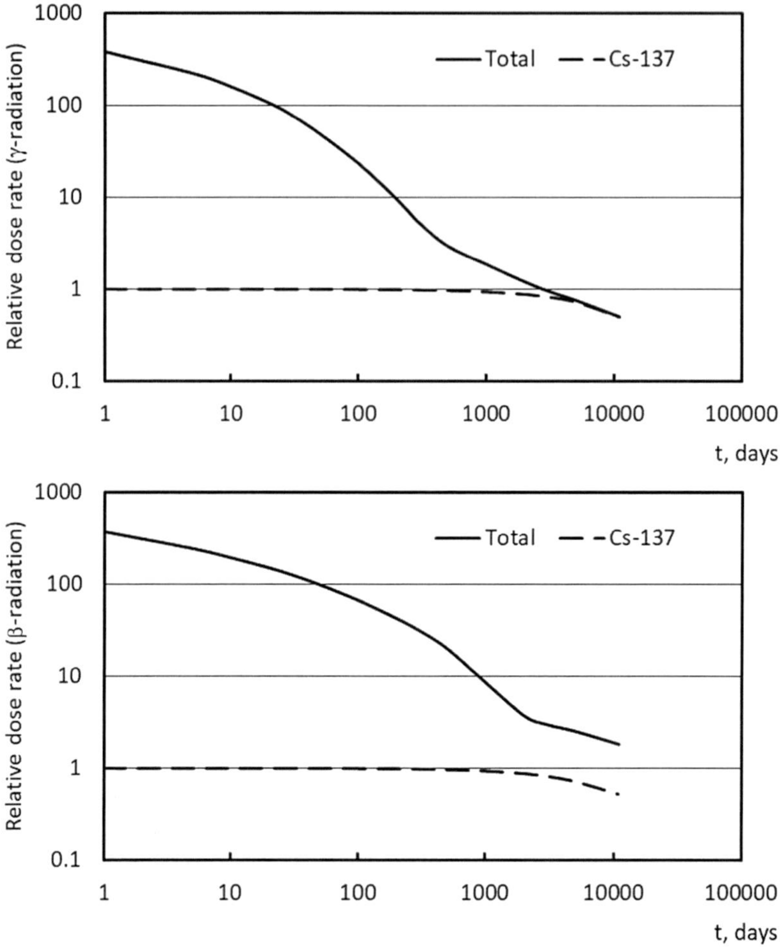

Fig. 6.9 The modeled dynamics of the dose rates of γ-radiation and β-radiation in the vicinities to the ChNPP. Solid lines show the total dose rates formed by all radionuclides released from the ChNPP normalized to the dose rate formed by ^{137}Cs in the moment of release (dash line). Calculated for $^{137}Cs \rightarrow {}^{137m}Ba$ decay with the average energies of γ- and β-radiation per decay 564 and 249 keV, respectively. *t* denotes the time after the release (Modified from Yoschenko et al. (2018b))

Forum Report (IAEA 2006), the doses to the pine needles in the zone of lethal damage were estimated as 80–100 Gy; 80% of the doses were received during the first month after deposition. β-radiation from radionuclides on the plant surfaces contributed up to 90% to the total doses.

According to our own observations in the Chernobyl zone (Yoschenko et al. 2018b), in 2016, there were almost no Scots pine trees older than 30 years in the areas where current dose rates exceeded 20 μGy h^{-1} (corresponds to current ^{137}Cs deposition levels above 20 MBq m^{-2}). Figure 6.9 indicates that the total γ- and

Fig. 6.10 Survivor Scots pine tree with a strongly deformed crown in Red Forest (Chernobyl exclusion zone, Ukraine) (Photo taken by authors in the spring of 2016)

β-radiation dose rates in the initial period after the accident in this area may have exceeded 10 mGy h^{-1}. We found only about ten such trees within an approximately 5 km^2 area; all these trees had well-expressed deformations of their crowns (Fig. 6.10).

Although the accuracy of the dose estimates in Table 6.10 may be debatable, this in no way diminishes the overall value of the cited studies: they clearly demonstrated large-scale impacts of the Chernobyl accident on forest ecosystems and identified Scots pine and European spruce (*Picea abies*) as the most radiosensitive tree species in that region. Deciduous tree species (birch, alder, oak, etc.) were much less impacted, even in the zone where lethal damage to pine trees was observed. They mainly exhibited partial death of the tree crowns and certain morphological abnormalities (Kozubov and Taskaev 1990). Damage to birch and black locust leaves was observed on plots where the γ-radiation dose rate exceeded 4 mGy h^{-1} (Smirnov and Suvorova 1996).

After the Chernobyl accident, direct losses in the forest industry included the death of 25–50% of about ten million trees in the zones of damage I–III during the first 2 years (Table 6.10). Kozubov and Taskaev (2002) estimated that the total loss of wood in zones I and II during the first 4 years was 250,000 m^3. According to Kozubov and Taskaev (2002), the total area of the radioactively contaminated forests is about 32,000 km^2, which exceeds the total area (29,400 km^2) of the three most contaminated zones where ^{137}Cs deposition density was above 185 kBq m^{-2} (UNSCEAR 2000). Therefore, the authors may have included forests in the zone of enhanced radiological control in their estimate (Table 6.6); however, forestry in this zone was not banned and restrictions were established only for the utilization of mushrooms, berries, game, and medicinal plants.

6.4.2 Effects of Chronic Radiation

After the acute radiation stages in April–May 1986 in Chernobyl and March 2011 in Fukushima, forest ecosystems remain exposed to chronic radiation from the radionuclides deposited in the ecosystems. The dose rates of external radiation to plant species decrease due to radioactive decay, washing of deposited radionuclides or FPs from plant surfaces, their movement deeper into the soil profile, and their removal from ecosystems due to lateral runoff. In contrast, internal radiation dose rates may increase with accumulation of radionuclides in plant tissues due to root uptake.

The effects of chronic radiation on forest species in the Chernobyl zone were widely reported. Very important observations were carried out by Mitrochenko et al. (1999) with regard to this issue in the Red Forest. A new plantation of Scots pine was established in this area approx. 2 years after the accident; the pine seedlings were germinated in a clean area from seeds collected from nonirradiated parent trees and then planted in the Red Forest. Within several years, the frequency of morphological abnormalities in this population sharply increased. The morphological changes were similar to those reported by Kozubov and Taskaev (2002) for pine trees in the zones of sublethal and moderate damage in 1987 and included cancellation of apical dominance, loss of needles, dense foliage, andso on. Morphological abnormalities and genetic defects in plant species in the Chernobyl zone were reported by other researchers (e.g., Kal'chenko et al. 1993a, b; Shevchenko and Grinikh 1995; Geras'kin et al. 2003a, b, 2008, 2013; Fesenko et al. 2005a; Geras'kin and Volkova 2014). Most of these abnormalities disappeared in a few years after the accident, but the cancellation of apical dominance (replacement of the single trunk with several branches that leads to deformation of the tree's crown and to growth suppression) remained.

Fig. 6.11 Typical morphological abnormalities (cancellation of apical dominance) in Scots pine in the Chernobyl zone (Photo by authors, Red Forest, 51.385082N, 30.066914E, June 2005)

In 2005, large-scale observations of the morphological effects of chronic radiation on Scots pine were carried out on several populations that were planted after the accident in the Chernobyl zone and on a control population outside of the zone (Yoschenko et al. 2011; Kashparov et al. 2012). The observed morphological abnormalities in the studied populations were exclusively presented by cancellation of apical dominance (Fig. 6.11). The abnormalities' frequency correlated very well with the dose rate increasing from 0.06 in the control population to 0.8 and 1 at the dose rates of 150 μGy h^{-1} and 800 μGy h^{-1}, respectively. The dose rate dependencies for DNA damage were formulated as well. Independent of dose rate, the abnormalities formed mainly in young (4–8 years old) trees.

In the present time, in many cases, it is difficult to distinguish between the morphological abnormalities formed by radiation and other factors. For instance, twigs in many young pine trees are damaged (eaten) by moose whose population in the Chernobyl zone has sharply increased during the past years.

Similar morphological abnormalities and time patterns of their formation were found after the Fukushima accident by Yoschenko et al. (2016) in Japanese red pine (*Pinus densiflora*; the close relative to Scots pine) and by Watanabe et al. (2015) in Japanese fir (*Abies firma*; another coniferous species in the Fukushima zone). This confirms that the observed effect is not specific for Chernobyl; rather, this may be the specific response to radiation in conifer species. Further in-depth studies are needed to clarify its mechanisms.

Formation under chronic radiation conditions of morphological abnormalities in plants other than cancellation of the apical dominance remains disputable. For instance, Møller (1998) reported a correlation between fluctuating asymmetry parameters of leaves and density of ^{137}Cs contamination (40–172 kBq m^{-2}) for three plant species—*R. pseudoacacia*, *S. aucuparia*, and *M. perforate*. The correlation was found despite low levels and a relatively narrow range of external dose rate (0.1–0.3 μGy h^{-1}) at the experimental sites, which were located outside the exclusion zone. However, Kashparova et al. (2020) showed that the dose rate now causes no effect on the fluctuating asymmetry of birch leaves and Scots pine needles over a wide range of the external (0.1–40 μGy h^{-1}) and internal (0.1–273 μGy h^{-1}) radiation exposures.

Numerous studies in the Chernobyl zone were performed to identify the chronic radiation effects on Scots pine at the population, organism, and cytogenetic levels (e.g., Zelena et al. 2005; Geras'kin et al. 2011; Kuchma and Finkeldey 2011; Geras'kin and Volkova 2014; and many others). They revealed increased mutation rates, changes in gene expression and enzyme status, an increase of the isoenzyme polymorphism, and other effects. However, interpretation of the results requires distinguishing between the effects caused by chronic radiation and those remaining from the acute radiation period in populations that existed when the accident occurred. Moreover, the morphological abnormalities in the Scots pine populations formed mainly in the past, and it may be difficult to explain their formation based on changes observed at the cell and gene levels in the present period. The young populations of conifers in the Fukushima zone offer an opportunity to study comprehensively a sequence of events that occur at different levels of the organism as a response to chronic radiation and which finally brought the observed radiation effects.

6.5 Contaminated Forest as a Source of External and Internal Exposures

6.5.1 *Radioactivity Spreading During the Forest Fires and Doses to Personnel and Population*

In the period 1993–2013, more than 1100 wildfires of different kinds and scales were officially registered in the ChEZ, including in the most contaminated 10-km zone. Periods of fire danger mainly occurred in April–May and August. The largest fires happened in August 1992 at a total area of 17,000 ha of meadows and forests, including a crown fire at more than 5000 ha (Evangeliou et al. 2015).

In the absence of the traditional economic activity in the exclusion zone during 30 years, dangerous fuel material in the forests and meadows accumulated in large amounts. The presence of young trees on the edges of the forest stands and excessively high density of plants in the pine forests throughout the Chernobyl zone increase the risk of development of the comparable small surface fires into intensive crown fires. The existing equipment, structure, and location of fire departments in the exclusion zone are insufficient in such conditions as they cannot guarantee either a rapid response or effective firefighting in critical weather conditions. For example, a fire department with two or three old vehicles and a limited amount of fuel and with 5–7 firefighters is responsible for an area of more than 65,000 ha, while outside the Chernobyl zone, the typical fire department usually is responsible for 15–20 times smaller territory. Besides, approximately the third part of the Chernobyl zone territory is not covered by the fire detection system (there are no fire lookout towers), and almost 23,000 ha of forest is out of reach for fire engines and fire brigades (Kashparov et al. 2018).

After 1992, the large wildfires in the Chernobyl zone occurred in April and August of 2015 at the areas of almost 11,000 ha and 6000 ha, respectively (Evangeliou et al. 2016). In July 2016, the fire occurred at a much smaller area of approx. 300 ha; however, this fire was widely reported in mass media because it spread at the heavily contaminated territory of the Red Forest. The maximum airborne concentrations of radionuclides in the vicinity to the fire front were measured on July 16–17. They reached 0.40 Bq m^{-3} for ^{137}Cs, 1.0 Bq m^{-3} for ^{90}Sr, 0.002 Bq m^{-3} for ^{238}Pu, and 0.004 Bq m^{-3} for ^{239,240}Pu (Kyiv Regional State Administration 2017). A significant increase of the airborne radionuclide concentrations was also detected at the longer distances from the source of release. During the fire, the airborne concentrations of ^{137}Cs and ^{90}Sr ten-fold exceeded the monthly averaged levels in Chernobyl, Kopachi, and in the northern and eastern parts of the exclusion zone, and 25-fold exceeded those in the areas located to the west of the fire (Chystogalivka and Shepelychi). At the borders of the exclusion zone, the airborne radionuclide concentrations were 3–4 orders of magnitude lower than those near to the source of release. After the fire, the airborne concentrations decreased to their usual levels.

Forest fires at the radioactively contaminated territories result in the release of radioactive aerosols from the burning forest litter and biomass into the atmosphere. For the Chernobyl zone conditions, the parameters of the radionuclide release, transportation, and deposition were measured during the real time (Kashparov et al. 2000b) and controlled forest and grassland fires (Yoschenko et al. 2006a). In the latter case, the radionuclide total inventories at the experimental sites and their distributions between the ecosystem compartments were studied in detail before conducting the experiments, and the radioactive aerosols were sampled using the isokinetic aerosol samplers, multicascade impactors, and deposition plates that were installed before the beginning of the experiment along the expected trace of spreading of the smoke plume. Such design of the experiments enabled estimation of the fractions of the total radionuclide inventories in the ecosystem that were released from the burnt area to the atmosphere through the fitting of the modeling results to the measured values of the radionuclide airborne concentrations and depositions at the different distances from the source of release (Yoschenko et al. 2006b). It was found that during the forest fire, up to 4% of ^{137}Cs and ^{90}Sr and up to 1% of the Pu isotopes can be released from the burning area (mainly from the forest litter) according to the model calculations. In all observations, the radioactive aerosol deposition area was limited to the distances from several hundred meters to several kilometers from the source of release. Therefore, their atmospheric transport could be successfully modeled using the Gauss model (IAEA 1980) coupled with the model of the plume rise (Talerko 1990) adapted for "heavy" aerosol particles and accounting for the fuel (biomass and litter) characteristics for calculation of the initial parameters of the plume (Yoschenko et al. 2006b). The radioactive aerosol aerodynamic parameters ranged from less than 1 μm to greater than 25 μm; dispersal compositions differed for the measured radionuclides that reflected their different distributions in the ecosystem compartments. In general, it was shown that the wildfires in the Chernobyl zone did not result in any significant redistribution of radioactivity even on the local scale. With the radionuclide resuspension factors ranging from 10^{-8} to 10^{-7} m^{-1} (Yoschenko et al. 2006a), the fires are less intensive sources of release than, for example, agricultural works (10^{-6} m^{-1} based on the data by Kashparov et al. (1994a, b)). However, the radionuclide airborne concentrations within the aerosol deposition areas are three orders of magnitude higher than the background values and can reach some Bq m^{-3} for ^{137}Cs and ^{90}Sr, and mBq m^{-3} for $^{239+240}Pu$. The doses from inhalation to the personnel (e.g., to the members of the fire squads) in these areas can contribute some tenth parts into the total exposure (Yoschenko et al. 2006a).

Based on the parameters of the radioactive aerosol release obtained from the controlled fires, the model calculations were carried out (Khomutinin et al. 2007) for assessment of the potential consequences of the fires in the most contaminated forest stands in the Chernobyl zone (Fig. 6.12). The modeling shows that even under the most unfavorable weather conditions (wind directions to the settlements adjacent to the exclusion zone or to the places of location of the personnel inside the zone, stable atmosphere, etc.), the expected dose for personnel and population of the ChEZ will not exceed a few μSv, and will be less than 1 μSv to the population of the settlements outside the zone (Tables 6.11 and 6.12). The highest doses, up to 40 μSv, might be received by the personnel employed to extinguish the fires. In all cases, deposition of the resuspended radionuclides will not lead to any significant additional contamination of the territories inside and outside of the Chernobyl exclusion zone; less than 1% of the released activity can be transported outside the zone (Khomutinin et al. 2007).

According to the most conservative estimates by Evangeliou et al. (2016), the large fires in the Chernobyl zone in April and August of 2015 at the total area of about 15,000 ha could contribute less than 10 Bq m^{-2} to the existing depositions of ^{90}Sr and ^{137}Cs at the European territories outside the zone, and about 0.1 Bq m^{-2} to the existing depositions of $^{238+239}Pu$ and ^{241}Am. These quantities are much lower than the depositions after the global fallout, 1–2 kBq m^{-2} for ^{137}Cs and ^{90}Sr and approx. 50 Bq m^{-2} for ^{239}Pu. Radiological consequences of the fire in 26–29 April 2015 (Fig. 6.13) for the personnel of the Chernobyl zone were estimated by Kashparov et al. (2015a, b, 2018) based on the detail data on the actual spatial

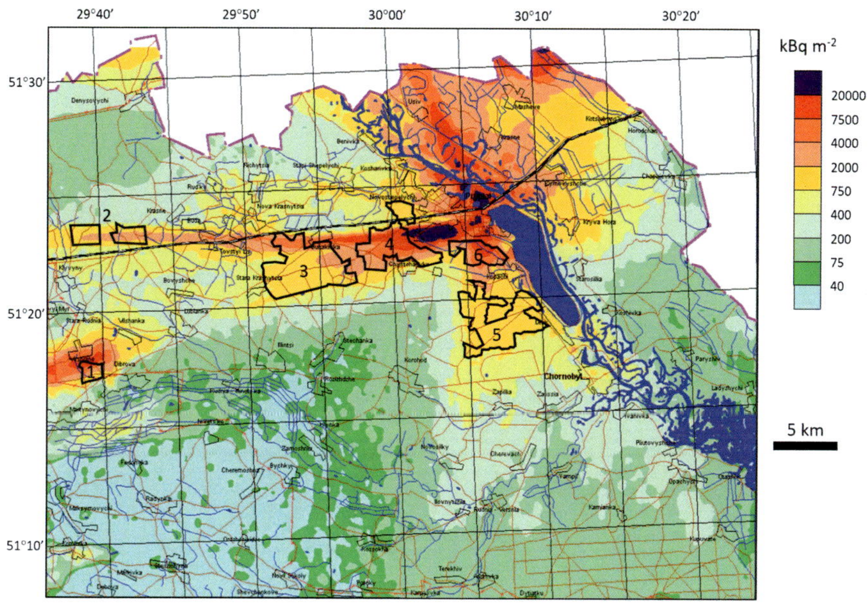

Fig. 6.12 Location of the most contaminated forest stands for assessment of the potential consequences of the fires: Vesniane (1), Krasne (2), Buryakivka (3), Chistogalivka (4), Kopachi (5), and Leliv (6). Background map: ^{137}Cs deposition in the exclusion zone (Fig. 6.2)

Table 6.11 Estimates of the maximum radionuclide deposition and the maximum doses from the radionuclide inhalation (50-year EED) to the population of the settlements adjacent to the exclusion zone (Khomutinin et al. 2007)

Settlement (distance from the fire)	Additional deposition density, Bq m^{-2}			Dose, μSv
	^{137}Cs	^{90}Sr	$^{239+240}Pu$	
Forest fire near village Buryakivka (3)				
Nivetske (17 km)	0.4	0.3	0.00005	1.0
Cheremoshna (18 km)	0.6	0.3	0.00004	0.9
Forest fire near village Chystogalivka (4)				
Cheremoshna (24 km)	0.4	0.2	0.00003	0.6
Nivetske (22 km)	0.5	0.2	0.00004	0.8
Forest fire near village Leliv (6)				
Dytiatky (24 km)	0.1	0.1	0.00001	0.3

distribution and radioactive contamination of the combustible materials within the grassland (6250 ha) and forest fire areas, classification of the fire type and weather conditions for each period of the fire, and on the parameters of the radionuclide releases obtained earlier in the controlled and real-time fires (Yoschenko et al. 2006a; Kashparov et al. 2000b). Those authors found that even such a large-scale

Table 6.12 Estimates of the maximum radionuclide deposition density and the maximum doses from the radionuclide inhalation (50-year EED) to the personnel of the exclusion zone and to the critical group (firefighters) (Khomutinin et al. 2007)

Location (distance from the fire)	Additional deposition density, Bq m^{-2}			Dose, μSv
	^{137}Cs	^{90}Sr	$^{239+240}$Pu	
Forest fire near village Buryakivka (3)				
Chernobyl (24 km)	0.2	0.01	0.00001	0.4
ChNPP (14 km)	1	0.5	0.0001	1.6
"Vector" (2 km)	2	1	0.0001	2.6
Firemen				22
Forest fire near village Chistogalivka (4)				
Chernobyl (19 km)	1	0.4	0.00008	1.3
ChNPP (8 km)	4	2	0.0003	4.6
"Vector" (6 km)	3	1	0.0003	3.9
Firemen				40
Forest fire near village Kopachi (5)				
Chernobyl (14 km)	1	0.6	0.0001	1.4
ChNPP (3 km)	4	3	0.0005	5.6
"Vector" (12 km)	1	0.8	0.0001	1.8
Firemen				24

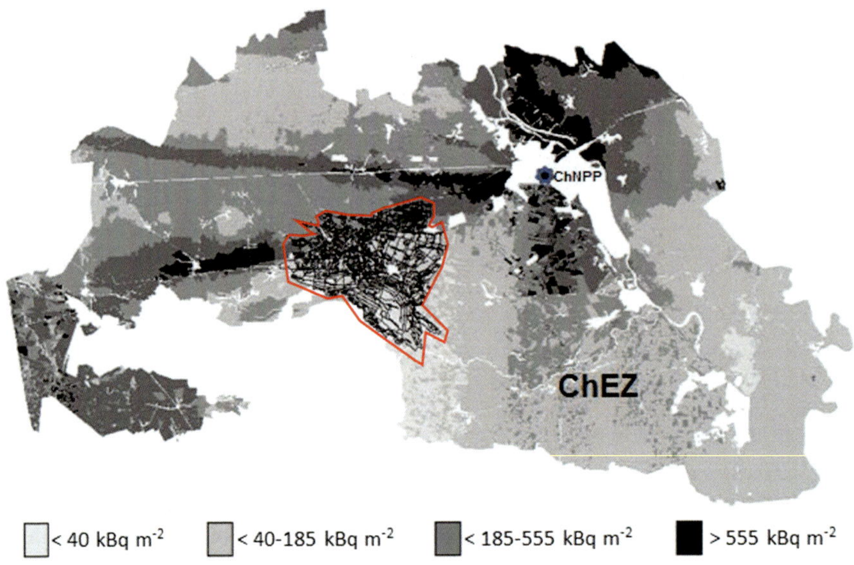

Fig. 6.13 Location of the large wildfire in the Chernobyl zone on April 26–29, 2015. Background map: ^{137}Cs deposition (Modified from Kashparov et al. 2015a, b, 2018)

fire did not result in significant doses even to the critical (most exposed) group of the personnel, firefighters. According to the calculations, the conservative expected value of the maximum total effective dose to the firefighters within the area of the fire in April 2015 was less than 42 μSv (assuming 10-h working day), being much lower than the control level of the effective dose for the personnel in the exclusion zone of 3000 μSv per year. The total doses were formed mainly due to external radiation (Fig. 6.14). During the forest fire, more than half of the internal dose could be contributed by inhalation of ^{90}Sr, while during the grassland phase of the fire, the contributions of ^{90}Sr, the sum of Pu isotopes, and ^{241}Am into the expected internal effective dose were approximately equal. The authors emphasize the necessity to take into account ^{241}Pu, a β-emitter whose contribution to the internal effective dose is currently comparable to the individual contributions of the α-emitting isotopes $^{238-240}$Pu (Kashparov et al. 2015a, b, 2018). The doses to the population outside the Chernobyl zone and the radionuclide depositions were negligible.

The possible impacts of the wildfires at the radioactively contaminated territories are not limited to the potential radiological consequences. For example, Dowdall et al. (2017) reported changes of the radionuclide chemical forms in soil after the fires at the territory of Belarus, which potentially can affect their mobility in the soil profile and bioavailability. Besides the loss of wood (e.g., the fire in the Chernobyl zone in August 2015 resulted in damage or loss of one million m^3 of wood (Nikonchuk 2015)), the wildfires may also impact fauna of the zone through destruction of habitat and food chains. Finally, even in the absence of any significant radiological consequences, public and mass media often consider the wildfires in the Chernobyl zone as the potential tread to health. It means that the fire situation in the Chernobyl zone must be improved through implementation of the measures aimed at prevention, early detection, and efficient suppression of fires (Zibtsev et al. 2011).

6.5.2 Contribution of the Forest and Forest Products to the Doses on Population Living in the Contaminated Areas: Forest Countermeasures

Forest ecosystems can significantly contribute to radiation exposure to the population at the territories outside the exclusion zone. Due to decreasing of ^{137}Cs migration rates in soil and accumulation of the radionuclide in litter and aboveground biomass, the highest ratio of dose rate at 1 m height to ^{137}Cs terrestrial contamination density, 1.7×10^{-3} (μSv h^{-1}) (kBq m^{-2})$^{-1}$, currently is found in pine forests; the ratios are lower in deciduous forests (1.3×10^{-3} (μSv h^{-1}) (kBq m^{-2})$^{-1}$) and at the meadows and arable lands (1.0×10^{-3} (μSv h^{-1}) (kBq m^{-2})$^{-1}$ and 0.6×10^{-3} (μSv h^{-1}) (kBq m^{-2})$^{-1}$, respectively).

An abnormal high radiocesium uptake into wild mushrooms and berries are observed at the radioactively contaminated territory of Ukraine outside the first and second zones (Balonov et al. (2018); Tables 6.13 and 6.14), and even at the

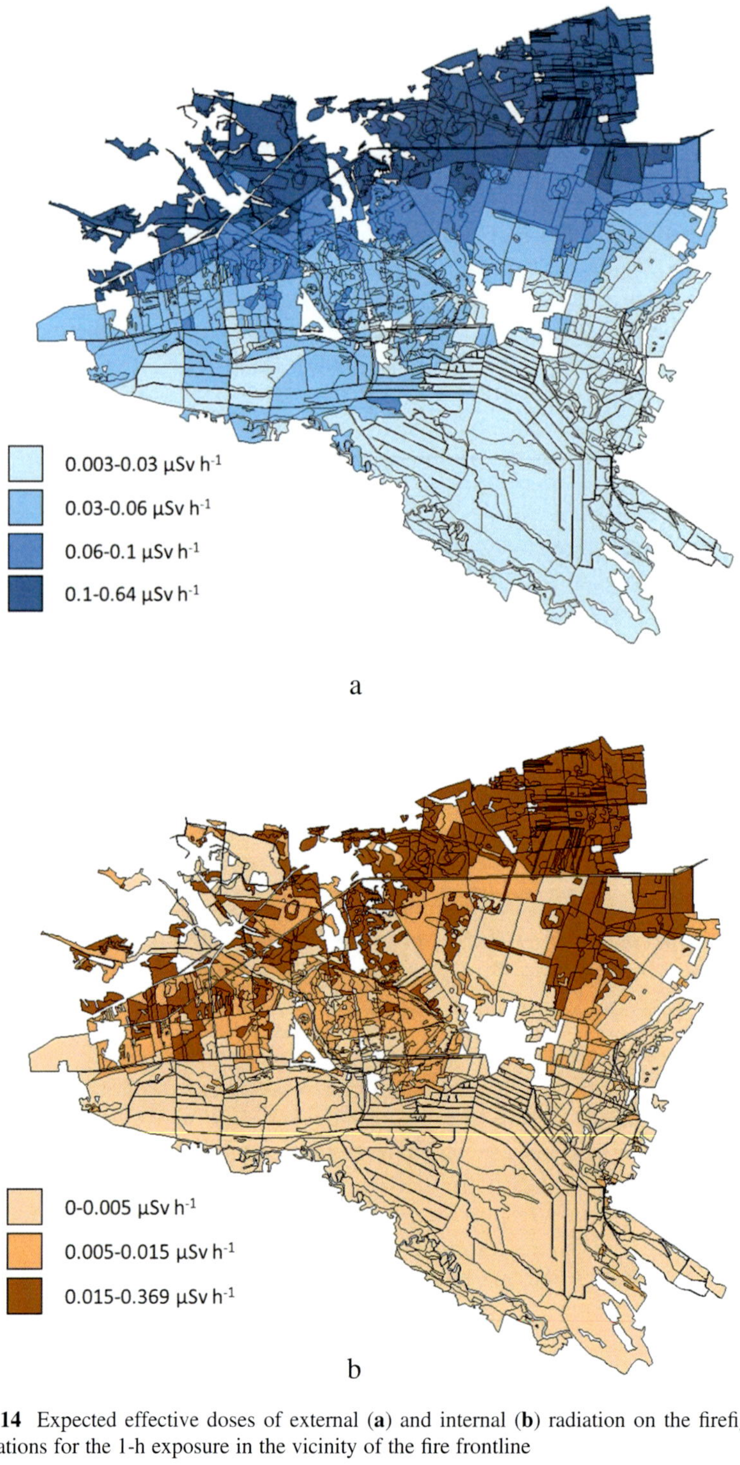

Fig. 6.14 Expected effective doses of external (**a**) and internal (**b**) radiation on the firefighters. Calculations for the 1-h exposure in the vicinity of the fire frontline

Table 6.13 Mean T_{ag} values of ^{137}Cs to mushrooms, (Bq kg^{-1}) (kBq m^{-2})$^{-1}$ (Kaduka et al. 2006)

Groups (species) of mushrooms by levels of ^{137}Cs accumulation	Soil groups			
	Sandy, sandy loam	Light and medium loam	Heavy loam, clay	Peaty
Severely accumulating: *Xerocomus badius, Xerocomus chrysenteron, Suillus variegatus, Lactarius* (all species)	30	10	2	40
Intermediately accumulating: *Leccinum, Bolétus edulis, Canthar̄ellus cibārius, Russula* (all species), *Tricholoma*	13	4	1	20
Poorly accumulating: *Armillaria mellea, Kuehneromyces mutabilis, Flammulina velutipes, Marasmius oreades, Gyromitra, Morchella, Agaricus* (all species), *Macrolepiota, Lycoperdon*	3	1	0.3	4

Table 6.14 Mean T_{ag} values of ^{137}Cs to wild forest berries, (Bq kg^{-1}) (kBq m^{-2})$^{-1}$ (Krasnov and Orlov 2004)

Species	Mean T_{ag} values	Range of T_{ag} values for different landscapes
Bilberry (*Vaccinium myrtillus* L.)	6.5	2.0–20.0
Great Bilberry (*Vaccinium uliginosum* L.)	10	9.4–15.6
Foxberry (*Vaccinium vitis-idaea* L.)	10	8.1–12.9
Cranberry (*Oxycoccus palustrix* Pers.)	13	8.9–16.6
Strawberry (*Fragaria vesca* L.)	3.8	2.0–5.8
Raspberry (*Rubus idaeus* L.)	2.6	0.8–6.6

relatively low contaminated areas radiocesium concentrations in mushrooms and berries exceed permissible levels.

In this reason, the major contribution to the dose of internal radiation to people living in the settlements of the Ukrainian Polissiya that are located near to forests is provided by wild mushrooms and berries (Fig. 6.15).

Fesenko et al. (2000) presenteda detailed analysis of the dynamics of contributions of forests into the doses on population and foresters in the Bryansk region of Russia. They investigated 15 forest areas ranked into three zones according to the ^{137}Cs deposition (>1480 kBq m^{-2}, 555–1480 kBq m^{-2}, and 185–555 kBq m^{-2}). In the studied area, the countermeasures were implemented ranging from restriction of the gathering of berries and mushrooms in the zone with the lowest deposition to complete exclusion from economic use of the forests in the first zone. In this study, mushrooms and wild berries in 1996 contributed 40–45% to the total internal dose; large contribution was also provided by the milk of cattle grazing at the forest pastures. In general, the radioactively contaminated forests could give 23–47% of the collective dose. Based on this estimate, the authors state that "agricultural

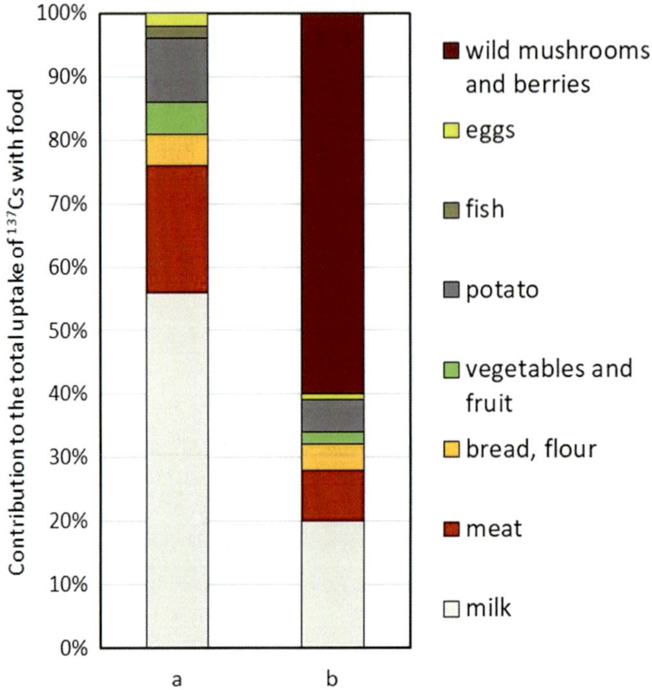

Fig. 6.15 Averaged contribution of different products to the peroral ^{137}Cs uptake by the rural population: human diet without (**a**) and with (**b**) forest mushrooms and berries (based on data by Kashparov et al. (2005) and Labunska et al. (2018))

countermeasures alone fail to attain the permitted level of exposure of the population. Therefore, serious attention should be drawn to forest countermeasures" (Fesenko et al. 2000).

The experience of the forest countermeasures and the decision-making framework for their optimization at the late stage after the Chernobyl accident were discussed in Fesenko et al. (2005b). The authors consider efficiency of application of restrictive countermeasures, optimized usage of forest and forest products, soil-based countermeasures, and animal-based countermeasures for the above-described forest area in the Russian Federation (Fesenko et al. 2000) using the set of criteria including reduction of individual dose to population and critical population group, feasibility and cost of implementation, cost of averted 1 man-Sv, ecological effects, and social acceptability. A detailed analysis of the exposures showed that the most efficient way to reduce the impacts of contaminated forests would be decreasing the exposure pathways formed by consumption of "forest milk" (for foresters as the critical group) and by gathering of mushroom (rest of population). The authors emphasized a needfor the application of countermeasures during the several-decades-long period. Instead of implementation of restrictive and soil-based countermeasures, at the late stage after the deposition, the most efficient strategy will be based on the optimization in forest and forest product usage. It might include limitation on gathering of mushrooms and wild berries to the species with low accumulation of radionuclides, mushroom processing before consumption (thorough washing in a large volume of water followed by boiling for 25–60 min with intermediate drain of water, etc.), application of Prussian Blue, and tree harvesting within the areas with low dose rates (Fesenko et al. 2005b).

The knowledge on the efficiency of the forest countermeasures was used for the preparation of guidelines for remediation strategies to reduce the radiological consequences of environmental contamination (IAEA 2012) and for the creation of the decision support software (Ulanovsky et al. 2011).

6.6 Conclusions

The Chernobyl accident caused severe impacts on forests and forestry at the large territories of Ukraine, Belarus, and RF. In the Chernobyl exclusion zone, the extremely high levels of acute radiation during the initial period after the accident killed pine forests close to the source of the release and formed zones of sublethal and moderate damage at greater distances; such severe effects, however, were not observed for broad-leaved tree species. In the exclusion zone, as well as in the second zone, that is, at the territories with ^{137}Cs deposition levels exceeding 555 kBq m^{-2}, the special regime has been implemented in forestry, which implies prohibition of commercial use of the forestry products and limitation of the working time. Restrictions in forestry were also established for the territories with the lower deposition levels (third and fourth zones), and only at ^{137}Cs deposition levels below 37 kBq m^{-2}, the forestry products may be used freely. Thus, the large forest area was partly or fully excluded from economic use. Being not decontaminated, Chernobyl forests, especially in the exclusion zone, will remain as long-term environmental repositories of radionuclides. Comprehensive studies are needed for prediction of the further dynamics of radionuclides in the forest ecosystems and for the understanding of the radiation effects on biota.

Important knowledge has been accumulated about the radionuclide cycling in the forest ecosystems and about the processes that redistribute radionuclides between the ecosystem compartments at the different stages after the deposition. Based on the knowledge, numerous models have been developed to predict the radionuclide dynamics in the forest ecosystems. However, attempts to apply the previously created models for prediction of the radiocesium dynamics in the Fukushima forests showed the need for their further development. Improved model prediction accuracy can be achieved through wider use of knowledge about plant physiology and nutrient cycles for determination of parameters of the radionuclide dynamics.

Observations in the Chernobyl zone enabled identification of the radiosensitive plant species. Besides the abovementioned effects of acute irradiation in the high doses, it was found that under chronic radiation conditions, at the comparable low dose rates, morphological abnormalities (cancellation of apical dominance) formed in young populations of Scots pine. A reliable correlation was established between the frequency of the trees with abnormalities in population and the radiation dose rate on the trees in population; the responses to chronic radiation also were found at the cell and molecular levels. After the Fukushima accident, similar morphological abnormalities were identified in young trees of Japanese red pine and Japanese fir; however, the mechanism by which this radiation effect forms has not yet been clarified.

Experimental studies, real-time observations in the Chernobyl zone, and modeling exercises showed that forest fires at the radioactively contaminated territories do not pose any significant radiological risks to the personnel of the zone and to the population outside; at certain conditions, inhalation of the resuspended radioactive aerosols may provide nonnegligible contribution into the doses on firefighters. However, the fires have severe consequences for flora and fauna of the zone. The outbreak of the large-scale fires in recent years clearly indicated the urgent need for improvement of forest management and modernization of the fire protection system. In addition, in many cases, media reports on the Chernobyl forest fires reveal the lack of adequate risk communication to the public without any opinion from experts being presented.

In the same time, the radioactively contaminated forests at the territories outside the first and second zones can form significant doses to population living in the adjacent villages, mainly due to consumption of the forest products (mushrooms, berries, and milk of the cows grazing at the forest pastures or being fed with hay harvested in forest). Reducing this exposure pathway at the late stage after the Chernobyl accident should be based mainly on the optimization in forest and forest products usage, and on informing the population.

Acknowledgments The authors' researches cited in this chapter were supported by the Japan Society for Promotion of Science (Grant Numbers 15K00563, 15H00968, and 15H04621), MEXT (24110007), Science and Technology Center in Ukraine (1992, 3674, p099, p170), and the National University of Life and Environmental Sciences of Ukraine, under the Ministry of Education and Science of Ukraine (project 110/90-f).

References

Abaturov A (1990) Osobennosti prostranstvennogo raspredeleniya radiatsionnogo porazheniya sosniakov vblizi ChAES (Specifics of the spatial distribution of radiation damages to the pine forests near to the ChNPP). In: Ryabov IN, Ryabtsev IA (eds) Biologicheskiye i radioekologicheskiye aspekti posledstviy avarii na Chernobyl'skoy AES (Biological and radioecological aspects of consequences of the accident at the Chernobyl NPP). USSR Academy of Sciences, Moscow, p 17

Abaturov Y, Gol'tsova N, Rostova N, Girbasova A, Abaturov A, Melankholin P (1991) Nekotoriye osobennosti radiatsionnogo porazheniya sosny v raione avarii na ChAES (Some peculiarities of radiation damage to pine in the zone of the accident at the Chernobyl NPP). Ecologiya (Ecology) 5:28–33

Balonov M, Kashparov V, Nikolaenko E, Berkovsky V, Fesenko S (2018) Harmonization of standards for permissible radionuclide activity concentrations in foodstuffs in the long term after the Chernobyl accident. J Radiol Protect. https://doi.org/10.1088/1361-6498/aabe34

Bunzl K, Schimmack W, Kreutzer K, Schierl R (1989) Interception and retention of Chernobyl USSR-derived cesium-134, cesium-137 and ruthenium-106 in a spruce stand. Sci Total Environ 78:77–78

Cole DW, Rapp M (1981) Elemental cycling in forest ecosystems. In: Reichle DE (ed) Dynamic properties of forest ecosystems. Cambridge University Press, Cambridge, pp 341–407

Davydchuk V (1994) Landshafty avariynoy zony i ih radiatsionnoye zagriazneniye (Landscapes of the accident zone and their radioactive contamination). In: Marinich A (ed) Landshafty Chernobyl'skoy zony i ih otsenka po usloviyam migratsii radionuklidov (Landscapes of the Chernobyl zone and their evaluation regarding to conditions of the radionuclides migration). Naukova Dumka, Kiev, pp 8–22

De Cort M et al (1998) Atlas of caesium deposition on europe after the Chernobyl accident, Rep. 16733. Office for Official Publications of the European Communities, Luxembourg

Dowdall M et al (2017) Investigation of the vertical distribution and speciation of ^{137}Cs in soil profiles at burnt and unburnt forest sites in the Belarusian Exclusion Zone. J Environ Radioact 175-176:60–69

Evangeliou N et al (2015) Fire evolution in the radioactive forests of Ukraine and Belarus: future risks for the population and the environment. Ecol Monogr 85(1):49–72

Evangeliou N et al (2016) Resuspension and atmospheric transport of radionuclides due to wildfires near the Chernobyl Nuclear Power Plant in 2015: an impact assessment. Sci Rep 6:26062. https://doi.org/10.1038/srep26062

Fesenko S, Voigt G, Spiridonov S, Sanzharova N, Gontarenko I, Belli M, Sansone U (2000) Analysis of the contribution of forest pathways to the radiation exposure of different population groups in the Bryansk region of Russia. Radiat Environ Biophys 39:291–300

Fesenko S et al (2001) Identification of processes governing long-term accumulation of ^{137}Cs by forest trees following the Chernobyl accident. Radiat Environ Biophys 40:105–113

Fesenko S et al (2005a) Comparative radiation impact on biota and man in the area affected by the accident at the Chernobyl Nuclear Power Plant. J Environ Radioact 80:1–25

Fesenko S, Voigt G, Spiridonov S, Gontarenko I (2005b) Decision making framework for application of forest countermeasures in the long term after the Chernobyl accident. J Environ Radioact 82:143–166

Geras'kin S et al (2003a) Genetic consequences of radioactive contamination by the Chernobyl fallout to agricultural crops. J Environ Radioact 66:155–169

Geras'kin S et al (2003b) Bioindication of the anthropogenic effects on micropopulations of *Pinus sylvestris*, L. in the vicinity of a plant for the storage and processing of radioactive waste and in the Chernobyl NPP zone. J Environ Radioact 66:171–180

Geras'kin S et al (2011) Effects of radioactive contamination on Scots pines in the remote period after the Chernobyl accident. Ecotoxicology 20:1195–1208

Geras'kin S et al (2013) Effects of long-term chronic exposure to radionuclides in plant populations. J Environ Radioact 121:22–32

Geras'kin S, Fesenko S, Alexakhin R (2008) Effects of non-human species irradiation after the Chernobyl NPP accident. Environ Int 34:880–897

Geras'kin S, Volkova P (2014) Genetic diversity in Scots pine populations along a radiation exposure gradient. Sci Total Environ 496:317–327

Goor F, Thiry Y (2004) Processes, dynamics and modelling of radiocaesium cycling in a chronosequence of Chernobyl-contaminated Scots pine (*Pinus sylvestris* L.) plantations. Sci Total Environ 325:163–180

Goor F, Thiry Y, Delvaux B (2007) Radiocaesium accumulation in stemwood: integrated approach at the scale of forest stands for contaminated Scots pine in Belarus. Environ Manag 85:129–136

IAEA (1980) Atmospheric dispersion in Nuclear Power Plant siting: a safety guide. International Atomic Energy Agency, IAEA. safety series no 50-SG-53. IAEA, Vienna

IAEA (1991) The international Chernobyl project. Technical report. Assessment of radiological consequences and evaluation of protective measures. Report by an International Advisory Committee. IAEA, Vienna

IAEA (2001) The international nuclear event scale (INES). User's manual, 2001 edn. Jointly prepared by IAEA and OECD/NEA. IAEA, Vienna

IAEA (2002) Modelling the migration and accumulation of radionuclides in forest ecosystems. Report of the Forest Working Group of the Biosphere Modelling and Assessment (BIOMASS) Programme, Theme 3. IAEA, Vienna

IAEA (2003) Assessing radiation doses to the public from radionuclides in timber and wood products. IAEA-TECDOC-1376. IAEA, Vienna

IAEA (2006) Environmental consequences of the Chernobyl accident and their remediation: twenty years of experience. Report of the Chernobyl Forum Expert Group 'Environment'. IAEA, Vienna

IAEA (2010) Handbook of parameter values for the prediction of radionuclide transfer in terrestrial and freshwater environments. Technical reports series no. 472. IAEA, Vienna

IAEA (2012) Guidelines for remediation strategies to reduce the radiological consequences of environmental contamination. In: Fesenko S, Howard BJ (eds) Technical reports series no. 475. IAEA, Vienna

IKEA (2011) IKEA of Sweden AB. Specification. Chemical compounds and substances. Spec. no: IOS-MAT-0010 Date: 2011-05-13 version no: AA–10911–10. Almhult, IKEA of Sweden AB

Izrael Y, Bogdevich I (eds) (2009) The Atlas of recent and predictable aspects of consequences of Chernobyl accident on polluted territories of Russia and Belarus (ARPA Russia-Belarus). Foundation "Infosphere" – NIA-Nature, Moscow-Minsk

Kaduka M, Shutov V, Bruk G, Balonov M, Brown JE, Strand P (2006) Soil-dependent uptake of ^{137}Cs by mushrooms: experimental study in the Chernobyl accident areas. J Environ Radioact 89:199–211

Kal'chenko V, Arkhipov N, Fedotov I (1993a) Mutagenez fermentnikh lokusov, indutsirovanniy v megasporakh *Pinus sylvestris* L. ioniziruyuschim izlucheniyem pri avarii na Chernobyl'skoi AES (Mutagenesis of ferment loci induced in *Pinus sylvestris* L. spores by ionizing irradiation associated with the ChNPP accident). Genetika (Genetics) 29:266–273

Kal'chenko V, Rubanovich A, Fedotov I, Arkhipov N (1993b) Geneticheskiye effecty, indutsirovanniye Chernobyl'skoy avariei, v polovykh kletkakh sosny obyknovennoi *Pinus sylvestris* L. (Genetic effects induced by the ChNPP accident in sexual cells of *Pinus sylvestris* L. species). Genetika (Genetics) 29:1205–1212

Kashparov V (2002) Assessment of the radiological situation resulted by the accidental release of fuel particles. Radioprotection 37:1061–1066

Kashparov V et al (1994a) Resuspension of radionuclides and the contamination of village areas around Chernobyl. J Aerosol Sci 25(5):755–759

Kashparov V et al (1994b) Inhalation of radionuclides during agricultural work in areas contaminated as a result of the Chernobyl reactor accident. J Aerosol Sci 25(5):761–766

Kashparov V et al (1996) Formation of hot particles during the Chernobyl Nuclear Power Plant. Nucl Technol 114:246–253

Kashparov V et al (1999) Kinetics of fuel particle weathering and ^{90}Sr mobility in the Chernobyl 30-km exclusion zone. Health Phys 76:251–259

Kashparov V et al (2000a) Dissolution kinetics of particles of irradiated Chernobyl nuclear fuel: influence of pH and oxidation state on the release of radionuclides in the contaminated soil of Chernobyl. J Nucl Mater 279:225–233

Kashparov V et al (2000b) Forest fires in the territory contaminated as a result of the Chernobyl accident: radioactive aerosol resuspension and exposure of firefighters. J Environ Radioact 51:281–298

Kashparov V et al (2001) Soil contamination with ^{90}Sr in the near zone of the Chernobyl accident. J Environ Radioact 56:285–298

Kashparov V et al (2003) Territory contamination with the radionuclides representing the fuel component of Chernobyl fallout. Sci Total Environ 317:105–119

Kashparov V et al (2004) Kinetics of dissolution of Chernobyl fuel particles in soil in natural conditions. J Environ Radioact 72:335–353

Kashparov V, Lazarev N, Polishchuk S (2005) Current problems of agricultural radiology in Ukraine. Agroecol J 3:31–41

Kashparov V et al (2012) Radionuclide migration in the experimental polygon of the Red Forest waste site in the Chernobyl zone – part 1: characterization of the waste trench, fuel particle transformation processes in soils, biogenic fluxes and effects on biota. Appl Geochem 27:1348–1358

Kashparov V et al (2015a) Evaluation of the expected doses of fire brigades at the Chornobyl exclusion zone in April 2015. Nucl Phys At Energy 16(4):399–407

Kashparov V et al (2015b) Radiological consequences of the fire in the Chernobyl exclusion zone in April 2015. Radiat Biol Radioecol 57(5):512–527

Kashparov V, Levchuk S, Zhurba M, Protsak V, Khomutinin Y, Beresford NA, Chaplow JS (2018) Spatial datasets of radionuclide contamination in the Ukrainian Chernobyl Exclusion Zone. Earth Syst Sci Data 10:339–353. https://doi.org/10.5194/essd-10-339-2018

Kashparova E, Levchuk S, Morozova V, Kashparov V (2020) A dose rate causes no fluctuating asymmetry indexes changes in silver birch (*Betula pendula* (L.) Roth.) leaves and Scots pine (*Pinus sylvestris* L.) needles in the Chernobyl Exclusion Zone. J Environ Radioact 211:105731

Kato H et al (2017) Temporal changes in radiocesium deposition in various forest stands following the Fukushima Daiichi nuclear power plant accident. J Environ Radioact 166:449–457

Kato H, Onda Y, Gomi T (2012) Interception of the Fukushima reactor accident derived ^{137}Cs, ^{134}Cs and ^{131}I by coniferous forest canopies. Geophys Res Lett 39:L20403

Khomutinin Y et al (2007) Assessment of radiological danger of hypothetical forest fire in the Chernobyl exclusion zone. Bulletin of ecological state of the exclusion zone and zone of an unconditional (obligatory) resettlement. Bull Ecol 1(29):28–33

Kozubov G, Taskaev A, Ladanova N, Kuzivanova S, Artemov V (1987) Radioekologicheskiye issledovaniya sosnovykh lesov v raione avarii na Chernobyl'skoi AES (Radioecological investigations of the pine forests in the region impacted by the accident at the Chernobyl NPP). Syktyvkar, Komi Branch of AS USSR

Kozubov G, Taskaev A (eds) (1990) Radiatsionnoye vozdeistviye na khvoiniye lesa v raione avarii na Chernobyl'skoi AES (Radiation influence to the pine forests in the zone of the accident at the Chernobyl NPP). Komi Branch of AS USSR, Syktyvkar

Kozubov G, Taskaev A, Fedotov I, Arkhipov N, Davydchuk V, Abaturov Y (1991) Karta-skhema radiatsionnogo porazheniya lesov v raione avarii nf Chernobyl'skoi AES (M 1:100000) s poiasnitel'noi zapiskoi (Schematic map of radiation damages of forests in the zone of the accident at the Chernobyl NPP (scale 1:100000) with the explanatory note). Komi Branch of AS USSR, Syktyvkar

Kozubov G, Taskaev A (2002) Radiobiologicheskiye issledovaniya khvoinikh v raione Chernobyl'skoi katastrofy (Radiobiological studies of coniferous species in the area of the ChNPP accident). Moscow, Design Information. Cartography

Krasnov V, Orlov A (2004) Radioecology of berries. Zhitomir, Volyn

Krasnov V, Orlov A, Buzun V, Landin V, Shelest Z (2007) Prykladnaya radioekologiya lesa (Applied forest radioecology). Zhytomyr, Polissiya

Kuchma N et al (1994) Radioekologicheskiye i lesovodstvenniye posledstviya zagriazneniya lesnykh ekosistem zony otchuzhdeniya (Radioecological and silvicultural consequences of contamination of forest ecosystems of the exclusion zone). Chernobyl, STC SPA Prypiat

Kuchma M, Arkhipov A, Bidna S (1997) Spetsializovana systema ekologo-lisivnychogo dogliadu za lisamy zony vidchuzhennia (Specialized system of ecological-silvicultural maintenance of forests of the exclusion zone). Chernobyl, Chernobyl Science and Technology Center of International Researches

Kuchma O, Finkeldey R (2011) Evidence for selection in response to radiation exposure: *Pinus sylvestris* in the Chernobyl exclusion zone. Environ Pollut 159:1606–1612

Kyiv Regional State Administration (2017) Regional report on the environmental conditions in Kyive region in 2016. https://menr.gov.ua/files/docs/Reg.report/ДОПОВІДЬ%20Київська%202016.pdf. Accessed 31 May 2018

Labunska I, Kashparov V, Levchuk S, Santillo D, Johnston P, Polishchuk S, Lazarev N, Khomutinin Y (2018) Current radiological situation in areas of Ukraine contaminated by the Chernobyl accident: part 1. Human dietary exposure to Caesium-137 and possible mitigation measures. Environ Int 117:250–259. https://doi.org/10.1016/j.envint.2018.04.053

Mamikhin S, Tikhomirov F, Shcheglov A (1997) Dynamics of ^{137}Cs in the forests of the 30-km zone around the Chernobyl nuclear power plant. Sci Total Environ 193:169–177

Melin J, Wallberg L (1991) Distribution and retention of cesium in Swedish Boreal forest ecosystems. In: Moberg J (ed) The Chernobyl fallout in Sweden, results from a research programme on environmental radiology. The Swedish Radiation Protection Project. Arprint, Lund, pp 467–475

Ministry of Emergencies of Ukraine, National Academy of Sciences of Ukraine (1996) Atlas of Chernobyl Exclusion Zone. Kartographiya, Kyiv

Ministry of Emergencies of Ukraine (2011) Twenty-five years after Chornobyl accident: safety for the future. National report of Ukraine. KIM, Kyiv

Ministry of Emergencies of Ukraine, All-Ukrainian Research Institute of Population and Territories Civil Defense from Technogenic and Natural Emergencies (2006) 20 years after Chornobyl catastrophe. Future outlook: National report of Ukraine. Atika, Kyiv

Ministry of Emergencies of Ukraine, Intellectual systems GEO Ltd (2008) Atlas. Ukraine. Radioactive contamination. http://radatlas.isgeo.com.ua. Accessed 21 Dec 2017

Ministry of Health of the Republic of Belarus (2000) Respublikanskiye dopustimiye urovni soderzhaniya tseziya-137 v drevesine, produktsii iz drevesiny i drevesnykh materialov, prochei nepischevoi produktsii lesnogo hoziaistva (RDU/LH-2001) HN 2.6.1.10-1-01-2001 (Republican permissible levels of cesium-137 in wood, products from wood and wood materials, other non-food products of forestry (RPL/F-2001) HN 2.6.1.10-1-01-2001). Ministry of Health, Minsk

Ministry of Health of the Russian Federation (1999) Permissible levels of caesium-137 and strontium-90 in forest products. Sanitary rules SR 2.6.1.759-99. Ministry of Health, Moscow

Ministry of Health of Ukraine (2005) Sanitary standard of specific activity of radionuclides ^{137}Cs and ^{90}Sr in wood and products of wood. Ministry of Health, Kyiv

Mitrochenko V, Kyrychenko O, Kuchma M (1999) Influence of ionizing radiation on the forest plantations. In: Principles of forest radioecology. Derzhkomlisgosp of Ukraine, Kyiv, pp 52–80

Møller AP (1998) Developmental instability of plants and radiation from Chernobyl. Oikos 81 (3):444–448

Myttenaere C, Schell WR, Thiry Y, Sombre L, Ronneau C, van der Stegen de Schrieck J (1993) Modelling of the ^{137}Cs cycling in forest: recent developments and research needed. Sci Total Environ 136:77–91

Nadtochy P, Malynovsky A, Mozhar A, Lazarev M, Kashparov V, Melnyk A (2003) Dosvid podolannya naslidkiv Chornobyl'skoyi katastrofy (sil'ske ta lisove hospodarstvo) (Experience of overcoming the consequences of the Chernobyl catastrophe (agriculture and forestry)). Kyiv, Svit

Nepyivoda V (2005) Forestry in the Chornobyl Exclusion Zone: wrestling with an invisible rival. J Forestry 103:36–40

Nikonchuk V (2015) Report for the round table "Exclusion zone: present and future" at SAUEZM on 13.11.2015. http://dazv.gov.ua/images/presentacii/Nikonchuk%20-%202015.11.13.pdf. Accessed 21 Dec 2017

Perevolotsky A (2006) Distribution of ^{137}Cs and ^{90}Sr in the forest biogeocenoses. RNIUP, Gomel

Shaw G (2007) Radionuclides in forest ecosystems. Radioact Environ 10:127–155

Shcheglov A, Tikhomirov F, Tsvetnova O, Kiyashtorin A, Mamikhin S (1996) Biogeochemistry of Chernobyl-derived radionuclides in forest ecosystems of the European part of the CIS. Radiat Biol Radiecol 36:469–478

Shcheglov A, Tsvetnova O, Klyashtorin A (2001) Biogeochemical migration of technogenic radionuclides in forest ecosystems. Nauka, Moscow

Shcheglov A, Tsvetnova O, Klyashtorin A (2014) Biogeochemical cycles of Chernobyl-born radionuclides in the contaminated forest ecosystems. Long-term dynamics of the migration processes. J Geochem Explor 144:260–266

Shevchenko V, Grinikh L (1995) Cytogenetic effects in *Crepis tectorum* populations growing in the Bryansk region 7 years after the ChNPP accident. Radiat Biol Radiecol 35:720–725

Smirnov E, Suvorova L (1996) Estimation and prediction of biological effects of radioactive contamination on the plant cover in the Chernobyl affected area. In: Taskaev A (ed) Effects of radioactive contamination on terrestrial ecosystems in the Chernobyl affected areas. Komi Branch of AS USSR, Syktyvkar, pp 27–37

Sombre L, Vanhouche M, Thiry Y, Ronneau C, Lambotte JM, Myttenaere C (1990) Transfer of radiocesium in forest ecosystems resulting from a nuclear accident. In: Desmet G (ed) Transfer of radionuclides in natural and seminatural environments. Elsevier, London, pp 74–83

Steinhauser G, Brandl A, Johnson TE (2014) Comparison of the Chernobyl and Fukushima nuclear accidents: a review of the environmental impacts. Sci Total Environ 470–471:800–817

Talerko N (1990) Calculation of radioactive admixture release from the accidental unit of Chernobyl power plant. Meteorol Hydrol 10:39–46

Teramage M, Onda Y, Kato H, Gomi T (2014) The role of litterfall in transferring Fukushima-derived radiocesium to a coniferous forest floor. Sci Total Environ 490:435–439

Thiry Y et al (2009) Impact of Scots pine (*Pinus sylvestris* L.) plantings on long term ^{137}Cs and ^{90}Sr recycling from a waste burial site in the Chernobyl Red Forest. J Environ Radioact 100:1062–1068

Tikhomirov F, Shcheglov A (1994) Main investigation results on the forest radioecology in the Kyshtym and Chernobyl accident zones. Sci Total Environ 157:45–57

Tikhomirov F, Sidorov V (1990) Radiation damage of forest in the ChNPP zone. In: Ryabov I, Ryabtsev I (eds) Biological and radioecological aspects of consequences of the accident at the Chernobyl NPP. USSR Academy of Sciences, Moscow, p 18

Ulanovsky A, Jacob P, Fesenko S, Bogdevitch I, Kashparov V, Sanzharova N (2011) ReSCA: decision support tool for remediation planning after the Chernobyl accident. Radiat Environ Biophys 50:67–83. https://doi.org/10.1007/s00411-010-0344-7

UNSCEAR (2000) Exposures and effects of the Chernobyl accident (annex J). United Nations, New York, NY

UNSCEAR (2008) Sources and effects of ionizing radiation (annex D). United Nations, New York, NY

Verkhovna Rada of Ukraine (1991) On the legal regime of the territories exposed to radioactive contamination in consequence of the catastrophe at the Chernobyl NPP. Bulletin of Verkhovna Rada 16, Kyiv

Watanabe Y et al (2015) Morphological defects in native Japanese fir trees around the Fukushima Daiichi Nuclear Power Plant. Sci Rep 5:13232. https://doi.org/10.1038/srep13232

Yoschenko V et al (2006a) Resuspension and redistribution of radionuclides during grassland and forest fires in the Chernobyl exclusion zone: part I. Fire experiments. J Environ Radioact 86 (2):143–163

Yoschenko V et al (2006b) Resuspension and redistribution of radionuclides during grassland and forest fires in the Chernobyl exclusion zone: part II. Modeling the transport process. J Environ Radioact 87(3):260–278

Yoschenko V et al (2011) Chronic irradiation of Scots pine trees (*Pinus sylvestris*) in the Chernobyl exclusion zone: dosimetry and radiobiological effects. Health Phys 101:393–408

Yoschenko V et al (2016) Morphological abnormalities in Japanese red pine (*Pinus densiflora*) at the territories contaminated as a result of the accident at Fukushima Dai-Ichi Nuclear Power Plant. J Environ Radioact 165:60–67
Yoschenko V et al (2017) Radiocesium distribution and fluxes in the typical *Cryptomeria japonica* forest at the late stage after the accident at Fukushima Dai-Ichi Nuclear Power Plant. J Environ Radioact 166:45–55
Yoschenko V et al (2018a) Radioactive and stable cesium isotope distributions and dynamics in Japanese cedar forests. J Environ Radioact 186:34–44. https://doi.org/10.1016/j.jenvrad.2017.09.026
Yoschenko V, Ohkubo T, Kashparov V (2018b) Radioactive contaminated forests in Fukushima and Chernobyl. J Forest Res 23:3–14. doi.org/10.1080/13416979.2017.1356681
Yoschenko V, Kashparov V, Ohkubo T (2019) Radioactive contamination in forest by the accident of Fukushima Daiichi Nuclear Power Plant: comparison with Chernobyl. In: Takenaka C, Hijii N, Kaneko N, Ohkubo T (eds) Radiocesium dynamics in a Japanese forest ecosystem. Initial stage of contamination after the incident at Fukushima Daiichi Nuclear Power Plant. Springer Nature Singapore Pte Ltd, Singapore, pp 3–22
Yoshida S et al (2002) Stable elements – as a key to predict radionuclide transport in the forest ecosystems. Radioprotection 37:391–396
Yoshida S, Muramatsu Y, Dvornik A, Zhuchenko T, Linkov I (2004) Equilibrium of radiocesium with stable cesium within the biological cycle of contaminated forest ecosystems. J Environ Radioact 75:301–313
Yoshida S, Watanabe M, Suzuki A (2011) Distribution of radiocesium and stable elements within a pine tree. Radiat Prot Dosim 146:326–329
Zelena L, Sorochinsky B, von Arnold S, van Zyl L, Clapham DH (2005) Indications of limited altered gene expression in *Pinus sylvestris* trees from the Chernobyl region. J Environ Radioact 84:363–373
Zibtsev S et al (2011) Wildfires risk reduction from forests contaminated by radionuclides: a case study of the Chernobyl Nuclear Power Plant Exclusion Zone. Paper presented at the 5th international wildland fire conference, Sun City, South Africa, 9–13 May 2011

Part IV
Behavior of Radionuclides in Aquatic Ecosystems

Chapter 7
Long-Term Dynamics of the Chernobyl-Derived Radionuclides in Rivers and Lakes

Alexei Konoplev, Volodymyr Kanivets, Gennady Laptev, Oleg Voitsekhovich, Olga Zhukova, and Maria Germenchuk

In memoriam: This chapter is dedicated to the memory of Dr. Anatoly Bulgakov, colleague and friend, who initiated and greatly contributed to the development of the approach presented here.

Abstract It has been shown that after the initial phase of the Chernobyl accident, for time periods longer than 1 year, the changes in activity concentrations of radionuclides, both dissolved and particulate, are determined by changes in the concentrations of the upper soil layer for rivers or top layer of bottom sediments for lakes and ponds. The advantage of the proposed diffusion-based approach to describe the time dependence of radionuclide concentrations in rivers and lakes is that the middle- and long-term phases after the accident can be described by the same equation with the same parameters. Datasets for Chernobyl contaminated rivers and lakes have been used to test the model. The proposed semiempirical diffusional model of radionuclide dynamics in surface waters provides a tool for long-term predictions of river contamination in the future not only for Chernobyl but also for Fukushima contaminated areas or any other nuclear accident.

Keywords Chernobyl · radionuclides · migration · transport · rivers · lakes

A. Konoplev (✉)
Institute of Environmental Radioactivity, Fukushima University, Fukushima, Japan

V. Kanivets · G. Laptev · O. Voitsekhovich
Ukrainian Hydrometeorological Institute, Kiev, Ukraine

O. Zhukova
Scientific-Practical Center of Hygiene, Ministry of Health of the Republic of Belarus, Minsk, Belarus

M. Germenchuk
Center for Nuclear and Radiation Safety, Ministry of Emergency Situations of the Republic of Belarus, Minsk, Belarus

A. Konoplev et al. (eds.), *Behavior of Radionuclides in the Environment II*,
https://doi.org/10.1007/978-981-15-3568-0_7

7.1 Introduction

Long-term dynamics of deposited radionuclides in the environment 33 years after the Chernobyl accident is becoming more and more important and relevant for understanding what is going on in Fukushima. Detailed analysis of Chernobyl data covering an extended time period can provide a basis for long-term prediction of changes in environmental radioactive contamination in Fukushima and any other contaminated areas (Beresford et al. 2016)

The Chernobyl contaminated area is located southwest of the East European Plain, partially in the Pripyat Polesye, bordering the Pre-Dnieper lowland to the east. The region is a plain with a temperate continental climate. The composition and physicochemical properties of released radionuclides were changing during the different stages of the accident. This resulted in the radionuclide composition in the fallout being nonuniform and depending on direction and distance from the Chernobyl Nuclear Power Plant (ChNPP) (Konoplev et al. 1992; Borzilov et al. 1994). Figure 7.1 presents the Dneper River basin (a) and watersheds of small rivers in the area surrounding the Chernobyl NPP (b), which were used for official prediction of radioactive contamination of Kiev reservoir and Dnieper river immediately after the accident (Borzilov et al. 1989).

Initial radioactive contamination of water bodies after the Chernobyl accident was relatively high owing to direct fallout onto the river and lake surfaces (Vakulovsky et al. 1994). The contamination of water bodies decreased sharply due to the fast processes of sedimentation, sorption, and fixation of radionuclides to sediments and sedimentation of particles to the bottom (Vakulovsky et al. 1994; Smith et al. 2005a,

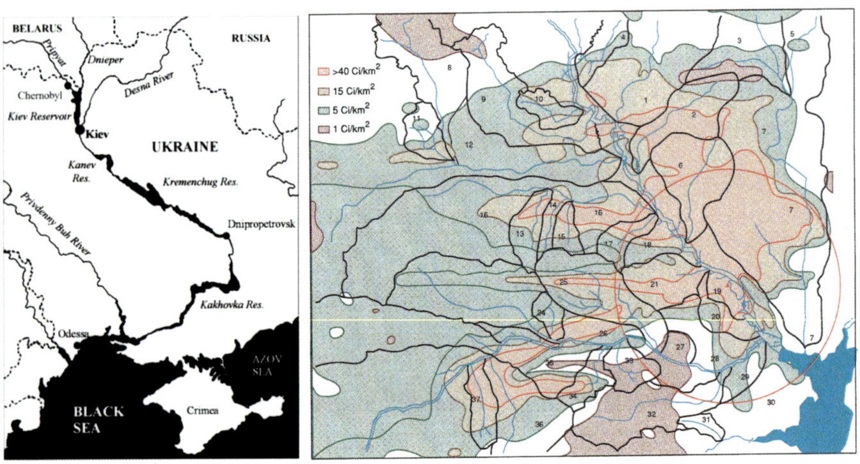

A B

Fig. 7.1 Dnieper River basin covering parts of Belarus, Russia, and Ukraine (**a**) (Smith et al. 2005a) and a historical sketchy map of the area surrounding the Chernobyl NPP (**b**) (red circle—30-km zone) with 36 watersheds of small rivers and ^{137}Cs deposition isolines as of 1988 (Ci/km^2)

b). After the Chernobyl accident, large areas of land were contaminated with radionuclides from the fallout. These areas became continuous sources of radionuclides to natural waters and aquatic ecosystems. Wash-off by surface runoff was the primary pathway for contamination of water bodies after the initial deposition (Borzilov et al. 1988, 1989; Konoplev et al. 1992, 1996a; Sansone et al. 1996; IAEA 2006).

The objective of this chapter is to review available data on changes of radionuclide concentrations in rivers and lakes of Chernobyl contaminated areas and put forward a quantitative approach to describe the long-term dynamics of radionuclides.

7.2 Empirical Modeling of Radionuclide Dynamics in Water

Temporal changes in dissolved radionuclide concentrations in surface runoff, rivers, and lakes were normally described using a series of exponential functions (Monte 1995, 1997; Smith et al. 2004, 2005a, b):

$$C_{\mathrm{w}}(t) = C^{0}_{\mathrm{w}1}\mathrm{e}^{-(\lambda+k_1)t} + C^{0}_{\mathrm{w}2}\mathrm{e}^{-(\lambda+k_2)t} + C^{0}_{\mathrm{w}3}\mathrm{e}^{-(\lambda+k_3)t} \qquad (7.1)$$

where $C_{\mathrm{w}}(t)$ is the current activity concentration of dissolved radionuclide; λ is the radionuclide decay rate constant; k_1, k_2, and k_3 are the empirically determined radionuclide-specific rate constants; and C^{0}_{w1}, $C^{0}_{\mathrm{w}2}$, and $C^{0}_{\mathrm{w}3}$ are the initial activity concentrations of a correspondent component of radionuclide in solution.

Application of the three-exponential function to describe dynamics of dissolved ^{137}Cs in Pripyat River using selected data from DB “RUNOFF” is presented in Fig. 7.2. It can be seen that the initial phase after the accident lasted less than 1 year. Henceforth, we will exclude the initial phase from consideration of long-term dynamics since changes during the initial phase are determined by different processes than those during the middle- and long-term phases.

In the case of Chernobyl, the analysis and prediction were focused on dissolved ^{137}Cs since it was predominant in the contaminated surface waters and controlling the radionuclide transport (Borzilov et al. 1989, 1994; Smith et al. 2004; Konoplev et al. 1996a). Data on mass particulate concentration of ^{137}Cs in surface waters were scarce. Luckily, in 2000–2002 through the French-German Initiative (FGI) on Chernobyl, data on radioactive contamination of rivers were collected and stored in the Data Base (DB) “RUNOFF” (Konoplev et al. 2002; Bulgakov and Konoplev 2004). This database includes data on the dissolved and particulate concentrations of ^{137}Cs and dissolved ^{90}Sr in 20 rivers flowing through the territory contaminated after the Chernobyl accident. The data were obtained in the period from 1986 to 2000 by

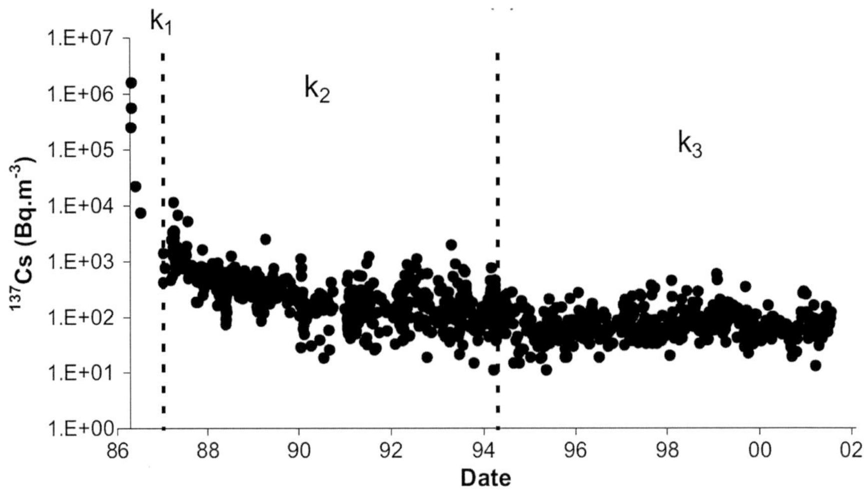

Fig. 7.2 Dynamics of ^{137}Cs activity concentrations in Pripyat River at Chernobyl for 1986–2002. The different phases in the exponential decline are illustrated by a dotted line and k_i values (Smith et al. 2005a)

the national monitoring programs with contributions by various organizations from Belarus, Russia, and Ukraine. After 2000, we used monitoring data mostly obtained by the Ukrainian Hydrometeorological Institute (UHMI) in Kiev (Ukraine) and by the Hydrometeorological Service of Republic Belarus in Minsk (Belarus). As far as we know, this is the first-time data on long-term dynamics of particulate ^{137}Cs in surface waters, which was reviewed and analyzed.

Figure 7.3 presents time dependences of annual mean particulate ^{137}Cs activity concentrations (C_s) in Pripyat River near Chernobyl (a) and in Dneper River near Nedanchichi (b) after 1987 and an approximation by an empirical two-exponential model:

$$C_s(t) = C_{s1}^0 \times e^{-\lambda_1 t} + C_{s2}^0 \times e^{-\lambda_2 t} \quad (7.2)$$

the parameters are presented in Table 7.1, where $\lambda_i = \lambda + k_i$.

The empirical model (Eq. (7.1)) based on the multiexponential description of radionuclide dynamics, however, does not seem to adequately reflect actual mechanisms underlying changes of radionuclide activity concentrations in water. The drawback of this approach is that it requires using a number of functions accounting for the short-, middle-, and long-term phases after the accident as well as parameters that are not known initially.

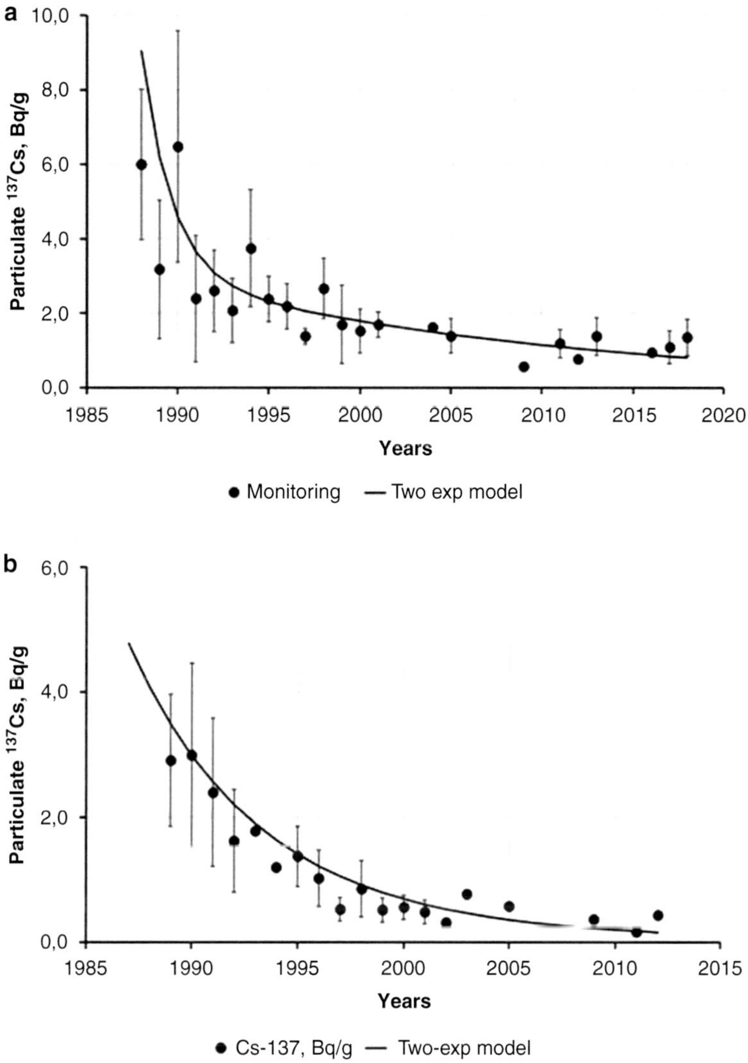

Fig. 7.3 Time dependence of annual mean particulate ^{137}Cs activity concentrations (black circles) in 1988–2018 for Pripyat River at Chernobyl (**a**), and in 1989–2012 for Dneper River at Nedanchichi (**b**) and the approximation by the empirical two-exponential model (line)

Table 7.1 Parameters used to approximate the time dependence of particulate ^{137}Cs in Pripyat River at Chernobyl and Dneper River at Nedanchichi by the two-exponential model (7.3)

	Two-exponential model (7.3)			
River (cross section)	C^0_{s1}, Bq g^{-1}	C^0_{s2}, Bq g^{-1}	λ_1, $year^{-1}$	λ_2, $year^{-1}$
Pripyat (Chernobyl)	20	3.3	0.60	0.044
Dneper (Nedanchichi)	5.0	0.6	0.17	0.068

7.3 Semiempirical "Diffusional Modeling" of Long-Term Dynamics of Radionuclides in Rivers

In the long term, a major source of suspended matter for surface runoff and rivers is the top layer of catchment soil. The initial period after fallout when the radionuclides deposited on the water's surface and were a major contributor to specific activity of river water is not considered here. The duration of this period, most likely, does not exceed one year, at least for the rivers flowing through the region, which was the worst contaminated after the Chernobyl accident. The ^{137}Cs transport by these rivers from May to December 1986 was several tenths of a percent of the total activity of radionuclide deposited on the catchment (Borzilov et al. 1993). Following, the radionuclide concentration in river water was determined by transport from the catchment, sorption–desorption, and resuspension–sedimentation in the system "water–suspended matter–bed sediments," as well as dilution by relatively clean groundwater and concentrating due to evaporation. For the long-term dynamics of radionuclides, wash-off is the determining factor (Bulgakov et al. 2002).

7.3.1 Dynamics of Radiocesium in the Topsoil Layer of the Chernobyl Exclusion Zone (ChEZ)

Radionuclide concentration in topsoil is decreasing over time due to its vertical migration to deeper layers (Prokhorov 1981; Konoplev and Golubenkov 1991; Ivanov et al. 1997; Konoplev et al. 2016). Long-term dynamics of ^{137}Cs vertical distribution in soil of the Pripyat River catchment is illustrated in Fig. 7.4. ^{137}Cs vertical profiles are presented for individual soil cores collected from two sites in ChEZ: (a) nearby the abandoned village Benevka (9 km NW of ChNPP) and (b) nearby the abandoned village Korogod (15 km S-SW of ChNPP). Data for the first 10 years after the Chernobyl accident were taken from the DB "RUNOFF" (Konoplev et al. 2002). Data for 2017 were obtained in the frame of Japan's Science and Technology Research Partnership for Sustainable Development (SATREPS) Project "Strengthening of the Environmental Radiation Control and Legislative Basis for the Environmental Remediation of Radioactively Contaminated Sites" (Konoplev et al. 2018). In both cases, the ^{137}Cs activity concentration in the topsoil layer can be seen to decrease significantly with time.

Figure 7.5 presents a characteristic vertical distribution of ^{137}Cs in soil of the Pripyat River catchment in 2017. It should be noted that the maximum ^{137}Cs activity concentration is still located in the topsoil layer.

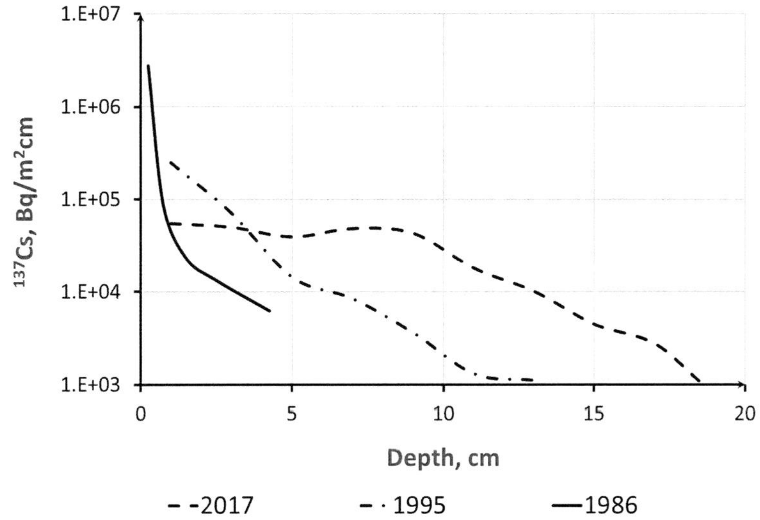

A

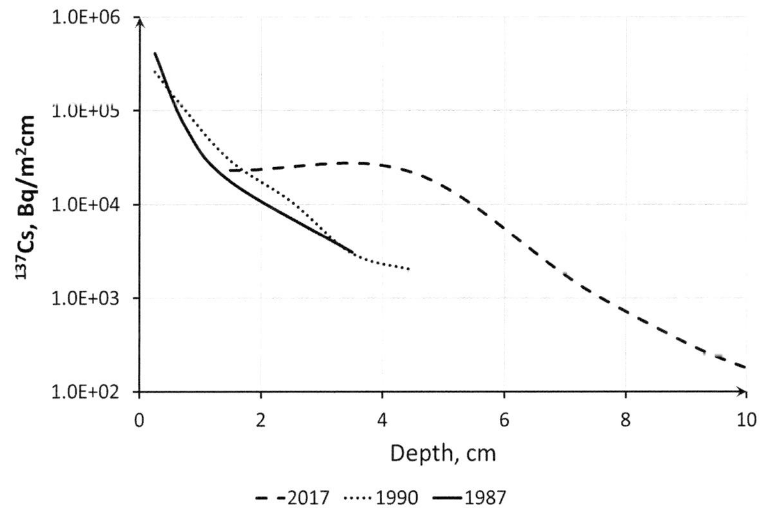

B

Fig. 7.4 Dynamics of ^{137}Cs vertical distribution in soil at Benevka (**a**) and at Korogod (**b**) in ChEZ for 1986–2017

7.3.2 Particulate r-Cs in Surface Runoff and Rivers

Vertical migration in catchment soils is a critical process responsible for a decline in particulate radionuclide concentration in surface waters in the long term. According to the convection–dispersion model (Crank 1975; Prokhorov 1981; Konoplev and

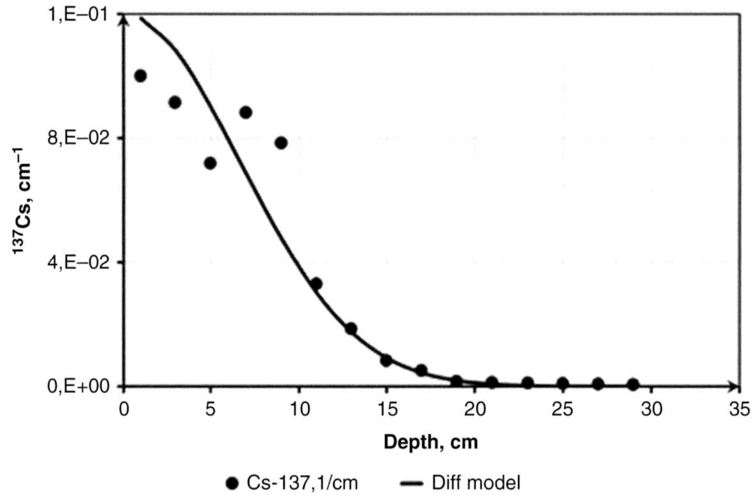

Fig. 7.5 ^{137}Cs differential distribution in the soil profile at Benevka in 2017: measurements (points) against "diffusion" model with $D_{\text{eff}} = 0.7$ cm^2/year (line). ^{137}Cs levels are presented as volumetric activity concentrations normalized by ^{137}Cs deposition (Bq cm^{-3}/Bq cm^{-2})

Golubenkov 1991; Konshin 1992; Ivanov et al. 1997; Bossew and Kirchner 2004), the change in radionuclide concentration in the topsoil layer C_s with time is given by the following equation (Bulgakov et al. 2002):

$$C_{\text{s}}(t) = \frac{\sigma}{\rho\sqrt{\pi D_{\text{eff}} t}} e^{-\left(\frac{v^2}{4D_{\text{eff}}}+\lambda\right)t} \tag{7.3}$$

where ρ is the average bulk density of soil on the catchment; D_{eff} is the effective dispersion coefficient; v is the effective velocity of convective transport; λ is the radioactive decay constant; σ is the radionuclide deposition density; and t is the time.

Time dependence of particulate radionuclide in surface runoff and river water can be described by this equation with the values of D_{eff} and v averaged over the catchment area.

Studies of radiocesium vertical migration in soils of river catchments show that, as a rule, its transport due to dispersion prevails over its convective transport (Konshin 1992; Konoplev et al. 1992; Ivanov et al. 1997). This is also confirmed by the shape of r-Cs profiles in soils presented in Fig. 7.5. Considering that

$$\frac{v^2}{4D_{\text{eff}}} \ll \lambda \tag{7.4}$$

Equation (7.1) can be simplified as follows:

$$C_s(t) = \frac{\sigma}{\rho\sqrt{\pi D_{\text{eff}} t}} e^{-\lambda t} = C_s^0 \frac{e^{-\lambda t}}{\sqrt{t}} \quad (7.5)$$

Particulate radionuclide concentration in surface runoff and rivers should follow Eq. (7.5) with parameters σ, D_{eff}, and ρ averaged over the catchment area. The advantage of the proposed semiempirical "diffusion-based approach" is that the middle- and long-term phases after a nuclear accident can be described by the same equation using the same physically based parameters, which can be estimated or determined by field or laboratory studies. Generally, decay corrected particulate r-Cs activity concentrations in surface runoff and rivers should follow the inverse square root of time function.

Figure 7.6 presents data on long-term dynamics of particulate ^{137}Cs in river water for two large rivers of the Chernobyl area: Pripyat and Dneper. Similar to Fig. 7.3, data for 1986–1999 were selected from DB "RUNOFF"; specifically, data obtained by RPA "Typhoon" (Obninsk, Russia, http://www.rpatyphoon.ru/) and UHMI (Kiev, Ukraine, https://uhmi.org.ua/) were used because they were most reliable. Data on particulate ^{137}Cs were available only for these two large rivers. Data were averaged on an annual basis for 1987–2001. For some years in the period of 2002–2017, only single measurements by UHMI were available and used. Data for the initial period after the accident were not used since it was impacted by initial deposition on the water surface.

The significant scatter in the experimental data on ^{137}Cs activity concentrations in river water, particularly in the first years after the accident, can be attributed to extremely nonuniform distribution of radionuclides on the catchment and occurrence of hot fuel particles as well as uncertainties associated with sampling, processing, and measurements (Bobovnikova et al. 1991; Konoplev and Bulgakov 1999).

As can be seen, both the proposed semiempirical "diffusional" model and the empirical two-exponential model account equally well for the general trend in the long-term dynamics of particulate ^{137}Cs in rivers. However, the empirical multiexponential model of radionuclide dynamics in rivers does not seem to adequately reflect the actual mechanisms underlying changes of radionuclide activity concentrations in water. The problem using this approach is that at least two functions must be used to account for the middle- and long-term phases after the accident, as well as parameters that are initially unknown (Konoplev et al. 2018). Just as importantly, long-term dynamics of radionuclide concentration on suspended matter can be described with only a single parameter $C_s^0 = \frac{\sigma}{\rho\sqrt{\pi D_{\text{eff}}}}$, the value of which is determined with physically meaningful and understandable characteristics such as average deposition on the catchment and the effective coefficient of radionuclide dispersion in catchment soils D_{eff}. Data on radionuclides D_{eff} in different soils are readily available in the literature (Bulgakov et al. 1991; Konoplev and Golubenkov 1991; Konoplev et al. 1992, 2016; Konshin 1992; Ivanov et al. 1997). Additionally, C_s^0 can be estimated based on monitoring data of radionuclide concentration on

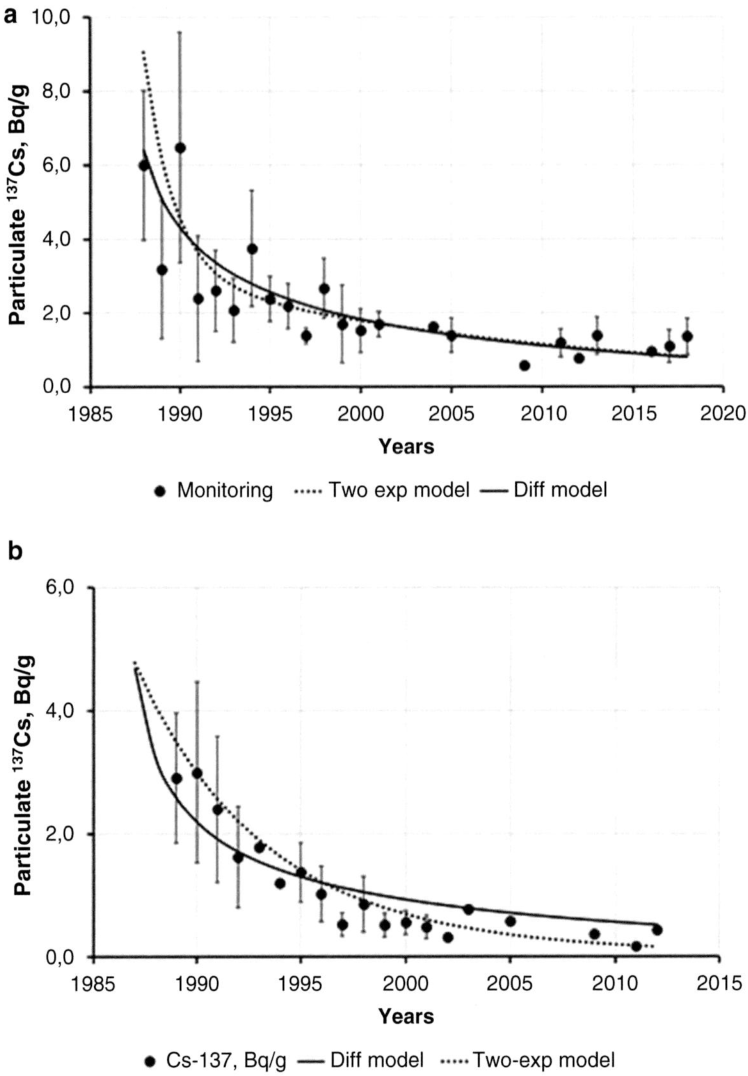

Fig. 7.6 Time dependence of annual mean particulate ^{137}Cs activity concentrations (points) in Pripyat River at Chernobyl (**a**) from 1989 (3 years after the accident) to 2017 and in Dneper River at Nedanchichi (**b**) for 1989–2012 against approximations by Eq. (7.5) (solid line) and by Eq. (7.2) with parameters from Table 7.1 (dotted line)

suspended matter in rivers when the initial phase is completed. Afterward, long-term predictions can be performed based on the "diffusional" model using Eq. (7.5). On the other hand, the two-exponential model is based on the application of four fitting parameters, which cannot be determined immediately after the accident and requires long-term field observations for their estimation.

7.3.3 Dissolved r-Cs in Surface Runoff and Rivers

Dissolved radionuclides in surface runoff and rivers are transferred from soil to water mostly due to cation exchange (Konoplev et al. 1992; Bulgakov et al. 1999). As this takes place, only the exchangeable fraction of radionuclide is involved in exchange with solution. In the course of time after the accident, the mobile r-Cs is fixed on a timescale of several months (Konoplev et al. 1992, 1996b; Konoplev and Bulgakov 2000). This means that for dissolved r-Cs the two major processes determining its dynamics in surface runoff and river water are relatively fast fixation of mobile (exchangeable) form and slow vertical migration in catchment soil. Fixation of r-Cs in soil can be treated as diffusion of Cs^+ cation into interlayers of the crystal lattice of clay minerals (Bulgakov and Konoplev 1996, 2001; Konoplev and Bulgakov 2000). Hence, if a radionuclide is deposited in mobile form (soluble or exchangeable) time dependence of its exchangeable form is described with fairly good accuracy by the following equation (Konoplev and Bulgakov 2000; Bulgakov et al. 2002; Bulgakov and Konoplev 2004):

$$\alpha_{\mathrm{ex}}(t) = \alpha_{\mathrm{ex}}(\infty)\left(1 + \frac{\delta}{\sqrt{t}}\right) \tag{7.6}$$

where $\alpha_{\mathrm{ex}}(t)$ is the current fraction of exchangeable radionuclide in soil; $\alpha_{\mathrm{ex}}(\infty)$ is the exchangeable fraction of the radionuclide at equilibrium; and δ is the soil specific kinetic parameter of radionuclide fixation (Konoplev and Bulgakov 2000):

$$\delta = \frac{l}{\sqrt{\pi D_{\mathrm{s}}}} \tag{7.7}$$

where l is the characteristic thickness of the diffusion layer in clay mineral crystal lattices and D_s is the radionuclide diffusion coefficient of the mineral solid phase.

It should be noted that both processes determining long-term dynamics of dissolved r-Cs in surface runoff and rivers, its faster fixation and slower decrease of its concentration in the topsoil layer due to vertical migration in soil, are diffusion based. In this regard, temporal changes of dissolved r-Cs concentration in surface runoff and/or river water can be described by the following equation (Bulgakov et al. 2002; Bulgakov and Konoplev 2004; IAEA 2006):

$$C_{\mathrm{w}}(t) = \frac{\alpha_{\mathrm{ex}}(\infty)\sigma}{K_{\mathrm{d}}^{\mathrm{ex}}\sqrt{\pi D_{\mathrm{eff}} t}}\left(1 + \frac{\delta}{\sqrt{t}}\right)\mathrm{e}^{-\left(\frac{v^2}{4D_{\mathrm{eff}}}+\lambda\right)t} \tag{7.8}$$

where $K_{\mathrm{d}}^{\mathrm{ex}}$ is the exchangeable distribution coefficient characterizing equilibrium distribution of the exchangeable fraction of radionuclide between the solid phase and solution (Konoplev and Bulgakov 2000a). $K_{\mathrm{d}}^{\mathrm{ex}}$ is the constant for sediments of similar properties and stationary water composition and does not depend on transformation processes.

$$K_d^{ex} = \alpha_{ex}(t) \cdot K_d^{tot}(t) = \alpha_{ex}(\infty) \cdot K_d^{tot} \tag{7.9}$$

where the total distribution coefficient $K_d^{tot}(t) = \frac{C_s(t)}{C_w(t)}$ and $K_d^{tot} = K_d^{tot}(\infty)$—its value at equilibrium.

For longer times when equilibrium between the exchangeable and nonexchangeable forms of r-Cs is reached and when condition (7.4) is valid, the equation can be simplified as follows:

$$C_w(t) = \frac{\alpha_{ex}(\infty)\sigma}{\rho K_d^{ex}\sqrt{\pi D_{eff} t}} e^{-\lambda t} = \frac{\sigma}{\rho K_d^{tot}\sqrt{\pi D_{eff} t}} e^{-\lambda t} = C_w^0 \frac{e^{-\lambda t}}{\sqrt{t}} \tag{7.10}$$

$$C_w^0 = \frac{C_s^0}{K_d^{tot}} \tag{7.11}$$

The ^{137}Cs long-term observations show that despite a relatively high variability of K_d^{tot} in the aquatic environment no temporal trend is seen in K_d^{tot} changes. This is illustrated in Fig. 7.7 presenting the time dependence of K_d for Chernobyl-derived ^{137}Cs at the cross-section "Chernobyl" in Pripyat River from 1990 to 2016 (a) and at the cross-section "Nedanchichi" in Dneper River from 1989 to 2012 (b).

For both rivers, the K_d value for ^{137}Cs is variable, and yet no trend has been detected. For Pripyat-Chernobyl $K_d(^{137}Cs) = (3.3 \pm 1.3) \times 10^4$ L/kg and for Dneper-Nedanchichi $K_d(^{137}Cs) = (6.4 \pm 2.2) \times 10^4$ L/kg. It is worth mentioning that K_d for Fukushima-derived ^{137}Cs in the rivers of Fukushima prefecture, including Abukuma River, is about an order of magnitude higher (Konoplev 2015; Takahashi et al. 2020).

To test the applicability of Eq. (7.8) for describing dynamics of dissolved r-Cs in small rivers of the Chernobyl 30-km zone: Uzh (Cherevach), Teterev (Loputki), and Irpen' (Kozarovichi) data from DB "RUNOFF" were used (Bulgakov et al. 2002). On the catchments of these rivers, fuel particles that could influence the mobility of r-Cs in soil and its availability for exchange with solution were insignificant in number.

Since mineral sandy loamy and loamy soils are predominant in the region, the value of parameter δ was taken to be 0.5 year$^{0.5}$, as was recommended for such soils by Bulgakov and Konoplev 2001. Figure 7.8 presents a comparison of monitoring data of dissolved ^{137}Cs in rivers Uzh, Irpen, and Teterev with the semiempirical "diffusional" modeling by Eq. (7.8).

Figure 7.9 shows the approximation by the simplified "diffusional" model (Eq. 7.10) to describe the time dependence of dissolved ^{137}Cs in small Sakhan River of the Chernobyl exclusion zone.

Figure 7.10 shows the time dependence of annual mean dissolved ^{137}Cs activity concentrations in Pripyat River at two cross sections—Mozyr, upstream of the most

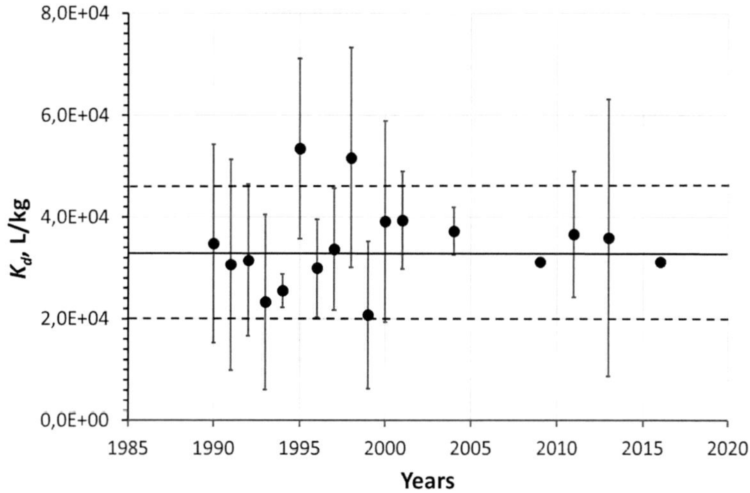

A

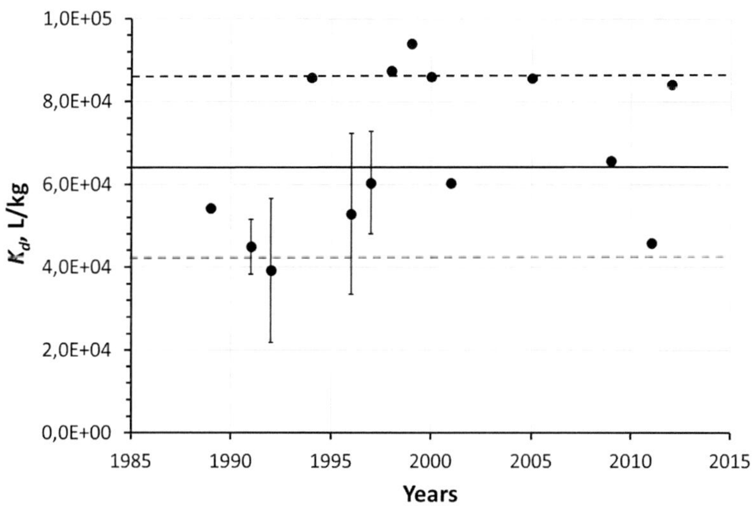

B

Fig. 7.7 Time dependence of annual mean values of K_d for ^{137}Cs in Pripyat River at cross-section Chernobyl (**a**) and Dneper River at cross-section Nedanchichi. The solid line corresponds to the mean K_d^{tot} value for 1990–2016 with the standard deviation (dash lines)

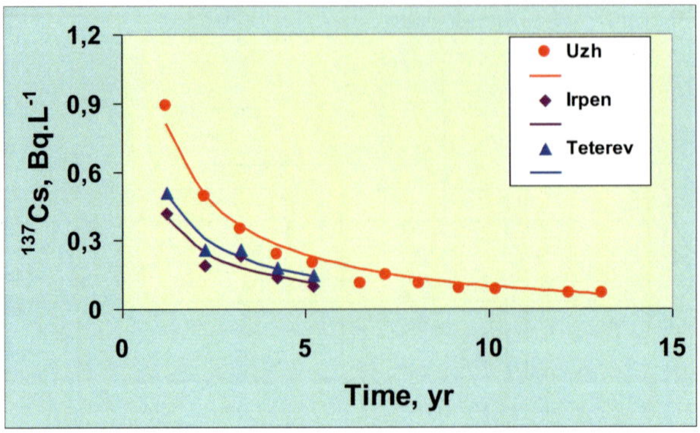

Fig. 7.8 Observation data on dissolved ^{137}Cs activity concentrations for three rivers in the southern sector of the Chernobyl contaminated area with low content of fuel particles: Uzh at Cherevach, Teterev at Loputki, and Irpen' at Kozarovichi against approximation by Eq. (7.8) (modified after Bulgakov et al. 2002)

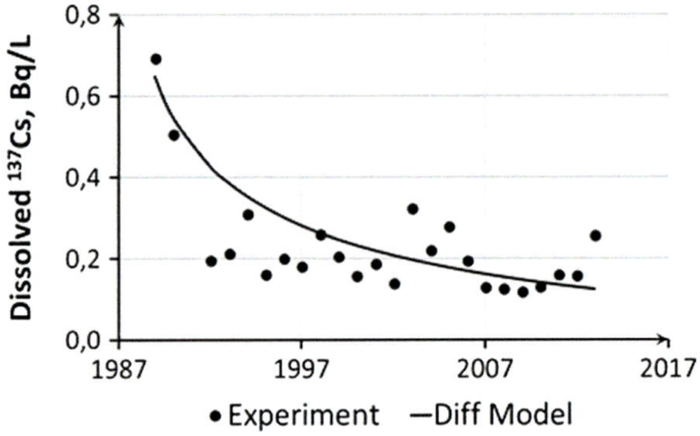

Fig. 7.9 Time dependence of annual mean dissolved ^{137}Cs activity concentrations (points) in Sakhan River at Novoshepelichi for 1989–2015 against the simplified approximation by Eq. (7.10) (line)

contaminated areas and Chernobyl—in comparison with the approximation by Eq. (7.6). In the case of Mozyr, data from DB "RUNOFF" for 1990–1999 and monitoring data of Hydrometeorological Service of Belarus Republic for 2000–2013 have been utilized.

The proposed approach to describe long-term dynamics can be applied not only for radiocesium but also for other radionuclides as well. Figure 7.11 presents the time dependence of dissolved ^{137}Cs and ^{90}Sr in Dneper River for 1990–2013 at the cross-section Rechitsa (Belarus) in comparison with the approximation by Eq. (7.6).

Figure 7.12 shows long-term dynamics of dissolved ^{137}Cs and ^{90}Sr in Pripyat River at Mozyr (1990–2013). Note that the dissolved activity concentrations of ^{137}Cs and ^{90}Sr in Pripyat River at this cross section are similar despite the lower

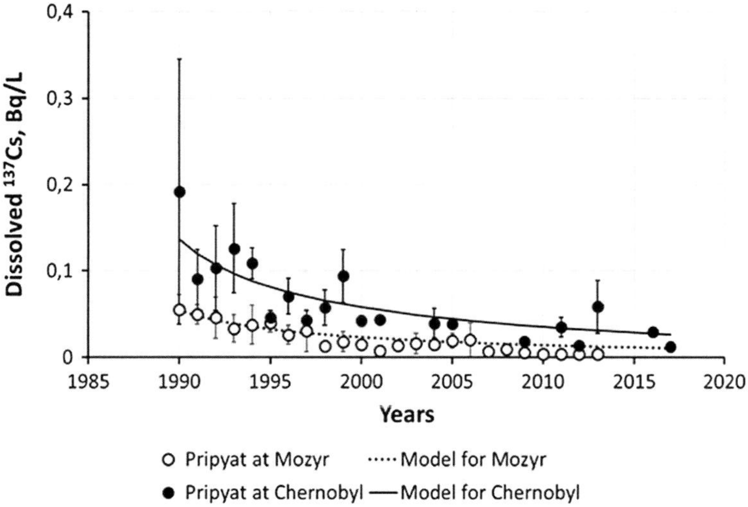

Fig. 7.10 Time dependence of annual mean dissolved ^{137}Cs activity concentrations (points) in Pripyat River at Mozyr and at Chernobyl against the simplified approximation by Eq. (7.10) (lines)

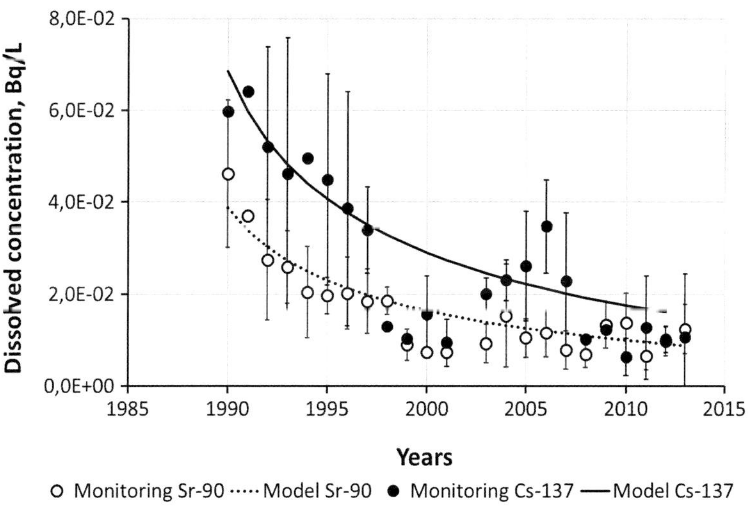

Fig. 7.11 Time dependence of annual mean values of dissolved ^{137}Cs and ^{90}Sr activity concentrations in Dneper River at cross-section Rechitsa (Belarus) against the simplified approximation by Eq. (7.10) (lines)

average deposition of ^{90}Sr on the catchment at Mozyr, $\sigma(^{90}\mathrm{Sr}) = 1.85$ kBq m^{-2}, as compared with ^{137}Cs, $\sigma(^{137}\mathrm{Cs}) = 51$ kBq m^{-2}. The lower K_d value for ^{90}Sr results in a larger portion of Sr transferring to the water relative to Cs. This can be approximated by Eq. (7.10) with the same value of $C_\mathrm{w}^0 = \frac{\sigma}{\rho K_\mathrm{d}\sqrt{\pi D_\mathrm{eff}}}$.

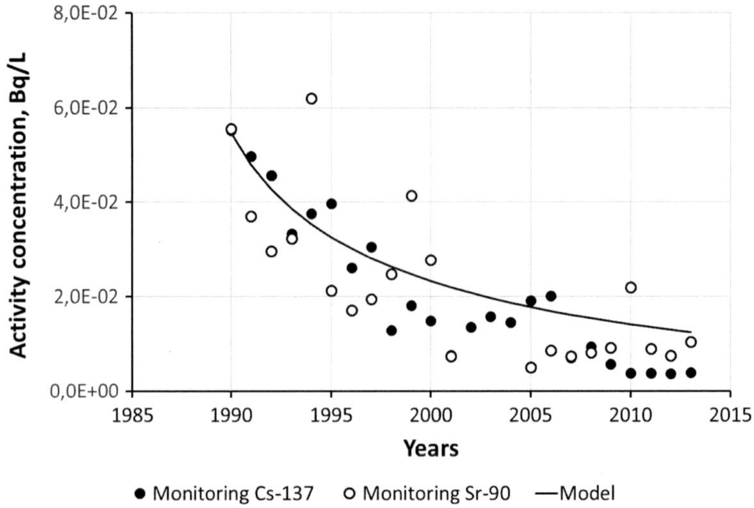

Fig. 7.12 Time dependence of annual mean values of dissolved ^{137}Cs and ^{90}Sr activity concentrations in Pripyat River at cross-section Mozyr (Belarus) against the simplified approximation by Eq. (7.10) (line)

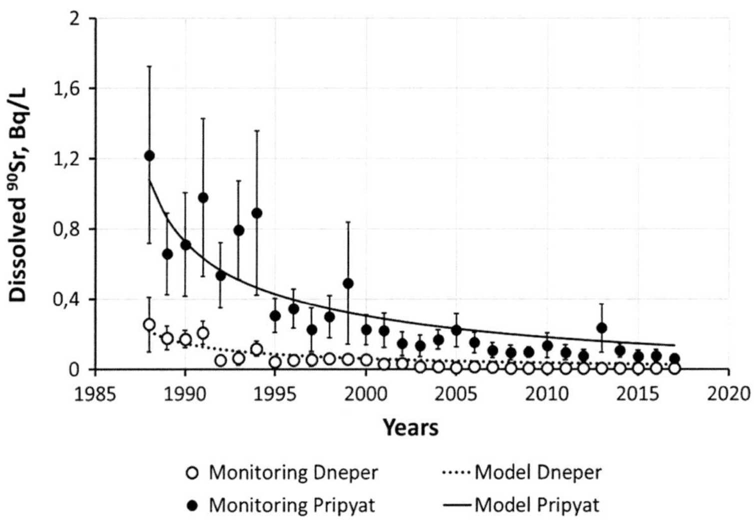

Fig. 7.13 Time dependence of annual mean dissolved ^{90}Sr in Pripyat River at Chernobyl and in Dneper River at Nedanchichi and their approximation by the "diffusional" model

Figure 7.13 illustrates the application of the proposed "diffusional" model to approximate the time dependence of dissolved ^{90}Sr in Pripyat River at Chernobyl and Dneper River at Nedanchichi in Ukraine for 1988–2017.

The analysis suggests that starting about one year after the initial deposition from the Chernobyl accident the changes in ^{137}Cs and ^{90}Sr activity concentrations, both

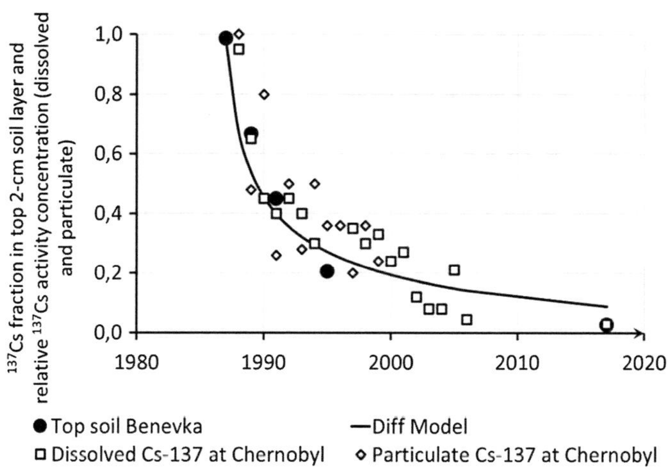

Fig. 7.14 Time dependence of ^{137}Cs inventory fraction in the top 2 cm of soil at Benevka on the Pripyat River catchment (black circles), its approximation by Eq. (7.3) and dynamics of annual mean concentrations of dissolved and particulate ^{137}Cs in relative units in Pripyat River at Chernobyl in 1987–2017

dissolved and particulate, are determined by changes in their concentrations in the upper soil layer. This is illustrated in Fig. 7.14 showing experimental data on temporal changes of the fraction of ^{137}Cs inventory in the top 2 cm of soil on the Pripyat catchment near the settlement Benevka in the Chernobyl exclusion zone, its approximation using Eq. (7.5) as well as dissolved and particulate ^{137}Cs in Pripyat River at the Chernobyl cross section. As can be seen, both the dissolved and particulate ^{137}Cs activity concentrations are generally described by the dependence above.

7.4 Reconstruction of Long-Term Dynamics of Chernobyl-Derived ^{137}Cs in River Using Its Vertical Distribution in Bottom Sediments of Dam Reservoir Accumulation Zones

Following the Chernobyl accident, many rivers in contaminated areas, regretfully, have not been subjected to systematic long-term monitoring. In view of this, to reconstruct time changes of particulate ^{137}Cs in rivers, Konoplev et al. (2019) suggested using the vertical distribution in sediment accumulation zones of dam reservoirs on contaminated rivers. The pattern of ^{137}Cs vertical distributions in sediment accumulation zones obtained for the central part of the Schekino reservoir on Upa River (Chernobyl contaminated areas of Tula region in Russia) showed that practically no vertical mixing of sediments occurred and the ^{137}Cs peak was well-defined, rather than diffused (Konoplev et al. 2019). Assuming that sediment accumulation rates have been uniform since the accident, layers of bottom sediments can be attributed to a certain time of sedimentation. With ^{137}Cs activity concentration in a specific bottom sediment layer corresponding to ^{137}Cs concentration on suspended matter at that moment, dynamics of particulate ^{137}Cs activity concentrations has been reconstructed for Upa River from 1986 to 2017 (Fig. 7.15) (Konoplev et al. 2019). The ^{137}Cs concentrations can be seen to have decreased by more than an order of magnitude over the time since the accident, with only minor changes during the last 15 years. It should be noted that the semiempirical "diffusional" model has provided a good representation of time dependency of particulate ^{137}Cs concentrations in Upa River reconstructed using radionuclide vertical distribution in bottom sediments of Schekino reservoir (Fig. 7.16).

Using a typical value of the distribution coefficient K_d for the rivers of the Chernobyl contaminated zone, activity concentrations of dissolved ^{137}Cs in Upa River over 30 years following the accident have been estimated (Konoplev et al. 2019). The obtained activity concentrations of dissolved ^{137}Cs in Upa River and their temporal trend were found to be consistent with the measurement data for 1987–1991 (Vakulovsky et al. 1994). In the open literature, data are available for these years only.

7.5 Long-Term Dynamics of Radionuclide in Lakes and Ponds

The presented diffusion-based approach to describe time changes of radionuclide activity concentrations can be extended to lakes and ponds. Like for rivers and open dam reservoirs, the most important mechanism for a decrease in lake water activity concentration in the initial stage of contamination is downward movement of radionuclides in the water column followed by adsorption and fixation in the top bottom sediment layer. For relatively small lakes, this stage can last for several weeks or months (Santschi et al. 1990) and will end when distribution of radionuclide in the "water—top sediment layer" system becomes close to equilibrium. Later dynamics of activity concentration in water is mainly determined by radionuclide migration to deeper layers of sediments and additional "spikes" of radioactivity to lake water as a result of surface runoff from the contaminated catchment.

In the case of a closed lake or pond, that is, when the effect of particulate and dissolved radionuclide runoff from lake/pond catchment can be neglected, particulate radionuclide activity concentration corresponds to its concentration in the top bottom sediment layer. Radionuclide transport in bottom sediments can be described by a simple diffusion equation (Prokhorov 1981; Bulgakov et al. 2002a). Therefore, radionuclide concentration in the top bottom sediment layer and hence particulate radionuclide concentration will follow an equation similar to (7.5) but with the value of D_{eff} relevant for bottom sediments (Bulgakov et al. 2002a).

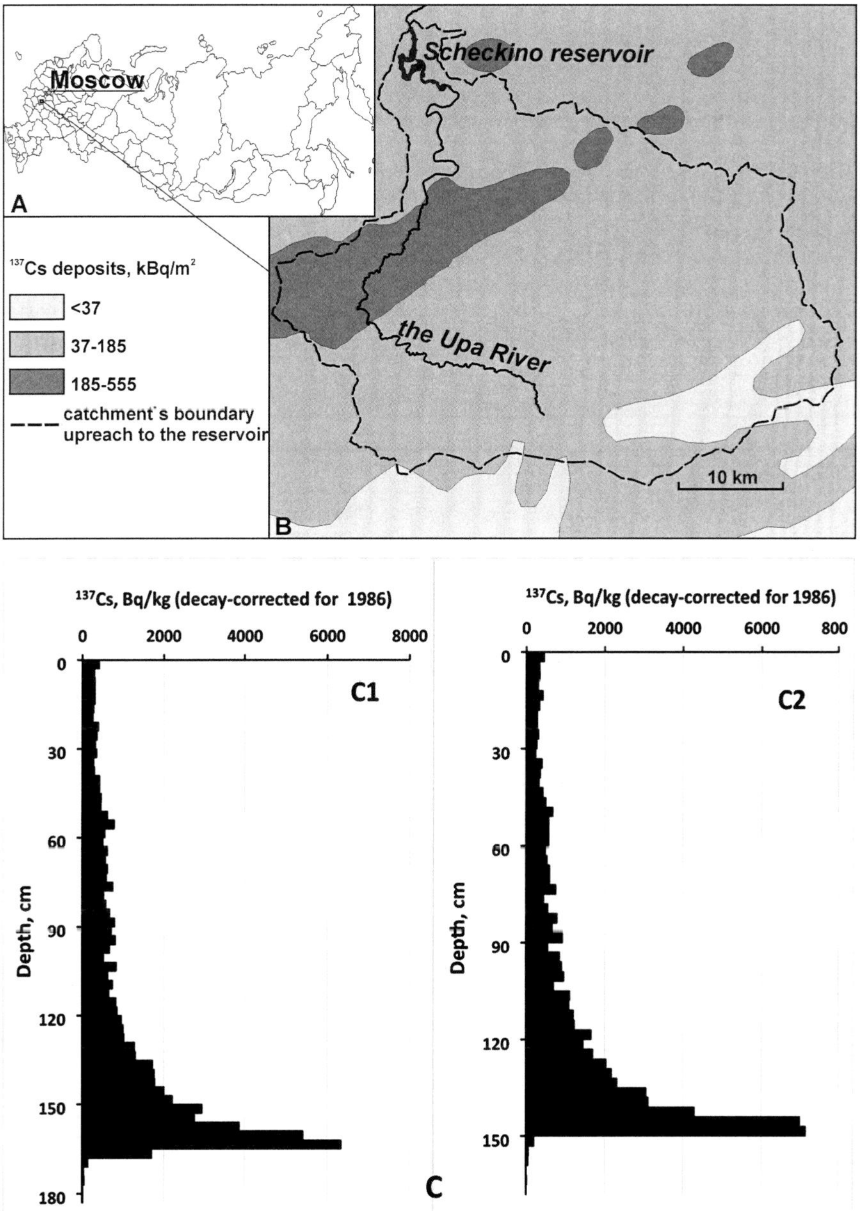

Fig. 7.15 Location of Upa River basin (**a**) and map of Chernobyl-derived ^{137}Cs deposition in Upa River catchment upstream Schekino reservoir, decay corrected for 1 May 1986 (**b**) and the vertical distribution of decay corrected ^{137}Cs in bottom sediments of Schekino dam reservoir cores (Konoplev et al. 2019)

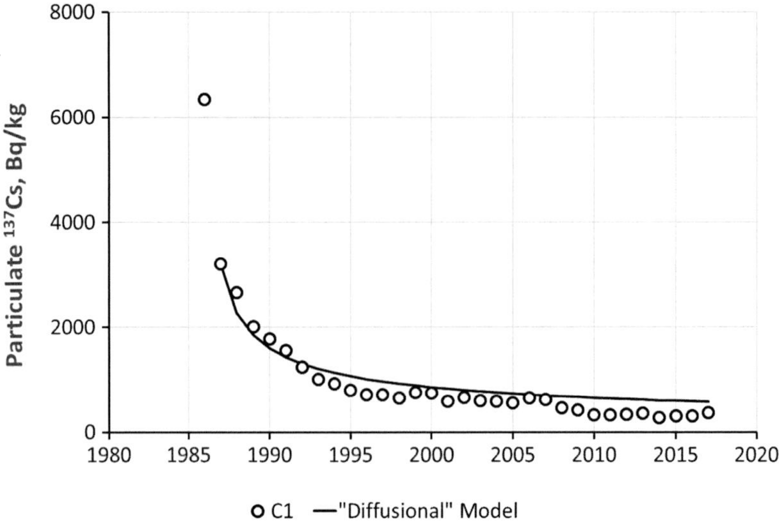

Fig. 7.16 Semiempirical modeling of reconstructed particulate ^{137}Cs in Upa River (line) based on the depth profile for bottom sediments (core C1, points) (Konoplev et al. 2019)

In the case of a lake with catchment soil as a predominant source of suspended matter, the time dependence of particulate radionuclide activity concentration should follow Eq. (7.5). It can be shown that in a general case for lakes or ponds particulate radionuclide activity concentration should follow the simplified equation:

$$C_s(t) = C_s^0 \frac{e^{-\lambda t}}{\sqrt{t}} \tag{7.12}$$

where C_s^0 is the fitting parameter, which can be estimated from monitoring data after the initial stage (a few years after the accident).

Like for surface runoff and rivers, dissolved radionuclide activity concentration in lakes or ponds should follow the simplified equation (Bulgakov et al. 2002a):

$$C_w(t) = C_w^0 \frac{e^{-\lambda t}}{\sqrt{t}} \text{ where } C_w^0 = \frac{C_s^0}{K_d^{tot}} \tag{7.13}$$

Equation (7.13) was tested against monitoring data for a closed lake Svyatoe in Bryansk region of Russia in 1992–2000 (Fig. 7.17), and the agreement between calculated and measured values was found to be reasonable (Konoplev et al. 1998; Bulgakov et al. 2002a).

This semiempirical "diffusional" model based on Eq. (7.13) was used to describe long-term dynamics of dissolved ^{137}Cs in most severely contaminated water bodies of the Chernobyl exclusion zone lakes Glubokoe and Azbuchin (Fig. 7.18). To test the proposed method of reconstruction of radionuclide long-term dynamics in lakes, cores of bottom sediments have been collected in 2018 on these two lakes.

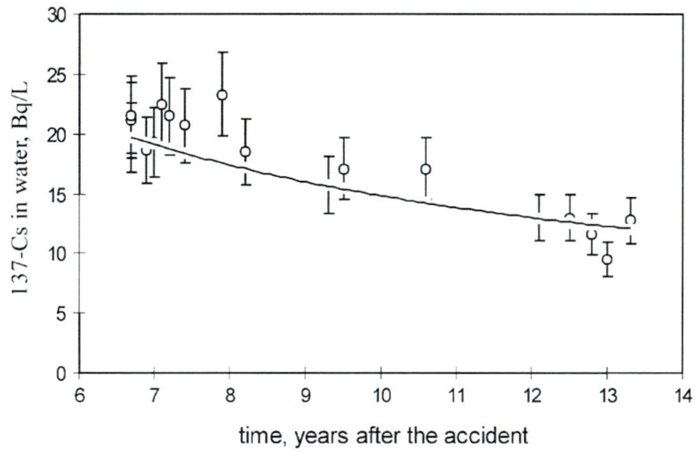

Fig. 7.17 Measured (points with standard deviations) versus calculated by the semiempirical "diffusional" model (line) dissolved ^{137}Cs activity concentration in Svyatoe Lake for 1992–2000 (Bulgakov et al. 2002a)

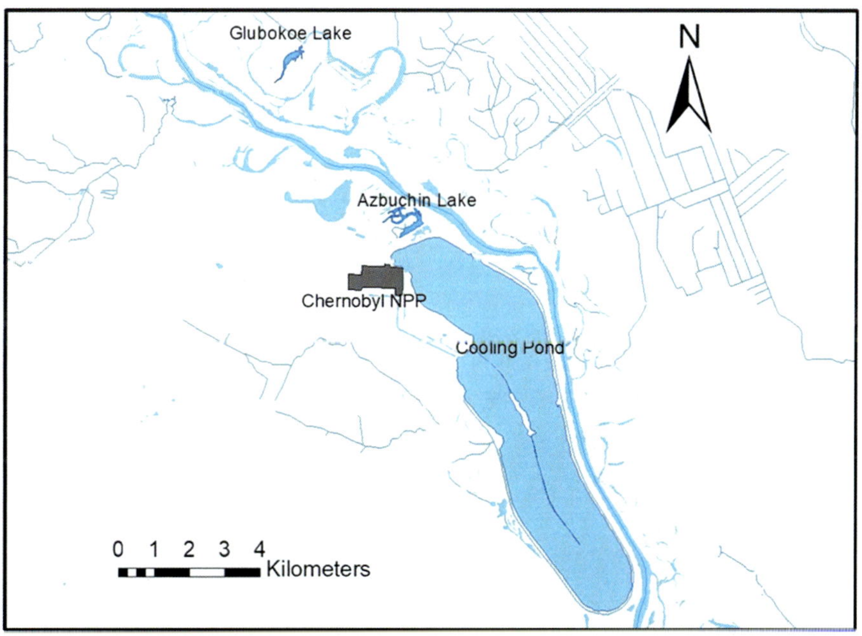

Fig. 7.18 Location of heavily contaminated lakes Glubokoe and Azbuchin in ChEZ

Figure 7.19 presents vertical distributions of ^{137}Cs, ^{241}Am, and ^{90}Sr (for Azbuchin) in bottom sediments of the lakes. The vertical distributions of the radionuclides ^{137}Cs and ^{241}Am strongly bound to particles in sediment accumulation

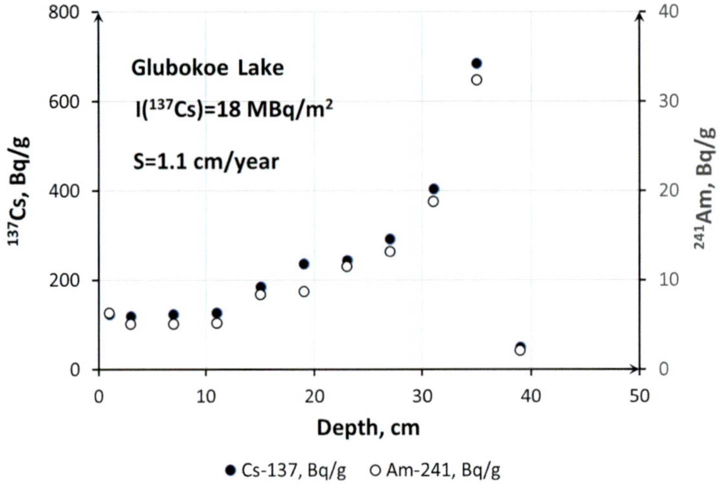

A

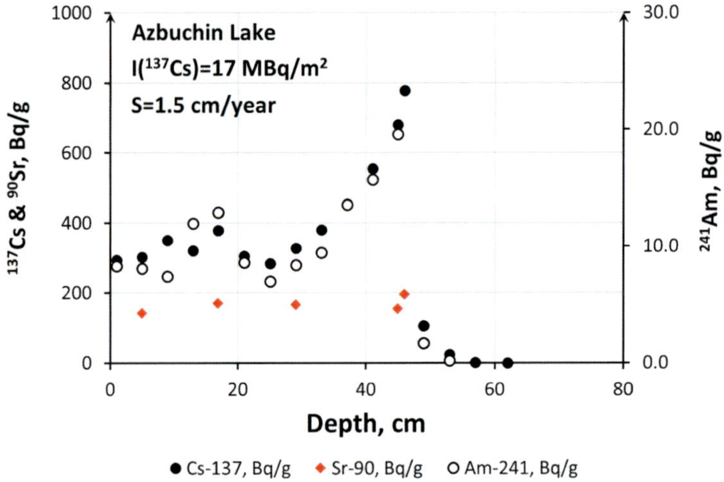

B

Fig. 7.19 Vertical distributions of the long-lived radionuclides in bottom sediments of Glubokoe Lake (**a**) and Azbuchin Lake (**b**). I is the radionuclide inventory in bottom sediments and S is the estimated average sedimentation rate

zones of the lakes (Fig. 7.18) suggest that practically no vertical mixing of sediments took place. Similar to the abovementioned Schekino dam reservoir on Upa River, layers of bottom sediments were matched with the time of sedimentation. This made it possible to obtain the dynamics of particulate ^{137}Cs activity concentrations from 1986 to 2018, indicating their decrease by about ten times over this time period. Next, based on particulate ^{137}Cs concentrations and using experimental values of the

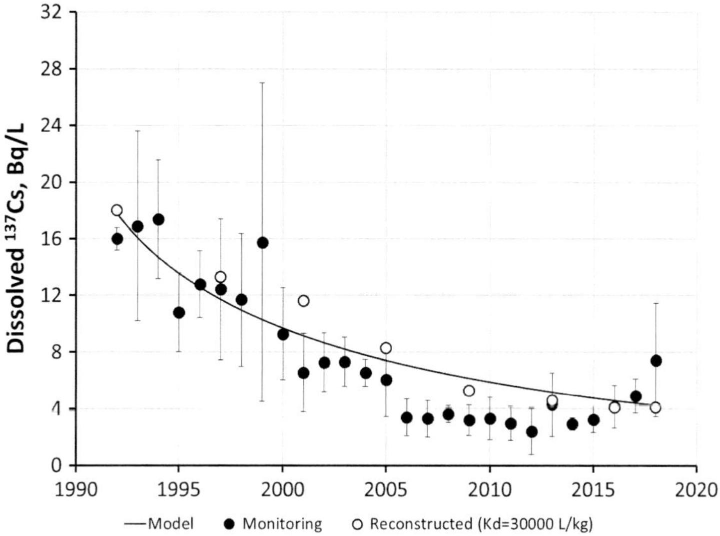

Fig. 7.20 Time dependence of annual mean activity concentrations of dissolved ^{137}Cs with standard deviations in Glubokoe Lake (black points) in comparison with reconstructed concentrations from the vertical distribution in bottom sediments (white points) and predictions by the semiempirical "diffusional" model (line)

distribution coefficient K_d, dissolved ^{137}Cs activity concentrations in the lakes Glubokoe and Azbuchin were estimated for 32 years after the accident.

Figure 7.20 presents a comparison of annual mean activity concentrations of dissolved ^{137}Cs in lake Glubokoe with reconstructed concentrations from the vertical distribution in bottom sediments and prediction based on the semiempirical "diffusional" model. The resulting reconstructed ^{137}Cs concentrations are in reasonable agreement with monitoring data and with prediction by the "diffusional" model. Similar agreement was found for lake Azbuchin.

7.6 Conclusions

The major process responsible for the long-term decline in Chernobyl-derived radionuclides in surface runoff and rivers is their vertical migration in catchment soils. Description of radionuclide decline in the catchment topsoil layer allows predicting, in essence, the long-term dynamics of radionuclide activity concentrations in surface runoff and rivers. The advantage of the proposed diffusion-based approach is that the medium- and long-term phases after the accident (more than 1 year for remote contaminated areas and 2–3 years for close areas) can be described by the same simple equation with a limited number of parameters.

For the first time, changes in particulate ^{137}Cs concentration in rivers with time after the Chernobyl accident have been closely scrutinized. The analysis of available data regarding the dynamics of particulate and dissolved ^{137}Cs and dissolved ^{90}Sr in surface water after the Chernobyl accident has demonstrated the benefits of semi-empirical description based on the diffusional model, in which case the same function with a limited number of physically meaningful parameters is used for the middle- and long-term phases after the accident.

When measurement data are not available, long-term dynamics of radionuclide concentrations in rivers and lakes can be reconstructed based on their vertical distribution in bottom sediments. The reconstructed time dependencies of particulate and dissolved ^{137}Cs activity concentrations in Upa River as well as lakes Glubokoe and Azbuchin were found to be described fairly well by the semiempirical diffusional model.

Thus, the proposed diffusion-based semiempirical model, especially its simplified version, can be advised for use in long-term predictions as part of emergency decision support for the purpose of developing a remediation strategy on radioactively contaminated sites.

Acknowledgments This research was partially supported by the Japan Society for the Promotion of Science, Grant-in-aid for Scientific Research (B) (18H03389) and by the Science and Technology Research Partnership for Sustainable Development (SATREPS), Japan Science and Technology Agency (JST)/Japan International Cooperation Agency (JICA).

References

Beresford N, Fesenko S, Konoplev A, Skuterud L, Smith JT, Voigt G (2016) Thirty years after the Chernobyl accident: what lessons have we learnt? J Environ Radioact 157:77–89

Bobovnikova TSI, Virchenko EP, Konoplev AV, Siverina AA, Shkuratova IG (1991) Chemical forms of long-lived radionuclides and their alteration in soils near the Chernobyl nuclear power station. Soviet Soil Sci 23(5):52–57

Borzilov VA, Konoplev AV, Revina SK, Bobovnikova TI, Lyutik PM, Shveikin YV, Scherbak AV (1988) An experimental study of the washout of radionuclides fallen on soil in consequence of the Chernobyl failure. Meteorol Hydrol 11:43–53. (In Russian)

Borzilov VA, Sedunov YS, Novitsky MA, Voszhennikov OI, Konoplev AV, Dragolyubova IV (1989) Physico-mathematical modelling of the washout of long-lived radionuclides from watersheds in the 30-km area around the Chernobyl Nuclear Power Station. Sov Meteorol Hydrol 1:5–13

Borzilov VA, Novitsky MA, Konoplev AV, Voszhennikov OI, Gerasimenko AC (1993) A model for prediction and assessment of surface water contamination in emergency situations and methodology of determining its parameters. Radiat Prot Dosim 50(2–4):349–351

Borzilov VA, Konoplev AV, Bulgakov AA (1994) Application of the Chernobyl experience in developing methodology for assessing and predicting consequences of radioactive contamination of the hydrosphere. In: Hydrological, chemical and biological processes of transformation and transport of contaminants in aquatic environment (proceedings of the Rostov-on-Don symposium, May 1993), vol 219. IAHS Publication, Wallingford, pp 157–167

Bossew P, Kirchner G (2004) Modelling the vertical distribution of radionuclides in soils. Part 1: the convection – dispersion equation revisited. J Environ Radioact 73:127–150

Bulgakov AA, Konoplev AV, Popov VE, Bobovnikova TI, Siverina AA, Shkuratova IG (1991) Mechanisms of the vertical migration of long-lived radionuclides in soils within 30-kilometers of the Chernobyl Nuclear Power Station. Soviet Soil Sci 23(5):46–51

Bulgakov AA, Konoplev AV (1996) Diffusional modelling of radiocaesium fixation by soils. Radiat Prot Dosim 64:11–13

Bulgakov AA, Konoplev AV, Shveikin YV, Scherbak AV (1999) Experimental study and prediction of dissolved radionuclide wash-off by surface runoff from nonagricultural watersheds. In: Linkov I, Schell WR (eds) Contaminated forests. Springer, New York, NY, pp 103–112

Bulgakov AA, Konoplev AV (2001) Diffusional model for radionuclide fixation in soils: a comparison with experimental data and other models. Geochem Int 39:191–195

Bulgakov AA, Konoplev AV, Kanivets VV, Voitsekhovich OV (2002) Modelling the long-term dynamics of radionuclides in rivers. Radioprot Colloq 37(C1):649–654

Bulgakov AA, Konoplev AV, Smith JT, Hilton J, Comans RNJ, Laptev GV, Christyuk BF (2002a) Modelling the long-term dynamics of radiocaesium in closed lakes. J Environ Radioact 61:41–53

Bulgakov AA, Konoplev AV (2004) Modelling the long-term dynamics of ^{137}Cs in rivers with different structures and types of watershed contamination. Meteorol Hydrol 1:34–45. (In Russian)

Crank J (1975) The mathematics of diffusion. Oxford University Press, London, pp 56–61

IAEA (2006) Radiological conditions in the Dnieper river basin. Assessment by an international expert team and recommendations for an action plan. IAEA, Vienna, p 185

Ivanov YA, Lewyckyj N, Levchuk SE, Prister BS, Firsakova SK, Arkhipov NP, Kruglov SV, Alexakhin RM, Sandalls J, Askbrant S (1997) Migration of ^{137}Cs and ^{90}Sr from Chernobyl fallout in Ukrainian, Belarussian and Russian soils. J Environ Radioact 35:1–21

Konoplev AV, Golubenkov AV (1991) Modelling vertical radionuclide migration in soil (as a result of a nuclear accident). Sov Meteorol Hydrol 10:62–68

Konoplev AV, Bulgakov AA, Popov VE, Bobovnikova TI (1992) Behaviour of long-lived Chernobyl radionuclides in a soil-water system. Analyst 117:1041–1047

Konoplev AV, Bulgakov AA, Popov VE, Popov OF, Scherbak AV, Shveikin YV, Hoffman FO (1996a) Model testing using Chernobyl data: I. Wash-off of Sr-90 and Cs-137 from two experimental plots established in the vicinity of Chernobyl reactor. Health Phys 70(1):8–12

Konoplev AV, Bulgakov AA, Hilton J, Comans R, Popov V (1996b) Long-term investigation of ^{137}Cs fixation by soils. Radiat Prot Dosim 64:15–18

Konoplev AV, Bulgakov AA, Zhirnov VG, Bobovnikova TI, Kutnyakov IV, Siverina AA, Popov VE, Virchenko EP (1998) Investigation of ^{90}Sr and ^{137}Cs behaviour in lakes Svyatoe and Kozhanovskoe in Bryansk region. Meteorol Hydrol 11:78–87. (In Russian)

Konoplev AV, Bulgakov AA (1999) Kinetics of the leaching of ^{90}Sr from fuel particles in soil in the near zone of the Chernobyl nuclear power plant. At Energy 86:136–141

Konoplev AV, Bulgakov AA (2000) Transformation of the forms of ^{90}Sr and ^{137}Cs in soil and bottom deposits. At Energy 88:56–60

Konoplev AV, Bulgakov AA (2000a) Exchangeable distribution coefficient of ^{90}Sr and ^{137}Cs in soil-water system. At Energy 88:158–163

Konoplev AV, Deville-Cavelin G, Voitsekhovich O, Zhukova OM (2002) Transfer of Chernobyl ^{137}Cs and ^{90}Sr by surface run-off. Radioprot Colloq 37(C1):315–318

Konoplev AV (2015) Distribution of radiocesium of accidentally origin between suspended matter and solution in rivers: comparison of Fukushima and Chernobyl. Radiochemistry 57 (5):471–474

Konoplev AV, Golosov VN, Yoschenko VI, Nanba K, Onda Y, Takase T, Wakiyama Y (2016) Vertical distribution of radiocesium in soils of the area affected by the Fukushima Dai-ichi nuclear power plant accident. Eurasian Soil Sci 49(5):570–580

Konoplev A, Wakiyama Y, Igarashi Y, Kanivets V, Laptev G, Obrizan S, Bogdan L, Balashevska Y, Nanba K, Onda Y (2018) Empiric and semi-empiric modelling of radionuclide long-term dynamics in the soil-water environment: Fukushima and Chernobyl. In: Arabi M,

David O, Carlson J, Ames DP (eds) Proceedings of the 9th international congress on environmental modelling and software, Fort Collins, CO, USA. Colorado State University, Fort Collins, CA

Konoplev AV, Ivanov M, Golosov VN, Konstantinov EA (2019) Reconstruction of long-term dynamics of Chernobyl-derived ^{137}Cs in the Upa River using bottom sediments in the Schekino reservoir and semi-empirical modelling. Proc Int Assoc Hydrol Sci 381:95–99

Konshin OV (1992) Mathematical model of ^{137}Cs migration in soil: analysis of observations following the Chernobyl accident. Health Phys 63(3):301–306

Monte L (1995) Evaluation of radionuclide transfer functions from drainage basins of freshwater systems. J Environ Radioact 26:71–82

Monte L (1997) A collective model for predicting the long-term behaviour of radionuclides in rivers. Sci Total Environ 201:17–29

Prokhorov VM (1981) Migration of radioactive contaminants in soils: physico-chemical mechanisms and modelling. Energoatomizdat, Moscow, p 99. (in Russian)

Sansone U, Belli M, Voitsekhovitch O, Kanivets V (1996) ^{137}Cs and ^{90}Sr in water and suspended particulate matter of the Dneper River-Reservoirs System. Sci Total Environ 186:257–271

Santschi PH, Bollhalder S, Zingy S, Luck A, Farenkothen K (1990) The self-cleaning capacity of surface waters after radioactive fallout. Evidence from European waters after Chernobyl, 1986–1988. Environ Sci Technol 24:519–527

Smith JT, Wright SM, Cross MA, Monte L, Kudelsky AV, Saxen R, Vakulovsky SM, Timms DN (2004) Global analysis of the riverine transport of ^{90}Sr and ^{137}Cs. Environ Sci Technol 38:850–857

Smith JT, Voitsekhovich OV, Konoplev AV, Kudelsky AV (2005a) Radioactivity in aquatic systems. In: Smith JT, Beresford NA (eds) Chernobyl catastrophe and consequences. Springer-Praxis, Berlin, pp 139–190

Smith JT, Belova NV, Bulgakov AA, Comans RNJ, Konoplev AV, Kudelsky AV, Madruga MJ, Voitsekhovich OV, Zibold G (2005b) The "AQUASCOPE" simplified model for predicting $^{89,90}Sr$, and $^{134,137}Cs$ in surface waters after a large-scale radioactive fallout. Health Phys 89:628–644

Takahashi Y, Sakaguchi A, Fan Q, Tanaka K, Miura H, Kurihara Y (2020) Difference in the solid-water distribution of radiocesium in rivers in Fukushima and Chernobyl. In: Kato K, Konoplev A, Kalmykov S (eds) Behavior of radionuclides in the environment I: Function of particles in aquatic system. Springer Nature, 115–150.

Vakulovsky SM, Nikitin AI, Chumichev VB, Katrich IY, Voitsekhovich OV, Medinets VI, Pisarev VV, Bovkun LA, Khersonsky ES (1994) Cesium-137 and strontium-90 contamination of water bodies in the areas affected by releases from the Chernobyl nuclear power plant accident: an overview. J Environ Radioact 23:103–122

Chapter 8
Distribution and Dynamics of Radionuclides in the Chernobyl Cooling Pond

Volodymyr Kanivets, Gennady Laptev, Alexei Konoplev, Hlib Lisovyi, Grygorii Derkach, and Oleg Voitsekhovych

Abstract As a result of the accident at the Chernobyl Nuclear Power Plant (ChNPP) in 1986, the Cooling Pond (CP) was heavily contaminated by radionuclides due to atmospheric fallout and discharges from the industrial site. Shortly afterward, most of the radionuclides were deposited to the CP bottom sediments and later served as a source for secondary contamination of the water column. By the end of the initial phase, ^{137}Cs and ^{90}Sr activity concentrations in the CP water began to decrease until decommissioning of the pond. This decrease of radionuclide activity concentrations in water was due to radioactive decay, sedimentation of radionuclides onto the bottom, dilution of pond's volume with water pumped from Pripyat River, and ^{90}Sr infiltration through the dam. Draining of CP began in May 2014, resulting in splitting the pond into three separate parts and the water surface area reduction by some 40% by the end of 2017. Later on, ^{90}Sr activity concentration in the water of the Northern and Western sectors increased by a factor of 3 at the end of 2017, while in the Southern sector the ^{90}Sr water activity concentration increased by less than 10–15%. In future, ^{90}Sr activity concentration in the CP water is expected to further go up due to dissolution of fuel particles at drained territories causing wash off of the radionuclide by surface runoff to residual lakes.

Keywords Chernobyl · Cooling Pond · Water level · Radionuclides · Speciation · Fuel particles · Dynamics · Sediments

---

V. Kanivets (✉) · G. Laptev · H. Lisovyi · G. Derkach · O. Voitsekhovych
Ukrainian Hydrometeorological Institute, Kiev, Ukraine
e-mail: kaniv@uhmi.org.ua

A. Konoplev
Institute of Environmental Radioactivity, Fukushima University, Fukushima, Japan

A. Konoplev et al. (eds.), *Behavior of Radionuclides in the Environment II*,
https://doi.org/10.1007/978-981-15-3568-0_8

8.1 Introduction

The cooling pond of the ChNPP is one of the major water bodies that were heavily contaminated as a result of the accident in April 1986. The CP was used as a reservoir of cooling water for Units 1–3 of the ChNPP until 2000 when the plant was finally shutdown and the CP became no longer required for operational purposes (IAEA 2006). However, the pumping station used for feeding the CP with water from Pripyat River was shut down in May 2014 based on the Feasibility Study (FS) for ChNPP CP decommissioning (Krasnov 2013). As a result, the water level began steadily declining and was reduced by 5 m by the end of 2017.

The water level is expected to decrease by 6–7 m (to natural equilibrium elevation marks) due to evaporation and water infiltration through the dam and pond floor. As a result, the once large reservoir will turn into a few lakes of smaller size, with a total residual area 27–38% of the original one. This process might take 4–6 years from the time of water pumping cessation. Eventually, the water level of residual lakes will apparently be controlled by the local groundwater level and water level in Pripyat River, as well as by hydrometeorological factors. As a consequence, three types of landscape zones have been formed: (1) permanently drained areas; (2) semi-wetland zone (transitional), which will be flooded in high-water period or stay dry in low-water period; and (3) lakes mostly bound to the old floodplain lakes and old river channel (Fig. 8.1).

The presented schematic map was constructed based on the latest data on the CP bed topography obtained by UHMI in 2016 when the actual coastline of the newly exposed "islands" was used for correction of all drawbacks of the older bathymetry map, presented in IAEA-TECDOC-1886 (IAEA 2019). All data presented herewith in, where bathymetry was a necessary component in calculations, have been thoroughly corrected using this up-to-date bathymetry map.

The total area of the pond bottom expected to become exposed is about 16 km^2 (70% of the total CP area), with the inventory being 25% of the total ^{137}Cs and 50% of the total ^{90}Sr in bottom sediments (Buckley et al. 2002).

Positive effects of CP drainage are expected to include (Krasnov 2013): (1) reduction of operational costs for maintenance of normal water level; (2) reduction of ^{90}Sr transport from the CP to Pripyat River; (3) as a result of decrease in groundwater level on territories adjacent to the CP, reduction of risk from groundwater contamination by radioactive wastes localized in TSRW; (4) reduction in liquid radioactive discharges that potentially could be generated due to inundation of the ChNPP premises by groundwater; and (5) minimization of engineering and environmental risks associated with a possible failure of the dam.

The main potential adverse impact of CP draining was assumed to be associated with wind resuspension and transport of radioactive particles (Krasnov 2013). Results of model simulation show that additional contamination of surrounding areas by ^{90}Sr will not exceed 0.05% of the current contamination level, even in case of extreme meteorological conditions (when wind 15 m/s lasting for 3 days could cause a wild fire on all vegetation cover of the exposed area of the CP). In

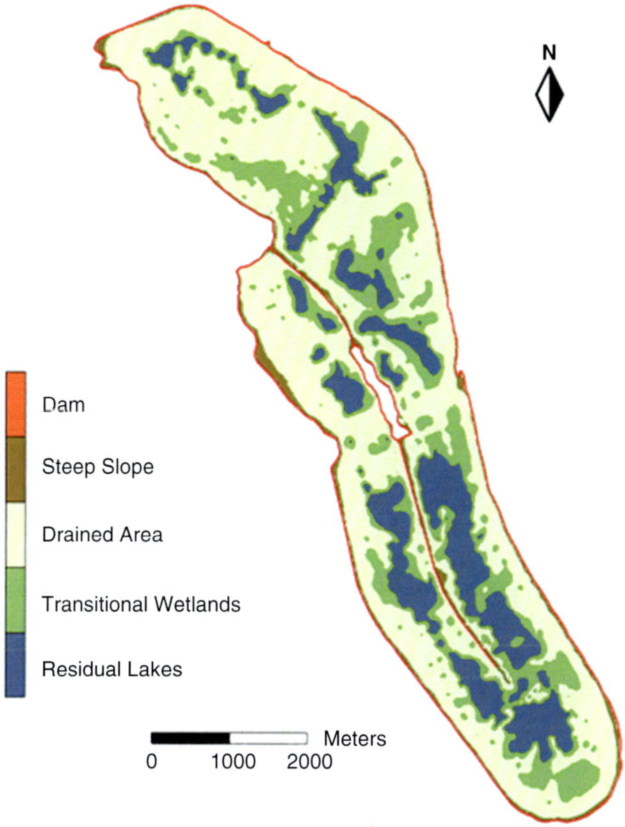

Fig. 8.1 Schematic representation of new landscape forms on the CP basin (based on 2016 updated bathymetry)

accordance with this simulation, an additional contribution to the 50-year lifetime effective dose equivalent due to inhalation for personnel is estimated to be 0.52 μSv for normal conditions and 3.0 μSv for extreme weather conditions, which is three orders of magnitude below the permissible doses (Krasnov 2013).

It is expected that in the course of a few years the exposed pond floor will be overgrown by grass and bushes typical of this geobotanical zone (Fig. 8.2).

Most of the radionuclides occurring in the CP, in particular ^{137}Cs, are associated with silt particles accumulating in deep-water areas due to the process of sediment focusing. Therefore, even when the water level goes down, these heavily contaminated deposits remain under water in residual lakes and are not affected by air resuspension.

According to Krasnov (2013), reduction in the CP water volume can cause deterioration of its overall ecological state, which in turn can result in changes in radioactive water contamination situation. The most likely changes are: (1) a general

Fig. 8.2 Self-sown vegetation on the exposed areas of the CP bottom

decrease in water ventilation (dissolved oxygen (DO) content) leading to establishment of anoxic reductive conditions in the near-bottom water layer and (2) a gradual increase in radionuclide activity concentration in water to the levels typical of floodplain water bodies in the near zone of the ChNPP.

Deterioration of the CP DO regime is believed to have a negative impact on aquatic organisms. The CP was dwelled by abundant fish, viable and differing in reproduction and feeding habits. Before the start of CP draw-down, the conditions for fish species diversity and reproduction were favorable: extensive parts of the bottom with varying depth and nutrients supply and food access. After the water level is lowered by 7 m, the residual lakes will, most likely, have a trough-like shape with steep littoral slopes and flatbed covered with a thick silt layer, which is not a favorable environment for flourishing higher aquatic vegetation, which serves as a basic diet for phytophagous fish. Actually, the question that remained unclear after the Feasibility Study (Krasnov 2013) is whether fish common for the CP are capable of adapting to new habitat conditions after water level reduction and whether new oxygen regime may lead to massive extinction ("excessive biomass effect").

In this regard, it is important to study changes in the thermo-oxygen and hydrochemical regimes in newly formed lakes. This is required for monitoring habitat conditions of aquatic organisms and predicting changes in the CP ecosystem, on the one hand, while hydrochemical conditions are controlling the behavior of radionuclides in the lake ecosystem. Hydrochemical observations are essential to support predictions of radionuclide activity concentrations in water.

In all components of the CP ecosystem, very high activity concentrations have been reported since integrated experimental studies after CP decommissioning began in 2014. These studies provide a basis for investigating the fate and transport of radionuclides in specific conditions of water level reduction. Such data, among other things, will be useful for parametrization and validation of lake models.

8.2 Characterization of Cooling Pond

Cooling Pond is an artificial reservoir created on the floodplain of Pripyat River (Fig. 8.3). In fact, a 10-km stretch of the Pripyat River was cut off to build this pond, the river flow being redirected to a new artificial riverbed. The construction of the CP was fully completed in 1981 prior to the commissioning of the third and fourth units of the ChNPP.

During the period of ChNPP operation, the composite topography of the bottom was contributing to the turbulence of circulatory flow, which promoted a more

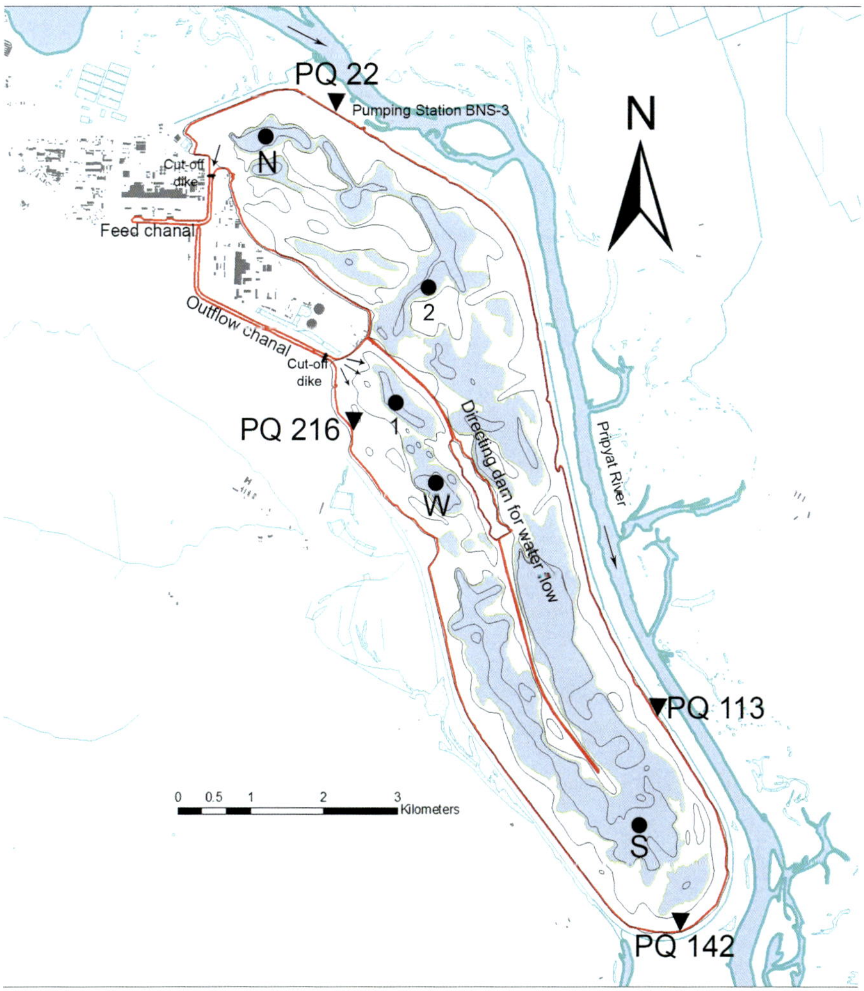

Fig. 8.3 Schematic drawing for the CP engineering design, bathymetry, and location of water sampling sites

efficient water cooling. The size and configuration of the pond were designed with a view to cool down four ChNPP units in power generation mode (Table 8.1). After the ultimate shutdown of the ChNPP in 2000, the water level was being maintained at 110.7 m above the sea level, which was about 7 m above the low water stage of Pripyat River. The high head pressure between the pond and Pripyat River, as well as high water permeability of the dam material, resulted in intensive filtration of water through the dam. The water level in the pond was maintained using a pumping station, supplying water from Pripyat River to compensate filtration and evaporation losses from the water surface.

Calculations of the CP water balance in different phases of its operation show that filtration losses have always been the main loss component of the balance (Table 8.2).

The equation of water balance takes the following form $Q_P + Q_C + P = E + F$, where Q_P is water supply from Pripyat River by the pumping station; Q_C—inflow of small creeks; P—atmospheric precipitation on the cooling pond surface; E—evaporation from the cooling pond surface; F—total filtration losses due to seepage through the enclosing dam and the pond bottom.

By estimates, before the ChNPP shutdown, the share of filtration losses was 62–80% of total losses in the water balance (Bugai et al. 1997; Filippov 1996). After the ChNPP shutdown, the water temperature in the CP seems to have decreased to the values typical of water reservoirs of the given climatic zone, evaporation from the water surface diminishing by a factor of 2, and the share of filtration losses increasing to 80% of all losses (Bugai et al. 2005).

Bugai and Skalsky (2002) calculated how long it would take the water level to decrease to equilibrium elevation in case of cessation of water supply from Pripyat River. For these calculations, a regional filtration model of the Chernobyl exclusion zone was used. The calculations were made based on two scenarios: "normal" and

Table 8.1 Cooling pond dimensions

Water level in the pond, m BHS	Length, km	Mean width, km	Surface area at normal water level, km^2	Volume, million m^3	Average depth, m	Maximum depth, m
110.7[a]	11.5	2.2	22.57	139	6.15	17.0[b]

[a]The water level in the pond from the bathymetric survey of 2001 used for estimating the pond dimensions

[b]The maximum depth recorded during the bathymetric survey of 2001

Table 8.2 Estimates of the water balance components for different phases of CP operation

	Input, 10^6 m^3/year			Losses, 10^6 m^3/year	
Source	Q_P	Q_C	P	E	F
Filippov 1996 (average values for 1982–1995)	155	0.6	13.4	45.6	123.7
Bugai et al. 1997 (for 1993)	122–143		16	46–51	76–92
Bugai et al. 2005 (for 2001)	120		14	19	101

"dry". The "normal" scenario accounts for mean annual meteorological and hydrological conditions of the exclusion zone, while the "dry" scenario provides for extreme conditions: minimal amount of precipitation, maximum evaporation, minimal infiltration feed, and minimum water levels in Pripyat River. It is assumed that the water level will continue to decline until a dynamic equilibrium is established: filtration losses from the pond toward Pripyat River and losses due to evaporation will be compensated by atmospheric precipitation and filtration inflow to the pond from the above-floodplain terrace of Pripyat River (Bugai and Skalsky 2002). The time of water level reduction in the CP to 104.7 m (pond averaged) is estimated to be about 100 months by the "normal" scenario and 40 months by the "dry" scenario (Bugai and Skalsky 2002).

The chemical composition of the CP water before its decommissioning was primarily controlled by the chemical composition of the Pripyat River water, since the volume of water annually pumped from the river to the pond was comparable to that in the pond. In fact, every year the entire volume of CP water was replaced by the water from Pripyat River.

In accordance with the classification of fresh waters, in terms of ion composition (Alekin 1948), the CP water belongs to bicarbonate category of calcium group and is characterized by moderate hardness. In the pond, unlike Pripyat River, no significant seasonal variations in ion composition are observed. After the ChNPP shutdown in 2000, there has been no change in the hydrochemical composition of the pond water, only that the content of sulphates and phosphorus was reduced, most likely, due to cessation of sewage discharge to the pond. The hydrochemical composition in 2012–2013 (before the beginning of CP decommissioning), was similar to that in 1991–2002 (Voitsekhovych et al. 2013).

The CP thermal and oxygen regimes after the ChNPP shutdown are determined solely by natural environmental factors and characterized by significant seasonal fluctuations of water temperature and oxygen concentrations, as well as by spring–summer direct and winter inverse temperature stratifications typical of large deep-water bodies in temperate climate zone.

8.2.1 *Features of Bottom Sediments and Their State Before CP Decommissioning*

Formation of bottom sediments in artificial reservoirs is a function of bottom relief, hydrodynamic activity, and sources of solid material. Being an artificial reservoir with a very rugged bottom relief, the ChNPP CP shows significant spatial variability of water-physical and mechanical properties of the bottom sediments. At transverse bathymetric cross sections, the bottom elements that are clearly discernible include: (1) the coastal shelf 50–100 m wide (depth 0–1.5 m); (2) underwater coastal slope (depth 1.5–5.0 m); (3) relatively flat surface of the river floodplain (depth 5–7 m); (4) the former channel of Pripyat River. The cooling pond is fringed with a wide strip of coastal shelf and underwater coastal slope composed of sandy material (70–200 m).

The soil cover of the river floodplain, before the pond had been created, consisted of sandy and sandy sod-podzolic, sod-weakly podzolic, and peat-marsh soils. The bottom areas, which had previously been river channel, were composed of fine and medium-grained sand. After filling the CP basin with water, the primary soil cover began to change. These changes consisted of physical and chemical transformation of primary soils; the topsoil layer was soaking and swelling; soil skeleton bulk weight was decreasing. Due to various currents (wind-wave, circulatory, and compensational types), the primary soil cover was subject to erosion. As a result, by 1989, the area of primary soils became reduced to 26% of the total area of the CP (Romanovsky et al. 1989). The surveys of bottom sediments in 1999 and 2001 showed that the bottom soil complex was composed of sand, silted sand, and silt. The primary soils remained only in some places at a depth of 5–7 m and were covered with 2–5 cm sandy silt layer (Buckley et al. 2002).

The hydrodynamic regime of the CP is not well understood. Over the time of the CP existence, only one survey of surface current velocities throughout the pond (Romanovsky et al. 1989) was conducted, which, however, did not provide a complete picture of the hydrodynamic situation in the reservoir. At the same time, a whole range of diverse factors should be kept in mind, such as complex relief of the bottom, significant horizontal and vertical temperature gradient (until the ChNPP shutdown in 2000), alignment of the pond longitudinal axis, and prevailing wind direction, all of which are responsible for complex and changeable structure of the currents and mixing of water masses in the reservoir.

Throughout the CP existence, the primary floodplain soils were eroded, silt particles on shallow areas were washed out, and re-deposited in silt accumulation zones (the old channel of Pripyat River, floodplain lakes, sites of former sand quarries, etc.). The processes of soil complex transformation were slowing down with time, and boundaries of areas with different types of bottom sediments became fixed.

Slow-down of sedimentation processes is directly indicated by a decrease in water turbidity with time. Occasional measurements during the vegetation season of 1982–1984 showed the water turbidity to be 8–24 g/m^3 at the content of organic matter 43–66% (Romanovsky et al. 1989). In November 1989 (minimal development of plankton), the spatial survey of water turbidity at seven cross sections of the CP with sampling at different depths was conducted. By the results of the survey, the average turbidity across the CP was 6.8 g/m^3 (Romanovsky et al. 1989). The measurements in deep-water areas in different seasons during 2002–2003 showed the turbidity not exceeding 3.5 g/m^3 (Voitsekhovych et al. 2013).

Before the pumping station shutdown in 2014, the main input to the CP sedimentation budget was made by suspended sediments from Pripyat River. The mean annual flux of river silt sediments (<0.05 mm) was calculated to be about 4740 tons (Kanivets and Voitsekhovych 2000). These sediments were deposited mainly in the Northern part of the CP. The total annual amount of silt material entering the pond is estimated to be 7100 tons dry mass (Table 8.3).

Table 8.3 Main inputs to the CP sediment budget, t/year (% of Σ) (Kanivets and Voitsekhovych 2000)

$Q_{Pripyat}$	$Q_{Airborne}$	$Q_{Phytoplankton}$	Σ
4740 (67)	1040 (15)	1320 (18)	7100

Table 8.4 Average particle size composition of bottom sediments at different depths of the CP (Voitsekhovych et al. 2013)

	Particle size, mm						
Depth interval, m	1–0.5	0.5–0.2	0.2–0.1	0.1–0.05	0.05–0.01	0.01–0.005	<0.005
0–3.7	25.5	50.5	18.8	2.4	1.6	0.4	0.7
3.7–7.0	14.5	42.0	28.1	6.6	5.6	1.3	1.9

Table 8.5 Silt thickness in different areas of the CP (Voitsekhovych et al. 2013)

Depth interval, m	0–3.7	3.7–7.0	7.0–10.0	>10
Area, km^2	3.56	16.4	2.2	4.0
Average silt layer thickness, cm	0	1.8	4.3	26.3[a]
Maximum silt layer thickness, cm	0	6.0	19.0	>98

[a]The average thickness of silt deposits at the depths more than 10 m may be greater than indicated, since the corer did not go through the entire silt layer in 33% of sampling events

By calculations of Kanivets and Voitsekhovych (2000), the total contribution of this solid material to the formation of silt deposits in deep-water areas was 10–15%. It was therefore reasonable to assume that there was another stronger source of silt material that should be autochthonous.

The thick silt deposits discovered in 1999–2002 were mainly made of particles that were believed to have been part of primary soils on the territory designated for the pond. By action of currents, silt particles were gradually washed out from primary soils at shallow and medium depths and re-deposited in the deep-water areas.

After reduction of water level by 7 m (to 103.70 m above the sea level), a total of 16 km^2, or 70% of the bottom area, is expected to become exposed to the air. Composition of bottom sediments to be denuded was studied before the beginning of CP decommissioning (Voitsekhovych et al. 2013). It was revealed that at depths 0–3.7 m, the bottom consisted of dense sandy deposit (Table 8.4). The content of silt particles (<0.01 mm) representing seasonal organogenic forms was, on average, 1.1%.

The bottom sediments at depths 3.7–7.0 m are silted sand with the average content of silt particles 3.2% (Table 8.4). Table 8.5 shows the averaged silt deposit thickness on sections with different depths.

Distribution of main types of bottom sediments over the CP bottom in 2001–2004 is schematically shown in Fig. 8.4. This sketch was drawn based on a digital model of the CP bottom relief. Some additional details concerning the state of bottom sediments are given in Sect. 8.5 describing radioactive contamination of the CP bottom sediments.

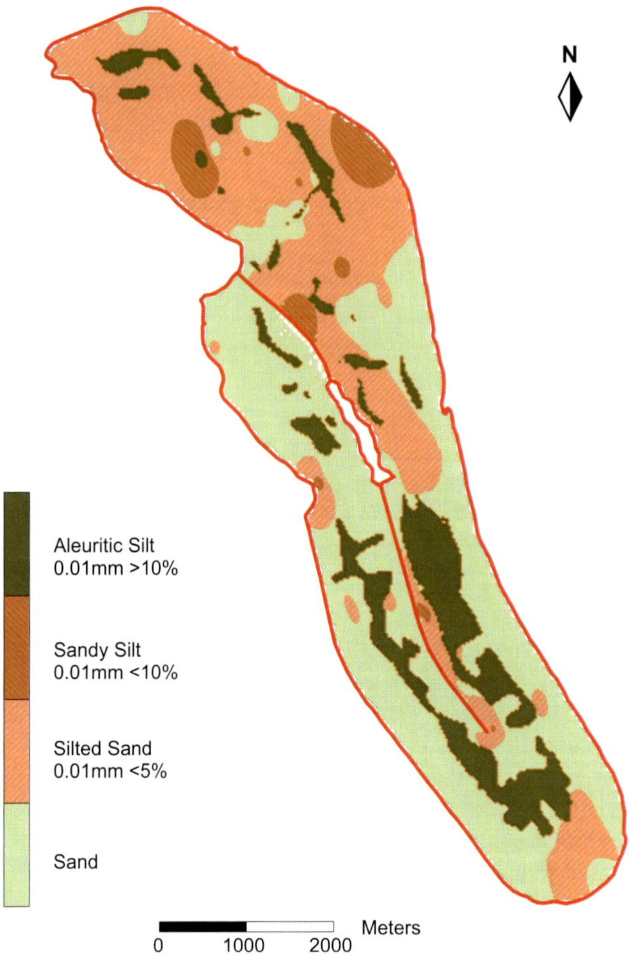

Fig. 8.4 Types of bottom sediments in the ChNPP CP (Voitsekhovych et al. 2013)

8.3 Sources of Radioactive Contamination

8.3.1 Pre-accidental Contamination

Prior to the accident from 1976 to April 1986, the CP contamination was associated with the operation of the ChNPP. The main source of contamination in the first 2–3 years after the ChNPP start-up were non-reusable waters discharged to the pond and the total annual activity of fission products being 37–370 GBq (excluding tritium). The contribution of ^{137}Cs was about 10% of the total activity (Kazakov 1995) and the input of other sources (flux of fission products and corrosion products with service water, deposition of aerosol emissions on the water surface) was an order of magnitude less. In subsequent years, discharges of radionuclides and aerosol emissions of the ChNPP were below the input due to the global fallout (Kazakov 1995). In 1981 and 1982, additional radioactive releases to the CP as a result of a leak (1981) and a radiation-related incident (1982) occurred, but information about scale and characteristics of these releases is not available in the open literature.

8.3.2 Contamination as a Result of the Accident in 1986

The radioactive contamination of the CP after the Chernobyl accident occurred within a short time due to deposition of radioactive materials on the water surface and discharge of highly contaminated water from the emergency cooling system and of water used to extinguish the fire at the NPP. The heaviest fallout onto the CP water surface occurred from April 29 to May 2, 1986 after the wind turning south and south-east, which coincided with the second and third phases of the accident evolution (IAEA 1993; Sivintsev and Khrulev 1995). The fallout composition was governed by physical processes in the accidental reactor. The main feature of the accidental release in the above-mentioned time period was that a major part of radionuclide occurred was incorporated in a hardly soluble matrix. Such matrices were primarily nuclear fuel debris, as well as compounds of zirconium, iron, silicon, and other elements. Some amounts of fission products entered the CP as adsorbed on aerosol carriers such as dust particles, soot, etc. (Ignatenko et al. 1989; Sobotovytch et al. 1989; Sobotovytch and Olshtynsky 1991; Rybalko et al. 1989; Meshalkin et al. 1989).

In 1986, a gamma-survey of the CP coastal zone was performed, supplemented by soil sampling followed by radioisotope analysis. The analysis results were extrapolated to the CP surface (Table 8.6). The total activity of gamma-emitting radionuclides deposited on the CP surface was estimated to be 13,300 TBq, including 159 TBq of ^{137}Cs (Egorov et al. 1989). The highest deposition level was in the Northern part of the CP and the lowest in the Southern part.

Egorov et al. (1989) estimated that about 120 TBq of ^{137}Cs was additionally released to the CP through the outlet channel, however, these estimates were not thoroughly explained or corroborated by actual measurements. At the end of May 1986, the ^{137}Cs inventory in the CP was approximately 281 TBq (Egorov et al. 1989), of which about 40% was in the water column (Table 8.7). Unfortunately, no data on ^{90}Sr in the CP was available for the accidental phase, as express analysis of ^{90}Sr was not organized at that time.

Table 8.6 Estimated radionuclide depositions on the CP surface as a result of the accident in 1986, TBq (based on Egorov et al. 1989)

^{144}Ce	^{131}I	^{134}Cs	^{137}Cs	^{95}Zr	^{106}Ru
2030	94	94	160	3050	1560

Table 8.7 Radionuclide inventories in the CP water column and bottom sediments at the end of May 1986, TBq (based on Egorov et al. 1989)

Radionuclide	^{144}Ce	^{131}I	^{134}Cs	^{137}Cs	^{95}Zr	^{106}Ru
Water	15	320	70	120	23	18
Bottom	1150	30	95	160	1590	320
Total	1165	350	165	280	1613	338

Table 8.8 Estimated ^{137}Cs and ^{90}Sr depositions onto the CP water surface in 1986, TBq

Data source	^{137}Cs	^{90}Sr
ARINPPO (Egorov et al. 1989)		
Atmospheric fallout	159	–
Discharge through the outlet channel	122	–
UHMI (G. Laptev) estimates based on UIAR maps for 1997	94 ± 30	65 ± 20
SHI surveys of bottom sediments in 1989 and 1991 (Shiklomanov 1992)	185 ± 44	33 ± 16
UHMI surveys of bottom sediments in 1999–2001 (Buckley et al. 2002)	218 ± 50	(45 ± 11)[a]
UHMI surveys until 2013 using 2016 updated bathymetry (decay corrected to 1 May 1986)	290 ± 70	52 ± 13

[a]This value allows for 10 TBq of ^{90}Sr discharged from the CP due to infiltration losses (Bugai et al. 1997, 2005)

Table 8.9 Location and time period of regular observations of radionuclides in the CP

Observation site (Fig. 8.3)	Inlet and outlet channels	PQ-22	PQ-113	PQ-142	PQ-216
Period of observations	1987–2017	1987–2017	1988–1992, 2001–2017	1987–2017	1987–2017

In 1989 and 1991, the ^{137}Cs and ^{90}Sr inventories in bottom sediments were estimated to be 170 and 29 TBq, respectively, based on the data obtained through the bottom sediments surveys (Shiklomanov 1992). In 1997, a large-scale study to update data about radionuclide depositions in the ChEZ, including territories adjacent to the CP, was carried out (Kashparov et al. 2000). Based on the results of this work, the maps of exclusion zone contamination with ^{137}Cs and ^{90}Sr allowed the calculation of ^{137}Cs/^{90}Sr ratio in soils, as of 1997. These maps were used by UHMI to estimate more accurate depositions of ^{137}Cs and ^{90}Sr onto the CP water surface in 1986 (Table 8.8).

The radionuclide deposition in the near zone of the ChNPP is thought to be similar to that of the reactor fuel (Meshalkin et al. 1989; Kashparov et al. 2000). The average ^{137}Cs/^{90}Sr ratio in the soils on the territory adjacent to the CP is estimated to be about 2, which is close to that in irradiated nuclear fuel, suggesting that the territory had been contaminated with fuel particles. At the same time, the ^{137}Cs inventory measured in the CP bottom sediments in 1989 and 2001 (Table 8.9) significantly exceeded the values calculated using the deposition maps, and the average $^{137}Cs/^{90}Sr$ ratio for the CP was found to be 5–6 (Kanivets and Voitsekhovych 2000; Buckley et al. 2002). These discrepancies can be explained by three reasons: (1) additional discharge of ^{137}Cs through the outlet channel (Egorov et al. 1989); (2) ^{90}Sr depletion in bottom sediments due to excess outflow of this radionuclide from the pond with seepage water (Bugai et al. 1997, 2005); (3) interception of ^{137}Cs condensation forms by presumably "mist cloud" developed over the CP's warmer surface after the accident in April 1986 (G. Laptev's hypothesis).

As to additional ^{137}Cs significant discharge to the CP mentioned above, it is questionable about its scale as well as amount quoted, since information about this event is available from a single publication (Egorov et al. 1989), with no details and quantitative characteristics provided. ^{90}Sr depletion of bottom sediments due to infiltration through the enclosing dam in 1986–2000 was estimated to be about 10 TBq, or 25% of the initial quantity (Bugai et al. 1997, 2005) and could not significantly influence the initial $^{137}Cs/^{90}Sr$ ratio in bottom sediments.

A hypothesis proposed by G. Laptev regarding the role of "mist cloud" hanging above the CP water surface in retaining ^{137}Cs condensation forms from the steaming damaged reactor in April–May 1986 sounds quite plausible, though has never been confirmed experimentally. The contamination level of the water column and bottom sediments all over the CP was evaluated for the first time in early June 1986 (Teplov et al. 1989). The results of this screening showed a marked excess of ^{137}Cs against refractory radionuclides (^{144}Ce for instance) in bottom sediments, with the abundance factor from 3 to 7. It is worth to note that the area most enriched in ^{137}Cs was the middle part of the "cold zone", quite far away from the outlet channel, while the bottom sediments in front of outlet channel were least containing excess ^{137}Cs. Given the forced circulation of water masses in the CP after the accident was temporarily terminated, the only mechanism of this enrichment formation could be by direct atmospheric fallout and not by discharges through the outflow channel. Overall, the pattern of $^{137}Cs/^{144}Ce$ enrichment factor measured in bottom sediments in June 1986 is very similar to modern estimates of $^{137}Cs/^{90}Sr$ enrichment factor (Voitsekhovych et al. 2013).

Thus, based on the above presented data, a total of 190–280 TBq of ^{137}Cs and 35–65 TBq of ^{90}Sr entered the CP in the initial phase of the Chernobyl accident (April–May 1986). Due to the lack of data about the activity of transuranic radionuclides in atmospheric fallout and bottom sediments, only rough estimates can be made regarding flux of radionuclides to the CP, using the known radionuclide ratios in the nuclear fuel matrix (^{90}Sr/TUE, ^{144}Ce/TUE).

In fact, timewise, the radioactive contamination of the CP can be considered as a pulse release. After the initial phase of the accident, the radionuclide inventory in the CP was only decreasing, which was due to radioactive decay and natural attenuation.

8.4 Radioactive Contamination of CP Water

8.4.1 Methods of Monitoring

Immediately after the accident, no regular observations were conducted at fixed stations, at times water sampling was organized by ad hoc groups such as Research and Design Institute of Power Technique (RDIPT, Russia, Moscow), ARINPPO, and others. Routine monitoring by Dosimetry Department of PA "Combine" (now SSE "Ecocenter") began in January 1987 for ^{137}Cs and as late as August 1989 for ^{90}Sr (Table 8.9). Sampling was organized at a number of locations around the CP perimeter (Fig. 8.3), as well as in the inlet and outlet channels separated from the pond by alluvial dams in 2013. Samples were collected from the CP bank and therefore characterized stagnant sections near the bank, not representing the entire water mass of the pond. In spite of limitations, such data are valuable, enabling a better understanding of general long-term trends in water radioactive contamination over the entire post-accident period.

Sampling frequency was at least once a month. In 1987–1992, water samples were not filtered and because of this only total dissolved and particulate ^{137}Cs was measured. Since 1993, samples began to be filtered and activity concentrations of dissolved and particulate ^{137}Cs were measured separately. A composite filter was used consisting of ashless paper filter ("blue ribbon" grade) and prefilter made of Petryanov textile FPP-15-1.5 (Tertyshnyk et al. 1980; Vakulovsky 1986; Makhonko 1990). It has been shown experimentally that the effective pore diameter of the composite filter is close to 0.45 μm (Kanivets 2010).

UHMI began regular research observations on CP in 2015 after the pumping station shutdown and the beginning of water level lowering. Since the previous studies showed that coastal water samples were not representative of the entire water body (Smith 2005), it was decided to conduct sampling at permanent vertical cross sections in deep-water areas. Sampling stations were assigned in the Northern (opposite the inlet channel), Western (opposite the outlet channel), and Southern parts of the pond (indicated as N, W and S on Fig. 8.3).

Sampling and measuring activities at the deep-water station included: (1) measurements of water temperature on different water column horizons starting from a depth of 1 m and then every 2 m, as well as in the near-bottom contact layer of water; (2) water sampling from the same horizons to measure pH, Eh, and EC; ^{137}Cs and ^{90}Sr activity concentrations; turbidity; oxygen and ammonium content; and overall hydrochemical composition of water; (3) sampling of bottom sediments to determine ^{137}Cs and ^{90}Sr activity concentrations, and to extract porous water for radionuclide and hydrochemical analysis.

Sampling was performed either from a motorboat or through a hole in the ice in winter (Fig. 8.5), using Molchanov bathometer (GR-18) with a built-in mercury thermometer (analogue of the Rutner's bathometer), 6 l of water from each horizon.

a) b)

Fig. 8.5 Water sampling on the CP: (**a**) from boat; (**b**) from ice

8.4.2 Long-Term Dynamics of Radionuclides in the CP

Before the ChNPP was put in operation (1972–1977), radioactivity in the CP was in conformity with the global levels resulting from nuclear weapon tests. At that time, the ^{137}Cs content in rivers and reservoirs at these latitudes was less than 1 Bq/m^3 and ^{90}Sr content—0–30 Bq/m^3 (Bochkov et al. 1983; Gedeonov et al. 1993). With the beginning of operation of the ChNPP power units, the water radioactivity in the CP increased manifold, conceivably due to discharges of non-reusable water in the first years of NPP operation and estimated to be 3.4–18.0 Bq/m^3 for ^{137}Cs in 1980–1981 (Kazakov 1995). In 1985, the ^{137}Cs activity in the pond water was found to be 30–37 Bq/m^3. This increase was presumably caused by a radioactivity leakage in September 1981 and an incident at the ChNPP in 1982, which led to the CP additional contamination by radionuclides ^{51}Cr, ^{137}Cs, ^{134}Cs, ^{90}Sr, ^{54}Mn, ^{60}Co, and ^{59}Fe. The ^{90}Sr activity concentration in the CP water before the accident in 1986 was 2–20 Bq/m^3 (Fomin et al. 1998).

Immediately after the accident of 1986, high radioactivity of the CP water was first mostly associated with short-lived fission products such as ^{131}I, ^{140}Ba, $^{103,106}Ru$, $^{141,144}Ce$, ^{95}Nb, ^{95}Zr, etc. (Table 8.10). However, as soon as August 1986, the radionuclides ^{137}Cs, ^{134}Cs, and ^{90}Sr became the major radionuclides of environmental concern.

During the period 1987–2013, the average annual water concentrations of ^{137}Cs and ^{90}Sr decreased almost by two orders of magnitude (Fig. 8.6). In the first years after the accident, the activity concentrations were decreasing very quickly

Table 8.10 Activity concentrations of gamma-emitting radionuclides in the CP water in 1986, Bq/l

Sampling site	^{131}I	^{95}Nb	^{95}Zr	^{103}Ru	^{134}Cs	^{137}Cs	^{140}Ba	^{141}Ce
	30 May 1986							
PQ43, North part	660	62		74	100	170	320	
PQ70, Central part	670			67	130	200	280	
PQ150, South part	1000	90		80	210	280	380	
PQ250, near inlet canal	4630	840	370	610	790	1570	660	440
	7 August 1986							
PQ70, Central part		7,3	19	31	62	206		

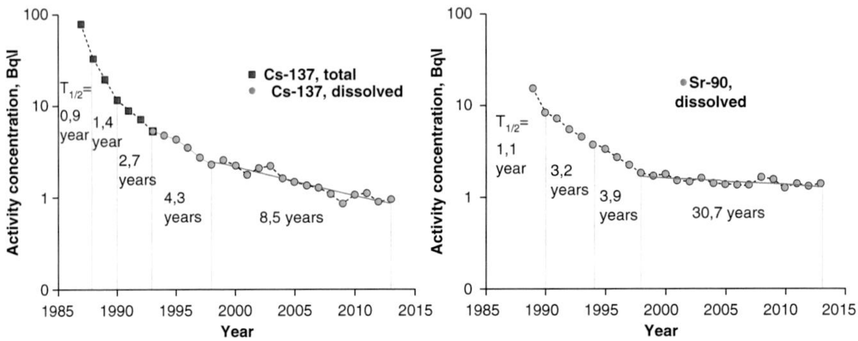

Fig. 8.6 Long-term dynamics of annual average activity concentrations of ^{137}Cs and ^{90}Sr in the CP water, data of coastal observations for 1987–2013 (Kanivets et al. 2014)

(environmental half-life being 0.9 year for ^{137}Cs and 1.1 years for ^{90}Sr). In 1998–2013, the environmental half-life was 8.5 years for ^{137}Cs, 30.7 years for ^{90}Sr (Kanivets et al. 2014).

The coastal observations show that the annual average activity concentration of ^{90}Sr was stabilized at 1.2–1.6 Bq/l since 2001 and starting from 2007, it has been higher than the activity concentration of dissolved ^{137}Cs, which was slowly and steadily decreasing until the beginning of 2015. Meanwhile, in other water bodies of ChNPP contaminated areas (Lake Glybokoe, Kyiv Reservoir), the activity concentration of dissolved ^{137}Cs in water is several times lower than ^{90}Sr. Therefore, the ratio ^{137}Cs/^{90}Sr = 0.6–1.6 in the CP water (Fig. 8.7) can be considered to be abnormal (Zarubin 2006).

8.4.3 *Spatial Non-uniformity of Contamination*

As shown by long-term observations, the lowest ^{137}Cs and ^{90}Sr activity concentrations before the pumping station shutdown (May 2014) were observed in the Northern part of the CP, probably due to the influence of dilution by “clean” water

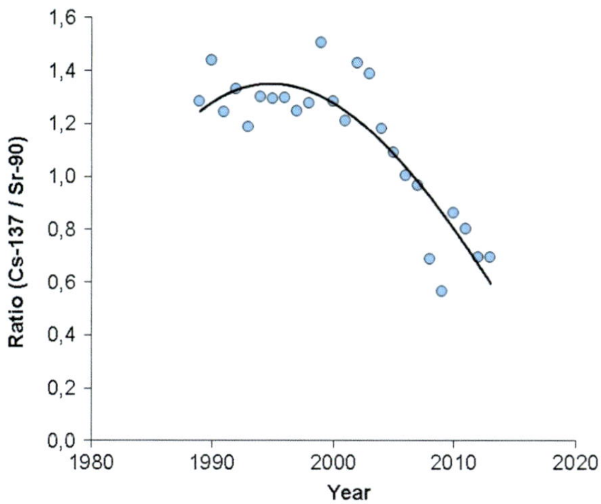

Fig. 8.7 Time change in $^{137}Cs/^{90}Sr$ ratio in the CP water in 1993–2013 (Kanivets et al. 2014)

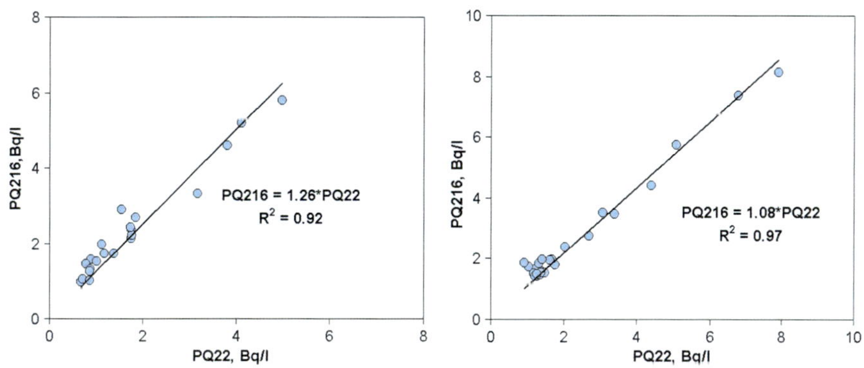

Fig. 8.8 Correlation between ^{137}Cs (left) and ^{90}Sr (right) activity concentrations in water near the pumping station (PQ22) and the outlet channel (PQ216) (Kanivets et al. 2014)

from Pripyat River. The ^{137}Cs and ^{90}Sr activity concentrations in Pripyat River were, respectively, 10- and 50-fold lower than in the pond.

Throughout the former "warm" part of the CP (from PQ142 to PQ216), the ^{137}Cs and ^{90}Sr water concentrations in 2003–2014 were approximately 25% and 8% higher than in the Northern part of the CP, where the diluting effect of Pripyat River was evident (Fig. 8.8).

Water exchange between the earlier existing "cold" and "warm" parts of the CP was prevented by a current-guiding dike. Obviously, the diluting effect of the Pripyat water did not go as far as the "warm" part. It can be assumed that radionuclide concentrations in the water and bottom sediments of this part of the CP were close to equilibrium values. The gradual decrease in radionuclide activity along the CP axis from the outlet channel mouth to the pumping station, possibly, can also be explained by a gradual decrease of radionuclide levels in bottom sediments in the same direction (Kanivets and Voitsekhovych 2000; Buckley et al. 2002).

8.4.4 Seasonal Variations of Radionuclide Activity Concentrations in Water

All the time since the accident, seasonal variations in dissolved ^{137}Cs activity concentrations were observed (Fig. 8.9) with a pronounced winter–spring minimum (February–April) and summer–autumn maximum (June–September). Concentrations, as a rule, are gradually increasing from April to August and then are gradually decreasing. The nature of this phenomenon has not been well understood until recently. The cyclic character of the changes in dissolved ^{137}Cs activity concentration is most likely to be associated with the CP temperature–oxygen regime characterized by stable summer–autumn temperature oxygen stratification, the presence of the thermocline, oxygen deficiency below the thermocline, and almost complete anoxia in the bottom water layer (Table 8.12).

It was suggested earlier (Nasvit 2002) that in the summer due to oxygen deficiency below the thermocline, nitrogen is reduced to ammonium by means of bacteria. Of all major cations, ammonium is the strongest competitor in exchange with ^{137}Cs selectively sorbed by micaceous clay minerals on frayed edge sites (FES) (Pirnach 2011a, 2011b; Konoplev 2020). An increase in ammonium concentration in the pore solution of bottom sediments and near-bottom water leads to remobilization of ^{137}Cs and its flux from the bottom to the water column. Results of hydrochemical observations on the pond in 1984–1987 (Romas 2002) and studies in 2002–2003 and

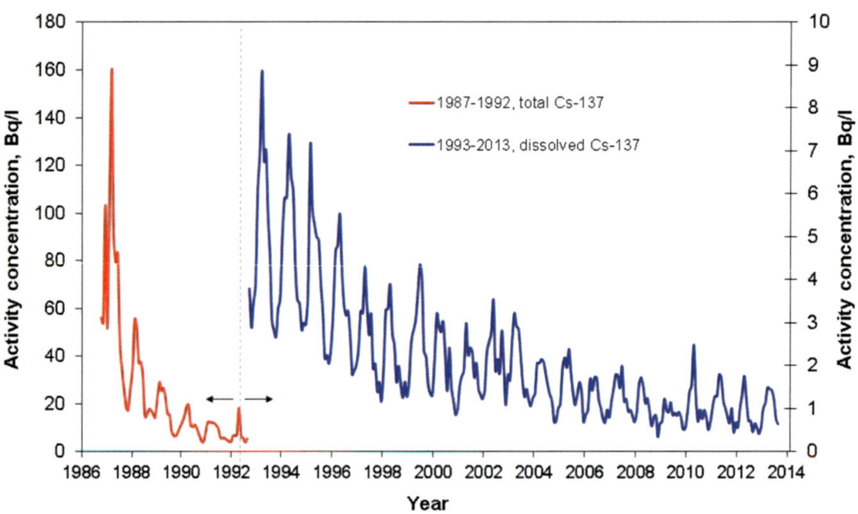

Fig. 8.9 ^{137}Cs in coastal water of the cooling pond (Kanivets et al. 2014)

2012–2017 (Voitsekhovych et al. 2016; Kanivets et al. 2014) have supported this hypothesis. The minimum concentration of ammonium in the contact layer of water and bottom sediments was usually observed in November–March and the maximum—in June–August (Table 8.11). By the end of August, the dissolved ^{137}Cs activity concentration in the CP water column was reaching the annual maximum (2.5–4.0 Bq/l in 2002–2003).

Subsequent to lasting calm hydrodynamic conditions, the dissolved ^{137}Cs activity concentration usually increases 1.5- to 2-fold from the water surface to the bottom (Table 8.12) and only after strong continuous storms or after autumn overturn of the water column, the dissolved ^{137}Cs concentration equalizes throughout the depth as a result of vertical mixing of water.

For ^{90}Sr, no seasonal variations in its activity concentration in water were detected; the ^{90}Sr vertical distribution in the water column was practically uniform in all seasons (see Table 8.12). The water balance indicates that the CP water mass was annually replenished, since the annual volume of supplied water was close to the CP water volume. Since the ^{137}Cs and ^{90}Sr activity concentrations in the Pripyat water were essentially lower than in the CP water, the source of water contamination is apparently the CP bottom sediments.

The difference in the dynamics of ^{137}Cs and ^{90}Sr concentrations in water stems from the difference in mechanisms of water contamination by each of these radionuclides. Based on available data, the source of CP water contamination with ^{137}Cs are silt deposits of deep-water areas of CP. The ^{137}Cs flux from the bottom to the water column is controlled by changes in ammonium concentration in the porous solution of the upper contact layer of silt deposits. Another evidence of the fact that deep-water silt deposits act as a major contributor of ^{137}Cs to water contamination is that ^{137}Cs concentrations in the water column in deep-water areas were always by about 30% higher than those near the shoreline (Table 8.13).

The ^{90}Sr concentrations in the deep-water areas, quite the reverse, have always been lower than near the shore (by 40% on average). The main source of water contamination with ^{90}Sr seems to be the shallow zone. This radionuclide might be coming from fuel particles randomly dispersed over the CP bottom as a result of initial fallout (Bulgakov et al. 2009). The high density of fuel particles precludes their trans-sedimentation from shallow areas to deep-water zones. In shallow water areas, fuel particles occur in water-permeable sediments, and the water here is saturated with oxygen, which seems to be conducive to the gradual destruction of the fuel matrix and ^{90}Sr dissolution in the CP water.

Overall, the long-term coastal monitoring by SSE "Ecocenter" and UHMI research in the deep-water areas in different seasons conducted before the beginning of the CP decommissioning have been demonstrated as follows:

1. Over the period 1987–2014, ^{137}Cs and Sr^{90} average annual concentrations in the CP water have decreased by nearly two orders of magnitude.
2. The average annual concentration of ^{90}Sr in the water became stabilized at the level 1.2–1.6 Bq/l since 2004 and was higher than the average annual concentration of dissolved ^{137}Cs starting from 2007. The most probable reason for such

Table 8.11 Vertical profile of water temperature (t °C), vertical distribution of oxygen and ammonium in the CP water column (Kanivets et al. 2014)

	October 2002			February 2003			May 2003			August 2003.		
Distance from bottom, m	t °C	O_2, mg/l	NH_4, mg/l	t °C	O_2, mg/l	NH_4, mg/l	t °C	O_2, mg/l	NH_4, mg/l	t °C	O_2, mg/l	NH_4, mg/l
"Cold" part of CP, depth 6.5 m, sampling station 1 (Fig. 8.3)												
5.5	13.0	9.8	0.35			0.12	19.0	10.4	0.41	24.1	7.8	0.25
2.0	13.0	10.1	0.44			0.10	17.2	8.8	0.31	23.5	4.2	0.27
0.05			1.7			0.18						0.49
−0.05[a]			1.05			0.43						1.95
"Warm" part of CP, depth 11 m, sampling station 2 (Fig. 8.3)												
9.0	13.5	10.0	1.4	1.6	17.2	0.02	19.9	11.4	0.26	23.5	7.04	0.12
2.0	13.4	9.7	0.29	3.1	10.6	0.03			0.4	16.0	0.9	0.43
0.05		8.7	1.05			0.07				13.8	0.11	0.71
−0.05[a]			3.86			0.41			1.12			2.35

[a]Pore solution in the upper 5-cm layer of bottom sediments

Table 8.12 Vertical distribution of the dissolved ^{137}Cs and ^{90}Sr activity concentrations in the CP water column (Kanivets et al. 2014)

	Activity concentration, Bq/l							
	October 2002		February 2003		May 2003		August 2003.	
Distance from bottom, m	^{137}Cs	^{90}Sr	^{137}Cs	^{90}Sr	^{137}Cs	^{90}Sr	^{137}Cs	^{90}Sr
"Cold" part of CP, depth 6.5 m, sampling station 1 (Fig. 8.3)								
5.5	2.8	1.2	1.41	1.25	1.41	0.97	2.5	1.0
2.0	2.8	1.2	2.54	1.43	2.7	1.1	2.0	1.1
0.05	3.2	1.6					5.5	1.5
−0.05∗	2.8	2.4		2.1			35.0	3.8
"Warm" part of CP, depth 11 m, sampling station 2 (Fig. 8.3)								
9.0	2.2	1.2	1.9	1.5	2.35	1.57	3.3	1.3
2.0	2.8	1.2	3.0	1.75	2.0	1.56	3.9	1,8
0.05	4.24	1.86	27.0	2.1	11.2	2.28	6.0	2.8
−0.05∗	31.6	7.8	33.3	3.2	33.0	2.2	38.0	4.0

[a]Pore solution of the upper 5-cm layer f bottom sediment

Table 8.13 Ratios of radionuclide activity concentrations in the water column of deep areas and near the shoreline, 2002–2003 (Kanivets et al. 2014)

	"Warm" part					
	^{137}Cs, Bq/l			^{90}Sr, Bq/l		
Month	Deep area	Shoreline	Deep/shore	Deep area	Shoreline	Deep/shore
October	3.2	2.2	1.5	1.4	2.0	0.7
February	3.5	2.5	1.3	1.8	2.5	0.7
May	2.2	1.6	1.4	1.6	1.8	0.9
August	3.9	3.6	1.1	1.7	3.0	0.6
Average			1.3	Average		0.7

stabilization is the resupply of radionuclides due to slow dissolution of fuel particles in the upper layer of bottom sediments in shallow areas. While ^{137}Cs is quickly fixed by micaceous clay minerals of suspended sediments, ^{90}Sr remains in the dissolved state. Before the beginning of CP decommissioning, such processes were running continuously resulting in compensation of ^{90}Sr infiltration losses and stability of its concentration in water.

3. The ^{137}Cs and ^{90}Sr concentrations in the Northern section of the CP (referred to as the "cold" part) in 1987–2013 were, respectively, 25 and 8% lower than in the Southern and Western sections ("warm" part) mainly because of dilution of the Northern section water by clean water from Pripyat River.
4. When the water column is stratified, the dissolved ^{137}Cs activity concentration in water is increasing from the water surface to the bottom, reaching the maximum in the bottom contact water layer. After the autumn "overturn" of water mass and in windy weather, the ^{137}Cs concentrations equalize in depth due to intensive vertical mixing. The vertical distribution of ^{90}Sr in the water column is practically uniform throughout the year.

5. All the time since the accident, seasonal variations in dissolved ^{137}Cs activity concentrations were observed with a pronounced winter–spring minimum (February–April) and a summer–autumn maximum (July–September) due to variations in ammonium concentrations in the pore water of bottom sediments and in the water column. Seasonal changes in ^{90}Sr concentration in the CP water are minor as a consequence of low variability of calcium and magnesium concentrations.
6. The ^{90}Sr activity concentration in the coastal water has always been 20–70% higher than that in the water column of deep-water zones due to high leaching rates of radionuclides from fuel particles in shallow areas. The dissolved ^{137}Cs activity concentration in deep waters was, on average, 30% higher than in coastal waters, since silt sediments of the deep-water areas are the main source of cesium re-mobilization from bottom sediments due to increased content of clay fraction and buildup of ammonium in the pore solution.

8.4.5 Changes in Radioactive Contamination of Water in the Initial Phase of the CP Decommissioning

8.4.5.1 Lowering the Water Level

In mid-May 2014, the pumping station was shutdown used previously for feeding the CP with water from Pripyat River to compensate water losses due to evaporation and infiltration through the enclosing dike, coastal slopes, and bottom. Afterward, the water level began to gradually decrease and should continue to decrease for several years until stabilization at natural equilibrium elevation.

In July 2015, the water level fell so much that the remains of the ChNPP first-stage enclosing dam (Fig. 8.10) emerged from the water and the CP became divided into three parts: Western, Northern, and Southern sectors, characterized by distinct temperature–oxygen regimes and radioactive contamination situations.

Fig. 8.10 View of the CP before decommissioning (left, photo by Olexander Sirota) and after the water level decline by 3 m (right, photo by Chornobyl Tour Ltd)

Monitoring of water level changes is the responsibility of the Chernobyl NPP. The main gauging station is located in the Northern part of the CP, and the second one is in the Southern part of the CP. Based on the time dependence of the water level in the Northern sector during 2014–2017 (Fig. 8.11), four stages of relatively constant rates of water level reduction can be identified (Table 8.14).

It is most likely that in the first and partly second stages of water level reduction, filtration losses were the most significant component of water balance. In the third stage, in view of the significant reduction in filtration losses, changes in water level were largely determined by evaporation and atmospheric precipitation. The rate of water level reduction in the CP Southern sector was higher than in the Western and Northern sectors because the filtration area is much larger (Bugai et al. 2005). In the CP Southern sector, the water level in August 2016 was 40 cm lower than in the Northern sector and at the end of 2017, this difference was as much as 80 cm. The

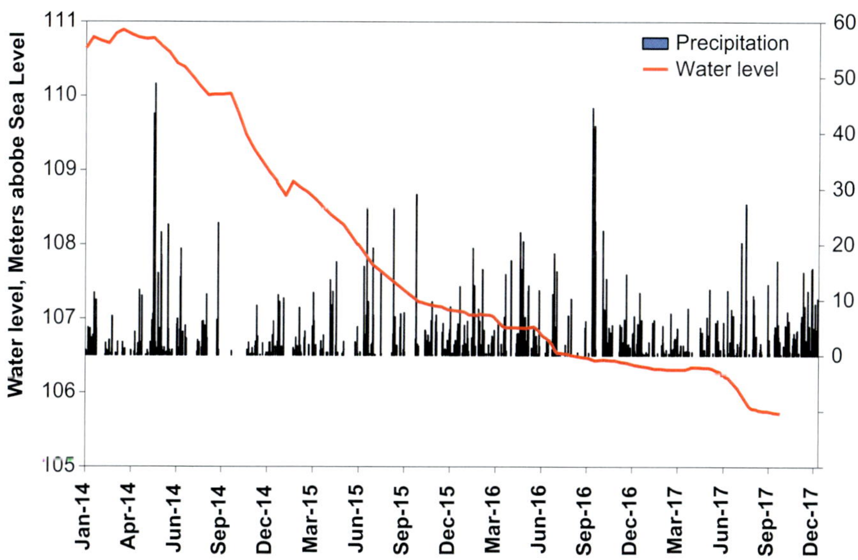

Fig. 8.11 Time dependence of the CP water level in 2014–2017 (ChNPP data) and atmospheric precipitation (Chernobyl meteostation data)

Table 8.14 Change in the rate of water level reduction in the Northern sector of the CP

Time period (stage)	Water level, m above the sea level		Water level reduction rate, mm/day
	Initial	Final	
15.05.2014–20.10.2015	110.78	107.23	6.80
21.10.2015–20.07.2016	107.23	106.54	2.55
21.07.2016–06.06.2017	106.54	106.28	0.08
11.08.2017–31.12.2017	105.78	105.71	0.05

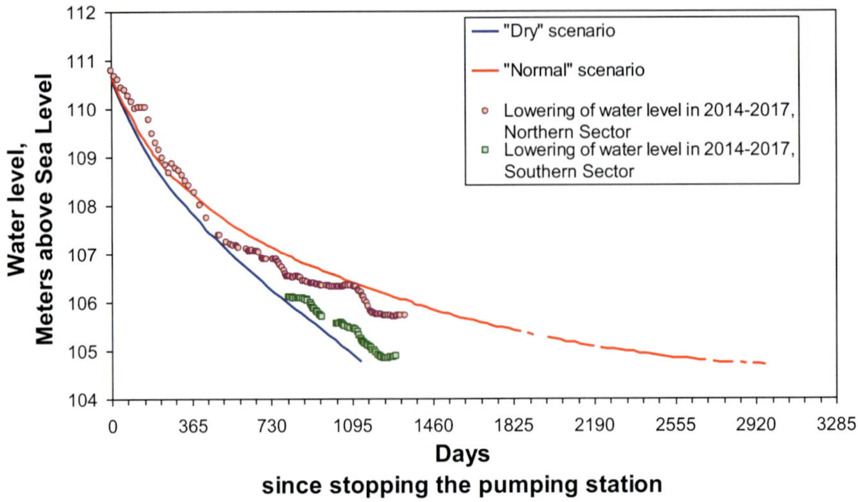

Fig. 8.12 Time changes in the CP water level in 2014–2017 compared with predictions under the dry and normal scenarios (Bugai et al. 2005)

rate of water level reduction in different parts of CP in 2014–2017 was consistent with IGS predictions made in 2001 (Fig. 8.12).

Overall, by the end of 2017, the CP lost 75% of its design volume. Losses were the largest in the Western and Northern sectors (80%) and smallest in the Southern sector (70%) because the average depth in this sector is greater. Meanwhile, the CP surface area was reduced by 40% only. Shallow areas of 1–1.5 m depth at that time occupied a considerable part (about half) of the reservoir.

In the spring of 2016, a lot of small islands appeared after snow and ice melting (Fig. 8.13). Until the end of summer 2016, the square area of these islands was expanding and they became covered with dense meadow grass and shrub, which prevents wind resuspension of soil particles. A good illustration of the dynamics of the CP water losses is provided by satellite images (Fig. 8.14) showing the evolution of the CP outlines until the spring of 2018.

8.4.5.2 Dynamics of Water Radioactive Contamination After the CP Decommissioning

It was expected that after cessation of water supply from Pripyat River to the CP and reduction in the pond volume, radionuclide activity concentrations, at the very least, should not be dropping. The monitoring data show that over three years since the beginning of the CP decommissioning, the ^{137}Cs activity concentrations in water have changed only slightly, whereas ^{90}Sr activity concentrations increased, on average, by 85% (Table 8.15).

Fig. 8.13 Exposure of the bottom of shallow areas after spring snowmelt of 2016 (above) and self-sown meadow vegetation on them in autumn 2016 (below) (photos by "Chernobyl zone—Radioactive TEAM")

In 2016, the difference in the ^{90}Sr activity concentrations in the Northern, Western, and Southern sectors, which became separated from each other from mid-2015, became apparent (Fig. 8.15). In general, by mid-2017, the ^{90}Sr concentrations in the Northern and Western sectors increased by 2.5 times and by 1.2 times in the Southern sector.

The growth of ^{90}Sr activity concentration in relatively shallow Northern and Western sectors is explained by the higher activity of this radionuclide in the upper contact layer of sandy-silt bottom sediments in the North-Western part of the CP in which nuclear fuel debris primarily deposited.

In the Southern sector, ^{90}Sr contamination of the bottom is much lower (Buckley et al. 2002), ^{90}Sr occurring mainly in deep-water sediments. There is practically no exchangeable ^{90}Sr in the upper contact layer of silt sediments (Voitsekhovych et al. 2016), so its flux from the silt bottom to the water column is very small. Just like

Fig. 8.14 Changes in the CP configuration from 2015 to the spring of 2018

Table 8.15 ^{90}Sr and ^{137}Cs mean annual activity concentrations in water for different sectors of CP based on data of coastal monitoring

	^{137}Cs				^{90}Sr			
Time	North	West	South	Average	North	West	South	Average
2011–2013	0.9	1.0	1.0	0.97	1.1	1.6	1.2	1.3
2015	0.9	1.1	1.0	1.0	2.0	1.7	1.2	1.5
2016	0.8	1.2	1.1	1.1	2.6	3.1	1.5	2.1
2017	0.65	1.3	1.0	1.0	2.7	4.0	1.5	2.4

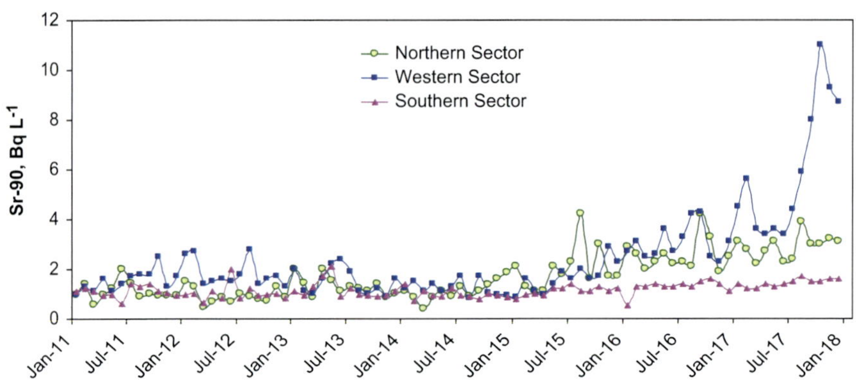

Fig. 8.15 Time dependence of ^{90}Sr activity concentration in different sectors of CP in 2011–2017, data of "Ecocenter" coastal monitoring

before the beginning of decommissioning, ^{90}Sr activity concentration near the shore remains higher than in the main water mass of deep-water areas (Figs. 8.16 and 8.17).

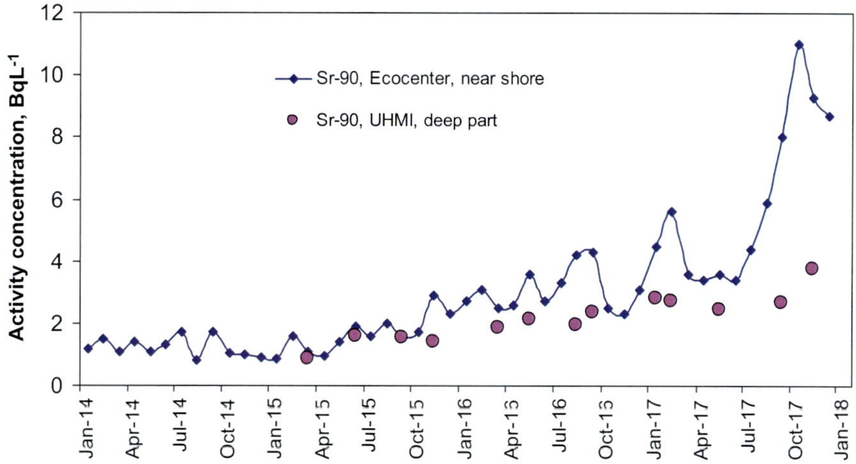

Fig. 8.16 Time dependence of ^{90}Sr activity concentration in water of the CP Western sector in 2014–2017, UHMI data (depth averaged values) and "Ecocenter" data (coastal monitoring)

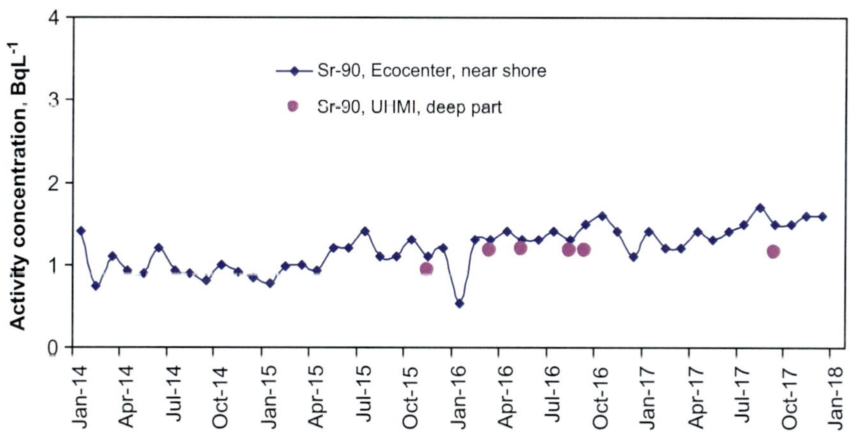

Fig. 8.17 Time dependence of ^{90}Sr activity concentration in water of the CP Southern sector in 2014–2017, UHMI data (depth averaged values) and "Ecocenter" data (coastal monitoring)

The ^{90}Sr water activity concentration in the CP Northern and Western sectors is expected to continue growing due to enhanced ^{90}Sr leaching from fuel particles on exposed sections of bottom sediments, as well as increased dissolution rate of fuel particles on the areas that became shallow as a result of lowering the water level, where oxygen content in the near-bottom water layer has increased.

Vertical distribution of ^{90}Sr in the water column is relatively uniform (Fig. 8.18) irrespective of seasonal changes in the hydrological, hydrophysical, and hydrochemical conditions. No major seasonal variations in ^{90}Sr activity concentration in water have been revealed.

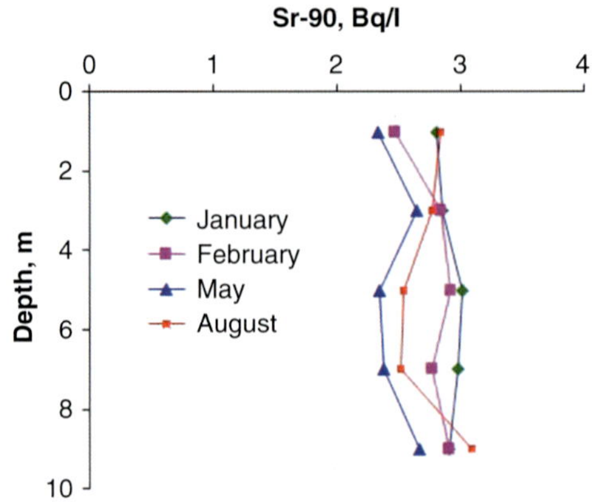

Fig. 8.18 Vertical distribution of ^{90}Sr activity concentration in the water column of the CP Western sector for different seasons of 2017

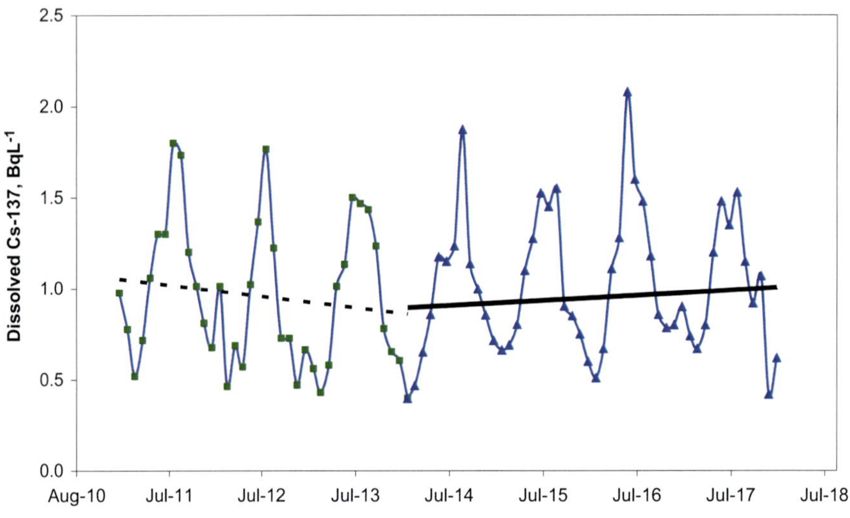

Fig. 8.19 Seasonal variations of monthly averaged dissolved ^{137}Cs activity concentrations in 2011–2017 (averaged over all coastal observation sites), SSE "Ecocenter" data

As in all previous years, in 2014–2017, seasonal variations in dissolved ^{137}Cs activity concentration in water were observed (Fig. 8.19) with a pronounced winter–spring minimum (February–April) and summer–autumn maximum (July–September). Seasonal variations in dissolved ^{137}Cs activity concentrations are associated with changes of physicochemical and biological processes in the CP, specifically: (1) significant temperature–oxygen direct stratification in deep-water areas in the summer (June to September); (2) occurrence of anoxic reducing conditions in the hypolimnion; (3) an increase in ammonium concentration in the pore solution of bottom sediments, and as a consequence, an increase of ^{137}Cs diffusion flux from bottom sediments to the water column.

Data obtained by UHMI in 2015–2017 made it possible to identify the main processes responsible for seasonal variations in ^{137}Cs vertical distribution in the water column. In the spring, when distribution of temperature and oxygen in the water column is uniform, at normal oxygen saturation of water and low ammonium concentration in the water column and in the pore solution of bottom sediments, the ^{137}Cs activity concentration in the pore solution is minimal (below 5 Bq/l), and hence diffusion from bottom sediments to water is low and the vertical distribution of this radionuclide in the water column is almost uniform (Fig. 8.20).

Subsequent to formation of strong direct temperature stratification at the beginning of summer (Fig. 8.21), the oxygen content below the thermocline begins to decrease because it is consumed for decomposing dead phytoplankton, and oxygen deficiency below the thermocline occurs to the point of complete anoxia in the near-bottom water layer at the end of summer. At the same time, nitrogen is reduced to its ammonium form and an increase of ammonium concentration in the contact "sediment-water" layer leads to ^{137}Cs remobilization and enhancing of ^{137}Cs diffusional

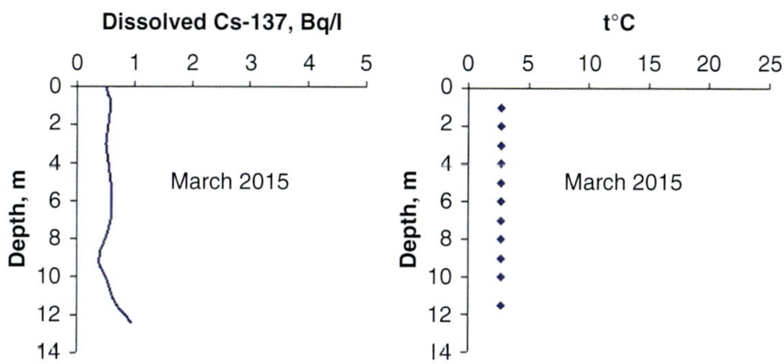

Fig. 8.20 Vertical distribution of dissolved ^{137}Cs under spring homothermal conditions in the water column of the CP Western sector, March 2015

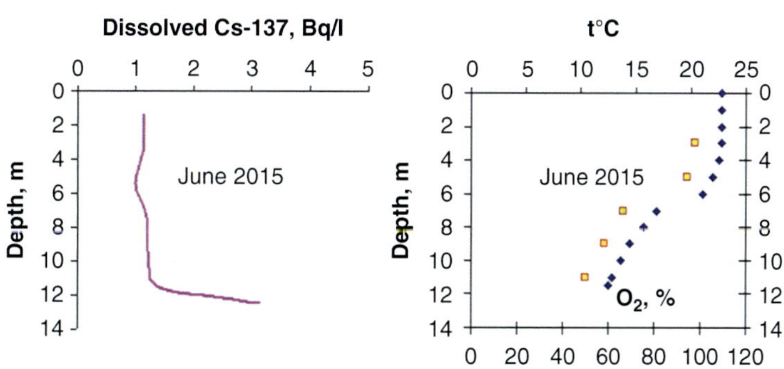

Fig. 8.21 Vertical distribution of dissolved ^{137}Cs, temperature, and oxygen in the water column of the CP Western sector, June 2015

flux from bottom sediments to the water column. As a result, the dissolved ^{137}Cs activity concentration from the water surface to the bottom increases linearly by the end of summer (Fig. 8.22).

In September, water cooling in the epilimnion leads to overturning of the water column and full vertical mixing. As a result of this seasonal phenomenon, the water column becomes homogeneous with respect to the characteristics such as temperature, oxygen content, dissolved ^{137}Cs activity concentration, etc. (Fig. 8.23).

The autumn increase in wind strength and duration causes enhancement of hydrodynamic processes in the CP. Due to resuspension of bottom sediments on shallow areas and transfer of fine particles by drift currents, the water column in deep-water areas is saturated with suspended sediments, their content increasing by 3–5 times in case of storms. In the process of sedimentation, dissolved ^{137}Cs is adsorbed and fixed by these particles and its activity concentration is significantly reduced and equalizes in depth by the end of autumn.

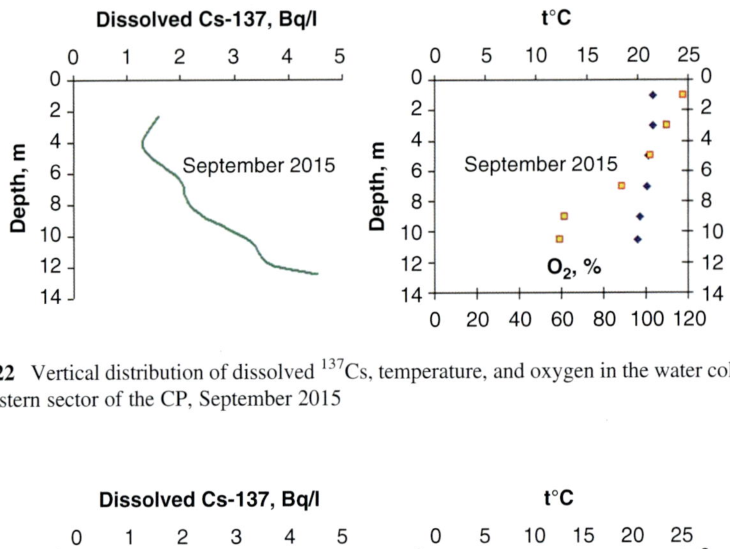

Fig. 8.22 Vertical distribution of dissolved ^{137}Cs, temperature, and oxygen in the water column of the Western sector of the CP, September 2015

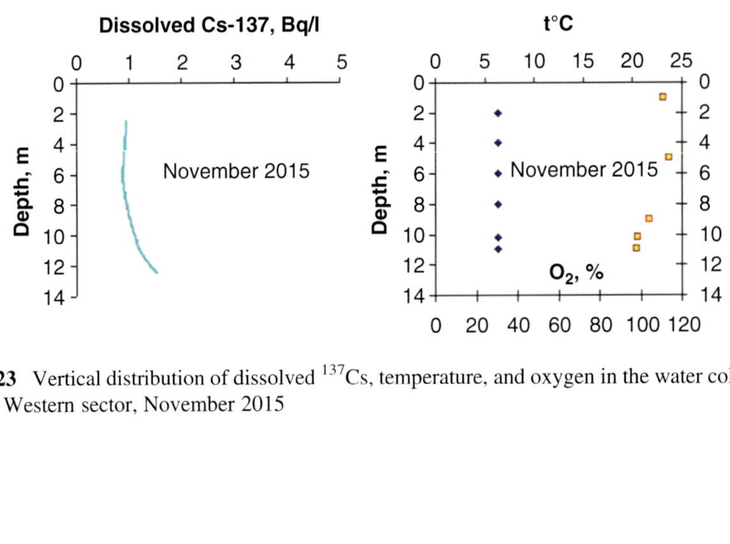

Fig. 8.23 Vertical distribution of dissolved ^{137}Cs, temperature, and oxygen in the water column of the CP Western sector, November 2015

Differences in the spatial and temporal changes of ^{137}Cs and ^{90}Sr water activity concentrations are associated with differences in mechanisms of water contamination by each of these radionuclides. The main source of ^{137}Cs in the CP water are sediments of the medium and large depth areas. The amount of ^{137}Cs exchangeable forms in silt bottom sediments in deep-water areas has not changed over the past 10–15 years and equals 1–2% of total radionuclide activity in bottom sediments, which is sufficient for ^{137}Cs continuous diffusional flux from the upper contact layer of bottom sediments to the water column.

It has been revealed that ^{90}Sr in bottom sediments mainly occurs as part of fuel particles. Practically no ^{90}Sr exchangeable forms have been found in silt sediments of deep-water areas (<0.3%, see Sect. 8.5), since anoxic reduction conditions in sediments are conducive to persistence of fuel particles. In the bottom sediments of shallow areas, the share of ^{90}Sr exchangeable form is 5–7% due to higher concentration of oxygen in water and, as a consequence, higher oxidation velocity of fuel particles. ^{90}Sr seems to be supplied by the dissolution of fuel particles occurring on shallow areas.

From 2016, the wash-off of ^{90}Sr mobile forms by surface runoff from the exposed bottom areas became a new source of contamination. Therefore, in future, ^{90}Sr water contamination in the lakes that will remain after the CP decommissioning will be primarily determined by wash-off of the radionuclide from the exposed bottom sections during heavy rain and snowmelt runoff events.

In 2015–2017, hydrochemical and water-physical indicators of water (content of main ions, electrical conductivity, pH, and Eh) did not differ from those observed previously. At the same time, changes in the reservoir oxygen regime occurred, specifically a decrease in oxygen concentration in the surface water layer (0–5 m, Fig. 8.24) and enhanced vertical oxygen stratification in the warm period of the year, with the occurrence of hypoxic or anoxic conditions below the thermocline.

In 2015, the CP oxygen regime was similar to the previous years: sufficient saturation of the entire water column with oxygen during the autumn and spring homothermy (100–110%), moderate oxygen direct stratification during ice cover period (70–100% in 0–6 m layer, 50–60% in 7–11 m layer), and strong oxygen direct stratification in the summer (80–90% in the epilimnion, 30–50% in the hypolimnion). In 2016, the oxygen conditions in the CP drastically changed. In 2015–2016 winter–early spring, the whole water column saturation with oxygen was higher than 100%. In mid-May 2016, however, significant oxygen stratification emerged and in early August, oxygen occurred only in the epilimnion (the upper 5-m layer). In the hypolimnion, anoxic conditions became established (Fig. 8.25) and continued until the water mass overturn in the mid-September 2016. The change in the CP oxygen conditions in 2016 was probably associated with: (1) splitting of the pond into three isolated parts and a decrease in the water surface area leading to decreased impact of wind on the CP and, as a consequence, a decrease in vertical mixing intensity in the water column and its aeration; (2) a significant decrease in water volume (30–35% from the design-basis volume) leading to concentration of aquatic organisms (plankton, fish, and dead biota).

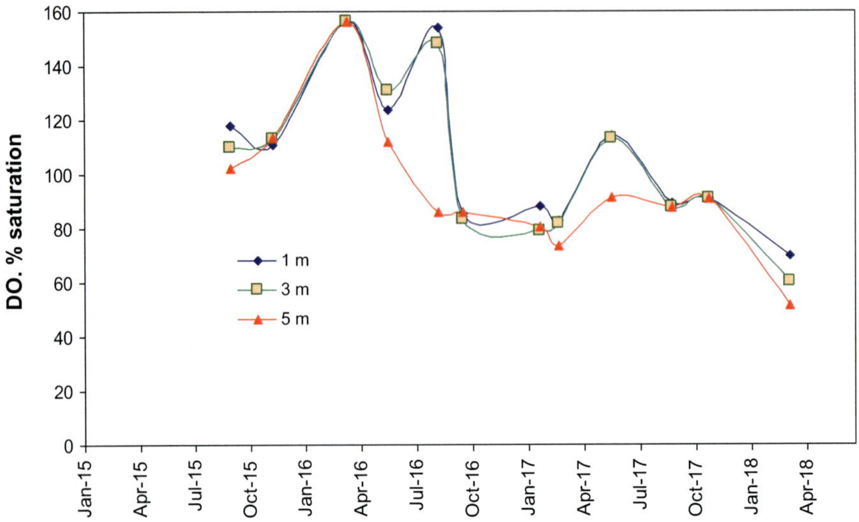

Fig. 8.24 Temporal change in oxygen water saturation, 2015–2017

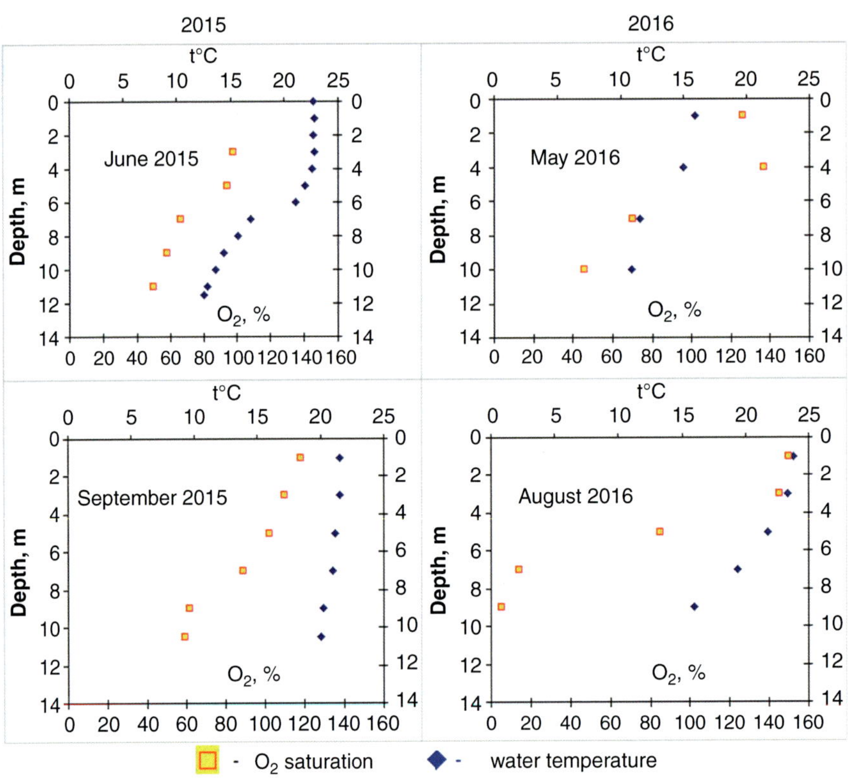

Fig. 8.25 Comparison of temperature–oxygen water stratification in 2015 and 2016

The oxygen conditions in the summer of 2016 might be termed adverse from the sanitary and hydrobiological standpoint. Moreover, oxygen-free reduction conditions were probably promoting ammonium generation in the water-bottom contact layer and in the pore solution, which might have led to an increase of ^{137}Cs diffusional flux from bottom sediments to the water column.

Lately, suspended sediment concentrations in water have been increasing, which, however, seems to be temporal. The observed increase in turbidity is associated with the CP current bathymetry, almost half of the pond having a depth of 1–1.5 m. At strong winds, fine mineral particles on shallow areas are resuspended and transported by wind currents, ending up in the water column on deep-water areas. In the process of sedimentation, these particles adsorb and fix dissolved ^{137}Cs and thereby transfer the radionuclide from water to the bottom. Before the water level reduction in the CP, the particulate ^{137}Cs fraction was 10–15%, increasing to 30–40% in 2016–2017 and occasionally being as high as 60–70%. This process seems to induce self-purification of the CP water, and this is the reason why an increase in ^{137}Cs concentration, expected based on its diffusion from bottom sediments to water, has never been observed to date.

A decrease in turbidity and an increase in dissolved ^{137}Cs activity concentration can be expected after the water level is reduced by 1 m and the bottom becomes exposed on shallow areas, which are the main supplier of suspended sediments until now.

In lakes to be left after completion of CP decommissioning, the following processes are anticipated to occur, influencing water contamination: (1) wind mixing of water mass will decrease due to reduction of water surface area, which will contribute to enhanced seasonal temperature–oxygen stratification and establishment of anoxic reduction conditions below the thermocline, since a significant portion of oxygen will be spent for decomposition of dead biota; (2) oxygen-free reduction conditions will facilitate persistence of fuel particles, but will have lead to an increase in ammonium concentration in the near-bottom water layer and enhanced ^{137}Cs diffusion from bottom sediments to the water column; (3) after stabilization of water level, some increase in dissolved ^{137}Cs activity concentrations in the water of new lakes due to diffusion from bottom sediments can be expected; (4) the fraction of ^{90}Sr mobile forms will increase due to dissolution of fuel particles on the exposed sections of the bottom. During rains and snowmelt, ^{90}Sr will be washed off by surface runoff to the lakes and it is likely that ^{90}Sr activity concentration in the lakes will be growing.

8.5 Radionuclides in Bottom Sediments

The first survey of bottom radioactive contamination throughout the CP was carried out in 1989 (repeated in 1991) by SHI with analytical support of SPE "Pripyat" (currently SSE "Ecocenter"). These surveys can be considered to be most representative, since sediment samples were collected at 100 locations in every case. In each sample, gamma-emitting radionuclides and ^{90}Sr were determined, after which ^{137}Cs and ^{90}Sr distribution patterns were obtained for the entire bottom area (Ci/km^2) of the pond and inventories of both radionuclides in bottom sediments were estimated (Shiklomanov 1992; Filippov 1996).

The merits of the conducted surveys were: (1) the surveys were performed a few years after the accident when the overall bottom contamination pattern was still close to that of primary fallout and (2) in the areas of intensive silt accumulation, the contaminated layer was not more than 10 cm, enabling through-thickness sampling. Conversely, the limitations are as follows: (1) ^{90}Sr analysis was not performed, ^{90}Sr activity was calculated using ^{144}Ce/^{90}Sr ratio in nuclear fuel, which could have caused additional uncertainty in deriving ^{90}Sr activity concentration and its inventory; and (2) distribution pattern of radionuclide inventories (Ci/km^2) on the bottom was drawn using regular interpolation between sampling points, with no account taken of the bottom relief. As a result, the obtained pattern of radionuclide distribution over the bottom area did not reflect the actual spatial pattern of contamination. It should, however, be noted that a few years later, results of this survey became outdated due to the active transformation of the bottom soil complex and essential change in radionuclide distribution pattern on the bottom.

In 1999–2004, the bottom sediment studies were continued within several international projects by UHMI and IHB NASU, with analytical support provided by SSPE "Ecocenter" (Kanivets and Voitsekhovych 2000; Buckley et al. 2002). In these studies, all bottom sediment samples were taken using corers, which permitted maintaining the integrity of collected bottom sediment cores. For core sampling in silt accumulation areas, a MacKereth minicorer (Duncan & Associates, UK) was used. A DTSh-3 rod sampler and automatic DA-3 sampler ("ECOTECHNIK", Russia) were used to collect sandy bottom sediments. A total of 122 core samples of bottom sediments were collected in which characteristics were determined such as: (1) physical properties of bottom sediments – bulk density, moisture content, loss on ignition, and particle size distribution in the top sediment layer; (2) radionuclide activity concentrations; (3) ^{137}Cs and ^{90}Sr speciation; (4) ^{137}Cs distribution among size fractions of sediments.

Later in 2011–2013, UHMI had carried out a series of the bottom sediment sampling mostly in order to verify ratio ^{137}Cs/^{90}Sr and ^{90}Sr/^{241}Am in areas which were not covered by all previous sampling campaigns at the CP (Krasnov 2013). Based on obtained data, features of bottom sediment dynamics, as well as vertical and lateral distribution of radionuclides in bottom sediment maps of ^{137}Cs and ^{90}Sr distribution over the bottom area, were drawn and inventories of each radionuclide were estimated. Along with this, a contribution of fuel particles to radioactive contamination of bottom sediments was assessed as well.

8.5.1 *Spatial Pattern of Bottom Radioactive Contamination, Radionuclide Inventories in Bottom Sediments*

Radioactive contamination of the bottom sediments in the CP was extremely non-uniform over the bottom area (in terms of inventory, activity concentration, and radionuclide ratio). For ^{90}Sr and transuranic radionuclides, the spottiness of primary depositions of fuel particles was the main reason for non-uniformity of their spatial distribution and hence the highest ^{90}Sr activity concentrations were detected in sediments with a high content of fuel particles, regardless of sediment type. For ^{137}Cs, spatial distribution non-uniformity was due to spatial variability of bottom sediment types. The highest ^{137}Cs activity concentration was found in silts primarily consisting of clay minerals and hence the largest ^{137}Cs inventories in bottom sediments occur in silt accumulation areas (former channel of Pripyat River, lakes and swamps on the former floodplain of Pripyat).

Throughout the pond existence, the bottom was subjected to transformation due to various types of currents (littoral, drift, and compensatory currents, large-scale horizontal and vertical circulations, and wave effects), resulting in bottom self-purification in erosion zones and enhancement of contamination on silt accumulation areas. The ^{137}Cs inventory on the bottom and silt layer thickness were closely correlated. For ^{90}Sr, such relationship is not so obvious, yet it is reasonable to suggest that ^{90}Sr inventory is higher on the areas of intensive silt accumulation as compared to those without silt.

Based on the results of radionuclide analysis of bottom sediment samples collected in 1999, 2001–2003, and later 2012–2013, schematic maps of the ^{137}Cs and ^{90}Sr bottom contamination were constructed using CP bathymetric map (Fig. 8.3) (Voitsekhovych et al. 2013). A special algorithm has been developed for geospatial interpolation analysis with consideration of different depth zones of the pond was used (Smith et al. 2009). Using the schematic maps of bottom contamination, total ^{137}Cs and ^{90}Sr inventories for the entire CP were calculated (Figs. 8.26 and 8.27), and distribution of these radionuclides among different depth zones was estimated (Table 8.16). It should be pointed out that the calculated inventories are conservative upper estimates and the actual values can be lower.

It can be seen from the schematic maps that a larger part of ^{137}Cs and ^{90}Sr depositions occur in the deep-water areas of the CP. More specifically, for both radionuclides, the highest depositions were found in the deep-water hollows of the former "warm" part of the CP, with the thickness of contaminated silt layer being more than 1 m in some parts even in 1999.

As indicated after the accident, the Northern part of the CP opposite to the intake channel (closest to the accidental reactor) was exposed to the highest contamination. At the same time, Egorov et al. (1989) estimated that about 120 TBq of ^{137}Cs was released through the outlet channel. The analysis of the schematic maps shown in Fig. 8.27 and current UHMI data indicate that in the initial stage of the accident, a large amount of ^{90}Sr and transuranic elements was discharged as embedded in fuel particles through the outlet channel to the Western sector of the CP because the

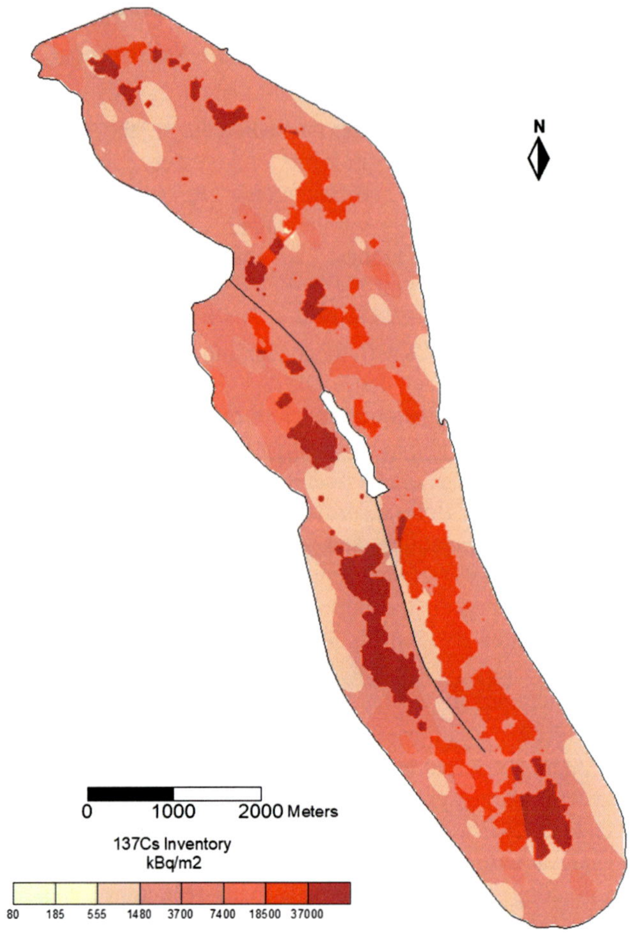

Fig. 8.26 Distribution of ^{137}Cs in the CP bottom sediments (modified after Voitsekhovych et al. 2013 using 2016 updated topography map)

bottom sediments of the Western sector (2 km^2) were found to be the most contaminated with these radionuclides. The Western sector of the CP has become a sort of trap for fuel particles released through the outlet channel.

So, the ratio $^{137}Cs/^{90}Sr$ in bottom sediments of the entire Western sector of the CP varies from 1,2–2 (sand deposits) to 3–6 (deep-water silt deposits), which indicates a significant content of fuel particles in bottom sediments of this part of the CP.

The bottom sediments in the Northern part of the CP between the pumping station (BNS-3) and the inlet channel also show a high content of ^{90}Sr (fuel particles), the $^{137}Cs/^{90}Sr$ ratio being 1.5–5.0. The highest values of $^{137}Cs/^{90}Sr$ ratio (up to 20 and on average 11–12) were found in the sediments along the Eastern part of the CP. As such, obtained data led us to assume that the amount of radioactivity discharged to the CP through the outlet channel was comparable to that due to atmospheric fallout

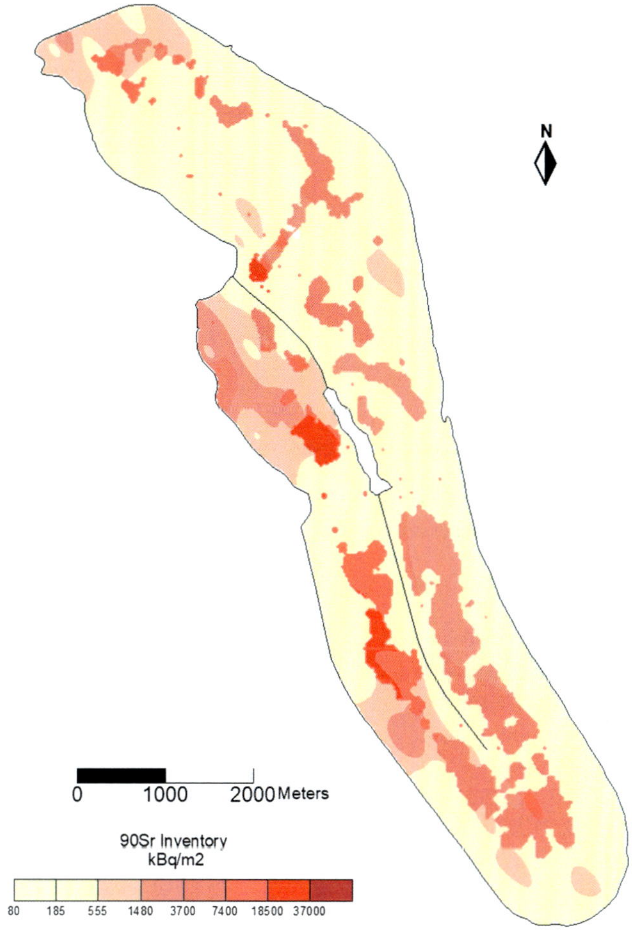

Fig. 8.27 Distribution of ^{90}Sr in the CP bottom sediments (modified after Voitsekhovych et al. 2013 using 2016 updated topography map)

Table 8.16 ^{137}Cs and ^{90}Sr inventories in the CP bottom sediments for different depth ranges in 2013 (calculated using 2016 updated topography map and decay corrected to 1 May 1986)

	Area		^{137}Cs		^{90}Sr	
Depth interval, m	km^2	%	TBq	%	TBq	%
0 3.5	3.50	16.2	13	4.4	3.3	6.2
3.5–7	13.50	62.5	54	18.6	10.5	19.8
7–10	2.65	12.3	81	27.9	14.2	26.7
10–12	0.97	4.5	65	22.4	11.9	22.4
>12	0.98	4.5	77	26.6	13.2	24.8
Total	21.7	100	290	100	53	100

onto the pond surface, or may even have exceeded the latter with respect to ^{90}Sr and transuranic nuclides.

Unfortunately, uncertainties involved in estimating ^{137}Cs and ^{90}Sr inventories in bottom sediments preclude drawing definitive conclusions about sources of radioactive contamination of the CP as a result of the 1986 accident. The main contributor to the uncertainties is the insufficient number of collected samples and hence radionuclide analysis data. To be more specific, uncertainties in estimating ^{137}Cs inventory are linked to the shortage of samples from silt accumulation areas, which could lead to overestimation of the total inventory of this radionuclide in the CP. On the other hand, an insufficient number of samples from the Western sector and North-Western part could lead to underestimation of total inventories of ^{90}Sr and transuranic nuclides.

8.5.2 Vertical Distribution of Radionuclides in Different Types of Bottom Sediments

Vertical distribution of radionuclides in bottom sediments was investigated in order to estimate deposition of radionuclides at each sampling point, reveal their distribution pattern in different types of sediments, estimate rates of silt accumulation, as well as to parameterize models for radionuclide vertical migration and predict secondary water contamination. The main types of the CP bottom sediments are sand, sandy silt, and silt (by the classification of Kudrin 1959). Underlying the silt and sandy silt layers are transformed primary soils (sandy loam and loam).

8.5.2.1 Bottom Sediments of Shallow Areas

According to the bathymetric map of 2001, the areas with depths of 0–7 m, expected to be dried up after the CP decommissioning, occupy about 75% of the bottom square area. *Sandy sediments* without silt particles occurred at the depths of 0–3.7 m. Layer-by-layer radionuclide analysis of sandy sediments showed that in the coastal zone along the CP perimeter, the depth of ^{137}Cs occurrence in sandy deposits was reaching 50 cm, which is indicative of essential wind-wave mixing of sandy deposits at depths up to 1–1.5 m (sand erosion and re-deposition in wave-cut zone, sediment flux along the coast during storms). At larger depths (outside the wave breaking zone), most of the contamination occurs in the upper 10–20 cm layer of sandy sediments. Due to low sorption properties of primary minerals (quartz and feldspar), sandy deposits are characterized by minimal levels of ^{137}Cs contamination. In 2001, the ^{137}Cs activity concentration in the top layer of sandy deposits was 4–17 kBq/kg (9 kBq/kg, on average) and ^{90}Sr—1.3–10 kBq/kg (2.5 kBq/kg, on average). The average ^{137}Cs deposition in sandy deposits for the entire CP was 1470 ± 450 kBq/m^2.

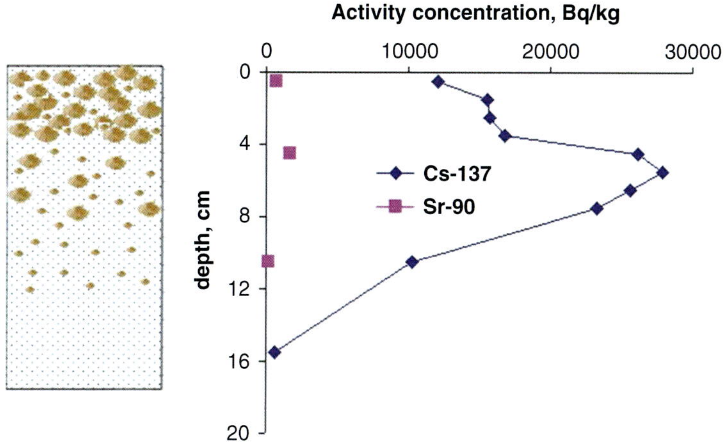

Fig. 8.28 Vertical distribution of ^{137}Cs and ^{90}Sr in sandy bottom sediments, the CP Eastern part. 0–4 cm: sand with abundant *Dreissena* shells (30%, dry weight). 4–8 cm: sand with individual shells and their fragments (7%, dry weight). 8–13 cm: sand with small shells fragments. 13–18 cm: sand

A typical vertical distribution of ^{137}Cs in the CP sandy sediments is shown in Fig. 8.28. The presented core sample was collected in 2017 on an island that emerged as a result of water-level reduction. The island surface contained deposits of *Dreissena* shells, the content of empty shells in the upper 4-cm layer of the sample being about 30%. At normal water level, the depth here was about 3.5 m. Most likely, over the time of the CP existence, the bottom in this place was exposed to wave impact, and the upper 4–5 cm layer of sediments from time to time was resuspended and mixed.

Abnormally high ^{137}Cs and ^{90}Sr activity concentrations were found in sandy deposits opposite to the ChNPP outlet channel (Fig. 8.29): an order of magnitude higher than the average values for sandy deposits of the entire pond. The $^{137}Cs/^{90}Sr$ ratio in the surface layer of sandy sediments of this zone was 0.6–1.5, which is indicative of high content of fuel particles. It also suggests that water discharges from the ChNPP operational site through the outlet channel was contributing to the CP radionuclide contamination after the initial atmospheric fallout.

In 1999–2004, *sandy silts* were widely spread on the areas with depths more than 5 m, the thickness of such deposits being 1–10 cm at depths 5.0–7.5 m. Underlying the silt layer were residues of primary soils. As a rule, radionuclides were evenly distributed throughout the silt layer (Fig. 8.30); the ^{137}Cs activity concentration in sandy silt was estimated to be100–300 kBq/kg.

^{137}Cs was found to penetrate sediments deeply (down to 12–15 cm) when sand was underlying the silt layer. With sandy loam or loam soil underlying silt, the radionuclide content at a depth of more than 8–10 cm from the sediment surface was the same as the pre-accidental level. The average density of bottom contamination with ^{137}Cs at a depth of 5.0–7.5 m was 2520 kBq/m^2. Differences in ^{137}Cs vertical

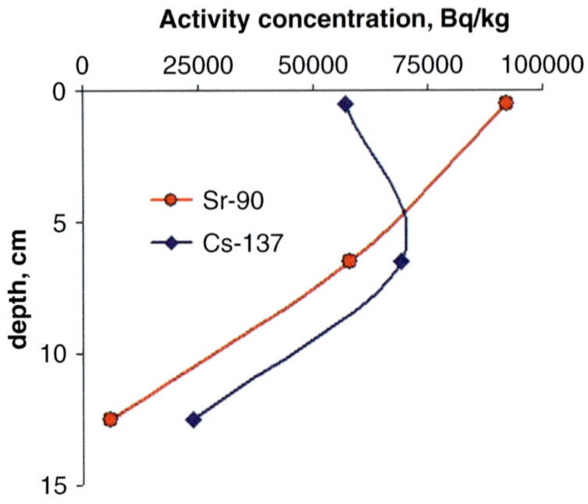

Fig. 8.29 Vertical distribution of ^{137}Cs and ^{90}Sr in sandy bottom sediments opposite to the outlet channel, depth 4.8 m

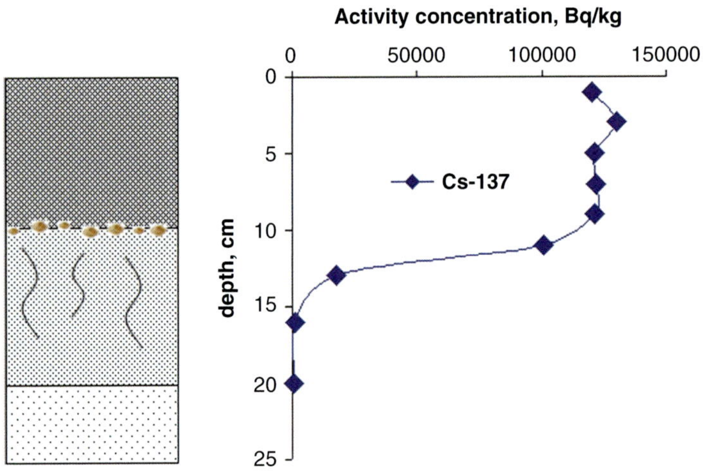

Fig. 8.30 Typical example of ^{137}Cs vertical distribution in the areas of silt particle trans-sedimentation, the CP North-Eastern part, depth 6 m. 0–10 cm: sandy silt. 10–20: primary sandy loam soil layer with remnants of meadow vegetation roots, on the surface – fragments of mollusk shells. 20–25 cm: sand

distribution depending on the type of underlying soils are shown in Figs. 8.31 and 8.32.

On bottom areas with a high content of fuel particles, ^{90}Sr, ^{241}Am, and $^{239,\,240}Pu$ were largely found in the upper layer of sediments (Fig. 8.33). As distinct from ^{137}Cs, these nuclides were penetrating into deposits to a lesser extent due to their incorporation in the fuel matrix.

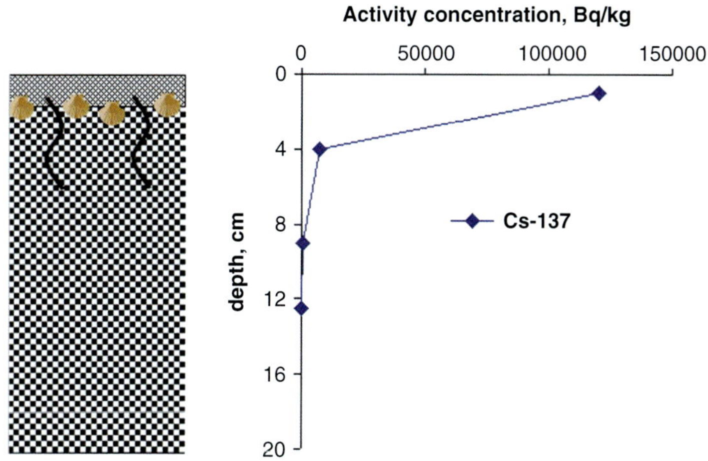

Fig. 8.31 Distribution of ^{137}Cs in silt covered loam, the CP North-Eastern part, depth 5.5 m. 0–1.5 cm: sandy silt with the mollusk shells. 1.5–13.5 cm: loam with remnants of meadow vegetation roots in the upper part of the layer

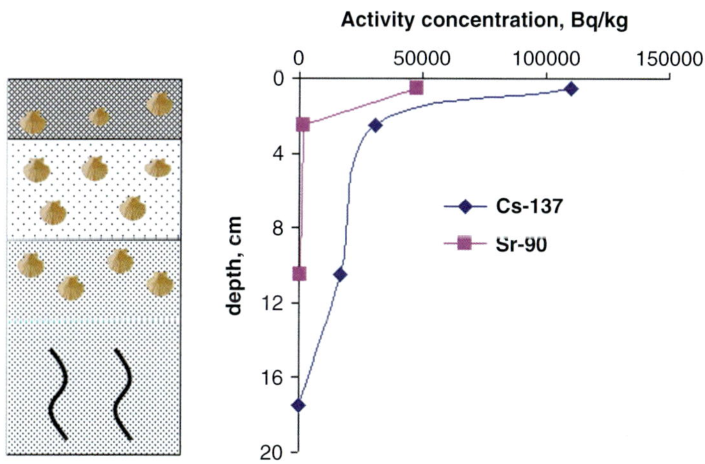

Fig. 8.32 Distribution of ^{137}Cs in silt covered sand and sandy loam, the CP North-Eastern part, depth 4.6 m. 1–3 cm: sandy silt with the mollusk shells. 3–8 cm: silted sand with the mollusk shells. 8–18 cm: sandy loam with the mollusk shells in the upper part of the layer and with remnants of meadow vegetation roots in the lower part of the layer

8.5.2.2 Bottom Sediments of the CP Deep Parts

Highest radionuclide activity concentrations were detected in silt deposits of deep-water zones, which are homogeneous and have no visible inclusions and impurities. The ^{137}Cs activity concentration in the top layer of these sediments throughout the

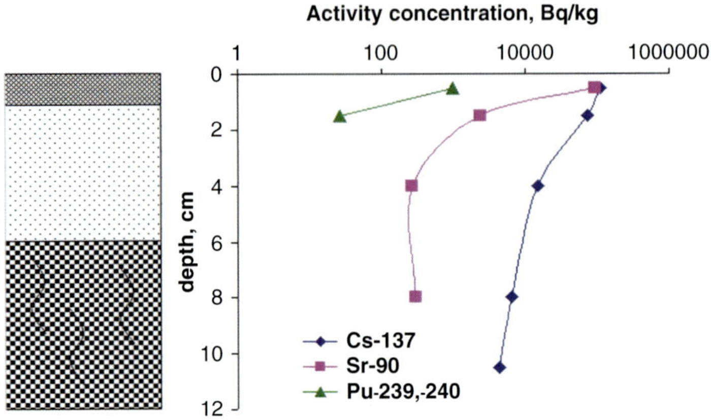

Fig. 8.33 Examples of the vertical distribution of ^{137}Cs, ^{90}Sr, and $^{239,\,240}$Pu on shallow water areas with a high content of fuel particles, the CP South-Western part, depth 6 m. 0–1 cm: sandy silt. 1–6 cm: small-grained sand. 6–12 cm: loamy soil with remains of the plant root system

CP was varying from 360 kBq/kg in the CP Northern sector to 500–700 kBq/kg in the Western and Southern sectors. The ^{90}Sr activity concentration in the upper layer of silt deposits was 40–160 kBq/kg. In the studies of 1999–2003, the maximum of ^{137}Cs deposition in deep-water silt sediments was determined to be 185 MBq/m^2, given 92 cm-thick silt layer, and the maximum of ^{90}Sr deposition was 35 MBq/m^2 (55 cm-thick silt layer).

Based on long-term observations, before the beginning of CP decommissioning, ^{137}Cs and ^{90}Sr activity concentrations in the CP water were steadily decreasing. One would think that activity concentrations in silt sediments should also gradually decrease from the layer with maximum activity (formed in 1986) to the surface layer. However, in most cases, the vertical profile of radionuclides in the silt layer on deep-water sections showed uniform distribution of activity in the layer formed after the accident (Fig. 8.34). This is another evidence that highly contaminated silt deposits in the CP deep-water areas were formed mainly due to transport of silt particles from shallow water areas, where these particles had contaminated during the period of initial fallout after the accident.

By the bathymetric map of 2001, the square area of deep-water bottom sections is about 4 km^2 (18% of the total design-basis CP bottom area). After the CP decommissioning, highly contaminated bottom sediments of these areas will remain under water.

8.5.3 Fuel Particles in the Bottom Sediment

As a result of the explosion and the fire at Unit 4 of the Chernobyl NPP in April–May 1986, about 3% of the irradiated nuclear fuel were released and dispersed in the form

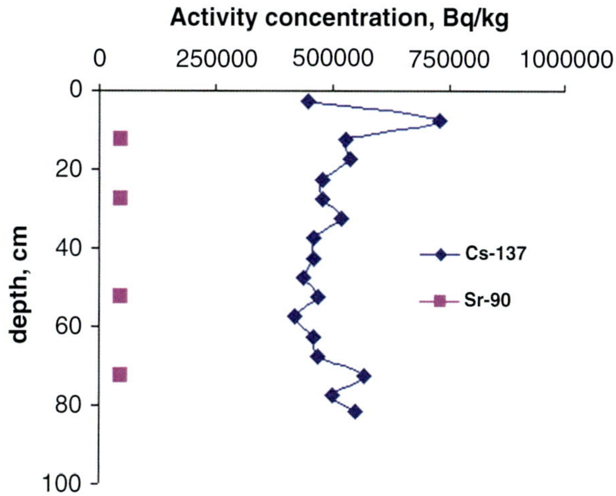

Fig. 8.34 Typical vertical distribution of ^{137}Cs and ^{90}Sr in silt sediments of deep-water zones, depth 11.0 m

of fuel particles. A dominant part of radionuclides deposited on the soil surface in the NPP vicinity was incorporated within these particles. Radionuclides identified in fuel particles represent a full spectrum of fission products: ^{90}Sr, ^{141}Ce, ^{144}Ce, ^{103}Ru, ^{106}Ru, ^{95}Zr, ^{95}Nb, ^{140}La, ^{99}Mo, ^{239}Np, ^{242}Cm, ^{244}Cm, ^{238}Pu, and other Cm and Pu isotopes, more volatile ^{131}I, ^{132}Te, ^{134}Cs, and ^{137}Cs (Ivanov et al. 1996; Smith et al. 2009; Konoplev 2020).

Radionuclides in the initial fallout of 1986 occurred mostly in the poorly soluble fuel matrix: only 25–30% of ^{137}Cs and 10–12% of ^{90}Sr in the fallout were in mobile form (water soluble and exchangeable) (Konoplev and Bobovnikova 1991; Bobovnikova et al. 1991).

In the course of time, degradation of the fuel particles caused radionuclide release from the uranium matrix. ^{137}Cs and ^{90}Sr showed markedly different geochemical behavior after being released from the fuel particles. ^{137}Cs was rapidly fixed by clay minerals and since 1988, the fraction of its mobile form was established at steady level, accounting for 2–10% of the total inventory in soils and bottom sediments. In contrast, ^{90}Sr mobile form constantly increased due to release from the fuel particles. In many soils, in particular acidic, the exchangeable form of ^{90}Sr approached the equilibrium level 80–90% (Bulgakov and Konoplev 2005).

Results of the studies carried out on the Cooling Pond have revealed that the fuel particles still occur in bottom sediments and account for a significant part of radioactivity. Fuel particles were directly determined using advanced beta-counting, high-resolution gamma-spectrometry, X-ray autoradiography, and other experimental methods. This allowed to corroborate that bottom sediments is favorable environment for preserving fuel particles from degradation by several reasons, among them prevailing of slightly alkaline conditions and low oxygen concentrations in sediment/water system, as well as minor annual gradient of temperatures, weathering, and electrostatic tension.

X-ray autoradiographical images (XRARI) taken from samples of bottom sediment show numerous "spots" developed by strong sources of radioactivity (Fig. 8.35). Dimensions of these spots are proportional to beta activity of the source particle, which is combined effect of major beta emitters long-lived fission products—^{137}Cs, ^{90}Sr, and its daughter ^{90}Y. The latter has much larger energy of beta spectra (2.2 MeV) and therefore is responsible for development of spot-image on autoradiography. Autoradiography can be used for not only visualization of the hot particles presence in a sample but also for quantitative assessment of beta activity of the sample. It is noticeable that recalculated beta activity of the hot particles from the XRARI satisfactorily correlated with the activity of ^{241}Am and ^{154}Eu measured by the method of high-resolution gamma-spectrometry, with the ratios comparable with theoretical estimates for ChNPP fuel (Fig. 8.36).

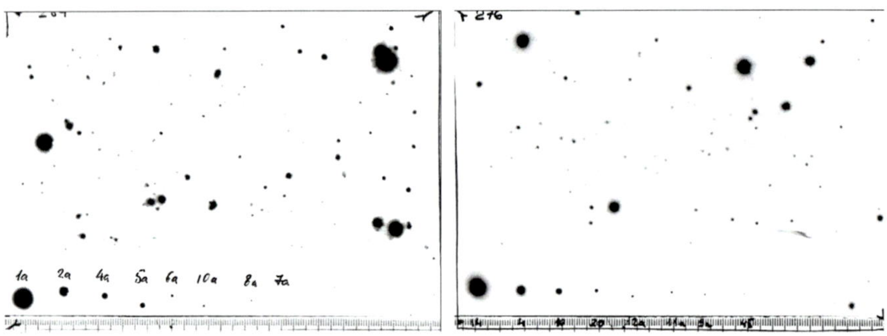

Fig. 8.35 X-Ray autoradiography images of hot particles in bottom sediments of the CP (aligned numbered spots in the footer of the image are calibration particles of known activity)

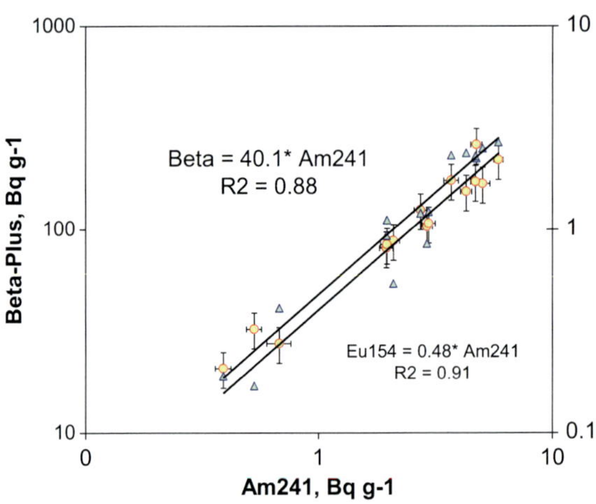

Fig. 8.36 Correlation between radionuclide activity concentration and β-activity in CP sediment samples

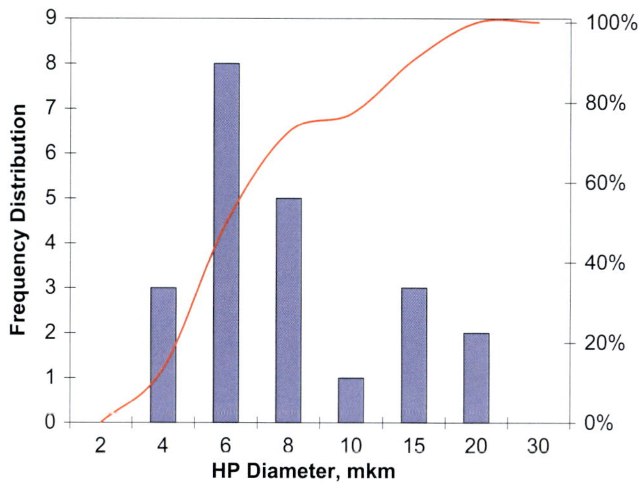

Fig. 8.37 Example of statistical distribution of hot particles equivalent diameters (obtained from autoradiography images shown above)

The spot diameter on autoradiography image can be converted to so-called equivalent diameter of the host-particle assuming it is a ball-shaped pure uranium oxide. Statistical distribution of the equivalent diameters calculated from the autoradiography images corresponds to a bi-modal log-normal type, with median size of 4.5–5.5 μm for the first set of small particles (75–80% of total number of detected particles) and 15 μm for the set of larger ones (Fig. 8.37). This distribution probably reflects different genesis of fuel particles: larger particles originated from the initial explosion and finer particles were formed at a later stage of the accident. It should be taken into consideration that these particles have different dissolution rates, which is important for modeling future state of the CP and newly formed lakes.

8.5.4 ^{137}Cs and ^{90}Sr Speciation in Bottom Sediments

Data about ^{137}Cs and ^{90}Sr speciation in bottom sediments are required, among other things, for prediction of radioecological consequences of CP decommissioning. In the CP bottom sediments, a significant part of ^{90}Sr still occurs embedded in fuel particles, whereas in soils of the Chernobyl exclusion zone, fuel particles have decomposed almost completely (Konoplev et al. 1992; Kashparov et al. 1999, 2000; Konoplev and Bulgakov 1999; Bulgakov et al. 2009). Exposure of the bottom to the air will lead to enhancement of the process of uranium dioxide (fuel matrix) oxidation to uranium trioxide, which is more water soluble as compared to dioxide. This will accelerate dissolution of fuel particles and remobilization of radionuclides. Mobility and bioavailability of ^{90}Sr in soils of drained areas will increase. After leaching from fuel particles, ^{137}Cs will be fixed quickly by micaceous clay minerals of sediments (Konoplev and Bulgakov 2000; Bulgakov et al. 2009).

After the Chernobyl accident, before the CP decommissioning, a number of studies of ^{137}Cs and ^{90}Sr speciation in sediments were carried out (Kanivets and Voitsekhovych 2000; Buckley et al. 2002; Bulgakov et al. 2004; Pirnach 2011b). In all cases, radionuclide speciation was analyzed by sequential extractions (Pavlotskaya 1974) and generally three chemical forms were extracted: exchangeable, non-exchangeable, and strongly bound.

In 2017 (the third year of CP decommissioning), UHMI determined ^{137}Cs and ^{90}Sr speciation in the upper 5-cm layer of deep-water silt sediments from the CP Western and Southern sectors and in suspended sediments from the Western sector. On the same year, they determined ^{137}Cs and ^{90}Sr speciation in sediments from drained bottom areas exposed to the air.

In 2012–2016, UIAR conducted a model experiment to study speciation changes in the CP sediments with time after draining the pond. Using the method of autoradiography, dimensions of fuel particles and their size distribution were determined (Protsak and Odintsov 2014; Protsak et al. 2017).

Earlier studies (Bulgakov et al. 2004) have shown that in the upper layers of bottom sediments, in which silt is predominant, the ^{137}Cs exchangeable fraction is minimum and equals to 3–7%, whereas in deeper parts with loamy sand and sand, this fraction is as large as 16–23%. In the upper 8-cm layers, the fraction of exchangeable ^{90}Sr varied from 1% to 10%. Similar exchangeable fraction of ^{90}Sr in the surface layer of the CP bottom sediments was found in 2001 (Buckley et al. 2002). Meanwhile, the ^{90}Sr exchangeable fraction in bottom sediments of PA "Mayak" reservoirs was determined to vary from 60 to more than 90% (Pavlotskaya 1974; JNREG 1997; Konoplev and Bobovnikova 1991; Konoplev 2020). The low content of ^{90}Sr exchangeable fraction in bottom sediments of the ChNPP CP is explained by the fact that the most part of ^{90}Sr here is embedded in fuel particles (Buckley et al. 2002; Bulgakov et al. 2009). It should also be noted that 1 M HCl tends to extract a larger fraction of ^{90}Sr (24–47%) as compared to ^{137}Cs (3–11%), which seems to be due to partial dissolution of the fuel matrix. A similar fraction of ^{90}Sr (20–25%) was extracted by 1 M HCl from the CP sandy-silt sediments in an experiment in 2012 (Protsak and Odintsov 2014). The authors estimated that about 70% of ^{90}Sr, 80% of ^{241}Am, and 90% of $^{239+240}Pu$ were embedded in fuel particles in the sediments under study in 2012. Bulgakov et al. (2009) believe that at least 90% of ^{90}Sr occurred in fuel particles in 1999–2002, assuming that the fraction of this radionuclide in fuel particles is equal to the fraction of its non-exchangeable form minus its fixed form fraction.

In 2001 in the upper contact layer of bottom sediments of shallow water areas (0–7 m), ^{137}Cs and ^{90}Sr exchangeable fractions were 2–6% and 0.5–3.0%, respectively, varying as a function of degree of sediment siltation for ^{137}Cs and degree of enrichment by fuel particles for ^{90}Sr. Between 1999 and 2002 in the deep-water areas with steady silt accumulation, the ^{137}Cs exchangeable fraction in the upper contact layer of silt deposits was 1.6–2.9% and ^{90}Sr exchangeable fraction—0.3–0.4%.

Fraction of exchangeable ^{137}Cs и ^{90}Sr in deep-water silt deposits did not change by 2017 and was 1.8–2.1% for ^{137}Cs and 0.2% for ^{90}Sr (Table 8.17). In September 2017, the $^{137}Cs/^{90}Sr$ activity ratio in the upper layer of deep-water silt sediments was

Table 8.17 ^{137}Cs and ^{90}Sr speciation in the upper layer of deep-water silt deposits and suspended sediments in 2017

Sample	Fraction, %						$^{137}Cs/^{90}Sr$
	^{137}Cs			^{90}Sr			
	1 M CH_3COONH_4	6 M HCl	Residue	1 M CH_3COONH_4	6 M HCl	Residue	
Bottom deposits, Western sector	2.1	90.3	7.6	0.2	93.9	5.9	2.5
Susp. sediments, Western sector (September)	1.6	93.3	5.1	0.7	87.3	11.9	2.9
Susp. sediments, Western sector (November)	1.7	92.7	5.6	0.3	94.4	5.4	3.2
Bottom deposits, Southern sector	1.8	85.0	13.2	0.2	87.9	11.9	6.2

2.5 in the Western sector and 6.2 in the Southern sector (Table 8.17), which suggests a higher content of fuel particles in the bottom sediments of the Western sector as compared to other sectors of the CP. Most likely, the main pathway of fuel particles to the CP was discharge of highly radioactive water from the accidental reactor cooling system to this sector of the CP in April–May 1986 (Egorov et al. 1989).

From 2002 to 2017, the silt deposit layer in the deep-water areas increased by 20–30 cm due to trans-sedimentation of silt particles from shallow-water areas. It looks like the trans-sedimentation process became more active after 2015 due to water level reduction in the CP and enhancement of wind-wave erosion of sandy-silt sediments that occurred at the depths of 5–7 m before the CP decommissioning.

Analysis of samples collected in 2017 has shown that in the CP Western sector deep water, the characteristics of suspended sediments were identical to those of bottom sediments, including total activity concentrations of ^{137}Cs and ^{90}Sr, ratio of ^{137}Cs and ^{90}Sr chemical forms, and ^{137}Cs/^{90}Sr ratio (Table 8.17). This confirms the above-mentioned hypothesis that silt deposits in the deep-water areas were formed by silt particles that had moved from surrounding shallow-water sections (trans-sedimentation).

Generally, the deep-water silts were quite uniform in terms of particle size and chemical composition in different parts of the CP. This seems to be the reason why the ratio of ^{137}Cs and ^{90}Sr chemical forms in different parts of the CP was similar. Differences were revealed only in the ^{137}Cs/^{90}Sr total activity ratio, which was presumably due to different content of fuel particles in silt deposits on each specific section.

Bulgakov et al. (2004), based on determining ^{137}Cs exchangeable form fraction in numerous samples of the top bottom sediment layer in 1999–2002, derived the following dependence by the least squares method:

$$\log \alpha_{\text{ex}}\left(^{137}\text{Cs}\right) = -1.06 \times \log\left[^{137}\text{Cs}\right] + 3.12 \tag{8.1}$$

where $\alpha_{\text{ex}}(^{137}\text{Cs})$ is the ^{137}Cs exchangeable form fraction in the top bottom sediment layer, %; [^{137}Cs] is the ^{137}Cs activity concentration in the top bottom sediment layer (Bq/g).

Relationship (8.1) can be explained by the fact that ^{137}Cs content in silt is much higher than in sand. Therefore, the higher the silt content, the higher ^{137}Cs activity concentration. What is more, ^{137}Cs is fixed by clay minerals of silt particles, and hence the higher the ^{137}Cs activity concentration, the lower the ^{137}Cs exchangeable fraction, as illustrated by Fig. 8.38.

Dependence of the exchangeable form fraction on total activity was also seen for ^{90}Sr. This is because that the greater quantity of fuel particles in sediments and the larger they are, the higher the sediment activity concentration and the smaller the ^{90}Sr exchangeable fraction.

In 2017, core samples were collected by UHMI at three locations on the CP drained bottom sections exposed to the air with a view to perform analysis of radionuclide speciation (Table 8.18). Two of the three cores were collected at the

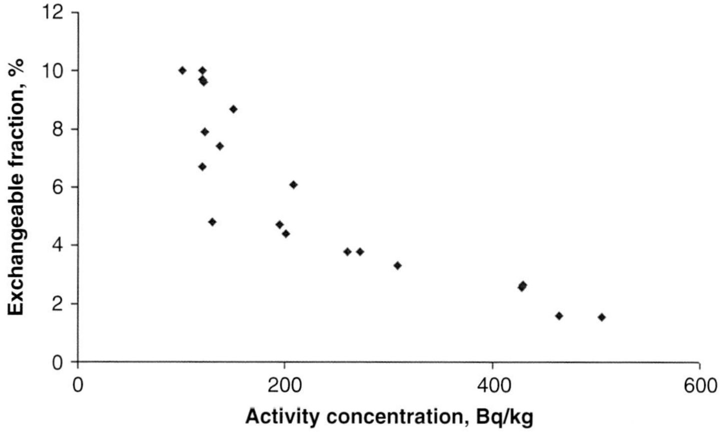

Fig. 8.38 Dependence of the ^{137}Cs exchangeable fraction on ^{137}Cs activity concentration in the CP upper layer of bottom sediments, 1999–2002

same sites where sampling was carried out in 2001. The ^{137}Cs exchangeable fraction in 2017 was found to be 2.5–6 times less than that in 2001, while the ^{90}Sr exchangeable fraction increased by 3–6 times. At the same time, this comparison may not be valid, since with the water level gradually going down, the type of bottom deposits was changing. For example, at the sampling point A, 30% of sediments in the upper 4-cm layer were Dreissena shells and their fragments, while only a small amount of shell material was found in sediments (<1%) in 2001. At the sampling point B, in 2017, the upper 8-cm layer was sandy loam, while in 2001, there was 7 cm of highly silted sand covered with 1-cm layer of sandy silt. It seems that with the water level being lower, silt particles were washed out and sediment density increased.

Another core was collected in the CP Northern part (sampling point C). The layered radionuclide analysis of these samples showed that at two locations, where clay particles occurred in soil (sampling points B and C), all ^{137}Cs and ^{90}Sr activity was contained to the upper 4-cm layer, and only in sandy sediments radionuclides penetrated to a depth of 10 cm (sampling point A).

It is commonly accepted today that dissolution of fuel particles will accelerate on the drained areas of the CP bottom, which by various estimates contain 70–90% of the radionuclide activity (Bulgakov et al. 2009; Protsak and Odintsov 2014). As a consequence, mobility and bioavailability of radionuclides there will be changing with time after CP decommissioning.

For ^{137}Cs, future changes will, most likely, be associated with transformation of clay minerals as a result of freezing-thawing and wetting-drying cycles, as well as additional fixation of ^{137}Cs exchangeable forms. After 30 freezing-thawing cycles, the ^{137}Cs exchangeable fraction in sediments was seen to have decreased by 1.5–2 times. A similar decrease was observed in the case of drying a sediment sample before determination of the exchangeable fraction (Bulgakov et al. 2004). Therefore,

Table 8.18 ^{137}Cs and ^{90}Sr speciation in the upper layer of exposed sediments collected in September 2017

Sample	Fraction, %						$^{137}Cs/^{90}Sr$	Dry density, g/cm^3
	Cs-137			Sr-90				
	1 M CH_3COONH_4	6 M HCl	Residue	1 M CH_3COONH_4	6 M HCl	Residue		
A	2.8	86.6	10.6	2.7	65.9	31.4	15	1.4
B	1.1	94.0	4.9	3.2	96.3	0.5	1.3	1.2
C	1.8	90.9	7.3	4.0	94.5	1.5	2.6	0.9

some decrease can be expected in ^{137}Cs mobility and bioavailability in soils (deposits) of drained territories.

On the other hand, the ^{90}Sr exchangeable fraction after 30 freezing-thawing cycles virtually did not change (Bulgakov et al. 2004). The same result was obtained earlier in an experiment with soils subjected to freezing cycles using liquid nitrogen. In this case, the average size of fuel particles was decreasing and their quantity increasing, possibly, as a result of destruction of aggregates consisting of nuclear fuel grains (Shalaeva et al. 1991). However, the ^{90}Sr exchangeability in this experiment did not change.

In the years ahead, an increase in dissolution rate of fuel particles will obviously be the most significant factor for changing ^{90}Sr chemical forms in drained bottom sediments. Due to anaerobic conditions and slightly alkaline medium of the CP bottom sediments, fuel particles tend to dissolve much slower than in soils occurring close to terrestrial systems.

The rate constant of fuel particle dissolution (k_l) in the CP bottom sediments was estimated by Bulgakov et al. (2009) based on data about ^{90}Sr speciation and ^{90}Sr budget in the CP (Weiss et al. 2000; Buckley et al. 2002). For CP bottom sediments $k_l \approx 0.02$ year^{-1}, and the corresponding half-time of fuel particles dissolution was estimated to be about 35 years. After drainage of the CP bottom sediments, k_l can increase up to 0.05 year^{-1} due to oxygen access and an increase in oxidation rate of the fuel matrix (UO_2). The authors (Bulgakov et al. 2009) assumed that later the pH of exposed sediments will be slowly decreasing because of overgrowth and microbiological activity on dried bottom. Besides, dissolution of fuel particles in remaining lakes is expected to slightly accelerate, since they will be better aerated due to depth reduction.

In a model experiment with the CP bottom sediments (Protsak et al. 2017), it was found that over 4 years of exposure of bottom sediment to the air, the ^{90}Sr exchangeable fraction increased from 1% to 10%, which is a substantial growth. At the same time, ^{137}Cs exchangeable fraction decreased by about 1.5 times, which is consistent with earlier estimates (Bulgakov et al. 2004).

Over 4 years of the above-mentioned model experiment (Protsak et al. 2017), the sediments' pH increased from 7.2 to 8.0 and at several dried bottom sites the pH was measured to vary from 7.6 to 7.9. This unexpected slight increase in pH was attributed by the researchers to the presence of a large amount of Dreissena shells on the dried bottom surface. In their view, calcium leaching from decomposed shells prevented anticipated acidification of drained bottom soils.

8.6 Conclusions

Prior to the Chernobyl accident, the presence of artificial radionuclides in the ChNPP CP was mainly associated with operating processes and were close to the levels of global radioactive contamination of water bodies in this geographical location.

The contamination of CP after the accident in 1986 occurred in two ways: (1) deposition of radionuclides from the damaged reactor onto the water surface and (2) discharge through the outlet channel of highly radioactive water from the reactor emergency cooling system and water used to extinguish the fire at the ChNPP (about 5000 m^3). By various estimates, a total of 190–280 TBq of ^{137}Cs and 35–65 TBq of ^{90}Sr entered the cooling pond as a result of the Chernobyl accident.

Due to sedimentation processes and sorption of dissolved radionuclides by the upper layer of bottom sediments, a major part of radioactivity was transferred to the bottom within a short time of the initial phase after the accident. In 2001, less than 0.2% of ^{137}Cs and about 1% of ^{90}Sr still remained in the CP water column.

The distinct feature of the CP radioactive contamination was enrichment of bottom sediments with ^{137}Cs (higher $^{137}Cs/^{90}Sr$ ratio), as compared with soils on territories adjacent to the CP. Presumably, the CP contamination with this radionuclide, to a degree, was due to fallout of condensational particles, which could facilitated by a high concentration of moisture above the warm water surface ("steam fog"), while ^{90}Sr and transuranic elements were mostly deposited as fragmented fine fuel particles.

The general pattern of the bottom radioactive contamination was changing over time in line with transformation of the bottom soil complex typical of man-made reservoirs. Due to various currents, highly contaminated silt particles were washed out from swelled primary soils and as a result of trans-sedimentation, these particles moved to the CP deepest sites, on which the silt layer was steadily increasing. While the thickness of the contaminated silt layer in 1989–1991 was not more than 15 cm (Shiklomanov 1992), in 1999, for some sites it exceeded 100 cm (Kanivets and Voitsekhovych 2000).

The source of secondary water contamination with ^{90}Sr was the CP shallow bottom areas made of highly permeable sands. Saturation of water with oxygen in these areas seems to contribute to dissolution of fuel particles and transfer of mobile forms of this radionuclide to the water column.

As far as ^{137}Cs is concerned, the source of secondary contamination of the CP water was highly contaminated deep-water silts. The ^{137}Cs exchangeable fraction there throughout the entire post-accidental period was 2–3%, which was sufficient for continuous diffusional flux of ^{137}Cs from the bottom to the water column. The ^{90}Sr exchangeable fraction in deep-water silt sediments has always been tenths of a percent, which is indicative of persistence of fuel particles in anoxic reducing conditions.

The exponential trends for reduction of dissolved ^{137}Cs and ^{90}Sr activity concentrations in the CP water have been observed since 1987. In 1998–2013, the environmental half-time of ^{137}Cs activity concentration reduction in water was 8.5 years and 30.7 years for ^{90}Sr (Kanivets et al. 2014). The annual average ^{90}Sr activity concentration in water has stabilized at 1.2–1.6 Bq/l since 2001 and from 2007, it began to exceed the dissolved ^{137}Cs activity concentration, which was slowly and steadily decreasing until the beginning of CP decommissioning. Starting from 2001, the $^{137}Cs/^{90}Sr$ ratio was decreasing from 1.6 to 0.6 in 2014, nevertheless the $^{137}Cs/^{90}Sr$ ratio in the CP water can be termed as abnormally high. The most likely reason for this high ratio is the low content of clay in the CP suspended matter and hence low sedimentation flux of ^{137}Cs from the water column to bottom sediments.

All the time since the accident, seasonal variations in dissolved ^{137}Cs activity concentrations were observed with a pronounced winter–spring minimum (February–April) and summer–autumn maximum (July–September). The cyclic character of changes in dissolved ^{137}Cs activity concentration seems to be associated with intra-annual cycles of physicochemical and biological processes in the CP. No seasonal variations in ^{90}Sr activity concentration in water have been detected.

The sustained long-term trend for a decrease in ^{137}Cs and ^{90}Sr activity concentrations until 2014 was due to stable hydrochemical and hydrological regimes in the CP and dilution of its water mass by Pripyat River water. Thus, before the beginning of CP decommissioning, the general radioecological situation for the pond was stable and predictable.

In mid-May 2014, the ChNPP began draining the CP and in July 2015, the pond was split into three separate sectors. By the end of 2017, the water level decreased by 5 m in the Northern and Western sectors and 6 m in the Southern sector. As a result, the CP lost 75% of its design-basis water volume; the water surface area being reduced by 40%. In 2017, about half of the remaining area was sections with a depth of 1–1.5 m.

During 2015–2017, the physical characteristics and macroion composition of the CP water did not change. At the same time, changes in the reservoir oxygen regime were seen such as a decrease in oxygen concentration in the surface water layer (0–5 m), enhanced vertical oxygen stratification in the warm time period of the year, with occurrence of hypoxic or anoxic conditions below the thermocline, and an increase in time length of these conditions.

When draining the CP, a significant increase in the suspended sediment concentrations in water was observed, which is explained by the current bathymetry of the CP, with almost half of the reservoir having a depth of 1–1.5 m. Given a strong wind, small mineral particles on these shallow areas are resuspended, transported by drift currents and saturate the water column in the deep-water zone. In the course of sedimentation in the water column, these particles adsorb and fix dissolved ^{137}Cs and settle to the bottom, bringing a significant part of this radionuclide from water to the bottom. This is strongly contributing to the CP water self-purification and weakening of ^{137}Cs diffusion flux from deep-water silt sediments to the water column.

In 2015, the ^{90}Sr activity concentration in the water of the Northern and Western sectors began to grow, and by the end of 2017, it increased by about 3 times. Over the same time period, the ^{90}Sr activity concentration in the water of the Southern sector increased by less than 10–15%. The increase of ^{90}Sr activity concentration in relatively shallow Northern and Western sectors is caused by the combination of several factors: (1) large ^{90}Sr inventories in the bottom sediments of these sectors; (2) wash-off of ^{90}Sr exchangeable forms from the surface of drained areas by surface runoff; (3) concentrating of this radionuclide due to significant reduction of the CP water volume, given no dilution of the CP water by clean river water.

It seems that on the drained bottom sections, the process of fuel particle dissolution has started. The results of the field model experiment (Protsak et al. 2017) indicate that the ^{90}Sr exchangeable fraction in drained sandy-silt deposits has increased from 1% to 10%. According to UHMI data, the ^{90}Sr exchangeable fraction in the upper layer of dry sandy and sandy loam deposits in the Western and Northern sectors in 2017 was 2.7–4%, being only 1–3% in 2001.

Most of the drained bottom surface becomes overgrown by meadow and shrub vegetation typical of this geobotanical zone, preventing wind erosion and atmospheric transport of radionuclides beyond the CP basin.

In future, the ecological situation in residual lakes is likely to deteriorate as a result of concentration of aquatic organisms, reduction of vertical mixing of water masses, and hence a decrease in the water column aeration. Effect of excessive biomass, which was indicated as a possibility in the feasibility study for CP decommissioning, can lead to degradation of existing aquatic biocenoses and a major deterioration in water quality of newly formed lakes.

Radioactive contamination of residual lakes in future will be controlled by hydrochemical and hydrological processes, as well as radionuclide wash-off by surface runoff from the drained sections of the bottom. Anoxic conditions are expected to facilitate generation of ammonium in the near-bottom water layer and in the pore solution of silt sediments, which is expected to lead to an increase in ^{137}Cs diffusional flux from bottom sediments to the water column. After final stabilization of water level and a decrease in water turbidity, a certain increase in dissolved ^{137}Cs activity concentration in the lakes can be expected. The ^{90}Sr activity concentration in residual lakes will be primarily determined by the rate of fuel particle dissolution and by wash-off of ^{90}Sr mobile forms to the lakes from adjacent drained sections of the bottom.

Acknowledgement This research was partially supported by the Japan Society for the Promotion of Science, Grant-in-aid for Scientific Research (B) (18H03389), and by Science and Technology Research Partnership for Sustainable Development (SATREPS), Japan Science and Technology Agency (JST)/Japan International Cooperation Agency (JICA), the State Emergency Service of Ukraine (projects 9/15, 12/18).

References

Alekin O (1948) Hydrochemical classification of the rivers of the USSR. Proc SHI 4(58):209–224. (in Russian)

Bobovnikova TI, Makhon'ko KP, Siverina AA, Rabotnova FA, Gutareva VP, Volokitin AA (1991) Physical-chemical forms of radionuclides in atmospheric fallout, and their transformations in soil, after the accident at the Chernobyl Atomic Energy Plant. Sov At Energy 71(5):932–936

Bochkov L, Vakulovsky S, Nikitin A, Tertyshnik E, Chumichev V (1983) About the content of cesium-137 in surface waters of land. Meteorol Hydrol 8:79–83. (In Russian)

Bulgakov A, Samokhvalova E, Popov V, Kanivets V (2004). Speciation of ^{137}Cs and ^{90}Sr in bottom sediments of the Chernobyl NPP cooling pond, 10 (Preprint in Russian)

Bulgakov A, Konoplev A (2005) Modeling of long-term transformation of 90Sr forms in soils. Eurasian Soil Sci 38(7):727–733

Bulgakov A, Konoplev A, Smith J, Laptev G, Voitsekhovich O (2009) Fuel particles in the Chernobyl cooling pond: current state and prediction for remediation options. J Environ Radioact 100(4):329–332

Buckley M, Bugai D, Dutton LMCD, Gerchikov M, Kashparov V, Ledenev A, Voitzehovich O, Weiss D, Zheleznyak M (2002) Drawing up and evaluating remediation strategies for the Chernobyl cooling pond, Final report, issue 01, Rep. no. C6476/TR/01, NNC Ltd, Manchester, p 90

Bugai D, Waters R, Dzhepo S, Skal'skij A (1997) The cooling pond of the Chernobyl nuclear power plant: a groundwater remediation case history. Water Resour Res 33(4):677–688

Bugai D, Skalsky A (2002) Results of hydrogeological modelling of ChNPP cooling pond. Rep. contr. no. C6476/D0427 AM 01, Geoexpert Ltd., Kiev, p 62

Bugai D, Skalsky A, Dzhepo S, Oskolkov B (2005) Experimental hydrogeological studies and filtration analyses for the Chernobyl cooling pond. Bull Eco Stat Chern Excl Zone 25(1):42–57. (in Ukrainian)

Egorov Y, Kazakov S, Leonov S, Staurin N (1989) The radiation state of the ChNPP cooling pond in May 1986. Proceedings of all union symposium on the consequence of the accident on the Chernobyl NPP. Chernobyl-88 6:75–84. (in Russian)

Gedeonov L, Gritchenko Z, Ivanova L, Orlova T, Tishkov V, Toporkov V, Prokopenko V (1993) Radionuclides of strontium and cesium in the water of the lower Danube in 1985–1990. At Energy 74(1):58–63. (In Russian)

Filippov A (ed) (1996) Long-term forecast of changes of water-salt and thermal regime, of radionuclide budget in the ChNPP Cooling Pond as a result of planned measures, Rep 4/94. NIVO, Chernobyl. (in Russian)

Fomin V, Antropov A, Oskolkov B, Dzhepo S, Bugai D, Skalsky A, Khersonsky E, Derevets V, Kononovych A, Drapeko V, Maksimenko A, Borozan A, Kubko Y (1998) The research plan of the characteristics of the ChNPP Cooling Pond as a source of radiation risks in order to collect data to justify measures for the decommissioning. Slavutych Laboratory of International Research and Technology, Slavutych, p 136. (In Russian)

IAEA (1993) Chernobyl accident: supplement for INSAG-1. Rep. of Intern. Consult. Groupe by the Nuclear Safety of IAEA. IAEA, Vienna, p 146

IAEA (2006) Environmental consequences of the Chernobyl accident and their remediation: twenty years of experience. IAEA, Vienna, p 166

IAEA (2019) IAEA-TECDOC-1886. Environmental impact assessment of the drawdown of the Chernobyl NPP cooling pond as a basis for its decommissioning and remediation. IAEA, Vienna, p 173

Ignatenko E, Komarov V, Zverkov V, Proskuryakov A (1989) Radiation situation within the 30-km one of the Chernobyl NPP. Proceedings of all union symposium on the consequence of the accident on the Chernobyl NPP. Chernobyl-88 1:4–11. (in Russian)

Yu Ivanov, Kashparov V, Sandalls J, Laptev G, Viktorova N, Kruglov S, Salbu B, Oughton D, Arkhipov N (1996). Fuel components of ChNPP release fallout: properties and behavior in the environment. In: Karaoglou A, Desmet G, Kelly GN, Menzel HG (eds) The radiological consequences of the Chernobyl accident, EUR 16544EN, Luxembourg, pp 173–177

JNREG (1997). Sources contributing to radioactive contamination of the Techa River and the area surrounding the 'Mayak' Production Association, Urals, Russia. Programme of the Joint Norwegian-Russian Expert Group for Investigation of radioactive contamination in northern areas, Østeras, p 134

Kanivets V, Voitsekhovych O (2000) Radioactive contamination of bottom sediments of the cooling pond of ChNPP. Proc UHMI 248:154–171. (in Russian)

Kanivets V (ed) (2010) Guideline to meteorological stations and posts. Issue 12, vol. 2 Observations for radioactive contamination of surface waters of land and sea waters. UkrHydromet, Kyiv, p 144. (In Ukrainian)

Kanivets V, Kireev S, Laptev G, Nasvit O, Obrizan S (2014) ^{137}Cs and ^{90}Sr in the water of the ChNPP cooling pond. Nucl Phys At Energy 15(4):370–379. (in Ukrainian)

Kashparov V, Oughton D, Zvarich S, Protsak V, Levchuk S (1999) Kinetics of fuel particle weathering and ^{90}Sr mobility in the Chernobyl 30- km exclusion zone. Health Phys 76 (3):251–259

Kashparov V, Lundin S, Khomutinin Y, Kaminsky S, Levchuk S, Protsak V, Kadygrob A, Zvarich S, Kovtun M, Zhurba M, Lanshyn V (2000) Contamination of the near zone territory of the Chernobyl accident with ^{90}Sr. Radiochemistry 42(6):550–559. (In Russian)

Kazakov S (1995) Management of the radiation state of cooling ponds of the nuclear power plants. Technic, Kyiv 191. (in Russian)

Konoplev AV, Bobovnikova TI (1991) Comparative analysis of chemical forms of long-lived radionuclides and their migration and transformation in the environment following the Kyshtym and Chernobyl accidents. Proceedings of seminar on comparative assessment of the environmental impact of radionuclides released during three major nuclear accidents: Kyshtym, Windscale, Chernobyl. Luxembourg, 1–5 Oct 1990, 1:371–396

Konoplev A, Bulgakov A, Popov V, Ts B (1992) Behaviour of long-lived Chernobyl radionuclides in a soil–water system. Analyst 117:1041–1047

Konoplev A, Bulgakov A (1999) Kinetics of the leaching of ^{90}Sr from fuel particles in soil in the near zone of the Chernobyl nuclear power plant. At Energy 86(2):136–141. (In Russian)

Konoplev A, Bulgakov A (2000) Transformation of ^{90}Sr and ^{137}Cs in soil and bottom sediments. At Energy 88(1):55–60. (In Russian)

Konoplev A (2020) Mobility and bioavailability of the Chernobyl-derived radionuclides in soil-water environment: review. In: Konoplev A, Kato K, Kalmykov SN (eds) Behavior of radionuclides in the environment II: Chernobyl. Springer Nature, Singapore, pp 157–193

Krasnov V (ed) (2013) Technical and economic feasibility study of the decommissioning of the Cooling pond of Chernobyl NPP, vol 1. IPS NPPs, Chornobyl. (in Ukrainian)

Kudrin V (1959) Classification and distribution of bottom sediments of the Rybinsk reservoir. Proc Inst Biol Reserv 4:25–37. (In Russian)

Makhonko K (ed) (1990) Guidance on the organization of the control of the environment state in the area of the nuclear power plant location, vol 264. Gidrometeoizdat, Leningrad. (In Russian)

Meshalkin G, Arkhipov A, Arkhipov N (1989) Physicochemical state of radionuclides in soils. Proceedings of all union symposium on the consequence of the accident on the Chernobyl NPP. Chernobyl-88 5(I):152–159. (in Russian)

Nasvit O (2002) Radioecological situation in the cooling pond of chornobyl NPP. In: Recent research activities about the Chernobyl NPP accident in Belarus. Kyoto, Ukraine and Russia. Research Reactor Institute, Kyoto University, pp 74–85

Pavlotskaya F (1974) Migration of radioactive products of global fallout in soils. Atomizdat, Moscow, p 215. (In Russian)

Pirnach L (2011a) Radioactive pollution of the Chernobyl cooling pond bottom sediments. I. Water-physical properties, chemical compound and radioactive pollution of pore water. Nucl Phys At Energy 12(1):86–93. (In Russian)

Pirnach L (2011b) Radioactive pollution of the Chernobyl cooling pond bottom sediments. II. Distribution of ^{137}Cs, ^{241}Am, ^{90}Sr in a solid phase. Nucl Phys At Energy 12(4):385–393. (In Russian)

Protsak V, Odintsov A (2014) Assessment of the forms finding of Chornobyl radionuclides in bottom sediments of cooling pond of ChNPP. Nucl Phys At Energy 15(3):259–268. (In Ukrainian)

Protsak V, Odintsov A, Khomutinin Y, Zhurba M, Prokopchuk N, Kashparov V (2017) Dynamics of physico-chemical forms of radionuclides in the bottom sediments of cooling pond of the ChNPP after their drying. 1. Model experiment. Nucl Phys At Energy 18(4):341–349. (In Ukrainian)

Romanovsky V, Khersonsky E, Vasilenko N, Voitsekhovych O, Abakumenko A (1989) The study of water and bottom sediments contamination, water masses dynamics and radionuclide transport in the Chernobyl cooling pond. Rep Contr. No.10/2. State Hydrological Institute, Leningrad. (in Russian)

Romas V (2002) Hydrochemistry of water bodies of nuclear and thermal power industry. Kyiv University, p 532. (in Ukrainian)
Rybalko S, Bondarenko G, Komarov V, Yu S, Tepikin V, Sobotovytch E (1989) Comprehensive investigation of the morphology and mattering composition of ChNPP emergency fallouts. Proceedings of all union symposium on the consequence of the accident on the Chernobyl NPP. Chernobyl-88 5(I):28–33. (in Russian)
Shalaeva T, Khitrov T, Cherkezyan V (1991) Stability of "hot particles". In: Geochemical migration routes of artificial radionuclides in the biosphere. Synopsis of V conference of IGCAC. IGCAC, Puschino, p 68. (In Russian)
Shiklomanov I (ed) (1992) Hydrological, thermal, chemical and radiation regimes of the ChNPP cooling pond. Rep State Hydrological Institute, Leningrad, p 129. (in Russian)
Sivintsev Y, Khrulev A (1995) Estimation of the emission of radioactive substances during the 1986 accident in the fourth power generating unit at the Chernobyl nuclear power plant (review of primary data). At Energy 78(6):390–401. (in Russian)
Smith J (ed) (2005) Radio-ecological study of the Chernobyl cooling pond and options for remediation (RESPOND). Final report, INTAS-2001-0556, Centre for Ecology and Hydrology (CEH), UK
Smith JT, Konoplev AV, Voitsekhovitch OV, Laptev GV (2009) The influence of hot particle contamination on models for radiation exposures via the aquatic pathway. In: Radioactive particles in the environment. Springer, Dordrecht, pp 249–258
Sobotovytch E, Bondarenko G, Olkhovyk Y, Komarov V (1989) Chemical speciations of fission products in Chernobyl accidental emissions. Proceedings of all union symposium on the consequence of the accident on the Chernobyl NPP. Chernobyl-88 5(I):56–67. (in Russian)
Sobotovytch E, Olshtynsky S (1991) Geochemistry of technogenesis, vol 228. Naukova Dumka, Kyiv. (in Russian)
Teplov P, Shaposhnikov B, Groshev I, Fedulov B, Yu Lebedeva, Yu Maslov, Shtverlov V, Kazakov V, Smirnov V, Krestianinov V, Demchenko S (1989). Survey of radiation conditions on chernobyl NPP after the accident. CHERNOBYL-88, Proceedings of the 1st all USSR scientific and technical meeting. Chernobyl, pp 40–133 (in Russian)
Tertyshnyk E, Nikitin A, Kabanov A, Ptakhin V (1980) A set of devices for separation of suspensions from water samples. Proc Inst Exp Meteorol 94(5):65–68. (In Russian)
Vakulovsky S (ed) (1986) Guidelines for the determination of radioactive contamination of water bodies, vol 78. Gidrometeoizdat, Moscow. (In Russian)
Voitsekhovych O, Kanivets V, Laptev G, Derkach G (2013) Preparation of the sections of the feasibility study on decommissioning of the cooling pond of ChNPP. Rep Contr. No. 100–13. Ukrainian Hydrometeorological Institute, Kyiv, p 116. (in Ukrainian)
Voitsekhovych O, Kanivets V, Kireev S, Laptev G, Obrizan S (2016) The state of the surface waters radioactive contamination. In: 30 years of the Chornobyl disaster (a collection of information and analytical reviews). Ukrainian Hydrometeorological Institute, Kyiv, pp 129–139. (in Ukrainian)
Weiss D, Larue P-J, Bogorinski P, Watermeyer V, Voitsekhovich O, Sobotovich E, Bugai D, Oskolkov B (2000) Collection and analysis of information and data related to the contamination of the Chernobyl cooling pond. Final report on CEC-contract no. B7-5350/99/6241/MAR/C2, Gesselenchaft fur Anlagenund Reaktorsicherhelt mbH: 199
Zarubin O (2006) Dynamic of the content of radionuclides in water of cooling-pond of Chernobyl NPP (1978–2004). Nucl Phys At Energy 17:73–80. (In Russian)

Chapter 9
Radioactivity of Aquatic Biota in Water Bodies Impacted with the Chernobyl-Derived Radionuclides

Ivan I. Kryshev, Tatiana G. Sazykina, and Alexander I. Kryshev

Abstract This chapter describes the dynamical behavior of radioactive contamination of aquatic biota in water bodies heavily impacted with the Chernobyl fallouts (Chernobyl Cooling pond, Dnieper river system, and Lake Kozhanovskoe). Specific characteristics of the radioecological situation in aquatic ecosystems are demonstrated for different time periods following the Chernobyl accident: early stage (within 1 year), intermediate stage (up to 5 years), and distant period (over 10 years). Special attention is given to dynamical processes of radionuclide distribution in age-structured fish populations and also transfers through food chains from nonpredatory to predatory fish. We present a modeling approach (ECOMOD methodology) and the results of modeling the dynamic processes of radionuclide migration in fish populations in the Chernobyl Cooling pond and Lake Kozhanovskoe.

Keywords Chernobyl · Aquatic biota · Radionuclides · Dynamic modeling · Contamination

9.1 Introduction

The most significant contamination of aquatic ecosystems occurred in the initial period of the Chernobyl accident: late April–early May 1986 (Izrael 1990; Kryshev 1992, 1995). Radionuclides entered water bodies through atmospheric fallouts and also with direct releases. In later periods, processes of washing out of radionuclides from catchment areas to water bodies became of considerable importance. In water bodies, radionuclides were subject to redistribution amongwater, bottom sediments, and aquatic biota. Migration, accumulation, and recirculation of radionuclides in aquatic biota, especially in food chains, play an essential role in determining the consequences of the Chernobyl accident for aquatic biota.

I. I. Kryshev · T. G. Sazykina · A. I. Kryshev (✉)
Research and Production Association "Typhoon", Obninsk, Kaluga region, Russia
e-mail: kryshev@rpatyphoon.ru

A. Konoplev et al. (eds.), *Behavior of Radionuclides in the Environment II*,
https://doi.org/10.1007/978-981-15-3568-0_9

This chapter starts with a brief outline of the theoretical background, which explains the biological reasons for radionuclide accumulation in living organisms, and the potential of different radionuclides to transfer through food chains (Sect. 9.2). An author's modeling approach (ECOMOD methodology) is presented intending to simulate dynamical processes of radionuclide accumulation in aquatic biota at different periods after the accidental contamination.

Section 9.3 describes the early period after the Chernobyl accident, which was characterized by a wide spectrum of accidental radionuclides in water bodies, associated with highly nonequilibrium processes of aquatic biota contamination. The dynamics of radionuclide accumulation in fish and other aquatic biota is given for the Dnieper cascade and the Chernobyl Cooling pond.

Section 9.4 outlines the dynamical processes of radioactivity redistribution in aquatic biota (mainly in fish) during the first five-year period (1986–1990) since the Chernobyl accident. The special attention is given to the dynamical processes of radionuclide transfer in age-structured fish populations and also migration through fish food chains from nonpredatory to predatory species.

Section 9.5 gives an example of a very slow decrease of radioactive contamination in fish, inhabiting lakes with slow water exchange and specific hydrochemical regime based on the investigation of Lake Kozhanovskoe, Bryansk region, Russia. The results of field studies are supported by modeling the dynamics of fish contamination.

Section 9.6 presents a brief summary of radionuclide levels in fish from water bodies, discussed in the previous sections, in the distant time periods following the radiation accident at the Chernobyl NPP (1986) (from one to three decades).

Section 9.7 gives a summary of lessons learned from the impact of radiation accidents on aquatic biota.

9.2 Theoretical Background for Migration of Radioactive Tracers in Aquatic ECOSYSTEMS

9.2.1 Radionuclides as Tracers of Chemical Elements

Aquatic radioecology as a science of radionuclide migration in aquatic ecosystems started in the 1920s with the discovery of high accumulation levels of some natural radionuclides (radium, uranium) in aquatic organisms (Vernadsky 1929). It is well known that living organisms have a fundamental capacity to accumulate some specific elements selectively from the environment, which are necessary for constructing the organic tissues of biological organisms and are essential elements for metabolism (C, P, N, O, H, Co, Zn, Fe, I, and others) (Lehninger 1982).

The elemental chemical composition of organisms is specific for each biological species and is kept, on average, constant to maintain optimal concentrations of elements in the tissues and organs (Bowen 1979). The elemental chemical composition of aquatic organisms differs considerably from the composition of natural waters and bottom sediments.

For biota, technogenic radionuclides are no more than radioactive tracers of similar stable elements. Bioaccumulation or sorption of a radionuclide by living organisms is not associated with its radioactivity, but just clearly demonstrates the difference between the concentration of the analogous stable element in the environment and within an organism. In physiological studies of the metabolism of humans and other organisms, radioactive tracers are widely employed as tracers, which follow the metabolism of analogous stable elements; in these studies, chemical features of 'radionuclide-stable element' are assumed to be identical (Design 1975).

Traditionally, the radionuclide accumulation by aquatic organisms is estimated in relation to radionuclide concentration in water using the radionuclide concentration factors:

$$CF = \frac{y}{X},$$

where CF is the concentration factor; y is the specific radionuclide activity of the organisms' tissues, Bq kg^{-1} dry weight or fresh weight; and X is the radionuclide concentration in filtered water taken from the habitat, Bq kg^{-1}. The 'concentration factor' theory was originated from observations of long-lived natural radionuclides in the environment and biota; therefore, it is based on a fundamental assumption about the equilibrium in the radionuclide distribution within the system 'aquatic media—organism'. For long-lived natural radionuclides, such equilibrium exists; the concentration factor for a natural radionuclide and a given organism in its habitat is constant and does not change with time.

The radionuclide concentration factors are not universal constants of aquatic organisms and can vary considerably depending on environmental conditions.

First of all, radionuclide concentration factors are strongly dependent on the concentration of stable analogous elements, which are available for uptake from the aquatic environment. Actually, stable and radioactive isotopes with identical or similar chemical properties form a common 'pool' of bioelements in the aquatic environment. Organism uptakes a radionuclide in parallel with its stable analog from an available pool of elements, and the share of the radionuclide in the uptake is equal to its relative share in the pool. In aquatic media with a low concentration of the stable analogous element, the quota of radionuclide uptake by the organism is relatively high; however, it is relatively low in a media with a high concentration of stable 'twin' element. That is why the radionuclide concentration factors for freshwater organisms are often higher than those for marine organisms.

Therefore, as the organisms accumulate stable bioelements (C, P, H, Co, Zn, and others) from the environment, they also consume their radioactive 'twins', such as ^{14}C, ^{32}P, ^{3}H, ^{60}Co, ^{65}Zn, and others.

Organisms consume not only the 'key' bioelements most essential for metabolism but also some other elements with similar chemical properties. Thus, Cs (and also ^{137}Cs, ^{134}Cs) is chemically similar to potassium, Sr (and ^{90}Sr) to calcium, and so on. Living organisms demonstrate partial discrimination in consuming elements, which are less suitable for metabolism. The substitute elements are consumed more actively when the concentration of main bioelements in aquatic media is low and is not sufficient for the organisms.

The physicochemical form of an element in aquatic media is of key importance for bioassimilation by living organisms. From the biochemical point of view, some chemical forms of a bioelement are more convenient for inclusion into metabolic processes than the others; that is why either preference or discrimination may be observed in biological uptake of different chemical forms of the same element.

Bioassimilation of bioelements and their radioactive 'twins' by organisms is an active multifactorial process, which depends on any factor influencing the normal course of biochemical reactions within the organism.

Many technogenic radionuclides are isotopes of heavy elements, such as ^{144}Ce, ^{106}Ru, and ^{239}Pu. Organisms do not need these chemical elements for metabolism, and those radionuclides are not actively bioassimilated but they are retained in organisms uptaken by passive adsorption. In aquatic media, heavy radionuclides usually exist in the form of colloids or suspensions, which are easily adsorbed on any submerged surfaces.

The ratio of surface/volume of small aquatic organisms, especially small planktonic species, is very large. Since the rate of adsorption is directly proportional to the surface area, organisms with a high surface/volume ratio show high concentration factors for adsorbing radionuclides. The surface/volume ratio for representatives of aquatic biota gradually decreases in the order 'bacteria—phytoplankton—zooplankton—crustaceans—fish'. It should be noticed that heavy radionuclides may be adsorbed not only on the external surface of organisms but also on the internal surfaces of organs, such as gills, intestine, kidney, and so on.

Therefore, two main mechanisms are responsible for radionuclide accumulation by organisms: active bioassimilation with incorporation of radionuclides into the tissues and passive adsorption on external and internal surfaces of the organism. In many cases, both bioassimilation and passive sorption contribute to the radionuclide uptake by aquatic biota.

Since the intensive development of the nuclear industry, technogenic radionuclides are becoming an important subject in radioecological studies. The old paradigm of 'equilibrium concentration factors' cannot be applied to the situations where releases of radionuclide to the environment vary with time and re-distribution processes of radionuclide between abiotic and biotic compartments are not in equilibrium.

Direct uptake of radionuclides from aquatic media is a major source of contamination of aquatic plants, plankton, and filtering organisms. Many other organisms, however, uptake radionuclides not only directly from water but also from a contaminated diet. The processes of radionuclide transfer via food chains are complicated; thus, the followings are the generalized statements obtained from experiments (Kryshev and Sazykina 1986; Cougthrey et al 1983):

- Radionuclides transformed into insoluble form within organisms such as ^{90}Sr and ^{45}Ca have low capacity for transfer via food chains; in many cases, concentration factors decreased from organisms of lower trophic levels (plants, mollusks) to fish and aquatic mammals.
- Radionuclides accumulated in soft tissues of organisms as analogues of major bioelements (e.g., ^{14}C, ^{40}K, ^{32}P as analogues of stable C, K, P) are easily transferred via food chains. Concentration factors for such radionuclides may increase in food chains from lower to higher trophic levels, following the accumulation of stable analogous elements.
- Radionuclides accumulated in organisms due to adsorption have a tendency of a general decrease in concentration factors in aquatic food chains 'phytoplankton—aquatic invertebrates—fish'. However, this effect is not only caused by trophic relations but also simply caused by the fact that larger organisms have lower surface/volume ratio compared with smaller organisms.
- The transfer of a radionuclide in the aquatic food chain is similar to the transfer of the stable analogous element. It is possible to predict the radionuclide transfer via a food chain using data on the elemental chemical composition of aquatic organisms.

9.2.2 Advantages of the Dynamic Modeling in Aquatic Radioecology: ECOMOD Methodology

Radioecological and ecological models have been based on different methodological assumptions. Classic radioecological models describe the radionuclide transfer in ecosystems using a linear compartment theory; these models use empirical accumulation coefficients for radionuclide distribution among the environmental compartments. In contrast, ecological models describing the dynamic fluxes of stable elements in ecosystems are solidly based on nonlinear differential equations. It is evident that the separation of models into two different classes is artificial and the transfer of stable and radioactive elements in aquatic ecosystems can be modeled on a unified methodological basis.

The theoretical aspects of an ecological approach to radioecological modeling based on ecological models are discussed (Kryshev and Sazykina 1986, 1990a, 2003; Sazykina 1994, 2000).

The unified ecological approach to radioecological modeling was developed in the period of 1978–1981 in Russia by T. Sazykina in the Ph.D. dissertation

(Sazykina 1981). Theoretical aspects of modeling the radionuclide migration in ecosystems were developed based on ecological models; mathematical models for radionuclides transfer in simple food chains were formulated. The methodology has developed further during the following years (Kryshev and Sazykina 1986; Kryshev and Sazykina 1990a, b; Sazykina 2000; Sazykina and Kryshev 1996; Kryshev and Ryabov 2000, 2005; Kryshev and Sazykina 2003; Kryshev 2000). This methodology called 'ECOMOD methodology' is intended to simulate dynamical and nonequilibrium radioecological processes in aquatic ecosystems.

In general, ecological models are based on the equations of biomass dynamics in populations of organisms or ecological groups of species. The ecological equations describe the transfer of biogenic elements from the environment to the food chains of organisms and back to the environment (ecological cycles of elements).

For an aquatic ecosystem, the dynamics of population biomass is described by the following general equation:

$$
\begin{aligned}
&\frac{\mathrm{d}M}{M\mathrm{d}t} = -\mathrm{BIOLOSS}(..,t) + \mathrm{BIOSYNT}(..,t);\\
&\mathrm{BIOLOSS}(\ldots,t) = \mathrm{MORTALITY}(..,t) + \mathrm{METAB}(..,t);\\
&\mathrm{BIOSYNT}(..,t) = \gamma \times \mathrm{FOOD}(..,t);
\end{aligned}
\qquad (9.1)
$$

where $M(t)$ is the biomass of the population of a given type of organisms (unit of mass per unit of water volume in a water body); BIOLOSS $(..,t)$ is the rate of biomass losses under metabolic processes METAB and mortality MORTALITY; BIOSYNT $(..,t)$ is the rate of biomass biosynthesis from diet; FOOD is the intrinsic rate of food consumption by population; γ is the bioassimilation coefficient; and METAB$(..,t)$ is the metabolic rate of existing biomass.

The terms 'BIOLOSS' and 'BIOSYNT' are actually complex formulas, which show the dependence of biological processes on the environment and on other organisms. The complex formulas for biological processes are described elsewhere (Alekseev et al. 1992; Jorgensen et al. 1978). Here the dependence on the set of parameters is shown as $(..,t)$.

Suppose that the specific concentration of chemical element "A" in the biomass is constant and equal Q_1^A (g/kg d.w.).

The diet of organisms contains a stable isotope of element 'A' with concentration Q_0^A (g/kg d.w.). The organisms assimilate the element 'A' from diet and use it for constructing their bodies. Therefore, the organisms concentrate the element 'A' from the diet with the concentration factor Q_1^A/Q_0^A.

Let us consider the transfer of the stable element 'A' between the biomass of population and environment.

The element 'A' is assumed to be a nonlimiting element for population, hence its content in the diet is sufficient to cover the needs of organisms, and using of the element to produce the new biomass from food is equal to $(Q_1^A/Q_0^A)\cdot\gamma\cdot\mathrm{FOOD}$. The element losses from the natural death of organisms and metabolism are equal to $Q_1^{\mathrm{A}} \times \mathrm{BIOLOSS}$.

The result of the equation for the stable element ‘A’ transferring in the population of biomass M is shown by the following form:

$$\frac{d\left(Q_1^{\mathrm{A}} \times M\right)}{M dt} = -Q_1^{\mathrm{A}} \times \mathrm{BIOLOSS}(\ldots, t) + \frac{Q_1^{\mathrm{A}}}{Q_0^{\mathrm{A}}} \times \gamma \times \mathrm{FOOD}; \qquad (9.2)$$

Assuming that a radioactive isotope of element ‘A’ is added to the diet (in the same chemical form as a stable isotope) with mass concentration q (g/kg d.w.) and corresponding activity X (Bq/kg d.w.). Then, the total concentration (pool) of element ‘A’ in the diet will be $(Q_0^{\mathrm{A}} v + q)$.

The share of radionuclide participating in the element uptake from bioassimilated diet is equal to its share in the total pool of element in the diet:

$$\mathrm{Share} = \frac{q}{Q_0^{\mathrm{A}} + q} = \frac{X}{Q_0^{\mathrm{A}} + X \times K} \approx \frac{X}{Q_0^{\mathrm{A}}}$$

where K is the conversion coefficient between mass and activity of radionuclide. For most important cases, $q \ll Q_0^{\mathrm{A}}$.

The rate of radionuclide removal from the biomass is proportional to the radionuclide concentration $y(t)$ (Bq kg^{-1} d.w.), already accumulated in the biomass.

Thus, Eq. (9.2) is transformed to describe the transfer a radioactive tracer in the biomass of population:

$$\frac{d(M \times y)}{M \mathrm{d}t} = \lambda_r \times y \quad y \times \mathrm{MORTALITY}(.., t) - \varepsilon_{\mathrm{A}} \times y \times \mathrm{METAB}(.., t) + \\ + \frac{X}{Q_0^{\mathrm{A}}} Q_1^{\mathrm{A}} \times \gamma \times \mathrm{FOOD}(.., t); \qquad (9.3)$$

where $y(t)$ is the specific activity of radionuclide in the biomass, Bq kg^{-1} d.w. and λ_{r} is the radioactive decay constant.

Taking into account the differential.

$$(M \times y)' = M' \times y + M \times y';$$

and using Eq. (9.1) for dM/dt, Eq. (9.3) is transformed to the differential equation for specific activity $y(t)$:

$$\frac{dy}{dt} = \gamma \times \mathrm{FOOD}(..,t) \cdot \left(\frac{Q_1^A}{Q_0^A} \times X - y\right) - \lambda_r \times y; \tag{9.4}$$

In the simplest case of $M(t) = \text{const}$, $X = \text{const}$, $y(0) = 0$, $\lambda_r \cong 0$, Eq. (9.4) has an analytical solution,

$$y_i(t) = X \times CF \times (1 - \exp(-p \times t)); \tag{9.5}$$

where $\mathrm{CF} = Q_1^A / Q_0^A$ is the equilibrium concentration factor for the element 'A' and $p = \gamma \times \mathrm{FOOD}(\ldots,t)$. Since $M(t) = \text{const}\ p = \varepsilon_A \cdot \mathrm{METAB}$.

Expression (9.5) is the simplest formula of the dynamics of radionuclide accumulation in biomass of the population. This relation provided by empirical values of CF and p is employed traditionally as an approximation of the results obtained from experiments, and also in compartmental radioecological models (Polykarpov 1966; Whicker and Schultz 1982). The advantage of the ecological approach is that the parameters of radionuclide transfer CF and p are described as ecological parameters and functions, which can be obtained from biological handbooks a priori.

In general, the ecological Eq. (9.1) together with the radiological Eq. (9.4) can describe a wide range of radioecological situations, for example, the radionuclide accumulation in nonequilibrium situations, transfer of radionuclide in some complicated food chains, and combined action of radiation and other environmental factors. Computer codes based on the ECOMOD approach have a module structure, including the following modules: 'Radionuclides in water-sediment components'; 'Population/ecosystem dynamics'; 'Radionuclides in aquatic biota'; 'Dose assessment for aquatic biota'. Depending on the specific task, some modules can be deleted from the computer program.

The module 'Radionuclides in water-sediment components' calculates the dynamics of radionuclides distribution in abiotic components of the aquatic system, in particular in water and in sediments. Any appropriate standard hydrological model can be used for this purpose. The module 'Population/ecosystem dynamics' calculates the biomass dynamics of populations in the aquatic ecosystem using any appropriate ecological model. The main module 'Radionuclides in aquatic biota' is intended to calculate the dynamics of radionuclide transfer between the aquatic media and food chains of aquatic organisms using the equations of the type (9.3) and (9.4). The output of the module 'Radionuclides in aquatic biota' is concentrations (activities) of radionuclides in the biomass of aquatic organisms (algae, zooplankton, several species of fish, etc.). The output data from this module serve as input data for the next module of 'Dose assessment for aquatic biota'; in this module, doses to reference organisms are calculated from data on water, sediment, and biota contamination with radionuclides. The ECOMOD modules make calculations for several radionuclides and several populations of aquatic biota simultaneously.

9.3 Accumulation of Radionuclides in Aquatic Biota During the Early Period Following the Chernobyl Accident

9.3.1 *Nonequilibrium Processes of Radionuclide Accumulation in Biota from Rivers and Reservoirs Within Highly Contaminated Chernobyl Area*

Immediately after the Chernobyl accident, a great variety of Chernobyl-driven radionuclides was observed in the components of the aquatic ecosystem of the Dnieper River, namely: ^{89}Sr, ^{90}Sr, ^{95}Zr, ^{95}Nb, ^{99}Mo, ^{103}Ru, ^{106}Ru, ^{131}I, ^{132}I, ^{132}Te, ^{134}Cs, ^{136}Cs, ^{137}Cs, ^{140}Ba, ^{140}La, ^{141}Ce, ^{144}Ce, ^{239}Np, and others (Accident 1986; Izrael 1990, 1996; Kryshev 1992). In the Kiev reservoir, radionuclide concentrations in water were up to 4 kBq/L (Accident 1986; Kryshev 1992). In this period, activities of short-lived nuclides were several orders of magnitude higher than those of long-lived radioisotopes of cesium and strontium. In water bodies, enrichment of suspended matter was observed with radioisotopes of tellurium, barium, molybdenum, ruthenium, cerium, zirconium, niobium, and neptunium. On May 1, 1986, the activity of ^{95}Zr and ^{95}Nb in suspended matter in the Kiev reservoir amounted to 97% of their total activity in water (suspended and dissolved fractions), whereas it was about 65% for ^{137}Cs and about 36% for ^{131}I (Izrael 1990; Kryshev 1992). As the short-lived nuclides decayed also were deposited to bottom sediments, contamination of water decreased distinctly. In June 1986, contamination of water was hundreds of times lower as compared to the maximum levels and was determined mainly by isotopes of Cs and Sr.

Table 9.1 presents the estimated concentration factors for radionuclides in aquatic organisms in June 1986 based on Izrael (1990) and Kryshev (1992). These

Table 9.1 The estimated radionuclide concentration factors (L/kg) in aquatic organisms of the Dnieper River (June 1986)

Radionuclide	Mollusks (f.w)	Aquatic plants (d.w.)	Fish (f.w.)
^{90}Sr	440 ± 130	240 ± 50	50 ± 12
^{95}Zr	2900 ± 800	20,000 ± 8000	190 ± 40
^{95}Nb	3700 ± 800	22,000 ± 11,000	220 ± 30
^{103}Ru	1000 ± 400	11,000 ± 5000	120 ± 50
^{106}Ru	750 ± 130	17,000 ± 3000	130 ± 40
^{131}I	120 ± 20	60 ± 30	30 ± 10
^{134}Cs	300 ± 120	2700 ± 900	300 ± 60
^{137}Cs	270 ± 90	3000 ± 1100	300 ± 70
^{140}Ba	2800 ± 800	3600 ± 1400	420 ± 210
^{140}La	2400 ± 300	3000 ± 1200	400 ± 190
^{141}Ce	3000 ± 800	20,000 ± 6000	500 ± 190
^{144}Ce	4600 ± 1500	24,000 ± 6000	900 ± 300

f.w. fresh weight, *d.w.* dry weight

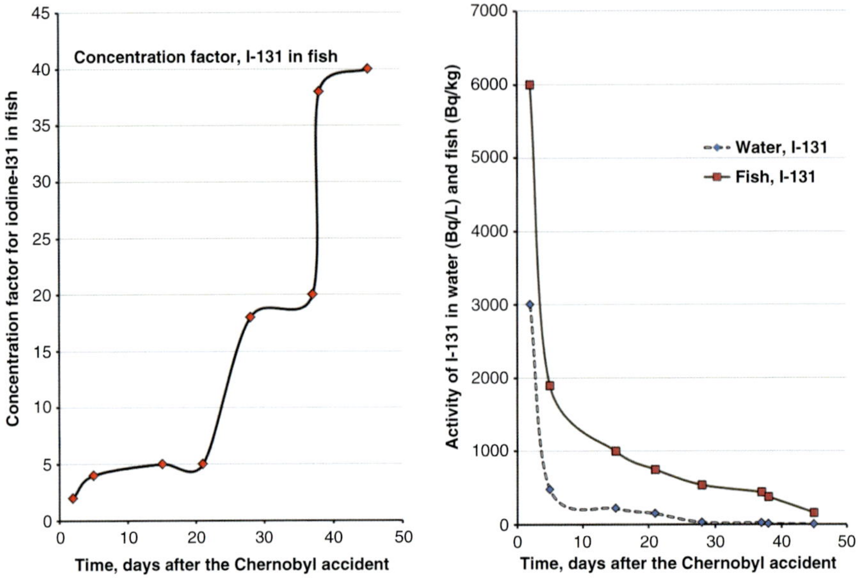

Fig. 9.1 Concentration factor for ^{131}I in fish from the Dnieper River, early period after the Chernobyl accident (left side) in 1986. Dynamics of ^{131}I activity in fish and water of the Dnieper River (right side) (Based on data from Kryshev and Ryazantsev (2000) and Kryshev and Ryabov (2000))

concentration factors differ essentially from the so-called 'equilibrium' values, which were measured in laboratory conditions, but practically never occur in nature. It is easily seen that for mollusks and aquatic plants the highest values of the concentration factors are typical for intensively adsorbed radioisotopes of Ce, Zr, Nb, Ba, La, and Ru. Radionuclides entering the organisms by the bioassimilation mechanism (I, Cs, and Sr) have lower values of the concentration factors. Since bioassimilation is a relatively slow process in June 1986, the concentration factors for ^{90}Sr, which selectively accumulated in mollusk shells, were still relatively small.

For fish muscles, the values of the concentration factors for intensively adsorbed radioisotopes were lower than those for mollusks and aquatic plants; this was due to lower body surface/weight ratio; bioaccumulation of biogenic radionuclides in fish was slower than that in mollusks and aquatic plants (Kryshev 1992, 1996).

Figure 9.1 (right side) shows the dynamics of ^{131}I concentration in fish from the Dnieper River during the early period after the Chernobyl accident (Kryshev and Ryabov 2000). The value of the ^{131}I concentration factor depends on several processes, including bioassimilation of ^{131}I in fish muscles, radioactive decay, and dilution of ^{131}I in river waters. The ^{131}I bioassimilation and self-purification of aquatic organisms occurred with some time lag. As a consequence, the radioactive contamination of fish muscles decreased much slower compared to the ^{131}I activity reduction in water. The concentration factor of ^{131}I in fish (Fig. 9.1, left side) was not constant, differing from the equilibrium considerably; thus, it could not be used for any predictions.

9.3.2 Distribution of Technogenic Radionuclides in Aquatic Biota of the Chernobyl Cooling Pond in the Early Period of Accidental Contamination

Heavy contamination of the Chernobyl Cooling pond was formed from highly radioactive accidental fallouts from 26 April through May of 1986.

The Chernobyl Cooling pond became one of the most heavily contaminated water bodies in the world. The Cooling pond was a suitable model for studying the dynamic processes of radionuclide migration and accumulation in closed aquatic systems (Kryshev et al. 1996, 1999; Kryshev 1992).

In the early period of the Chernobyl accident (first few months), very little information existed about biota contamination in the Cooling pond. In 1986, the dynamical ECOMOD model was employed to produce the first predictions of the radioactive contamination of the Chernobyl Cooling Pond ecosystem (Kryshev 1992). First calculations of the radioactive contamination of the Chernobyl Cooling pond ecosystem were made in May–August 1986 within a few months after the Chernobyl accident.

Estimations of accidental fallout of radionuclides to the Cooling pond were used as input information for model calculations. The available biological data on the Cooling pond ecosystem were taken into account. The radionuclide distribution in abiotic components of the pond was calculated using a simple 'water-sediments' compartmental model. The dynamics of radioactive contamination of aquatic biota was calculated for a number of radionuclides, and for some of them, adsorption processes were considered.

The following set of differential equations was used to describe the dynamics of i-th radionuclide activity (Bq/kg w.w.) in the biomass of ecosystem components (Kryshev and Sazykina 1990a, b):

Vegetation

$$\frac{\mathrm{d}y_k^{(i)}}{\mathrm{d}t} = \beta_k\left(T^0, \mathrm{LC}(t)\right) \times S(t) \times \left[-y_k^{(i)} + CF^{(i,\mathrm{stab})} \times X^{(i)}(t)\right] - \lambda^{(i)} y_k^{(i)}; \quad (9.6)$$

Consumers (zooplankton and fish)

$$\frac{\mathrm{d}y_j^{(i)}}{\mathrm{d}t} = -y_j^{(i)} \times \left(\lambda^{(i)} + \varepsilon_j^{(i)} + \frac{1}{M_j}\frac{\mathrm{d}_M}{\mathrm{d}t}\right) + \frac{Q_j^{(i,\mathrm{stab})}}{Q_{\mathrm{food}}^{(i,\mathrm{stab})}} \cdot \sum \gamma_m \times \mathrm{FOOD}_m \cdot y_m^{(i)}; \quad (9.7)$$

Adsorption on the surface of organisms

$$\frac{\mathrm{d}y_{\mathrm{sorp}}^{(i)}}{\mathrm{d}t} = \left(q_{\mathrm{sorp}}^{(i)} \times \frac{X^{(i)}}{Q_{\mathrm{stab}}^{(i)}} - q_{\mathrm{des}}^{(i)} \times y_{\mathrm{sorp}}^{(i)} \right); \tag{9.8}$$

where indexes $k = 1, 2$ refer to phytoplankton and macroalgae and indexes $j = 1, 3$ refer to zooplankton, nonpredatory fish, and predatory fish; $y_k^{(i)}$, $y_j^{(i)}$ are the activity concentrations of ith radionuclide in the biomass of ecosystem components; M_j is the biomass of the jth ecosystem component; $\beta_k(T^0, \mathrm{LC}(t))$ is the coefficient of biomass production by photosynthesis as a function of water temperature T^0 and light condition $\mathrm{LC}(t)$, which in turn is a specific function of the latitude and season; $S(t)$ is the concentration of dissolved limiting nutrient in the aquatic medium; the values of water temperature, light conditions, and nutrient concentration in the aquatic medium were inserted to the computer code as input data; $\mathrm{CF}^{(i,\mathrm{stab})}$ is the coefficient of accumulation of stable element, which is analogous to ith radionuclide by vegetation; $X^{(i)}(t)$ is the activity of the ith radionuclide in the aquatic medium at the time t; the values of $X^{(i)}(t)$ are calculated in the hydrological block of the model; $\lambda^{(i)}$ is the radioactive decay constant; $\varepsilon_j^{(i)}$ is the coefficient of metabolic losses of the ith element; $Q_j^{(i,\mathrm{stab})}$, $Q_{j,\mathrm{food}}^{i,\mathrm{stab}}$ are the concentrations of the stable analogous element in the biomass of aquatic organism and in its diet; $\Sigma\gamma_m\ \mathrm{FOOD}_m \times y_m^{(i)}$ is the intake of the ith radionuclide by an organism from its diet; $y_{\mathrm{sorp}}^{(i)}$ is the activity adsorbed on the unit of the surface of the organism (Bq/cm^2); q_{sorp}, $q_{\mathrm{des}}^{(i)}$ are parameters of sorption and desorption of the ith radionuclide and its stable 'twin'; and Q_{stab} is the concentration of a stable analogue of the ith radionuclide in the aquatic medium.

The ECOMOD calculations demonstrated a very complicated character of radionuclides accumulation and redistribution in aquatic biota from the Chernobyl Cooling pond in the first months after the accident. Phytoplankton accumulated radionuclides rapidly. Due to the short individual life span of algal cells, the activity of phytoplankton followed, in general, the activity of water.

Radionuclide activities in fish demonstrated much more complicated dynamical features—there was no equilibrium of any radionuclides; instead, bioaccumulation of radionuclides in fish tissues occurred despite the decrease of radionuclides in water. Obviously, the methodology of equilibrium concentration factors did not work in situations of accidental acute contamination of a water body.

Figure 9.2 shows the complicated dynamics of radionuclides activity concentrations in fish from the Chernobyl Cooling pond in the first months after the radiation accident of 1986.

Table 9.2 shows the model predictions compared with measured averaged concentrations of some radionuclides taken from Kryshev (1992). As seen from Table 9.2, the ECOMOD model predictions were in good agreement with measurements even for an extremely nonequilibrium radiological situation, which was observed in the Chernobyl Cooling Pond in the early days after the heavy radiation accident.

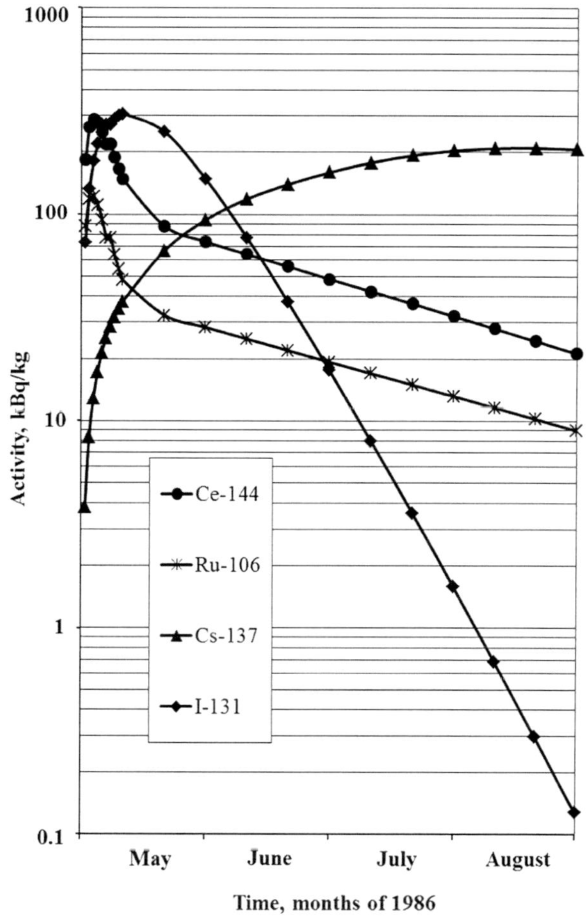

Fig. 9.2 The dynamics of radionuclides activity concentrations in fish from the Chernobyl Cooling pond, early period of the Chernobyl accident; model calculations (ECOMOD model) in 1986 (The figure is modified from Kryshev and Sazykina (1990a, b))

9.4 Radionuclide Migration in Aquatic Ecosystems Within the 5-Year Period After the Chernobyl Accident (1986–1990)

9.4.1 Dynamics of Radionuclide Concentration Factors in Aquatic Biota from Water Bodies of the Chernobyl Area

In the years following the Chernobyl radiation accident, radioecological consequences for biota were largely associated with long-lived radionuclides ^{137}Cs and ^{90}Sr. These long-lived radionuclides contributed only a few percentages to the contamination in the early period after the accident when dominating contaminants were numerous short-lived radionuclides. Following the decay of short-lived

Table 9.2 Model predictions and measured activity concentrations of radionuclides in the components of the Chernobyl NPP cooling pond, early period after the accident (May–August in 1986), kBq/kg w.w.

Component	^{131}I	^{137}Cs	^{90}Sr	^{144}Ce
30 May 1986				
Water				
Measured numbers	1.7 ± 1.4	0.4 ± 0.2	0.04 ± 0.026	0.2 ± 0.13
Model prediction	3.0	0.2	0.018	0.12
Fish:				
Measured numbers	30 ± 20	50 ± 30	0.3 ± 0.2	90 ± 50
Model prediction	50	85	0.2	115
Algae:				
Measured numbers	300 ± 200	150 ± 80	16 ± 10	800 ± 500
Model prediction	400	120	9	850
15 June 1986				
Fish				
Measured numbers	10 ± 6	80 ± 50	0.5 ± 0.3	110 ± 50
Model prediction	26	115	0.3	140
Algae				
Measured numbers	100 ± 60	140 ± 60	15 ± 7	400 ± 200
Model prediction	150	100	9	500
18 July 1986				
Algae				
Measured numbers	–	110 ± 60	15 ± 8	150 ± 70
Model prediction	9	80	9	100
7 August 1986				
Algae				
Measured numbers	–	100 ± 60	12 ± 6	110 ± 70
Model prediction		50	8	50
Fish				
Measured numbers	–	110 ± 60	0.9 ± 0.4	–
Model prediction		160	0.6	–

radionuclides, the contribution of long-lived radionuclides to the total activity was growing and ultimately became dominant.

A distinguishing feature of the Chernobyl accident from a biogeochemical standpoint was a substantially nonequilibrium character of biogenic migration of long-lived radionuclides over 3–5 postaccidental years. As a result of biogenic accumulation and biofiltration, radionuclides initially deposited in bottom sediments and dispersed in water were involved in biogeochemical cycles within contaminated water bodies.

The assessment of radionuclide concentration factors in aquatic biota was made on the basis of experimental data on the specific activity of radionuclides in fish, mollusks, aquatic plants, and water of the Chernobyl Cooling pond and Dnieper River (1986–1990). The measurements of the radionuclide compositions of about

10,000 samples (water and biota) were used for assessment (Kryshev 1992; Vakulovsky 1986).

Standard methods were used for water sampling and measurements (Vakulovsky 1986). Fish were caught with fishing nets. Samples of aquatic plants (algae) and other aquatic organisms were rinsed to remove particles on surfaces. As a rule, water or sediment was collected at the same time and place as biota. Collected samples of aquatic organisms were burned at a temperature of not more than 350 °C. The activities of gamma-emitting nuclides were determined by standard gamma-spectrometric methods with lead protection using highly sensitive semiconductor detectors and multichannel pulse analyzers. The activity of ^{90}Sr was determined by radiochemical methods from daughter nuclide ^{90}Y. The activity of the latter was measured with the low-background beta radiometer. The activities of radionuclides were calculated in aquatic plants for dry weight, whereas in mollusks (tissue and shell) and fish (muscles) with respect to the fresh weight of the organisms.

Tables 9.3 and 9.4 show radionuclide concentration factors estimated for aquatic organisms in Dnieper River and the Chernobyl Cooling pond obtained from the analysis on the observation data for the period 1986–1990.

For fish in the Cooling pond of the Chernobyl NPP, the processes of ^{137}Cs accumulation were nonequilibrium during the first five years following the Chernobyl accident. For most species of fish, maximal values of the concentration factors in muscles occurred in 1988 for silver carp in 1989 and for goby in 1990. For fish from the Dnieper River, the dynamics of ^{137}Cs concentration factors in fish muscles was similar to that for fish in the Cooling pond. However, the values of ^{137}Cs concentration factors for fish from the Dnieper water system were 2–3 times lower than those for fish from the Cooling pond.

The processes of ^{137}Cs accumulation in other organisms (aquatic plants and mollusks) were also nonequilibrium, that is, concentration factors changed over the time. For mollusks, however, the variability of the ^{137}Cs concentration factors with time did not change much with time as was found for fish. In general, the equilibrium with the environment is established rapidly for small organisms and more slowly for larger organisms because of differences in metabolism.

The ^{90}Sr accumulation in fish and mollusks from the Dnieper River increased with time during the first 3 years after the accident (Table 9.4). In 1986, the values of the concentration factors of ^{90}Sr and ^{137}Cs in mollusks were practically identical. Three years later, the concentration factors of ^{90}Sr for mollusks were 6 times greater than those of ^{137}Cs.

Thus, the analysis shows that a nonequilibrium situation remained in the contaminated water bodies over a long period after the accident. According to the assessments based on observational data of 1986–1990, during the 5-year period since the accident, a radioecological equilibrium had not been established in the aquatic ecosystems.

Table 9.3 Dynamics of the ^{137}Cs concentration factors in aquatic organisms within the 5-year period after the Chernobyl accident (muscles, f.w.)

	1986	1987	1988	1989	1990
Cooling pond of the Chernobyl NPP					
Nonpredatory fish (mixed)	720 ± 160	1300 ± 270	1900 ± 280	1600 ± 240	1000 ± 230
Carp	470 ± 190	840 ± 170	2100 ± 500	1300 ± 240	1100 ± 300
Bream	440 ± 160	1700 ± 400	1600 ± 380	1200 ± 500	860 ± 400
Roach	520 ± 200	1000 ± 170	1060 ± 320	–	800 ± 300
Silver bream	1100 ± 400	1700 ± 800	2100 ± 950	–	570 ± 210
Goldfish	1200 ± 500	850 ± 500	2600 ± 900	–	930 ± 210
Silver carp	670 ± 140	1700 ± 300	2100 ± 360	2400 ± 400	860 ± 410
Goby	660 ± 300	1400 ± 340	1600 ± 260	–	2100 ± 860
Predatory fish (mixed)	360 ± 150	2900 ± 400	8000 ± 1000	6000 ± 1300	5000 ± 900
Perch	860 ± 190	3400 ± 800	8400 ± 3600	6100 ± 2200	4300 ± 1400
Pike–perch	140 ± 60	2900 ± 700	7900 ± 2000	5800 ± 1800	5700 ± 1400
Pike	100 ± 40	2500 ± 340	7800 ± 1000	–	–
Mollusks	120 ± 30	240 ± 100	200 ± 80	–	–
Aquatic plants (dry weight)	3300 ± 1000	5100 ± 1800	10,600 ± 5000	11,000 ± 3400	10,400 ± 4000
The Dnieper River					
Nonpredatory fish (mixed)	700 ± 120	1000 ± 200	540 ± 160	520 ± 110	400 ± 100
Bream	760 ± 280	1300 ± 600	600 ± 300	460 ± 180	340 ± 120
Roach	750 ± 130	800 ± 300	450 ± 200	600 ± 220	500 ± 200
Silver bream	570 ± 100	900 ± 400	560 ± 300	500 ± 150	400 ± 150
Predatory fish (mixed)	200 ± 80	1700 ± 400	1800 ± 500	1300 ± 300	1500 ± 400
Perch	210 ± 130	1500 ± 300	1700 ± 500	1500 ± 600	1400 ± 500
Pike–perch	220 ± 100	1400 ± 200	1400 ± 600	1000 ± 340	1600 ± 600

Pike	140 ± 100	2200 ± 1100	2400 ± 1100	1400 ± 600	1500 ± 700
Mollusks	300 ± 140	500 ± 220	390 ± 170	320 ± 90	400 ± 130
Aquatic plants (dry weight)	3200 ± 1400	5200 ± 2200	6000 ± 3000	2600 ± 700	2000 ± 600

Based on data from Kryshev (1992); Vakulovsky (1986)
Latin names of fishes: carp (*Cyprinus carpio*); bream (*Abramis brama*); roach (*Rutilus rutilus*); silver bream (*Blicca bjoerkna*); goldfish (*Carassius carassius*); silver carp (*Hypophthalmichthys molitrix*): goby (*Cottus gobio*); pike (*Esox lucius*), pearch (*Perca fluviatilis*); pike–perch (*Stizostedion lucioperca*)

Table 9.4 Dynamics of the ^{90}Sr concentration factors in aquatic organisms of the Dnieper River after the Chernobyl accident

Component	1986	1987	1988	1989
Fish (muscles, f.w.)	40 ± 12	70 ± 23	90 ± 26	115 ± 30
Bream	70 ± 40	60 ± 40	50 ± 16	120 ± 50
Roach	40 ± 10	50 ± 20	120 ± 60	200 ± 50
Silver Bream	35 ± 10	90 ± 50	120 ± 40	140 ± 60
Perch	40 ± 14	60 ± 30	130 ± 60	80 ± 10
Pike–perch	24 ± 6	15 ± 6	70 ± 40	80 ± 30
Pike	50 ± 18	130 ± 50	75 ± 15	70 ± 30
Mollusks (muscles, f.w.)	440 ± 210	700 ± 300	1000 ± 360	1700 ± 700

Based on data from Kryshev (1992)

9.4.2 *Migration of ^{137}Cs Through Food Chains in Aquatic Ecosystems*

Considerable differences were found in ^{137}Cs concentration factors for predatory and nonpredatory species of fish. The ^{137}Cs concentration factors in the Chernobyl Cooling pond were 4–5 times higher for predatory fish than those for nonpredatory fish since 1988. A similar relationship had already been observed in the Cooling pond before the Chernobyl accident. The ^{137}Cs concentration factors for predatory fish were high ($K = 2300 \pm 600$), whereas for nonpredatory fish, they were consistently lower ($K = 700 \pm 300$) even in 1979–1983. This is ascribable to the manifestation of the effect of trophic levels (Kryshev and Ryabov 2000). Such effect was not observed for ^{90}Sr because this radionuclide was transformed into insoluble form in bones/shells of organisms and insoluble ^{90}Sr had low capacity for transfer via food chains. The values of the ^{90}Sr concentration factors in fish and mollusks increased with time during the first three years after the accident. The differences in accumulation between radioactive strontium and cesium can be explained by their different roles in organisms: cesium is accumulated in soft tissues; strontium is gradually deposited within bones/shells in insoluble form with a very slow rate of depuration.

Figure 9.3 demonstrates the concentration factors (CF) of ^{137}Cs for fish from the Chernobyl Cooling pond after the radiation accident based on measurement data from Table 9.3. The calculated concentration factors for different fish species had a definite dynamical behavior during the five postaccidental years, which were the most important for assessment.

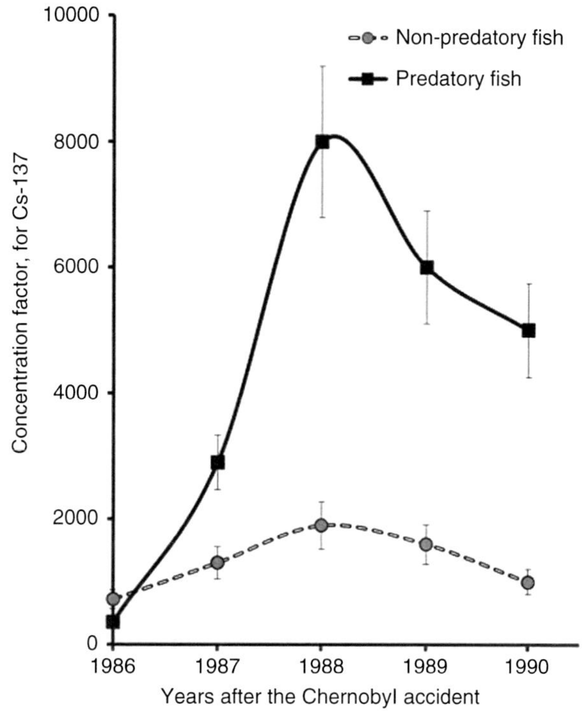

Fig. 9.3 Values of ^{137}Cs concentration factors (CF) for nonpredatory fish (mixed) and predatory fish (mixed) from the Cooling pond of the Chernobyl NPP (1986–1990) (Based on measurement data from Table 9.3)

9.4.3 Dynamics of Radionuclide Distribution in Age-Structured Fish Populations (Size Effect)

In a natural aquatic ecosystem, a population of each fish species is represented by a set of individual fish of different age classes. Each age class is characterized by a specific growth rate, the amount and assortment of the food components, and the activity of metabolic processes. Therefore, the rates of accumulation and elimination of radiocaesium in fish are specific for each age class.

An interesting and important phenomenon called the 'size effect' was observed in many water bodies contaminated by the Chernobyl fallouts (Evans 1988; Hakanson 1991; Kryshev 1992; Kryshev and Ryabov 2000). Usually, a 'positive size effect' has been observed—older fish were more contaminated with ^{137}Cs than younger members of the same species. More rarely, mostly during the first year following the accident, a 'negative size effect' was observed—the specific activity of ^{137}Cs in older fish was lower than that in young fish of the same species. The size effect was particularly pronounced during the second year following the accidental contamination.

The dynamics of the size effect in radionuclide accumulation by fish from the Chernobyl Cooling pond was modeled using the ECOMOD methodology (Kryshev 2000; Kryshev and Ryabov 2000).

In the case of accidental contamination of a water body, it seems convenient to divide the population of each species of fish into discrete age classes and to express the fish age as $\tau = \tau_0 + t$, where τ_0 is the age of the fish at the time of the accident, and t is the elapsed time since the occurrence of the accident. It is assumed in the model that the major part of the radiocaesium enters the fish via contaminated food.

The basic equation for the intrinsic radionuclide activity in fish was written as follows (Kryshev and Ryabov 2000):

$$\frac{\mathrm{d}y}{\mathrm{d}t} = -\left(\lambda_r + \varepsilon_A \frac{W}{M} + \frac{1}{M}\frac{\mathrm{d}M}{\mathrm{d}t}\right) \times y + \frac{Q_1^A}{Q_0^A} \cdot \left(\frac{1}{M}\frac{\mathrm{d}M}{\mathrm{d}t} + \varepsilon_A \frac{W}{M}\right) \times X, \quad (9.9)$$

where y is the activity concentration of the radionuclide in the muscle of the fish, in Bq/kg; λ_r is the radioactive decay constant; ε_A is a proportionality coefficient between the elimination rate of the biogenic element and the general rate of metabolism W; $M = M(\tau)$ is the fish weight averaged over a particular age class, in kg; Q_1^A is the concentration of the biogenic element in fish, Q_0^A is the concentration of the biogenic element in the fish diet, and X is the concentration of radionuclide in the fish diet.

Equation (9.9) was written for each discrete generation of fish, and the set of equations was solved numerically.

The specific increase of fish biomass and the value for specific metabolic rate are related to each other by the 'energy balance equality' (Ivlev 1962; Winberg 1956):

$$\gamma \times F = \frac{1}{M}\frac{\mathrm{d}M}{\mathrm{d}t} + \frac{W}{M} \quad (9.10)$$

where γ characterizes the degree of food assimilation in the gastrointestinal tract (GIT) of the fish and does not depend on the radionuclide; F is equal to the total amount of food consumed by the fish during the time period divided by the average biomass of fish during this period.

The assortment of feedstuffs consumed by fish is strongly dependent on fish size (age) as well as on the trophic position of the fish in the food chain. The contamination of fish food is estimated separately for each fish species and each age class.

If a fish feeds on small organisms like plankton, y_{food} can be estimated using the constant equilibrium concentration factor $\mathrm{CF}_{\mathrm{food}}$ (Kryshev and Sazykina 1990a, b, 1995). This simplification is justified since an equilibrium in the exchange of radionuclides between plankton and water is established rapidly because of a large surface/volume ratio of plankton for interaction with the environment. Then, $X = \mathrm{CF}_{\mathrm{food}} \times C_w(t)$, where $C_w(t)$ is the activity of the radionuclide in water, and Eq. (4.2) takes the form

$$\frac{\mathrm{d}y}{\mathrm{d}t} = -ay + G \times C_w(t) \quad (9.11)$$

where $a = \lambda_r + \varepsilon_A(W/M) + (dM/Mdt)$ and $G = (Q_1^A/Q_0^A) \times (\varepsilon_A(W/M) + (dM/Mdt)) \times CF_{food}$.

The specific activity of radiocaesium in predatory species of fish is determined from the system of equations of types (9.9) and (9.11), where the activity in the prey fish is calculated by the first equation:

$$\begin{aligned}\frac{dy_{prey}}{dt} = &-\left(\lambda_r + \left(\varepsilon_A \frac{W}{M}\right) + \left(\frac{dM}{Mdt}\right)_{prey}\right) y_{prey} + \\ &+ \left(\frac{Q_1^A}{Q_0^A}\right)_{prey} \times \left(\left(\varepsilon_A \frac{W}{M}\right)_{prey} + \left(\frac{dM}{Mdt}\right)_{prey}\right) CF_{food} \times C_w(t)\end{aligned} \tag{9.12}$$

$$\begin{aligned}\frac{dy_{pred}}{dt} = &-\left(\lambda_r + \left(\varepsilon_A \frac{W}{M}\right)_{pred} + \left(\frac{dM}{Mdt}\right)_{pred}\right) y_{pred} \\ &+ \left(\frac{Q_1^A}{Q_0^A}\right)_{pred} \times \left(\left(\varepsilon_A \frac{W}{M}\right)_{pred} + \left(\frac{dM}{Mdt}\right)_{pred}\right) y_{prey}(t)\end{aligned} \tag{9.13}$$

In Eqs. (9.12) and (9.13), the index *prey* is related to the prey fish, index *pred* is related to the predatory fish, and CF_{food} is the average concentration factor for organisms, which are consumed by prey fish.

Two fish species from the Chernobyl cooling pond were selected for model testing: silver carp (*Hypophthalmichthys molitrix*) as a representative nonpredatory fish and pike–perch (*Stizostedion lucioperca*) as a representative predatory fish. Silver carp from the Cooling pond of the Chernobyl NPP feeds mainly on chironomid larvae, oligochaetes, planktonic crustaceans, mollusks, and detritus, whereas pike–perch is a piscivorous predator from the second year of life. Data for average observed annual biomass growth rates of fish from the Chernobyl NPP cooling pond were taken from Kazakov (1995). The dynamics of the ^{137}Cs activity in water of the Chernobyl NPP cooling pond $C_w(t)$ were estimated earlier (Sazykina and Kryshev 1996; Kryshev et al. 1999).

The numerical values of the model coefficients, used in the calculations, are summarized in Kryshev and Ryabov (2000). In the early period after the radioactive contamination of the water body, the concentration factors in the food of fish-benthic feeders may depend on fish size. For example, large individual carp feeds on benthic organisms predominantly from deeper layers of the bottom sediments (5–8 cm) where radionuclide has not migrated. Therefore, different numerical values were selected for the concentration factors in the food of carp in the first year following the accident depending on fish age.

The calculations were performed for the 5-year period (1986–1990) after the accidental contamination. The model predictions were made for fish from the generation of 1978 up to the generation of 1988. The model predictions were compared with the experimental data (Kryshev and Ryabov 2000; Kryshev 2002).

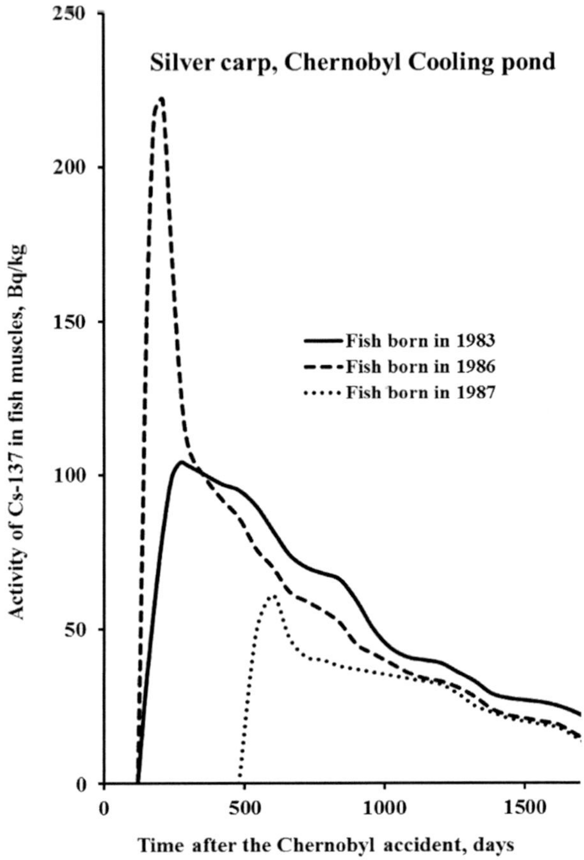

Fig. 9.4 Model reconstruction of the dynamics of fish contamination with ^{137}Cs following the Chernobyl accident. Water body—Chernobyl Cooling pond. Fish species—nonpredatory silver carp of different age generations (1979–1988). Curves with symbols show the dynamics of ^{137}Cs in fish of each age generation (The figure is modified from Kryshev and Ryabov (2000))

The calculated ^{137}Cs activity concentrations in the muscles of silver carp of different generations are presented in Fig. 9.4; calculations were made by A. Kryshev using Eqs. (9.9)–(9.11). Carp specimen of younger age classes accumulated radiocaesium more actively and reached a maximal level of contamination earlier than older carps. However, following the decrease of ^{137}Cs activity in the water with time, young silver carps lost radiocaesium more quickly than the older specimen. Due to more intense metabolism, ^{137}Cs activity in young fish followed rather rapidly the changes in the food contamination. Radiocaesium accumulation by fish of older age groups was slow because of their low metabolic activity. Thus, ^{137}Cs removal from older fish was slow. This was due to the fact that the rate of biological removal ε was proportional to the rate of overall metabolism in the organism divided by the weight (W/M), which in turn decreased with increasing weight of fish (e.g., for carp $\varepsilon \sim (W/M) \sim M^{-0.2}$).

Different types of size effects were observed in different periods after the accidental contamination. During the first year of the accident, carp specimens of the younger age groups were highly contaminated compared to the older carps; that

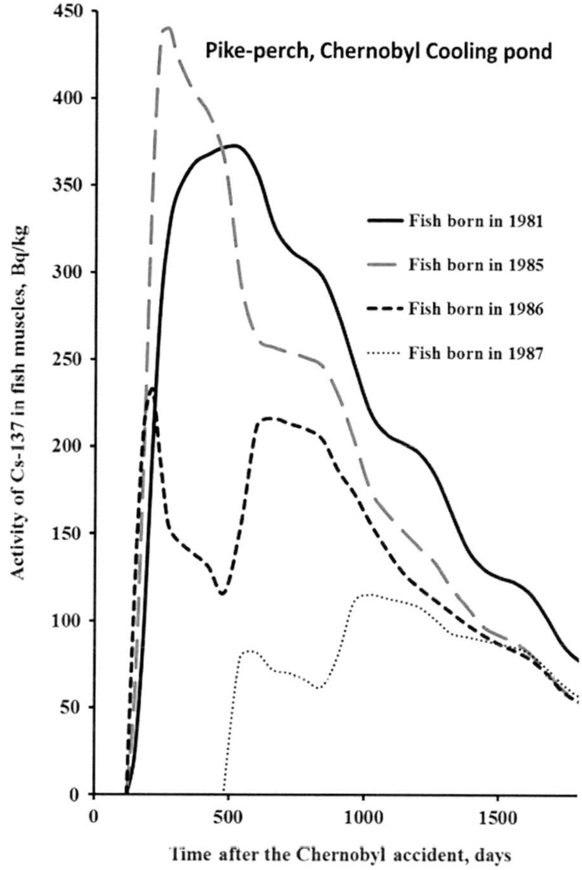

Fig. 9.5 Model reconstruction of the dynamics of fish contamination with ^{137}Cs following the Chernobyl accident. Water body—Chernobyl Cooling pond. Fish species—predatory fish pike–perch of different age generations (1981–1988). Curves with symbols show the dynamics of ^{137}Cs in fish of each age generation (The figure is modified from Kryshev and Ryabov (2000))

is, a negative size effect was observed. An additional contribution to this phenomenon was due to the fact that silver carp of younger age was feed on more contaminated food. Beginning from the second year after the accident, older carp specimens were contaminated to a greater extent than younger silver carps. Along with differences in the rates of accumulation and removal in the different age groups of fish, a sound reason for a positive size effect in radiocaesium accumulation was a decrease in the water activity and hence in the food of fish. The fish born within a year after the accident or later were less contaminated with ^{137}Cs.

Figure 9.5 shows the calculated ^{137}Cs dynamics of the activity concentration in the muscles of predatory fish—pike–perch of different generations; calculations were made by A. Kryshev using Eqs. (9.9)–(9.12). Although the observed effects, on the whole, were similar to those for silver carp, the size effect for predatory fish was more pronounced at all stages after the accident. In the first year following the accident, a negative size effect was enhanced by the fact that pike–perch of the first year of life feeds as nonpredatory fish. After a change to a predatory type of feeding, the secondary maximum of contamination takes place. The maxima in ^{137}Cs activity

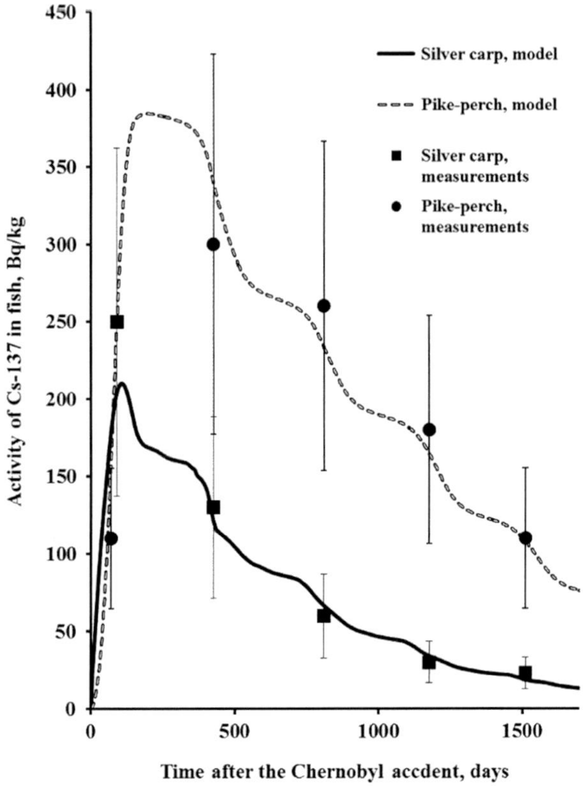

Fig. 9.6 Comparison of model predictions with data of measurements: activity of ^{137}Cs activity in fish muscles, silver carp (nonpredatory), and pike–perch (predatory) from the Chernobyl Cooling pond (1986–1990). Bars show measurement values with uncertainties (The figure is modified from Kryshev and Ryabov (2000))

in pike–perch came later than in carp due to the delay caused by the passage of ^{137}Cs through the trophic chain (the 'effect of trophic levels'). The change from a negative size effect to a positive one occurred, both in pike–perch and carp, beginning in the second year after the accident, but in pike–perch it occurred more smoothly.

Figure 9.6 shows the comparison of calculated ^{137}Cs activity concentrations in the muscles of silver carp and pike–perch with measurements based on data from Kryshev and Ryabov (2000). The dynamical model, based on ECOMOD methodology, describes the radionuclide migration in fish in good agreement with real data.

9.4.4 Model Testing Scenario: 'Chernobyl Cooling Pond'

The 'Chernobyl Cooling pond' scenario was developed in the International Atomic Energy Agency by the Post Chernobyl Data Working Group of the International BIOMOVS II Programme (Biospheric Model Validation Study, Phase II) (BIOMOVS II 1999; Kryshev et al. 1996). The objective of the 'Chernobyl Cooling pond' Scenario (Scenario CP) was to test models for radioactive contamination in aquatic ecosystems. The Scenario CP was based on data from the Chernobyl Nuclear Power Plant (NPP) cooling pond, which was contaminated in 1986 as a result of damage to Unit 4 of the NPP.

Being one of the most heavily contaminated water bodies in the world, the Cooling pond was a suitable model for studying the dynamic processes of radionuclide migration and accumulation in closed aquatic systems (Kryshev 1992). Dynamic processes worthy of detailed investigation include ^{137}Cs accumulation in different trophic levels, the accumulation and removal of ^{137}Cs from fish flesh, and the dependence on the size of the accumulation of ^{137}Cs by various species of fish. The Scenario CP is supplied with the test data of measurements including ^{137}Cs concentrations in water, sediments, and biota for the 5-year period following the Chernobyl accident.

The calculation endpoints of the Scenario are in two stages. In Stage I of the Scenario, the endpoints are model predictions of the dynamics of ecosystem contamination with ^{137}Cs, which was a dominant contaminant in the Cooling pond ecosystem over a long time period. For this stage of the Scenario, model predictions can be tested using the test data.

In Stage II, the endpoints are dose and risk assessment calculations both for aquatic biota and for humans. This stage provides modelers with the opportunity to compare dose and risk estimates made with different dosimetric models using the same input information. Of particular interest is the assessment of doses received in the first two years after the accident, when short-lived radionuclides contributed largely to doses.

The scenario 'Chernobyl Cooling pond' is available at the International Atomic Energy Agency and also the scenario description and model intercomparison are given in publications (BIOMOVS II 1999; Kryshev et al. 1996).

9.5 Long-Term Contamination of Fish Populations with Radionuclides of the Chernobyl Origin

The Chernobyl accident had a pronounced effect on the content of ^{137}Cs in aquatic ecosystems not only in the vicinity of the accidental release site but also across the whole contaminated area (Atlas 1998; Kryshev 1992, 1995, 1996; Kryshev and Sazykina 1994; Kryshev and Ryazantsev 2000; Vakulovsky et al. 1994, 2000; Severe 2001).

Lakes are typical critical components of the natural environment exposed to long-term radioactive contamination. Radioactive contamination of lakes was observed in contaminated territories of Belarus, Russia, Ukraine, Sweden, Finland, Norway, Germany, the United Kingdom, and other countries.

9.5.1 *Lake Kozhanovskoe, Bryansk Region, Russia*

The specific features of radionuclide behavior in lakes, demonstrating high levels of ^{137}Cs in the lake ecosystem, are discussed in this section using as a model example of Lake Kozhanovskoe. Lake Kozhanovskoe is located in Bryansk Region of Russia at a distance of about 300 km to the northwest from the Chernobyl NPP. The contamination of Lake Kozhanovskoe and its catchment area occurred in 1986 as a result of radioactive fallouts of the Chernobyl origin. Lake Kozhanovskoe is located on the territory of 'cesium spot' with contamination with ^{137}Cs in the range of (0.6–1.5) $\times$ 10^6 Bq/m^2. In Russia, Lake Kozhanovskoe is one of the water bodies that ismostly contaminated with ^{137}Cs of the Chernobyl origin.

A radioecological survey of water bodies in the Bryansk Region carried out in 1990–1992 revealed an abnormally high content of ^{137}Cs in the muscles of fish from Lake Kozhanovskoe, which was comparable to the ^{137}Cs content in fish from the Chernobyl Cooling pond in the first years after the accident, and was about 2 times higher than fish contamination from the Chernobyl Cooling pond in 1993. In 1990–1994, ^{137}Cs activities in lake fish were 5–28 kBq/kg, which were higher than the permissible Russian standard by over an order of magnitude (Ryabov 2004; Vakulovsky et al. 1994). In 1998–1999, the ^{137}Cs activity was within the range 5–11 Bq/L in lake water and 7–12 kBq/kg in lake fish (Vakulovsky et al. 2000).

At the same time, the concentration of ^{137}Cs in the Iput River in the immediate vicinity of Lake Kozhanovskoe was approximately 100 times lower than in Lake Kozhanovskoe.

9.5.1.1 Description of Lake Kozhanovskoe

Lake Kozhanovskoe is located at Novozybkov district in Bryansk Region of Russia (52° 47′N, 31° 40′E) in a central part of the East-European flatness. The climate is moderately continental. Winter is rather mild and snowy. Mean annual precipitation is 600–650 mm/year. Vegetation season is about 180–220 days. The lake is frozen in winter.

Lake Kozhanovskoe is a shallow water body; its surface area is about 6 km^2 and its average depth is 1.5 m. The bottom of the lake is covered with thick silt sediments, which on further waterlogging turn into a peat bog. The main type of soil in the catchment area of the lake is peat boggy soil (90%).

Lake Kozhanovskoe is a weakly flowing water body. The characteristic time of water exchange is about 1.3 years. The inflow to the lake is through a bypass channel (max 0.5 m^3/s) and the total intensity of outflow from the lake is about 0.25 m^3/s.

Water in Lake Kozhanovskoe is characterized by low content of potassium ions ([K] = 1.4–2 mg/L).

The peculiarity of ^{137}Cs behavior in Lake Kozhanovskoe is a very slow decrease of the radionuclide activity in water. Fixing the ability of bottom sediments with respect to ^{137}Cs is relatively weak. According to the data of measurements in 1993–1994, the decrease of ^{137}Cs activity concentration in the lake water did not exceed 10% per year, which was comparable with the decrease of ^{137}Cs activity concentration due to radioactive decay (about 3%). As a result, the concentration of ^{137}Cs in the lake water was of the same order of magnitude as in the lakes of the near zone of the Chernobyl NPP.

The decrease in the radiocaesium concentration of water in Lake Kozhanovskoe is determined mainly by the lake flowage and removal of the radionuclide by effluent streams. The radionuclides loss caused by the lake flowage, which is partially compensated by the wash-off of radiocaesium from the contaminated catchment area.

Consequently, the radiocaesium contamination of lakes with a high content of organic matter in the bottom sediments represents the considerable radioecological risks.

Lake Kozhanovskoe is a eutrophic water body with considerable development of large aquatic plants (macrophytes). Beaches of the lake are spongy, overgrown by aquatic vegetation.

The littoral area of the lake is overgrown with a limnetic cane—Scirpus lacustris. In summer, the surface of the lake is coated by the freshwater soldier (*Stratiotes aloides*); in winter, these plants are immersed to the bottom of the lake. Phytoplankton does not play a considerable role in lake productivity, being superseded by the great development of aquatic plants.

9.5.1.2 Fish Contaminated with ^{137}Cs in Lake Kozhanovskoe

Ichthyophauna of Lake Kozhanovskoe is represented by 10–11 species of fish. Goldfish (*Carassius auratus gibelio)* is a predominant species among nonpredatory fishes. Pike (*Esox lucius)* dominates among predatory fishes. The share of goldfish and pike is about 80%.

The data on the radioactive contamination of the Lake Kozhanovskoe are presented in Table 9.5.

Table 9.5 Concentration of ^{137}Cs in water and fish from the Lake Kozhanovskoe (1990–1999)

Date	Water, Bq/L	Goldfish, kBq/kg	Pike, kBq/kg
1990	–	15 ± 1(7)	–
1992	14 ± 2 (4)	21 ± 1(2)	–
1993	10 ± 4 (14)	12 ± 3(45)	35 ± 14(15)
1994	8 ± 2 (6)	11 ± 4(74)	28 ± 13(8)
1995	7.5 ± 1 (4)	15 ± 2(2)	31 ± 7(5)
1996	5.7 ± 1 (1)	–	–
1997	–	–	16 ± 1(2)
1998	5.1 ± 1 (1)	7 ± 4(18)	24 ± 5(8)
1999	5.3 ± 2 (2)	3 ± 2(14)	12 ± 3(1)

The concentration of ^{137}Cs in different fish species taken from the lake was high in 1990–1999 and varied depending on their body size and trophic levels. According to the data of 1993, the highest specific activity was observed in large-sized pike (70 kBq/kg, fresh weight), whereas the lowest one was observed in goldfish and roach (about 5–8 kBq/kg, fresh weight).

In 1993–1995, the concentration of ^{137}Cs of fish taken from Lake Kozhanovskoe remained practically unchanged. In 1998, the concentration of ^{137}Cs in predatory species of fish from Lake Kozhanovskoe remained high and was, on average, 25 kBq/kg, fresh weight, whereas in nonpredatory species, it was 10 kBq/kg, fresh weight.

High fish contamination levels in Lake Kozhanovskoe were determined by a high concentration of ^{137}Cs in water, hence in food of fish, and a low content of K^+ ions in water. In 1998, Lake Kozhanovskoe was registered as a radioecological preserve.

9.5.1.3 Long-Term Model Predictions of Fish Contamination in Lake Kozhanovskoe

Calculations of ^{137}Cs activity in fish from Lake Kozhanovskoe were made using the fish module of the ECOMOD model (Kryshev and Ryabov 2005).

Goldfish (*Carassius auratus gibelio*) was selected as a dominant nonpredatory fish species in Lake Kozhanovskoe. Pike (*Esox lucius*) was selected as a dominant predatory fish species.

Calculations of ^{137}Cs activity concentrations in nonpredatory fish (goldfish, *Carassius auratus gibelio*) from Lake Kozhanovskoe were made in two sets. The first set was performed for the time period 1993–1995 years for which the most detailed experimental information existed on ^{137}Cs activity concentrations in different components of the lake. The second set of calculations was performed for a longer period, covering 1990–1999 years.

The goldfish generation of 1990 was chosen for the ECOMOD model testing; for this generation, the most detailed information existed on the radiocaesium contamination.

The calculations of ^{137}Cs activity concentrations in goldfish for the long-term period 1990–1999 were performed with parameters validated from the first calculations; the results are shown in Fig. 9.7 together with the existing data of observations. A comparison with data of observations shows that the ECOMOD model provided good results for the entire 10-year period.

Calculations of ^{137}Cs activity concentrations in predatory fish (pike, *Esox lucius*) from Lake Kozhanovskoe were made in two sets. The first set was performed for the years 1993–1995, and the second set of calculations was performed for the 10-year period (1990–1999). The pike generation of 1989 was chosen for the ECOMOD model testing.

Model predictions for pike were in good agreement with the existing data of observations (Fig. 9.8).

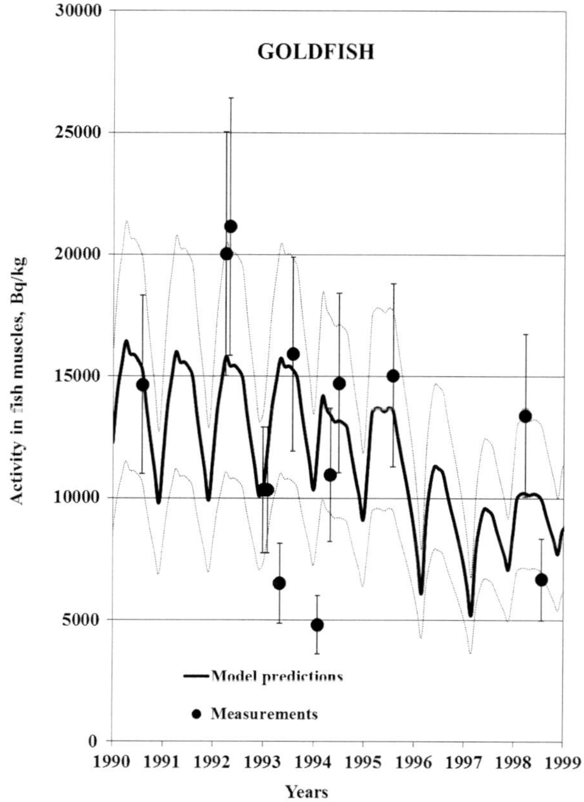

Fig. 9.7 Dynamics of ^{137}Cs activity in nonpredatory goldfish from Lake Kozhanovskoe, Bryansk region, Russia. Model predictions are shown by a thick line; bars with uncertainties show the results of measurements (The figure is modified from Kryshev and Ryabov (2005))

A comparison of observational data on the ^{137}Cs dynamics in lake fish with the model predictions demonstrated that the dynamic approach employed by the ECOMOD model was successful. This dynamical approach in radioecological modeling can be recommended for the reconstruction and prognosis of ^{137}Cs dynamics in fish living in natural water bodies contaminated as a result of a radiation accident.

9.6 Fish Contamination with ^{137}Cs in the Distant Periods After the Chernobyl Catastrophe

9.6.1 Rivers and Reservoirs

Fish in the Pripyat River (the near zone of the Chernobyl NPP) was characterized in 1997–2006 by the following average levels of ^{137}Cs activities in muscles: silver carp—155 Bq/kg (range 10–372 Bq/kg); pike—229 Bq/kg (range 42–547 Bq/kg); and pike–perch—190 Bq/kg (range 10–471 Bq/kg) (Gudkov et al. 2008).

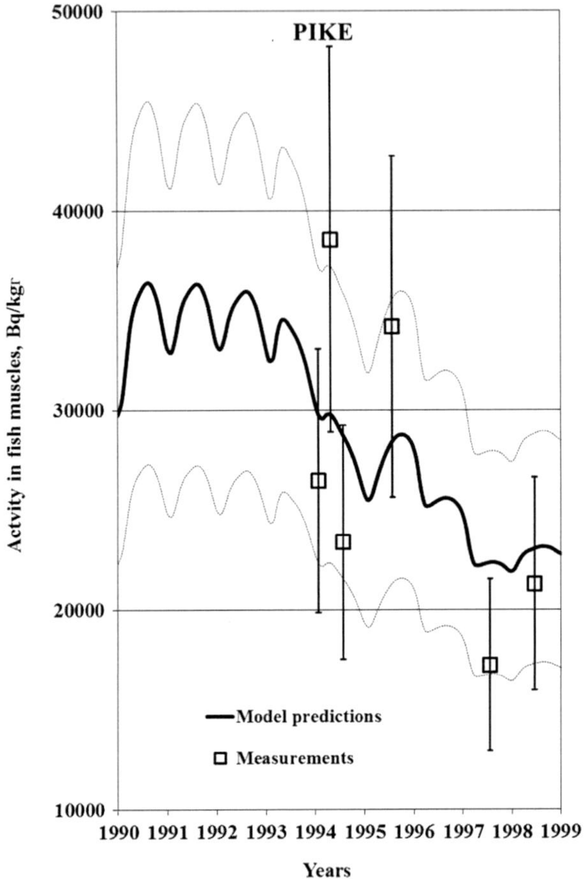

Fig. 9.8 Dynamics of ^{137}Cs activity in pike (predatory fish) from Lake Kozhanovskoe, Bryansk region, Russia. Model predictions are shown by a thick line; bars with uncertainties show the results of measurements (The figure is modified from Kryshev and Ryabov (2005))

In Kiev reservoir (2000–2006), average activities of ^{137}Cs in fish muscles were below the permissible levels (150 Bq/kg in Ukraine): nonpredatory fish—28.7 Bq/kg (range 5.6–62.3 Bq/kg); and predatory fish—52.4 Bq/kg (range 27.3–105.8 Bq/kg) (Environmental consequences 2006; Kaglyan et al. 2011).

9.6.2 Lake Kozhanovskoe

In Lake Kozhanovskoe, levels of fish contamination with ^{137}Cs are slowly decreasing with years following the Chernobyl accident. At the remote period (2009), reported data from the survey were the following: roach—1892 Bq/kg fresh weight (max. 3981 Bq/kg); and bream—1478 Bq/kg fresh weight (max. 3351 Bq/kg) (Polyakova and Pelgunova 2014).

9.6.3 Chernobyl Cooling Pond

According to the data of biological surveys (1997–2013), average activities of ^{137}Cs of fish taken from the Chernobyl Cooling pond were the following: silver carp—1900 Bq/kg (range 1300–2500 Bq/kg); pike–perch—8600 Bq/kg (range 7400–9740 Bq/kg); and perch—9300 Bq/kg (range 7500–10,500 Bq/kg) (Gudkov et al. 2014, 2008).

Since the years 2012–2014, the Chernobyl Cooling pond was being gradually drained with water volume decreasing, and the original aquatic ecosystem is being replaced with terrestrial plants (Thirty years 2016).

9.7 Lessons Learned from the Impact of the Radiation Accident on Aquatic Biota

Analysis of radioecological consequences of the Chernobyl accident on aquatic biota makes it possible to formulate the following general conclusions:

1. During the early stage following the radiation catastrophe at NPP (within 1 year), numerous short-lived radionuclides were of major importance with a considerable contribution to biota contamination. In the early period, the methodology of 'equilibrium concentration factors' could not be used; dynamical radioecological models were successfully employed for predicting and reconstructing of radioecological processes.
2. Intermediate postaccidental stage lasted about 3–5 years following the accident. Short-lived radionuclides gradually decayed during this period, and long-lived biologically active radionuclides became dominant contaminants in ecosystems. In aquatic ecosystems, the intermediate period was characterized with redistribution of radionuclides between ecosystem components with gradual transfer through food chains and involvement into biochemical cycles. Dynamic models simulate the processes of radionuclide redistribution in fish populations and aquatic food chains.
3. Distant time periods are characterized with a decrease of radioactive contamination in all components of aquatic ecosystems. Rivers and water bodies with great water exchange demonstrated a rapid decrease in contamination levels, associated with decreases of radionuclides content in aquatic biota. In contrast, lakes and other water bodies with slow exchanging of waters remain contaminated over long periods of time. In some lakes, a peculiar hydrochemical composition of bottom sediments and water may result in abnormally high concentration factors of biogenic radionuclides in fish and other biotas, which may exist for decades after the initial contamination.

References

Accident at the Chernobyl NPP and its Consequences Information (1986) Prepared for the IAEA expert meeting (25–29 August 1986, Vienna). SCAE of the USSR, Moscow, p 1986

Alekseev VV, Kryshev II, Sazykina TG (1992) Physical and mathematical modelling of ecosystems. Hydrometeoizdat, St.-Petersburg, p 367. (in Russian)

Atlas (1998) The Atlas of contamination with ^{137}Cs after the Chernobyl accident. CEC, Luxemburg

BIOMOVS II (1999) Assessment of the consequences of the radioactive contamination of aquatic media and biota. Model testing using chernobyl data. BIOMOVS II technical report no. 10, Scenario CP. Stockholm, Swedish Radiation Protection Institute, p 182

Bowen HJM (1979) Environmental chemistry of the elements. Academic Press, London

Cougthrey PJ, Thorne MC, Jackson D (eds) (1983) Radionuclide distribution and transport in terrestrial and aquatic ecosystems, vol 1–3. CRC Press, Rotterdam

Design (1975) Design of radiotracer experiments in marine biological systems. Technical report series № 167. IAEA, Vienna

Environmental Consequences of the Chernobyl Accident and Their Remediation (2006) Twenty years of experience. Report of the Chernobyl forum expert group 'Environment'. IAEA, Vienna

Evans S (1988) Accumulation of Chernobyl-related ^{137}Cs by fish populations in the biotest basin, northern Baltic Sea. Studsvik Rep STUDSVIC/NP-88/113:1–30

Gudkov DI, Kaglyan AE, Kireev SI, Nazarov AB, Klenus VG (2008) The main radionuclides and dose formation in fish of the Chernobyl NPP Exclusion Zone. Radiat Biol Radioecol 48 (1):48–59. (in Russian)

Gudkov DI, Protasov AA, Scherbak VI, Kaglyan AE, Nazarov AB, Diachenko TN, Silaeva AA (2014) Evaluating the current state and radionuclide contamination levels in biocenosis components of the Chernobyl NPP cooling pond. Proceedings of Scientific and Practical Conference "Radioecology-2014", 24–26 Apr, 2014 (in Russian)

Hakanson L (1991) Ecometric and dynamic modeling exemplified by cesium in lakes after Chernobyl. Lect Notes Earth Sci 35:158

Jorgensen SJ, Friis HB, Henriken J, Mejer HF (eds) (1978) Handbook of environmental data and ecological parameters. ISEM, Vaerlse, p 1162

Ivlev VS (1962) The method for calculation of food amount consumed by growing fish. Biologia vnutrennih vodoyomov Pribaltiki, pp 132–138 (in Russian)

Izrael YA (ed) (1990) Chernobyl: radioactive contamination of the natural environments. Gidrometeoizdat, Leningrad, p 296

Izrael YA (1996) Radioactive fallout after nuclear explosions and accidents. Gidrometeoizdat, St.-Petersburg, p 356

Kaglyan OE, Gudkov DI, Klenus VG, Kuzmenko MI, Shiroka ZO, Tkachenko VO, Melnik MK, Pomortseva NA, Yurchuk LP, Nazarov OB (2011) Current state of radioactive contamination of fresh-water fish in Ukraine. Rep Natl Acad Sci Ukraine 12:164–169. (in Ukrainian)

Kazakov SV (1995) Management of radiation conditions of water cooling reservoirs of nuclear power plants. Kiev, Technika, p 192. (in Russian)

Kryshev AI, Ryabov IN (2000) A dynamic model of ^{137}Cs accumulation by fish of different age classes. J Environ Radioact 50(3):221–233

Kryshev AI, Ryabov IN (2005) Model for calculating fish contamination with ^{137}Cs and its application for Lake Kozhanovskoe (Bryansk Region). Radiat Biol Radioecol 45(3):338–345. (in Russian)

Kryshev AI (2000) Dynamical model of radionuclides migration in fresh-water ecosystem. Ph.D. dissertation, Moscow State University, Department of Biophysics

Kryshev AI (2002) Modelling the accumulation of ^{137}Cs by age-structured fish population. Radioprotection 37(C1, 2):627–632

Kryshev II (ed) (1992) Radioecological consequences of the Chernobyl accident. Nuclear Society, Moscow, p 142
Kryshev II, Ryazantsev EP (2000) Ecological safety of the nuclear energy complex of Russia. Izdat, Moscow, p 354. (in Russian)
Kryshev II, Sazykina TG, Hoffman FO, Thiessen KM, Blaylock BG, Feng Y, Galeriu D, Heling R, Kryshev AI, Kononovich AL, Watkins B (1999) Assessment of the consequences of the radioactive contamination of aquatic media and biota for the Chernobyl NPP cooling pond: model testing using Chernobyl data. J Environ Radioact 42(2–3):143–156
Kryshev II, Sazykina TG, Kryshev AI (2003) The Chernobyl accident and aquatic biota. In: Scott M (ed) Modelling radioactivity in the environment. Elsevier, Amsterdam, pp 391–416
Kryshev II, Sazykina TG, Ryabov IN, Chumak VK, Zarubin OL (1996) Model testing using Chernobyl data. II. Assessment of the consequences of the radioactive contamination of the Chernobyl NPP cooling pond. Health Phys 70:13–17
Kryshev II, Sazykina TG (1990a) Modelling the dynamics of exposure doses to hydrobionts following the accidental contamination of a cooling pond. Water Resour 6:85–88. (in Russian)
Kryshev II, Sazykina TG (1990b) Simulation models for ecosystem dynamics under anthropogenic effects of heat electric generation and nuclear power plants. Energoatomizdat, Moscow, p 183. (in Russian)
Kryshev II, Sazykina TG (1995) Assessment of radiation doses to aquatic organism's in the chernobyl contaminated area. J Environ Radioact 28(1):91–103
Kryshev II, Sazykina TG (1986) Mathematical modelling of radionuclide migration in aquatic ecosystems. Energoatomizdat, Moscow, p 151. (in Russian)
Kryshev II, Sazykina TG (1994) Accumulation factors and biogeochemical aspects of migration of radionuclides in aquatic ecosystems in the areas impacted by the Chernobyl accident. Radiochimica Acta, 66/67, 381–384.
Kryshev II (1995) Radioactive contamination of aquatic ecosystems following the Chernobyl accident. J Environ Radioact 27(3):207–219
Kryshev II (1996) Radioactive contamination of aquatic ecosystems in the areas of nuclear power plants and other nuclear facilities in Russia. Radiochim Acta 74:199–202
Lehninger AL (1982) Principles of biochemistry. Worth Publishers. Inc., New York, NY
Polyakova NI, Pelgunova LA (2014) Estimation of the modern radioecological conditions of the freshwater ichthyofauna from reservoirs at Russia and Ukraine. Proceedings of international conference on biological effects of low dose ionizing radiation and radioactive pollution on environment. Syktyvkar, Russia. 17–21 Mar 2014 (in Russian)
Polykarpov GG (1966) Radioecology of aquatic organisms (translation from Russian). Reinhold, New York, NY, p 314
Ryabov IN (2004) Radioecology of fish in water bodies within the zone impacted by the Chernobyl accident. Association of scientific publications KMK, Moscow, 215 p. (in Russian)
Sazykina TG, Kryshev AI (1996) Radioecological model for transport of radionuclides and exposure to organisms in aquatic ecosystems. In: International symposium on ionising radiation, Stockholm, 20–24 May 1996. Proceedings, vol. 2, pp 503–510
Sazykina TG (2000) ECOMOD – an ecological approach to radioecological modelling. J Environ Radioact 50(3):207–220
Sazykina TG (1981) Migration of radioactive isotopes in closed aquatic ecosystems. Chapter IV in Ph.D. dissertation (physical and mathematical sciences, speciality: biophysics). Moscow State University, Department of Physics (in Russian)
Sazykina TG (1994) Modelling the transfer of fission products in aquatic ecosystems. In: Proceedings of the fourth international conference on the chemistry and migration behavior of actinides and fission products in the geosphere. Charleston, SC, 12–17 Dec, 1994, pp 727–731
Severe (2001) Severe radiation accidents: consequences and countermeasures. Izdat, Moscow, p 752
Thirty Years (2016) Thirty years of the chernobyl catastrophe. Chernobyl Center. Publishing House KIM, Kiev, p 169. (in Russian and Ukrainian)

Vakulovsky SM (ed) (1986) Methodical recommendations for determining radioactive contamination of aquatic samples. Gidrometeoizdat, Moscow, p 76. (in Russian)

Vakulovsky SM, Gaziev YAI, Nazarov LE, Tertyshnik EG, Uvarov AD, Martynenko VP, Petrenko GI (2000) Radioecological monitoring of environment in Bryansk region in 1998–1999. Proceedings of international conference "radioactivity after nuclear explosions and accidents". Moscow, 24–26 Apr 24–26, 2000. Vol. 2. Gidrometeoizdat, St.-Peterburg, pp 19–24

S.M. Vakulovsky, A.I. Nikitin, V.B. Chumichev, I.Yu. Katrich, O.A. Voitsekhovich, V.I. Medinets, V.V. Pisarev, L.A. Bovkum, E.S. Khersonsky, (1994) Cesium-137 and strontium-90 contamination of water bodies in the areas affected by releases from the chernobyl nuclear power plant accident: an overview. Journal of Environmental Radioactivity 23 (2):103–122

Vernadsky VI (1929) On the accumulation of radium by living organisms. Rep Acad Sci USSR 2:33–34. (in Russian)

Whicker FW, Schultz V (1982) Radioecology: nuclear energy and the environment, vol 1–2. CRC Press, Boca Raton, FL

Winberg GG (1956) Rate of metabolism and food requirements of fish. Fish Res B Can Transl Ser 194:1–239

General Conclusions

The Chernobyl accident was the most severe nuclear accident in history. Its long-term effects continue to be studied, even though numerous projects were completed and books published.

Long-lived radionuclides released as a result of the Chernobyl accident were largely embedded in fuel particles also called "hot" particles. These particles per se were insoluble in water; however, their disintegration and oxidation led to leaching of radionuclides and hence changes in their mobility and bioavailability. The timescale of fuel particles disintegration in soil was estimated to be from 1 to 10 years and even more. Most of fuel particles in soils have already disintegrated by now.

Modeling of radionuclide re-entrainment and consequent redistribution in space and over time is essential for estimating human inhalation doses. The tools for doing this include resuspension factor models, a mass loading approach, and mathematical models of radionuclide atmospheric transport, and dispersion based on the semi-empirical equation of turbulent diffusion. Using results of measurements undertaken in the Chernobyl exclusion zone, several models have been verified and validated.

Assessment of potential environmental risks of nuclear power is impossible without good understanding about behavior of radionuclides in the environment. In this respect, prediction of mobility and bioavailability of radiocesium and parameterization of its behavior in the soil–water environment were facilitated by development of the concepts of frayed edge sites (FES) capacity and radiocesium interception potential (RIP) for soils and sediments, which was a major step forward in elucidating the mechanisms underlying radiocesium retention in soils and sediments. With respect to transformation of radiocesium and radiostrontium, a model was developed accounting for leaching from fuel particles, sorption-desorption by ion exchange, fixation, and remobilization. Radiocesium in soil and sediments was found to be retained through two distinct mechanisms: highly selective exchangeable sorption by micaceous clay minerals and fixation.

It was revealed after the Chernobyl accident that natural attenuation processes associated with vertical and lateral migration of radionuclides in soils and sediments played an important role. The most significant lateral migration of Chernobyl-

A. Konoplev et al. (eds.), *Behavior of Radionuclides in the Environment II*,
https://doi.org/10.1007/978-981-15-3568-0

derived ^{137}Cs is observed within the arable slopes. Most of the sediment-associated ^{137}Cs transported outside the arable lands are redeposited along the bottom parts of cultivated slopes and in the bottoms of dry valleys. The bottoms of dry valleys and especially artificial reservoirs located in the upper reaches of the fluvial network are the main sink of Chernobyl-derived ^{137}Cs deposition. It is worth noting that even in the areas with a high proportion of arable land, no more than 0.3–0.5% of the total inventory of Chernobyl-derived ^{137}Cs was transported with river sediment yield over the past three decades.

Important knowledge has been accumulated about radionuclides cycling in forest ecosystems and the processes redistributing radionuclides between the ecosystem compartments at different stages after deposition. Numerous models have been developed to predict radionuclides dynamics in forest ecosystems. Experimental studies, real-time observations in the exclusion zone and modeling exercises have shown that forest fires in contaminated territories did not pose any significant radiological risks to personnel of the zone and the public outside. Under certain conditions, inhalation of resuspended radioactive aerosols can give a non-negligible contribution to doses for firefighters, while for flora and fauna in the zone the fires do have severe consequences.

As regards contamination of water bodies, vertical migration of radionuclides in the catchment soils was found to be a key process responsible for their decline in surface runoff in the long term. Long-term dynamics of activity concentrations in surface runoff and rivers can actually be predicted based on describing radionuclide decline in the catchment topsoil layer. Relying on the analysis of changes in the particulate and dissolved ^{137}Cs and dissolved ^{90}Sr in surface water, a semi-empirical diffusional model was proposed enabling the use of the same function for middle- and long-term phases after the accident with a limited number of physically meaningful parameters.

The Chernobyl NPP Cooling Pond (CP) merits special attention. It was contaminated in 1986 by deposition of radionuclides onto the water surface and discharge through the outlet channel of highly radioactive water from the reactor emergency cooling system and water used to extinguish the fire at the ChNPP. The dissolved ^{137}Cs activity concentration in CP shows seasonal variations with a pronounced winter–spring minimum (February–April) and summer–autumn maximum (July–September), which seems to be associated with intra-annual cycles of physicochemical and biological processes in CP. In May 2014, decommissioning and draining of CP was started, leading to splitting of the pond into three separate parts and reduction of the water surface area by some 40% by the end of 2017. As a result, the ^{90}Sr activity concentration in water in the Northern and Western sectors increased by a factor of 3 by the end of 2017, while in the Southern sector the ^{90}Sr water activity concentration increased by less than 10–15%. In future, the ^{90}Sr activity concentration in the CP water is expected to further go up due to dissolution of fuel particles on the drained territories causing wash-off of radionuclide by surface runoff to residual lakes.

Radioactive contamination of biota was varying at different stages after the accident. In the early stage (within 1 year), short-lived radionuclides were predominant, but their decay in the following 3–5 years resulted in prevalence of long-lived radionuclide in the ecosystems. In aquatic ecosystems, the intermediate period was characterized by redistribution of radionuclides between ecosystem components, with their gradual transfer through food chains and involvement into biochemical cycles. In the long term, radioactive contamination was decreasing in all components of aquatic ecosystems. Rivers and water bodies characterized by significant water exchange showed a rapid decrease in water contamination levels and radionuclide concentrations in aquatic biota. On the contrary, lakes and other water bodies with slow exchange of waters remain contaminated over extended periods of time. In some lakes, peculiar hydrochemical composition of bottom sediments and water can result in abnormally high concentration factors of biogenic radionuclides in fish and other biota, persisting for decades after the initial contamination.

In case of a radiological accident, diverse radioecological information about contaminated territory is required to calculate doses for the population, to identify territories of higher hazard from the viewpoint of absorbed doses, and to design a framework for comprehensive countermeasures. It is also crucial to understand the mechanisms of radionuclide migration in the environment and transfer through the food chain in order to predict actual radiation hazard, and to conduct emergency measures such as population and cattle evacuation and iodine prophylaxis. The mitigation of the Chernobyl accident consequences has demonstrated that the countermeasures for crops and livestock were highly effective, and applied timely. This has been confirmed by independent data from Ukraine, Belarus, and Russia, as well as international projects of the European Commission, FAO, and IAEA. It is essential that the gained experience of mitigating negative consequences of the accident be widely disseminated and included in guidelines, so that in case of a nuclear accident the practical knowledge be immediately used.

As the years passed after the accident, new knowledge has emerged enabling improvement of *decision support systems* which are key for emergency preparedness and emergency response in the event of a nuclear accident. The results of the studies described in the book, in our view, can serve this important goal. Also, the book provides experimental data and long-term environmental monitoring data that are crucial for validation of radioecological models. The presented data regarding radionuclide behavior in the environmental compartments can be used for developing and enhancing a strategy for remediation and rehabilitation of contaminated territories.